DESERTS AND
DESERT ENVIRONMENTS

Environmental Systems and Global Change Series

Series Editors: Professor Antony Orme (UCLA), Professor Olav Slaymaker (University of British Columbia), and Dr Tom Spencer (University of Cambridge)

The development of this new series of advanced undergraduate and graduate textbooks has been stimulated by three widely recognized trends in the teaching of earth and environmental sciences at university level.

Firstly, the systems approach is now well established in university physical geography and earth/environmental science curricula around the world, at both undergraduate and graduate levels.

Secondly, concerns about the pace and extent of global change have increasingly informed – and given an urgent social relevance to – a wide range of course offerings in these subjects.

Lastly, implicit in the environmental systems approach is the importance of integrating findings and methodologies from a wide range of disciplines, including ecosystems science, geomorphology, hydrology, geophysics, oceanography, climatology, archaeology, and environmental planning.

The ESGC Series is explicitly designed to reflect these educational trends. It is an ambitious new venture resulting from the merging of two existing publishing initiatives – Blackwell's Environmental Systems and Pearson's Understanding Global Environmental Change Series – and its objectives may be simply stated:

- to create an awareness and understanding of the way key environmental systems operate and interact;
- to explore the pace and extent of global (and regional) environmental change and to show how environmental systems respond to change over a variety of scales in time and space;
- to attract students from a range of disciplines and to encourage students to think in new ways that transcend traditional discipline boundaries;
- to underline the relevance of these studies to social/environmental problems, and to encourage students to bring a scientific approach to solving such problems.

Books in the series are aimed at advanced undergraduates and graduates taking degree courses in physical geography, earth science, environmental science, ecology, and archaeology. Titles in the series will have an international relevance, with examples and case studies taken from varied environments around the world.

1 *The Cryosphere and Global Environmental Change*, Olav Slaymaker and Richard E.J. Kelly
2 *Deserts and Desert Environments*, Julie Laity

Forthcoming
3 *The Pace of Environmental Change*, Antony Orme
4 *Oceans and Global Environmental Change*, Tom Spencer
5 *Water in a Changing World*, John Pitlick, James Wescoat, and Harihar Rajaram

DESERTS AND DESERT ENVIRONMENTS

Julie Laity

WILEY-BLACKWELL

A John Wiley & Sons, Inc., Publication

This edition first published 2008, © 2008 by Julie Laity

Blackwell Publishing was acquired by John Wiley & Sons in February 2007. Blackwell's publishing program has been merged with Wiley's global Scientific, Technical and Medical business to form Wiley-Blackwell.

Registered office: John Wiley & Sons Ltd, The Atrium, Southern Gate, Chichester, West Sussex, PO19 8SQ, UK

Editorial offices: 9600 Garsington Road, Oxford, OX4 2DQ, UK
The Atrium, Southern Gate, Chichester, West Sussex, PO19 8SQ, UK
111 River Street, Hoboken, NJ 07030-5774, USA

For details of our global editorial offices, for customer services and for information about how to apply for permission to reuse the copyright material in this book please see our website at www.wiley.com/wiley-blackwell

Library of Congress Cataloguing-in-Publication Data

Laity, Julie.
    Deserts and desert environments / Julie Laity.
        p. cm.
    Includes bibliographical references and index.
    ISBN 978-1-57718-033-3 (pbk. : alk. paper)   1. Deserts.   2. Geomorphology.   I. Title.
    GB611.L25 2008
    551.41′5–dc22

2008015536

A catalogue record for this book is available from the British Library.

Set in Meridien 9.5 on 11.5 pt by SNP Best-set Typesetter Ltd., Hong Kong
Printed in Singapore by Markono Print Media Pte Ltd

CONTENTS

PREFACE

The desert environment, owing to it complexity, is best appreciated through a long period of discovery. Its nature and beauty are subtle and illusory. Over time, the observer sees a very complex environment: a mosaic that retains the imprint of the past and offers glimpses into the present. It is my hope that this book will be a reference source and provide a coherent and informative basis for desert study.

On a practical level, the desert is of interest because humans increasingly exploit these regions, as technology allows us to expand into a realm that hitherto posed strong limitations. Given the recent explosive population growth in arid areas, the increasing problems of environmental degradation, and a need to make informed decisions in desert management, this book fills a niche for those seeking to understand the desert environment as a whole: its climate, hydrology, geomorphology, and basic biology. Throughout the book, there are specific references to human interactions and environmental concerns.

I hope that the ideas and information presented herein play a role in informing management decisions, as the desert recovers very slowly from harm. In particular, it must be recognized that water is a precious resource, which should not be exploited to the extent that the natural biological environment is impaired. As case studies in this book show, only a few decades of mismanagement for short-term gain can destroy an ecosystem that has evolved over thousands of years.

From an academic perspective, the book is concerned with the nature, origin, and evolution of desert environments. In recent years, increasingly detailed studies, the incorporation of new dating techniques, and an expansion of the global scope of scientific inquiry have helped us to better understand the desert environment. The approach in *Deserts and Desert Environments* differs from that of many other excellent books by exploring broader themes, discussing in depth global deserts, providing more details on both the surface and subsurface

climatic environment, considering all aspects of the hydrologic environment, and examining geomorphic linkages to the biological environment. Thus, the book aims to provide a comprehensive view of deserts, through intertwining themes, and exploring the interactions between the geologic, geomorphic, biological, climatic, and hydrologic spheres.

The book emphasizes many basic principles, and discusses some details in more depth. Examples are given from many areas of the world and, where appropriate, more extensive case studies are presented. The book may be used as a reference or as a text for a class in desert environments or desert geomorphology. By and large, the chapters in the book can be considered independently and read in any order.

Following an introduction to the nature of deserts and the imprint of past climates in Chapter 1, the book explores aspects of physical geography. The second chapter introduces the reader to the deserts of the world. Although this chapter can be read later in order, its placement near the beginning of the book provides an overview of the great diversity of desert landscapes and gives a sense of the different foci of research. Chapter 3 provides an essential introduction to desert climatology. This chapter explores weather data, the surface boundary layer, the effects of urbanization on desert climatology, the variability of climatic influences in both time and space (El Niño Southern Oscillation forcing and shifting desert boundaries), and the impact of climate on the biological environment. The fourth chapter reviews the hydrologic framework (precipitation, interception, evapotransiration, infiltration, and groundwater), explores surface runoff and flooding, and briefly reviews the water resources available to humans. Although lakes are rare in modern deserts, they were much more widespread in the past, and Chapter 5 examines palaeolakes and lakebed surfaces, as well as contemporary lakes, such as Mono Lake. In an environment in which vegetation is sparse, surface runoff infrequent, and salts are common, weathering processes become important, and this theme is considered in Chapter 6. Additionally, the nature of hillslope processes in deserts is examined. Chapter 7 introduces desert soils and surfaces, emphasizing how these differ from their counterparts in more humid environments. Water as a geomorphic agent is explored in Chapter 8, where both subsurface and surface processes are examined, as well as some of the most significant fluvial landforms and landform assemblages. Aeolian processes are examined in two chapters (9 and 10), which are subdivided into sections on basic wind processes, landforms of accumulation (dunes), landforms of erosion, and desert dust. Dust has received increasing attention in recent years, as it is globally disseminated by the winds, affecting not only deserts, but environments as far away as the Amazon basin. Anthropogenic factors in sand destabilization and dust production are also discussed in these chapters. Chapters 11 and 12 examine how the biological community responds to the limitations of water, high temperatures and radiation, and the presence of salt in desert soils; and emphasizes the important hydrologic and geomorphic impacts of plants and animals. Finally, in Chapter 13, the human dimension of deserts is developed more fully, with particular attention drawn to issues of desertification and problems associated with excessive groundwater withdrawal and depletion of surface water.

One of the goals of this book is to provide the readers with imagery designed to enhance their understanding of global deserts. High-quality maps and annotated orbital imagery illustrate the spatial dimensions and regional topography of deserts. Many of the maps and graphs in this book were provided by David Deis, whose exemplary attention to design and detail is evident throughout. Some of the photographs in this book were graciously provided by Robert Howard, Tony and Amalie Orme, Steve Adams, and Lloyd Laity.

At the end of the book, there is a long list of references to enable the reader to further explore the topics discussed. The research in arid lands continues to expand rapidly, and it has not been possible to read or cite all of this work. During the process of writing this book, I have had the very great pleasure of exploring many new themes and ideas in desert research. I realize that I may have insufficiently cited some work or missed important contributions to the field, and for that I apologize to the authors.

I would like to sincerely thank an anonymous reviewer who provided useful comments on the content and organization of the book. Freelance editor Nik Prowse oversaw the editing, proofreading, and indexing stages of the book and provided invaluable assistance and a push to the finish line.

I am fortunate to have a desert environment so close to my home, which has made it possible to explore large areas over many decades. Throughout this time, many people have been instrumental in inspiring or supporting me. My initial interest in geography was stimulated by my parents. My father drove us around large areas of Australia, Canada, and the United States and, with camera in hand, encouraged my interest in photography. My mother never found a pebble, a feather, or a flower that was not her friend, and always walked with her eyes trained on the ground. Although she had no formal training, her enthusiasm for the beauty of nature was my strongest influence. The desert was formally introduced to me by Antony Orme of UCLA, who first encouraged me in my academic pursuits, and later provided long-term inspiration through his own lifetime of enthusiasm for geomorphic study. During many of my desert travels, I have been accompanied by graduate students and my family, who were invariably cheerful and useful companions. In particular, I would like to thank Tim Boyle, Aaron Davis, Mark Kuhlman, and Alaric Clark. They have helped to fix flat tires, cajole transmissions, help other stranded motorists, and dig and tow vehicles out of the sand. My daughter, Kelsey Laity-D'Agostino, who played in sand dunes while in diapers and suffered greatly from saltating sand until she grew sufficiently tall, matured to be an asset to my field explorations, able to quickly erect a tent in the darkest and windiest conditions. Furthermore, her zeal for learning sustains my own interests. My husband, Saverio D'Agostino, is the glue that holds my life together. Over the years, he has served as field assistant, mechanic, laboratory technician, horse caretaker, fellow scientist, and chief cook and bottle washer. This book would not have been possible without his extraordinary generosity and the many cups of tea he has provided.

Julie Laity

1

INTRODUCTION: DEFINING THE DESERT SYSTEM

1.1 Defining the Desert System

Deserts and semideserts are the most extensive of the Earth's biomes, occupying more than one-third of the global land surface. Of this area, approximately 4% is classified as extremely arid (hyperarid), 15% arid, and about 14.6% semiarid (Meigs 1953, 1957). In total, about 49 million km² are affected by aridity. If dry-subhumid areas are included in the classification, then drylands comprise about 47% of the Earth's land surface (United Nations Environment Program 1992). "True" deserts are considered to be the warm hyperarid and arid regions, and semiarid and dry-subhumid regions the desert fringes. Collectively, the dry areas of the world occupy more land than any other major climatic type.

1.1.1 Physical, Biological, and Temporal Components

Deserts are characterized by their great aridity and may share in common features of climate, weather, geomorphology, hydrology, soils, and plant and animal life. However, defining a desert is not a simple matter, as witnessed by the many attempts at a systematic characterization based on aspects of climate (precipitation, evaporation, and temperature), geomorphic features, and flora and fauna. Although high temperatures, winds, and shifting sands may be present in some deserts, they are not components of all arid environments. In an attempt to define a desert, Shreve (1951) describes regions of "low and untimely distributed rainfall, low humidity, high air temperatures, strong wind, soil with low organic content and high content of mineral salts, violent erosional work by water and wind, sporadic flow of streams and poor development of nominal dendritic drainage." Although this definition is a good fit for many North American deserts, it poorly constrains others. For example, the Atacama and Namib Deserts, with their low average temperatures and high coastal humidities, do not conform to Shreve's description. However, it has proven difficult to arrive at a universally accepted definition, perhaps because deserts themselves show considerable individuality, and because of the existence of continuous transitions between the different types of deserts.

Considered from a biological standpoint, deserts may be considered to be areas where the availability of water is low. "True" deserts result from a deficiency in the amount of precipitation received relative to water loss by evaporation. For organisms, aridity may be a relative condition, as the amount of water available may be a function of several interacting variables, including precipitation, temperature, soil texture, groundwater seepage, and slope and aspect. Furthermore, some organisms obtain their moisture from fog or dew, a source of water that is not regularly measured.

Climate, vegetation, and fauna have all been used to delimit desert boundaries. Deserts may be divided into categories based on their temperature (hot; temperate; coastal) or moisture (hyperarid; arid; semiarid) characteristics. In many instances, the drier areas of the Earth are simply divided into two groups: arid and semiarid (or, synonymously, desert and semidesert; or desert and steppe). Plant and animal components are commonly incorporated in desert classification systems. Vegetative criteria of Shreve (1942) include floristic content, physiognomy and life forms, and structure and social organization. Herpetofauna and climate were used to set boundaries for the Chihuahuan Desert (Morafka 1977); and herpetofauna and plants for the eastern Sonoran Desert boundary (Lowe 1955).

The delimitation of desert areas is difficult, particularly the location of the outer boundaries. This is the case even for relatively well-studied deserts in populated areas. For example, the boundaries assigned to the Sonoran Desert vary widely according to the criteria used by individual researchers (MacMahon & Wagner 1985). Additionally, desert boundaries are often considered as shifting zones of transition rather than lines clearly demarcated by climate or by abrupt changes in species or associations. Transitional boundaries may result from human impact or from decadal climatic fluctuations. Satellite imagery has allowed an annual examination of fluctuating boundaries, most notably the southern boundary of the Sahara (Tucker et al. 1991; Nicholson et al. 1998).

Understanding desert climates is essential because of strong linkages between climate, biological processes, and geomorphology. Indeed, studies of desertification suggest that human-induced ecological and geomorphic changes may induce climate change, or at least prolong natural drought episodes. These themes are explored more fully in Sections 3.4.2, 3.4.3, and 13.2.

Although all deserts are characterized by aridity, other climatic factors, such as temperature and humidity or season of precipitation, show considerable variation. The following climatic characteristics are common to many interior tropical and subtropical deserts: (1) high summertime temperatures, (2) an excess of potential evaporation over precipitation as a result of high temperatures, wind, and clear skies, (3) high variability of precipitation totals, distribution, and intensity, (4) a more prominent role for wind than in other zones, (5) clear skies prevailing over 70% of the time, and (6) low humidity (commonly 15–30% for inland deserts and as low as 5% in the Saharan Desert). Winter temperatures show large variations from place to place, largely a reflection of continentality and latitude. For coastal deserts, conditions are considerably different than for interior deserts, as proximity to cool ocean currents and the occurrence of frequent fogs give rise to cooler maximum and average temperatures and very high relative humidities for areas immediately adjacent to the sea.

As may be inferred from the difficulties in defining and delimiting deserts, they do not present a homogeneous landscape. Long-term differences in climatic, tectonic, biological, and geologic history cause deserts to be individually distinct. Each tends to have a unique assemblage of landscape elements and processes. Geomorphologically, the North American deserts are dominated by the erosional and depositional effects of surface water; the eastern Sahara by aeolian processes; and the Atacama by extreme aridity, barren landscapes, saline deposits, and mass wasting that is enhanced by earthquake activity (Oberlander 1994). Deserts also vary tremendously in their tectonic settings. The stability and age of

Australian deserts contrasts sharply with the youth and tectonic instability of arid North America. Furthermore, unique landscape assemblages bear the imprint of climatic change over long time periods.

The Earth's climate has changed profoundly during the Quaternary and Tertiary, so many desert landscapes are palimpsests; that is, composed of relict elements produced under the influence of past climates and modern elements formed in the present climatic regime. Thus, it is impossible to understand modern desert environments without a consideration of earlier climatic, hydrologic, tectonic, geomorphic, and biological conditions. Deserts are superb repositories of these past legacies, as aridity and, in some cases, relative inactivity of the surface, act to preserve landscape assemblages. As noted by Williams (1994, p. 644), "one intriguing outcome of the polygenetic nature of desert landscapes is...the frequent juxtaposition of very old elements of the landscape with others that are very new." This book aims to examine the many forces that have shaped and continue to influence desert landscapes, and to provide a broad appreciation of how the legacy of the past informs the processes of the present.

1.2 EVOLUTION OF DESERTS

1.2.1 GLOBAL CONSIDERATIONS

The arid regions of the world, other than those in the high Arctic, owe their origin to climatic, topographic, and oceanographic factors that prevent the incursion of moisture-bearing weather systems. Although the causes of aridity are discussed separately below, it should be noted that most deserts are arid because of a combination of factors. For example, west of the Andes in the Peru–Chile desert, aridity is a result of subtropical atmospheric subsidence, reinforced by upwelling of the cold Humboldt Current, and by the Andean rainshadow.

1.2.1.1 Subtropical high-pressure belts

The world's arid and semiarid regions are mainly subtropical in distribution, covering about 20% of the Earth's land surface (Glennie 1987) (Fig. 1.1). As shown in Fig. 1.1, the Equator is flanked to the north and south by Hadley cells, each composed of a rising branch in the rainy equatorial zone, and a descending branch near 30° North and South in the arid subtropical zones. In the equatorial zone, opposing

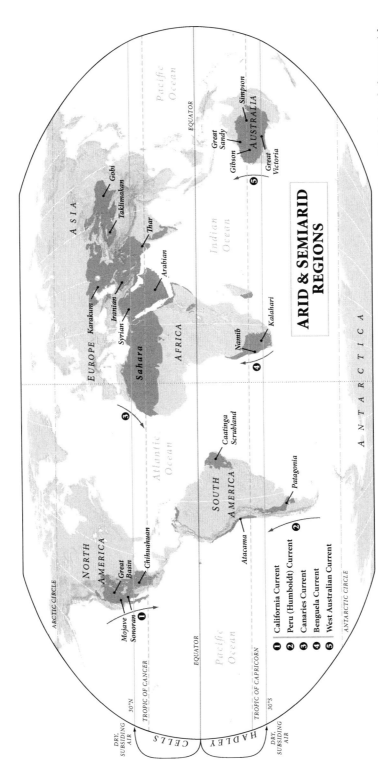

Fig. 1.1 Map of the global deserts. The world's deserts are largely subtropical in distribution. Cold currents are labeled and shown with arrows.

winds converge in the *intertropical convergence zone* or ITCZ, and feed air into the rising branches, favoring convective rainfall. The descending branch forms zones of elevated sea-level pressure, referred to as the *subtropical high-pressure belt* (McIlveen 1992). Although localized vertical motions of the atmosphere occur in this zone of subsidence, larger-scale vertical motion is suppressed by persistent thermodynamic stability, leading to a general lack of rainfall. Relative humidities throughout the troposphere are very low in the zone of subsidence, and only rarely do active disturbances penetrate to break the normal aridity.

The subtropical high-pressure belt is broken up into anticyclonic cells, so that the associated subsidence is discontinuous and aridity is not present in all longitudes. In the Northern Hemisphere, air moving clockwise around the equatorial sides of cells brings moist air to the eastern continental margins including the Caribbean, East Africa, and south and central China. Although subsidence dominates North India and Pakistan for much of the year, the summer monsoon brings 3 months of abundant rainfall except in the far northwest.

1.2.1.2 Continental interiors

Continental interiors are arid owing to the distance from the sea and sources of moisture. In central Asia, for example, a great mountain arc prevents the deep incursion of moisture-laden air of the summer monsoon. In winter, a vast high-pressure cell develops, with very dry subsiding air.

Interior deserts have a much greater range in annual temperature than those closer to a coast. Despite the high latitudes of Asian deserts, the summers are extremely hot, with maximum temperatures in excess of 38°C at all lowland stations. Winters are very cold, with minima ranging from −30 to −50°C (Fig. 1.2). The Great Basin Desert of North America also experiences great annual ranges of temperature and freezing conditions and snow in winter. The average temperature in January is −2°C. Nonetheless, summer temperatures in the southern Great Basin can be very hot (Laity 2002). The highest temperature ever recorded in the USA (57°C) was in Death Valley.

1.2.1.3 Polar deserts

Low levels of solar radiation at the poles result in very cold temperatures. Because of this, the atmosphere contains little moisture for precipitation, and although precipitation may be frequent, it is very light, with the depth of precipitable water not exceeding 10 mm at any time (Hidore & Oliver 1993). In Antarctica, mean annual precipitation ranges from 51 mm on the plateau to as much as 510 mm at some peninsular locations. Relative humidity values may be as low as 1%, with both humidity and cloud cover decreasing inland. Winds are also a predominant factor in polar deserts, making blizzards and drifting snows a common occurrence.

For reasons of space, polar deserts and periglacial environments cannot be addressed in this book. It is interesting to note, however, that they share several geomorphic processes in common with warm deserts, particularly as the paucity of vegetation cover allows the free sweep of the wind. Thus, aeolian features such as dunes and ventifacts are shared by both desert types.

1.2.2 REGIONAL CONSIDERATIONS

1.2.2.1 Cold-current influences

Cool coastal deserts form adjacent to cold ocean currents on the western margin of continents. They are often long and narrow in form, and may be bounded to the east by north–south-oriented mountain ranges. The climate is greatly moderated by the proximity to cold waters and is characterized by rainlessness, fog and dew, and cold temperatures. Coastal deserts include the Atacama, along the coast of Chile and Peru and adjacent to the Humboldt Current; the Namib on the coast of southwest Africa along the Benguela Current (Fig. 1.3); and the desert along the Pacific coast of Baja California, Mexico, adjacent to the California Current. Other deserts with cold-current influences are the coastal Sahara in northwest Africa, the Arabian Peninsula and Horn of Africa, and the western coast of Australia (Warner 2004). Subsidence from the subtropical high-pressure belt reinforces the effect of the cold coastal waters. Orographic barriers, such as the Andes Mountains, may prevent the incursion of moisture bearing systems from continental interiors.

The movement of water from polar latitudes to low latitudes, with associated upwelling of deep cold waters, produces cold currents. Warm air from the high-pressure cells is cooled by contact with the water and layers of fog form. The air that crosses the land is foggy (relative humidity at or near 100%) and chilled nearly to the temperature of the water, normally from 15 to 18°C. The affected layer is thin

Fig. 1.2 The deserts of continental interiors are arid owing to their distance from the sea and sources of moisture. Such deserts have a much greater range of temperatures than those close to the coast. Winters in the Asian deserts can be very cold, with minima ranging from −30 to −50°C. In this image, the shrinking Aral Sea appears in 2002, filling with seasonal ice, and the deserts of the Kyzylkum and Karakum (also known as Kara Kum) to the southeast and south of the lake, respectively, are blanketed in snow. The diversion of freshwater inflows to the saline Aral Sea for agriculture has led to a considerable loss of lake volume and quality. Source: NASA MODIS Rapid Response Team, NASA/GSFC. See Plate 1.2 for a color version of this image.

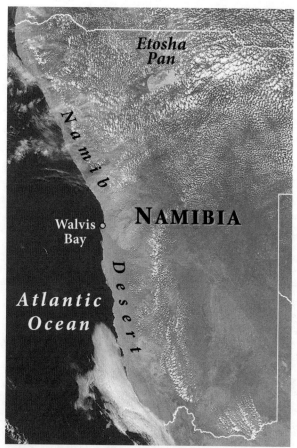

FIG. 1.3 Cool coastal deserts form adjacent to cold ocean currents on the west coasts of continents. The climate is moderated by the proximity to cold waters, which tend to impede convection. Moisture is largely provided by fog, shown here along the southern coast of the Namib Desert. True desert conditions with intense aridity occur in the Namib Desert, a strip 80–150 km wide along the Atlantic coast. To the east, on the right side of the satellite image, is the inland Kalahari Desert. Source: NASA Aqua/MODIS sensor. See Plate 1.3 for a color version of this image.

(150–600 m), and above these levels, hot and dry subsiding subtropical air prevails, causing a temperature inversion.

Rainfall amounts are very low because: (1) the air aloft lacks moisture, (2) the layer of cold air near the surface impedes convection, and (3) the moist surface air is too small in volume to provide an adequate moisture source. Condensation, however, occurs nightly and all exposed objects are wetted.

The seasons have little impact, with winter temperatures averaging only 3–6°C below those of summer. During the winter the sea breeze may weaken somewhat, but in general conditions vary little throughout the year.

1.2.2.2 Rainshadow effect

When air crosses mountain barriers, it rises on the windward side, and subsides on the lee side. Subsidence prevents convection and causes adiabatic heating that results in a pronounced drying effect. Thus, many of the world's arid areas lie to the lee of mountainous regions. In Australia, the Great Divide and other mountains on the east coast lie in the path of the prevailing southeasterly trade winds, creating a rainshadow effect that accentuates the aridity of the central continent. Aridity in the Patagonian Desert of South America results from the Andes mountain range, which blocks rain-bearing westerly air masses and gives rise to strong, dry adiabatic winds and dust storms. Little moisture results from cold air masses from the South Atlantic. Cool winters and mild summers characterize the Patagonian Desert, with temperatures decreasing southward. Similarly, the dry belts of Canada and the USA owe their origin to rainshadow effects resulting when westerly winds cross high-elevation mountains aligned north–south. Lying in the rainshadow of the Sierra Nevada-Cascade chain, the Great Basin Desert receives from 100 to 300 mm of precipitation annually (Fig. 1.4).

1.2.2.3 Edaphic environments

Edaphic deserts result in large measure from the influence of the soil. For example, in the Kalahari Desert, high evaporation rates and sandy soils that absorb rainfall produce a region that lacks surface water despite a rainfall range from 200 mm in the south to 500 mm in the north. Edaphic factors enhance the apparent aridity of this region, which is also fostered by continentality and atmospheric subsidence.

1.3 INDICES OF ARIDITY

Defining the margins of deserts and the boundary between arid and semiarid regions is difficult. Indices to determine aridity are based on rainfall alone, on water balance (the relationship that exists in a given area between precipitation (P), losses due to evapotransparaton (ET), and changes in storage (S)), on soil type, or on vegetation.

FIG. 1.4 The subsidence of air to the lee of mountain barriers creates deserts in the rainshadow. In this image, the Sierra Nevada of California, which rises to over 4400 m, creates a rainshadow to the west in the Owens Valley. The decline in rainfall on the lee slopes can be seen as a decrease in vegetation. The Olancha Dunes are a small dune field formed in the southern section of Owens Valley. Vegetation within the dunes probably grows by exploiting groundwater. Alluvial fans can be seen along the basal slope of the Sierra.

On the basis of moisture indices, deserts are commonly divided into three categories: hyperarid, arid, and semiarid. Hyperarid regions have at least 12 consecutive months without rainfall and no regular seasonal cycle of rainfall. The Western Desert of Egypt and the Atacama Desert are hyperarid. Arid regions receive between 25 and 200 mm of rainfall annually, whereas semiarid lands have between 200 and 500 mm (Grove 1977). Semiarid grasslands are generally referred to as steppes, although this term is often ill-defined.

Most climatic systems used to define aridity are based on the concept of water balance. An example is the Thornthwaite Moisture Index (I_m) (Thornthwaite 1948).

$$I_m = \frac{s - 0.6d}{e} \times 100$$

In this equation, s is the sum of monthly surpluses of precipitation above the estimated potential evaporation; d is the sum of monthly deficits in precipitation; and e is the estimated annual potential evaporation based on mean monthly values of temperature, with an adjustment for season of rainfall, and including a factor for soil-moisture storage. Thornthwaite moisture value indices of 0 to −20 are considered dry subhumid; values of −20 to −40 are semiarid (200–500 mm precipitation); and values of −40 to −56 are arid (25–200 mm precipitation). Below −57, the region is considered hyperarid (generally <25 mm annual precipitation, with no seasonal rhythm of precipitation, and at least 12 consecutive rainless months) (Meigs 1953). Meigs (1953) further subdivided deserts according to whether there was no marked season of precipitation, summer rain, or winter rain.

Because evaporation or evapotranspiration increase proportionately with temperature, it is impractical to use a fixed precipitation figure for desert boundaries. Furthermore, the effectiveness of precipitation is strongly influenced by the season of the year (the higher temperatures of summer cause more rapid evaporation than the cooler conditions of winter), and a seasonality factor is therefore essential.

The United Nations Environment Program (1992) classification of drylands is similar to that of Thornthwaite, but uses meteorological data from 2000 meteorological stations over a fixed time period (1951–1980) rather than the full length of available record. It includes dry-subhumid areas in its classification, as these regions experience certain climatic characteristics of semiarid areas.

Whereas precipitation and temperature are easy to measure and have been recorded at stations for many years, evaporation and transpiration are more difficult to measure and the records are of shorter duration and less plentiful.

Furthermore, in a determination of Thornthwaite's index, evaporation cannot be considered a function of temperature alone. Other factors include the amount of soil moisture, soil type and texture, wind velocities, atmospheric pressure, relative humidity, plant cover, and land use. These latter contributions are more difficult to determine.

There are several problems with climatic measures of aridity. The effect of climate on plant growth, human occupation, and geomorphological process is not dependent on absolute amounts of precipitation. The temporal distribution of rain throughout the year is significant. For example, cold-season rainfall may be more effective for plant growth (if temperatures are sufficiently high) as less moisture is lost by direct evaporation than with hot-season rainfall. Another important factor is the variability of rainfall from year to year in both time and space. Additionally, human activities in the twentieth century have resulted in the expansion of arid surface conditions into semiarid environments, principally as a result of a decrease in vegetation cover.

Another means of determining the extent of arid lands is to determine the extent of certain soil types. Pedocals, poorly drained soils with free calcium in the profile, occur where evaporation exceeds precipitation in both middle and low latitudes. The distribution of these soils suggests that 43% of the Earth's surface is arid. By contrast, climatic and vegetation assessment suggests 35–36%.

1.4 Desert surfaces

Popular representations employ images of dunes to invoke the desert. However, sand covers only about 20% of desert surfaces, most of it in vast sand sheets and sand seas. In North Africa, such sand seas are referred to as ergs. Nearly 50% of desert surfaces are stone pavements or regs (Walker 1986). Pebbles (4–64 mm in diameter) surface most, but larger clasts, in the cobble range (64–265 mm), are also found. Hamadas are relatively level plains that are covered by boulders and exposed bedrock. In the Sahara, nearly horizontal Mesozoic marine and nonmarine sedimentary formations form extensive sandstone and limestone hamada plateaus (Williams 1994). A

classic example is the Gilf Kebir in the hyperarid desert of southeastern Libya and southwestern Egypt. The remaining areas of desert surface are composed of bedrock outcrops, soils, and fluvial features including playas, alluvial fans and oases.

There are significant differences in the percentage cover of different surface types from region to region. In the Sahara, sand seas and regs are the dominant landform classes, covering 22 and 21% of the surface respectively, whereas alluvial fans cover less area (14%) (Ballantine et al. 2005). By contrast, in the southwestern USA, sand seas cover a mere 0.6% and alluvial fans 31% (Clements et al. 1957). The Peru–Chile desert also lacks extensive sand seas, being primarily a rocky desert (Berger & Cooke 1997).

1.5 Tectonically stable and unstable deserts

The tectonic setting of deserts plays a role in determining the relative and absolute relief of the region, affecting sediment sources and sinks, influencing timescales of relative stability or instability and, over the long term, the degree of aridity as mountain systems rise and block incoming moisture. Arid zones provide an important setting for tectonic study, as, owing to the lack of vegetation, features such as folds and faults are plainly visible on both remote imagery and the ground. In the Peru–Chile desert, for example, fault scarps, thrust sheets, and diverted drainage are clearly evident owing to active seismicity and very low denudation rates (Berger & Cooke 1997).

Tectonism provides a framework for basins and a source of renewed energy through uplift. The scale and alignment of relief features plays a critical role in climate, creating zones of enhanced rainfall and rainshadows, funneling winds, and providing numerous microclimatic environments. Tectonism also plays an important biological role, as fault-related springlines provide a source of water in many deserts.

Five types of tectonic setting are identified: cratons (shield and platform areas), which have shown relative stability since the late Tertiary; active continental margins, associated with Cenozoic orogenic belts, in a compressional setting, with thrust and transcurrent faulting; older, Phanerozoic orogenic belts, in which there is some reactivation of existing fault zones; inter-orogenic basin and range and inter-cratonic rift zones, with "pull-apart" basins in an extensional tectonic setting; and passive continental margins (Rendell 1997). Examples of deserts in these settings are shown in Table 1.1. The Saharan Desert is of such

TABLE 1.1 Tectonic settings and desert examples.

Tectonic setting	Contemporary deserts
Cratons	Kalahari
	Australian desert
	Saudi Arabia
Active continental margins and Cenozoic orogenic belts	Atacama
	Sahara
	Thar
	Sinai-Negev
	Arabian-Zagros
Older orogenic belts	China
	Sahara
Inter-orogenic, inter-cratonic	North American deserts: Mojave, Great Basin, Sonoran, Chihuahuan
	Sahara (Afar, Ethiopia)
	Monte Desert
Passive continental margin	Namib
	Patagonian

After Rendell (1997).

physical extent that it is included in more than one category.

These tectonic settings play a major role in determining the character of deserts. In the American southwest, the Basin and Range Province is distinguished by a distinctive physiography of narrow mountain ranges separated by broad, sediment-filled basins, within which are numerous playas (Leeder & Jackson 1993). Within this region the continental lithosphere has been stretched and thinned. Whereas many of the rocks exposed in this region are ancient, the landscape itself is modern, having largely acquired its form from extension during the late Miocene to Holocene (Baldridge 2004). Consequences of crustal stretching include faulting and volcanism. In 1887, a magnitude 7.2 earthquake in northern Sonora, Mexico, caused 50 km of ground breakage and developed a 3 m fault scarp (DuBois & Smith 1980). If an earthquake of similar magnitude were to occur today, it would have severe consequences for the heavily populated areas of southern Arizona. Volcanism is also widespread in the American southwest, although recent activity (the past few thousand years) has been largely confined to isolated, relatively quiet effusions of basaltic lavas. The high elevations and complex orographic conditions of the American west cause a patchy mosaic of climate and vegetation zones. Attempts to reconstruct palaeoclimates suggest very complex spatial patterns of hydrologic

and vegetation change that reflect local topographic influences.

By contrast to this largely modern landscape, Mesozoic and Cenozoic uplands and lowlands in the Saharan desert are a legacy of a Precambrian and early Phanerozoic tectonic inheritance. In Australia, certain landscape elements are exhumed surfaces that developed well before the onset of aridity at about 0.91 million years ago (Ma) (Williams 1994).

The tectonic environment profoundly affects fluvial and lacustrine systems. In the stable tectonic settings of shield and platform deserts, such as Australia, arid alluvial fans are poorly developed (Mabbutt 1977a). In more active tectonic zones, hydrologic systems are altered when surface uplift increases erosion and changes the rate of sedimentary supply. Tectonic activity sometimes deflects, alters, or blocks drainages. The drainage history of the Mojave River in the southwestern USA has been difficult to decipher and remains controversial owing to tectonic disruptions during the Holocene (Brown et al. 1990). In Algeria, the 1980 El Asnan earthquake blocked the Chelif River and produced a 5 km^2 lake (King & Vita-Finzi 1981).

1.6 DESERTS OF THE PAST

Aridity has existed on the surface of the Earth since Precambrian times, as shown by desert dune and evaporite sediments in the rock record (Glennie 1987). The earliest indicators of widespread dry conditions are 1.8 billion-year-old dune sediments in the Northwest Territories of Canada (Ross 1983). It is interesting to consider that prior to the colonization of the Earth's surface by plants, wind may have been a more potent agent of geomorphological change than it is today.

The location and extent of arid zones have shifted though time owing to changes in continental configuration, position, and relief. The onset of modern aridity can be traced back millions of years, with the development of drier conditions 80 Ma in the Namib and 150 Ma in the Atacama (Hartley et al. 2005). Other deserts, such as those of northern Africa and Australia, probably developed during the Tertiary (Thomas 1997). At a global level, it is thought that aridity was in place by 3 Ma (Williams 1994), related to declining temperatures in the late Cenozoic associated with the movement of continents towards their modern latitudes. As more land moved poleward, snow and ice expanded and global albedo increased, causing global cooling, increases in latitudinal temperature gradients, and changes in global circulation as a consequence. The lowering of sea level during glacials also increased continentality, with the land area of Australia increasing by one-fifth (Williams et al. 1993). Although the first signs of aridity may be associated in some regions with events in early Cenozoic time, the system entered its most dramatic phase in the Quaternary, with strong oscillations in climate.

During the late Quaternary, all deserts experienced climate change, with alternations between more humid or more arid conditions, depending on location and time. Additionally, deserts expanded or contracted along their margins. In Africa, Australia, and Asia, the peak of late Quaternary aridity is thought to have occurred during the Last Glacial Maximum (LGM) (about 18–21 ka, where ka means thousand years ago), when deserts expanded up to five times their present extent (Sarnthein 1978; Stokes et al. 1997). The eastern Sahara, for example, was hyperarid (<10 mm annual rainfall) and unoccupied by humans. This was a period of intense aeolian erosion, during which many drainage systems were obliterated or inverted, and the region became pockmarked with deflation basins (Haynes 2001). By contrast, the deserts of the American southwest experienced cooler and wetter conditions (Fig. 1.5)

as the massive Laurentide Ice Sheet that covered much of Canada and the northeastern USA caused the westerly winds and jet stream to enter at more southerly latitudes than today. As a consequence of increased effective moisture, several hundred lakes formed in the southwestern deserts and woodlands expanded (Benson et al. 1990, 1995; Orme 2002). Thus, climate changes are not always globally synchronous in deserts.

In order to fully assess the role of tectonism and climate change in the development of arid-zone landscapes, a clear chronological framework is necessary. However, many challenges remain in dating sedimentary deposits and surface exposures. The preservation of organic carbon (for ^{14}C dating) is very poor in arid environments and radiometric dating of geomorphic features in deserts is very difficult. There are, as yet, limited applications of other techniques, many of which are suitable only when favorable geomorphic/geologic settings are present. Luminescence techniques have found widespread favor for the investigation of aeolian deposits (Munyikwa 2005), particularly as they have undergone methodological innovations, such as the development of optically stimulated luminescence (OSL), which yields more precise dates for construction of high-resolution chronologies. OSL dates represent the last sunlight exposure event before deposition. Buried

FIG. 1.5 During the Quaternary, all deserts experienced climate change. In the American southwest, cooler and wetter conditions were experienced. Geomorphic indicators of a more humid climate include palaeoshorelines, shown here at Shoreline Butte in Death Valley. In this image, the butte is covered in wildflowers owing to an exceptionally wet winter season.

mineral grains, such as quartz and feldspar, gain energy from decaying radioactive isotopes. The energy is stored as trapped electrons within crystal lattice defects: as these traps are light-sensitive, they can be zeroed by exposure to sunlight. The burial age is calculated by measuring the stored energy (palaeo-dose), in conjunction with an evaluation of the annual energy-accumulation rate (Munyikwa 2005). Environments that have been dated by luminescence include loess terrain (Yu et al. 2007), desert dunes (Stokes et al. 1997; Kar et al. 2001; Twidale et al. 2001), and lunettes (Dutkiewicz & Prescott 1997; Lawson & Thomas 2002; Stone 2006). OSL dates have been instrumental in showing that dune processes may be different or more complex than previously thought (Hollands et al. 2006; Tooth 2007). Luminescence dates are increasingly being used to determine the timing of landscape response to late Quaternary climate changes (Munyikwa 2005). To date, however, there remains considerable uncertainty in the results, and it is unclear whether they reflect actual differences in geomorphic events and timing between locations, or differences in sampling strategies, laboratory practices, or improvements in luminescence dating procedures over the years (Tooth 2007). The cosmogenic nuclide age-determination technique has been applied to alluvial deposits (Repka et al. 1997), but there is a large age uncertainty and the practice is costly (Liu & Broecker 2008). Varnish microlamination dating is a correlative age-determination technique used on surfaces where rock varnish has formed, such as alluvial fans and exposures of desert pavement (Liu & Broecker 2008).

The use of geomorphic features as climatic indicators involves the application of a uniformitarian approach, using modern analogs as indicators of past climate. Palaeolake shorelines, karst and fossil-spring deposits, lake muds, palaeosols, wadi gravels, and shorelines are used to indicate wetter conditions, whereas degraded and fossilized sand dunes, yardangs, or deflation zones suggest more widespread aridity. Some of these features, including shorelines and fossil dunes, may be difficult to decipher or be masked at the ground by vegetation, and are more apparent in remotely sensed imagery.

Notable fossil dune systems include those of Africa: south of the Equator, relict dunes are widespread in Botswana, Angola, Zimbabwe, and Zambia (Thomas & Shaw 1991; Stokes et al. 1997) and may extend as far north as the Congo rainforest; whereas, north of the Equator fossil dunes extend south into the savan-

nah and forest zone of West Africa. These ancient dunes are considered indicative of the nature and extent of past periods of greater aridity (Goudie 1996). Aeolian dune systems of the Mega Kalahari (2.5×10^6 km^2) of southern Africa indicate that there were many cycles of dune activity associated with Quaternary glacials. Chronologies established with optical dating techniques suggest significant events at approximately 95,000–115,000, 41,000–46,000, 20,000–26,000, and 9000–16,000 years before present (BP) (Stokes et al. 1997). Within this region there were latitudinal gradients, with the northeastern desert margin having more episodic activity than the southwestern desert core, where aridity was more sustained.

Relict dune systems, yardangs, and ventifacts may also be used to infer palaeocirculation patterns (R.S.U. Smith 1984; Thomas & Shaw 1991; Laity 1992, 1994; Corbett 1993; Stokes et al. 1997; Lancaster et al. 2002; Kocurek & Ewing 2005). In Australia, there is a vast anticlockwise whorl of longitudinal dunes, which covers approximately 40% of the continent and represents 38% of the Earth's aeolian landscape (Wasson et al. 1988). Individual dunes are up to 300 km in length and 10–35 m in height. During the period of peak aridity (18–20 ka), the Australian desert expanded on its southern margin and the dunes were active. Today, vegetation is insufficient to inhibit sand movement, but the modern environment is less windy than in glacial times, and the majority of the dunes are no longer mobile (Ash & Wasson 1983).

Whereas ancient aeolian features indicate drier periods in the past, relict lacustrine phenomena are evidence of formerly more humid conditions. During humid phases, bodies of water appeared in even the driest parts of the Sahara, and pre-existing lakes, such as Lake Chad, enlarged. Evidence of lake expansion (lacustral phases) includes fossil shorelines, calcareous deposits, and archaeological sites (Williams 1982). In Australia, raised shoreline features around the southern margin of Lake Eyre and in the Murray Basin of southeast Australia (Stone 2006), and deep-water lacustrine facies in stratigraphic sequences, are used to infer major lake expansions during the last interglacial period. It is thought that the arid belt of Australia may have disappeared completely during the more humid phases of the last 400,000 years (Croke 1997). In the Darb el Arba'in Desert of the eastern Sahara, rainfall increased from LGM values of less than 10 mm year^{-1} to perhaps 100–600 mm year^{-1} during the period between 9800 and

4500 years BP (Haynes 2001). This pluvial period, which followed Younger Dryas time (the Younger Dryas was a brief cool period, from 11.5 to 13 cal ka BP), consisted of three wet phases, and resulted in the deposition of playa muds in deflation basins. Palaeolake systems are discussed more fully in Sections 5.2.3 and 5.3.

Additional advances in our understanding of past climates have come from an examination of ocean sedimentary cores, which retain information on aeolian and fluvial inputs, and from palynological and biological studies, which provide important data on the expansion and contraction (in both area and altitude) of deserts. In the central Saharan Desert, for example, the Nile crocodile (*Crocodylus niloticus*) is presently found only in pools in the Tibesti Mountains. As the crocodile could not migrate across the Saharan desert, its isolated presence inland attests to past pluvial conditions (Goudie 1996).

Faunal and floral evidence indicate that wetter conditions than today characterized much of Egypt and the northern Sudan during the early Holocene. The moist conditions fostered seasonal grassy plains and shrubs and trees of Sahelian-Sudanian affinities, with vegetation concentrated in wadis and around lakes and springs. Rhinoceros bones have been found at Merga in the Sudan, elephant and giraffe bones at Abu Ballas in western Egypt, and hippopotamus, ostrich, hartebeest, and gazelle from Dahkla, western Egypt (Nicoll 2004). In addition to allowing a broad interpretation of environmental conditions, biological indicators often allow constraint of temperature and rainfall values during earlier times. For example, pollen studies, and isotopic datasets for the middle–late Holocene in the eastern Sahara suggest that rainfall isohyets shifted northward by as much as 5° in latitude from their present position (Ritchie & Haynes 1987; Abel & Hoelzmann 2000; Rodrigues et al. 2000).

Archaeological evidence of human occupation also provides valuable information on changing desert climates. Chronologies are generated from ^{14}C (radiocarbon) age determination of organic-rich materials associated with archaeological sites, including hearth materials, eggshells, animal dung, and wood. Patterns of human occupancy suggest that rapid hydroclimatic change in Egypt and northern Sudan from the late Pleistocene through the late Holocene may have driven cultural innovation, settlement, and migration. Dwindling water supplies after 6000 years BP caused human migration towards more reliable water and may have aided in the for-

mation of hierarchical societies in the overpopulated Nile Valley (Butzer 1959; Nicoll 2004). In the Atacama Desert of northern Chile, campsites intercalated between debris flows suggest that even in the hyperarid mid-Holocene period there were extreme, short-lived hydrologic events. Major storms had a recurrence interval of 500–1200 years, whereas moderate storms returned about every 100–200 years (Grosjean et al. 1997). These extreme but rare events highlight the nonlinear nature of climate and landscape evolution in arid regions, and the difficulty of survival in such areas.

The final phase of desert climatic history began with the origins of humans in Africa some 2.5 Ma, and their migration to other arid regions such as Australia and the Americas during the late Pleistocene or early Holocene. From that time forth, desert habitats and climate were unwittingly modified by human activity, with contributions to desertification or to the local expansion of desert margins. This aspect of the climatic history of deserts is further treated in Chapter 13.

1.7 CHANGING HUMAN PERSPECTIVES ON DESERTS

The human population in deserts increased greatly in the twentieth century, with an estimate of at least 900 million people living in arid zones (Thomas 1997). In Africa, 49% of the population lives in arid areas of the continent (United Nations Environment Program 1992).

As the population increases, and resources are more severely stressed, the human perception of deserts continues to evolve. In some areas, twentieth-century technological innovations, including large-scale water-transfer projects, have allowed very high population densities. In the southwestern USA, Las Vegas, Nevada, is the fastest-growing city in the country. The citizens of this arid area, living in air-conditioned homes and businesses, are divorced from the heat and aridity of the surrounding landscape. In most areas of the world, however, the limitations of arid regions still impose serious constraints on human activities and occupations.

The continuing growth of population in deserts raises a host of social issues and new problems. One of the most significant concerns is the availability and quality of surface and ground water. Legal arguments concerning the allocation of surface water (e.g. the Colorado and Tigris/Euphrates River basins) affect both intranational and international relationships. Many communities rely heavily or exclusively on groundwater, most of which was recharged during earlier, wetter climatic periods, and the depletion of this resource has resulted in large-scale ground subsidence, the loss of surface water in streams, and the demise of ecosystems. Groundwater that was stored during earlier geological periods is being rapidly depleted and wells have become progressively deeper and brackish within the timescale of human memory, as seen in the New Valley, Egypt (Nicoll 2004). In China, the sustainability of groundwater reserves is a subject of impassioned debate as development plans for this resource in the arid basins of northwestern China are produced. Urban development and irrigated agriculture often rely on shallow groundwater reserves that are recharged from mountains or uplands surrounding the basins. Once new economic development is in place, the exploitation of groundwater increases rapidly, with consequent drops in the water table and loss of desert vegetation (Chen & Xue 2003).

There are many issues of geologic concern in deserts that have a uniquely human dimension. These include the disposal of hazardous and non-hazardous wastes, soil stability, flooding hazards, deforestation, and the extraction of hydrocarbon and mineral resources. Desertification, the degradation of drylands, is a multidimensional issue with biophysical and socio-economic linkages. It affects human welfare insofar as it influences the capability, sustainability, vulnerability, and resilience of the land and, ultimately, the carrying capacity of arid and semiarid regions (Geist 2005). Although some progress has been made in understanding desertification since the issue first reached the world stage in the 1970s, our knowledge is better at a local than a global level. The dynamics of landscape modification, the interconnections of natural systems, and the impending impact of global warming, need to be better understood if future human tolls, such as civil and nationalistic wars, famine, and the stagnation and decline of economies, are to be averted.

2

DESERTS OF THE WORLD

2.1 INTRODUCTION: THE EXTENT OF GLOBAL ARIDITY

The dry climates of the continents occur in five great zones separated by oceans or the wet equatorial regions. Of the total surface area of arid climate, Africa is home to 36.7%, Asia 31.7%, North America 12%, Australia 10.8%, and South America 8.8% (Meigs 1957; Warner 2004). The combined North Africa–Eurasia province is larger than all the remaining dry areas of the world: the arc of deserts sweeps from the Atlantic coast of Africa, through the Arabian Peninsula and into southern Asia (Fig. 1.1). It incorporates the Sahara Desert, which stretches across northern Africa to the Red Sea, as well as arid regions extending eastwards through Arabia to Pakistan and India and Central Asia, including the Arabian, Karakum, Lut, Thar, Taklimakan, and Gobi Deserts. The Somali-Chalbi deserts of east-central Africa might also be considered part of this system.

Within the broader arid zones are many individual deserts with distinctive characteristics. Their nature is determined by physiography, ecozone, geology, edaphic characteristics, and aridity, and these factors affect their mapped extent. For example, the size of the Karoo Desert in South Africa depends on whether it is defined by ecological characteristics, by topography, or by geology (the Karoo System) (Shaw 1997).

Our knowledge of the world's deserts is spatially unequal. The North American southwest, for example, is a relatively small area and has received a great deal of study, whereas other much larger regions have only received superficial examination. Isolated deserts remain poorly described and understood, owing either to their inaccessibility or to political turmoil that precludes scientific study.

The following descriptions of the world's deserts summarize some of the essential characteristics of each region, including climate and climatic history, areal extent, vegetation, elevation, geology, and geomorphology. More detailed information on other significant topics – for example, dust generation or specific landforms – is presented in the relevant chapters.

2.2 GLOBAL DESERTS

2.2.1 AFRICA

The largest arid areas of Africa straddle the Tropics of Cancer (the Saharan Desert) and Capricorn (the Kalahari, Karoo, and Namib Deserts). The Somali-Chalbi Desert is found on the eastern horn of Africa. The southern sector of the island of Madagascar also has a warm semiarid climate. The largest area of hyperarid desert in the world (69%) is found in Africa, and collectively Africa is home to just over one-third of the Earth's arid lands (United Nations Environment Program 1992). The great age of African arid regions is supported by the very high biodiversity of arid-adapted taxa.

The plant and animal communities of Africa have evolved to reflect the considerable heterogeneity in climate and landscape. Many contemporary distribution patterns are explained by plate-tectonic-driven shifts in the latitudinal position of the continent. The climatic history of Africa has fashioned the phytogeographical, zoogeographical, and ecological characteristics of the deserts. A wide range of taxonomic groups has been fostered in this ancient continent, including the ancestors of modern humans. Vegetation is a clue to a wide range of habitat factors and is an indication of dependent animal species. Biomes (large-scale mapping units that represent reasonably homogeneous areas and demonstrate close ties between the biota) include semiarid and arid formations, Mediterranean-type vegetation along the north coast, and limited montane regions. In the semiarid and arid biomes the length of the dry season becomes a major ecophysiological constraint.

Biomass and net primary productivity are limited and much of the biomass is concentrated beneath the ground for most of the year. It rarely exceeds a few tonnes per hectare, and productivity after rainy periods normally does not exceed 0.5 t ha^{-1} year^{-1} (Meadows 1996). Productivity is highly variable from year to year, driven by and responding rapidly to precipitation events.

Despite the environmental constraints of aridity, there is considerable species richness. The arid and semiarid zones of northern Africa support more than 3000 plant species, about 10% of which are endemic (Le Houérou 1986). The Namib, Kalahari, and Karoo Deserts, and the area around the Horn of Africa and southern Arabian Peninsula (Somalia, Ethiopia, and Yemen), show high to extremely high biodiversity (Jürgens 1997). There is little connection between the northern and southern deserts of Africa, with only eight species shared between the Saharan and Namib Deserts. Jürgens (1997) analyzed these disjunct populations and concluded that they represented fragmentation of a once continuous arid belt running from the Namib to the western Sahara via East Africa. The flora of the Namib is considered young, responding to the development of the Benguela Current in the middle Miocene. Nonetheless, this region is very diverse, with more than 7000 plant species, over 2000 of which are succulents (Meadows 1996).

The Mediterranean shrublands of North Africa have evolved in a climate with a warm, dry summer and a cool, rainy winter. The vegetation is dominated by evergreen sclerophyllous shrubs and trees that form open shrublands to closed-canopy woodlands (Meadows 1996). Canopy height and percentage cover vary and are in large measure a function of soil nutrients. Fire is an additional ecosystem constraint. Owing to a long history of human occupation, human impact is more significant here than in other regions (Le Floc'h et al. 1990). Elephant, rhinoceros, and zebra were plundered during the Roman Empire and became extinct in northern Africa. The landscape has also been altered by the invasion of introduced alien plant species (Le Floc'h et al. 1990).

The mountains of the Sahara exist as an archipelago in the desert. The Tibesti-Jebel Uweinat montane xeric woodland ecosystem is an island of higher biodiversity. It consists of the Tibesti Mountains in northern Chad and southern Libya, formed of seven inactive volcanoes, reaching as high as 3415 m; and a second smaller area, the Jebel Uweinat, located at the intersection of eastern Libya, southwestern Egypt, and northwestern Sudan, with peaks reaching almost 2000 m in elevation. Although the rainfall in this region remains low, it is somewhat

more regular than in the surrounding Sahara. Owing to the higher elevation, mean temperatures are also somewhat lower, with a mean maximum of 20°C at the highest elevations (Cloudsley-Thompson 1984). The vegetation varies according to elevation and aspect, but includes numerous trees such as palm species, *Acacia*, and *Tamarix*. The northern slopes support some wetland species, including *Juncus maritimus* and *Phragmites australis* (White 1983). Most vegetation is concentrated in wadis (dry streambeds), which receive runoff from surrounding mountain areas. The montane ecoregion sustains populations of several important Saharan large mammals, including the addax (*Addax nasomaculatus*) and scimitar-horned oryx (*Oryx dammah*) (Hilton-Taylor 2000). Palms and *Hibiscus* spp. are remnant tropical and Mediterranean plant species, occurring because the climate was once wetter and there was a more continuous connection between ecosystems. Fossils and rock paintings indicate that the Nile crocodile (*Crocodylus niloticus*) used to occur across the breadth of the Sahara, but it is now limited to a few specimens in canyon pools (de Smet 1998).

In semiarid parts of Africa, and most notably the Sahel, the increase in land exploitation in combination with erratic, unreliable rainfall and sparse and unproductive vegetation has caused considerable land degradation. This has produced a potentially unstable ecological situation that is the subject of much debate (see Chapter 13).

2.2.1.1 North Africa: the Saharan Desert and the Sahel

The Saharan Desert, with an area of 9,149,000 km^2, is larger than the continental USA (Tucker & Nicholson 1999) (Fig. 2.1). It spans 5000 km in an east–west direction and 1500 km from north to south. The Mediterranean Sea forms the northern boundary and the southern boundary is a zone referred to as the Sahel. Permanent sources of water include the Nile River, which flows northward from well-watered areas in the south to the Mediterranean Sea, the Niger River along the southwestern edge of the Sahara, smaller rivers that drain from the Atlas Mountains into the desert, and oases, which constitute less than 2000 km^2 in area (Warner 2004).

The Sahara is a region of extreme aridity, high temperatures, low humidity, and strong winds. El Azizia, Libya, maintains the highest global recorded temperature of 58°C (136.4°F). The northeast trade winds prevail across the Sahara and are also known

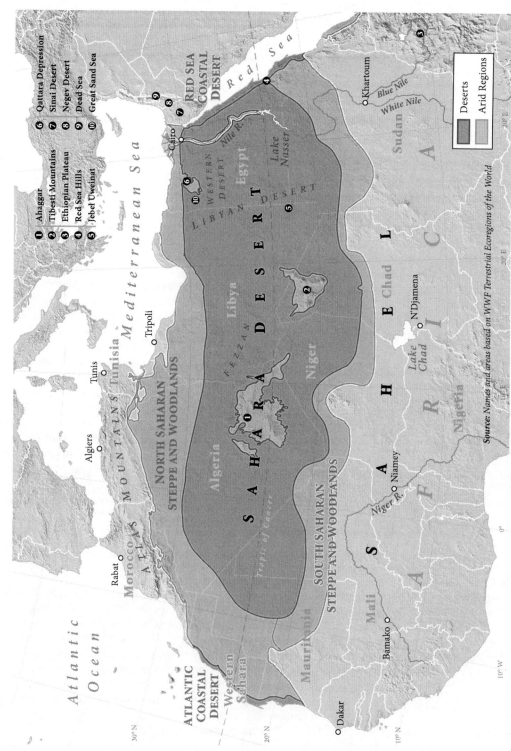

Fig. 2.1 Map of the arid areas of northern Africa.

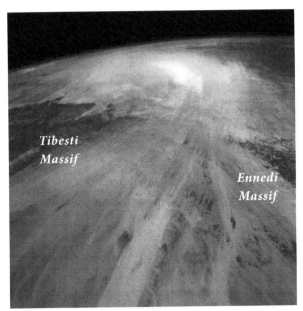

Tibesti Massif

Ennedi Massif

FIG. 2.2 The northeast trade winds prevail across the Sahara and are referred to as the *harmattan*. These dust-generating winds extend to the intertropical convergence zone. This NASA astronaut photograph shows the winds moving between the Tibesti and Ennedi Mountains towards the Bodélé Depression in the Borkou region. In the foreground are vast sand streaks and in the background are light clouds of dust. Yardang systems flank the Tibesti Mountains (dark massif to the center left). See Plate 2.2 for a color version of this image.

as the *harmattan* (Fig. 2.2). At the surface, these dust-bearing winds extend to the ITCZ. The amount of dust generated is greatest in winter owing to the enhanced pressure gradient and longer trajectory of the airflow.

The core desert has a mean annual precipitation of 100 mm or less. Winter rainfall penetrates south of the Atlas Mountains in the northwest Sahara, and into northern Ethiopia and the Red Sea Hills to the east. Summer rainfall affects the southern Sahara as far north as Tibesti (Shaw 1997). Precipitation is influenced by the large-scale migration of the ITCZ, which is found at about 5°N in January and 15–20°N in July, and the westerly winds that bring polar-front rains to the northern coasts. As a result, there are three distinct rainfall regimes: (1) the northern Sahara, with a Mediterranean climate, and a winter rainfall season, (2) the central Sahara, where rainfall is very scarce and irregular, but enhanced locally by mountains, and (3) the southern Sahara, where

rainfall in summer is related to the position of the ITCZ, with summer season convection related to monsoonal flow from the Gulf of Guinea (see Fig. 2.1). The central Sahara is the most arid sector as it is south of the limit of polar-front rains and north of the typical limit of monsoonal flow. However, moisture occasionally penetrates this region and mountains orographically enhance rainfall. The Tibesti and Ahaggar Mountains have a maximum observed rainfall of 300 mm, and the Atlas Mountains 500 mm (Warner 2004).

The late Ordovician glaciation affected roughly half of the area now encompassed by the Sahara, extending over northern and western Africa and into the Arabian Peninsula, and leaving as evidence striations, tillites, and erratics (Sheehan 2001). During the Cenozoic, the Atlas Mountains of the northwestern coast developed as the African plate moved northwards. Volcanic activity in the Tibesti region produced volcanic calderas and clusters of volcanic necks. The first evidence of aridity is in the Oligocene (Sarnthein 1978) and, from the Pliocene to the present, Saharan and Arabian climates have alternated between subhumid and arid (Williams 1994). Periods of greater moisture are recorded by fluvial landforms, soil development, the appearance of lakes and marshes, changes in flora and fauna, tufa deposition and archaeological sites. Arid phases are evidenced by lake desiccation and deflation, fragmentation of fluvial systems, and increased aeolian activity (Oberlander 1994; Haynes 2001; Nicoll 2004). Evidence of major climatic change includes large-scale fossil fluvial systems in regions where aeolian aggradation currently prevails. Ancient river courses, buried beneath up to 5 m of sand, have been identified from Shuttle Imaging Radar (SIR-A) imagery in the southwestern part of the Western Desert of Egypt near the border with Sudan (McCauley et al. 1982; Burke & Wells 1989).

Much of the land surface of the Sahara is occupied by sandy sheets (15%) and sand seas or *ergs* (22%), but gravel (*reg* in the western Sahara or *serir* in the eastern desert; 21%) and boulder plains (*hamada*) are also very widespread (Ballantine et al. 2005). Africa is relatively stable in comparison to other continents, and sedimentary erosion surfaces and ancient basins give it relative flatness. Mountains comprise 500,000 km² of the Sahara. The Atlas Mountains (elevations of 2000–3000 m) and the Tibesti and Hoggar massifs (altitudes near 3000 m) contribute flood flows onto the surrounding plains. Plateaus, formed of sedimentary strata, are commonly bounded by escarpments or cuestas, with some strata showing evidence of karst activity during wetter periods marked by spring activity and higher groundwater tables (Mainguet 1972). The dry valleys of the Libyan Desert of Egypt are considered to be expressions of former groundwater activity (Peel 1941). There are also large closed basins, the most famous being the Qattara Depression of western Egypt, which is 134 m below sea level (BSL).

Aeolian processes dominate in the Sahara (Fig. 2.3). Mainguet et al. (1980) consider the Sahara a single vast aeolian unit, divided into sectors where either sand transportation or deposition dominates. Sand-transport zones are characterized by landforms of erosion, including wind-abraded ridge and swale systems, yardangs, and zones of deflation. Sand seas or ergs represent the depositional sector. More than 20 large depocenters of sand are found in the Saharo-Arabian desert, separated by regs or hamadas. These sand seas are notable for their great mean thickness (20–100 m) (Wilson 1973), with individual mega-dunes up to 300 m in height (Breed et al. 1979). Most ergs are located within topographic depressions: of 12 sand seas in the Western Desert of Egypt, 10 occur in basins (El-Baz 1998). Orbital imagery has revealed that the sand seas are composed of many dune types and transitional forms (Breed et al. 1979). The sand appears to be cycled from one erg to another over intervening surfaces, eroding enormous fields of bedrock yardangs in some areas, particularly to the west of the Nile in Egypt and in the Tibesti region of Libya and Chad, where the systems sweep in a broad arc around the mountains, the sand flowing in a generally clockwise direction.

At its southern boundary, between about 16 and 17°N, the Sahara merges into the semiarid landscape of the Sahel (Monod 1986; Tucker et al. 1991). The Sahelian zone is an area approximately 400–600 km in width (between the approximate latitudes of 12–18°N) that stretches from the Atlantic coast to the Horn of Africa (Fig. 2.1). Its rainfall is characterized by a strong north–south gradient and high inter-annual variability, and is largely derived from the northward penetration of the West African Monsoon in the boreal summer, giving rise to lines of organized convective disturbances (Rowell & Milford 1993). The northern limit of the Sahel is often drawn at the 200 mm year^{-1} isohyet and its southern at 600–700 mm year^{-1} (Nicholson 1978). The landscape is largely located at altitudes between 200 and 500 m above sea level (ASL), and the topography is level or gently undulating. Large areas are covered by fossil sand dunes, with activation occurring only in areas of disturbance (Le Houérou 1980). Other geomorphic features include clay pans, floodplains, and fossil valleys. Much of the drainage system is relict in nature and modern stream flow is short-lived and endoreic. The Niger, Senegal, and Nile Rivers carry water through the Sahel from source regions external to it (Le Houérou 1980). The vegetation consists largely of annual grasses and scattered bush steppe in the north, merging gradually to the south into savannahs with perennial grasses and scattered trees. The vegetation is strongly seasonal, and almost all woody species are deciduous. The potential evapotranspiration of the region averages about 2000 mm year^{-1}. Rain-fed cultivation is extensive in the southern Sahel. Traditionally, the Sahel was a pastoral zone with nomadic herding in the north and more settled herding and agriculture in the south (Le Houérou 1980). However, the population has more than doubled in the past 40 years, pushing sedentary agriculture into less suitable areas of the central Sahel.

2.2.1.2 North Africa: the Somali-Chalbi Desert

The semiarid to arid Somali-Chalbi Desert (Fig. 2.4) is separated from the Saharan Desert by the Ethiopian Plateau, with elevations that exceed 4000 m. The Somali-Chalbi desert is unique in that it is a near-equatorial east-coast desert. Its most arid areas lie near the coasts of the Gulf of Aden and the Indian Ocean, where annual rainfall is less than 100 mm.

There are several reasons for the lack of rainfall. The Ethiopian Plateau and the high terrain of Kenya block incoming summer moisture. Cold water upwelling along the coast suppresses convection, and the movement of maritime air masses over the continent is not common (Warner 2004).

Southwesterly winds associated with the summer monsoon represent one of Earth's strongest and most-sustained winds, with gale-force winds common. These winds produce dust storms along the northeast coast and Gulf of Aden (Warner 2004). Sand dunes are generally limited to coastal areas.

2.2.1.3 Southern Africa: arid Madagascar

The southwestern coastal strip (≈50 km wide) of the island of Madagascar is semiarid (Fig. 2.5), with a mean annual rainfall of approximately 350 mm, and no pronounced wet or dry seasons. In some years, rain may only fall in a single month of

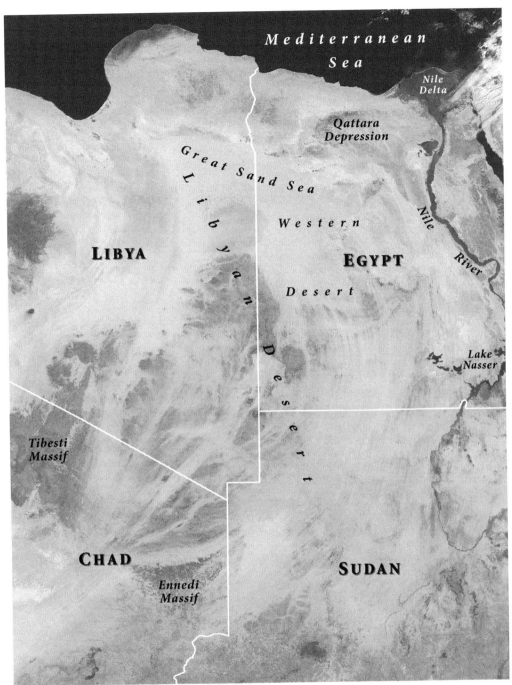

Fig. 2.3 Aeolian processes dominate the geomorphology of the Sahara Desert. The landscape can be divided into sectors of transportation and deposition. Sand seas or ergs represent the depositional sector. This photograph shows the Libyan Desert, where sand can be seen to extend from the north to the south in a broad clockwise arc that represents the flow of the northeast trades. The Nile River carries water through Sudan and Egypt from distant equatorial sources. Large areas of relict fluvial landscape are presently covered by sand. Evidence of previous volcanic activity in the Sahara can be found in the dark-toned Tibesti Massif in northeastern Chad. Source: NASA image. See Plate 2.3 for a color version of this image.

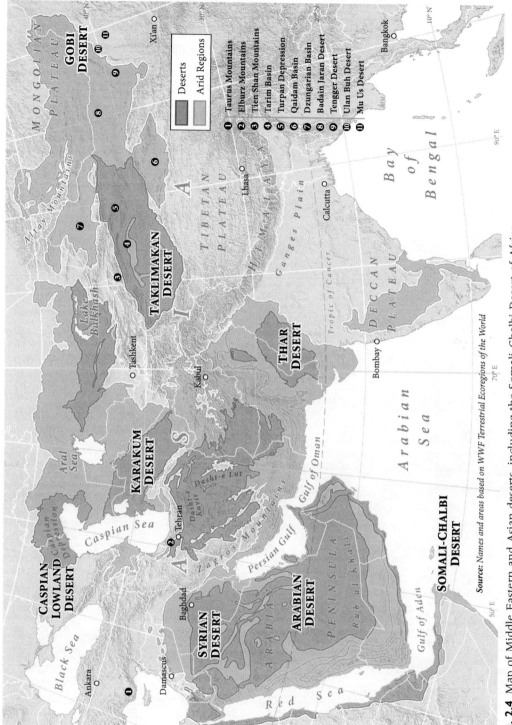

Fig. 2.4 Map of Middle Eastern and Asian deserts, including the Somali-Chalbi Desert of Africa.

Legend:

Deserts

Arid Regions

❶ Taurus Mountains
❷ Elburz Mountains
❸ Tien Shan Mountains
❹ Tarim Basin
❺ Turpan Depression
❻ Qaidam Basin
❼ Dzungarian Basin
❽ Badain Jaran Desert
❾ Tengger Desert
❿ Ulan Buh Desert
⓫ Mu Us Desert

Source: Names and areas based on WWF Terrestrial Ecoregions of the World

GOBI DESERT

MONGOLIAN PLATEAU

Xi'an

TAKLIMAKAN DESERT

Altay Mountains

Lake Balkhash

Tashkent

Aral Sea

TIBETAN PLATEAU

Lhasa

H I M A L A Y A

Ganges Plain

Calcutta

Bay of Bengal

Bangkok

Tropic of Cancer

DECCAN PLATEAU

Bombay

THAR DESERT

Kabul

KARAKUM DESERT

Dasht-e Lut

Dasht-e Kavir

Tehran

Caspian Sea

Caspian Depression

CASPIAN LOWLAND DESERT

Aral Sea

Zagros Mountains

Persian Gulf

Gulf of Oman

Arabian Sea

ARABIAN PENINSULA

Rub al Khali

ARABIAN DESERT

SOMALI-CHALBI DESERT

Gulf of Aden

Red Sea

SYRIAN DESERT

Baghdad

Damascus

Ankara

Black Sea

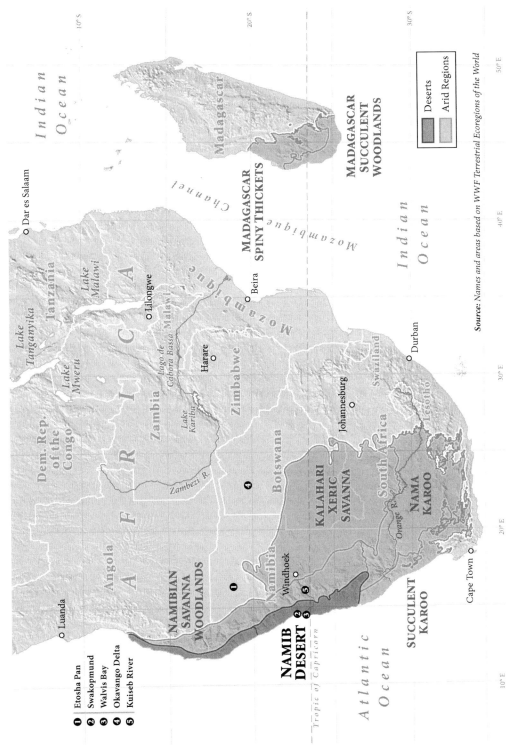

Fig. 2.5 Map of the southern African desert provinces, including the Karoo, Kalahari, and Namib Deserts.

Source: *Names and areas based on WWF Terrestrial Ecoregions of the World*

thunderstorms. The skies are generally clear and soil temperatures are high (50–70°C). The high relative humidity near the coast leads to dew and fog formation (Rauh 1986). Owing to its maritime position, arid Madagascar has one of the highest mean minimum temperatures and smallest mean annual temperature differences between summer and winter of the Earth's deserts.

The Madagascar area shifted from a more mesic to a more arid climate between 3000 and 2000 years BP, with increased fire frequency and disturbance beginning about 1900 years BP, possibly associated with human settlement of the area (Burney 1993). The modern vegetation is xerophilous bush and thickets. The region has high diversity, endemism, and biomass, with succulents dominating many parts of the landscape. Trees reach up to 15 m high, and beneath these is a dense zone of thorn bushes and succulents, up to 3 m high. The leaves are shed at the beginning of the dry season and the green cortices of the shoots perform photosynthesis (Rauh 1986).

Humans have destroyed much of the native vegetation. In the late Holocene, the region's entire endemic megafauna became extinct, the probable result of both human and climatic factors. Large mammals, so characteristic of arid and semiarid mainland Africa, are absent from Madagascar.

2.2.1.4 Southern Africa: the Karoo, Kalahari, and Namib Deserts

The southern African province consists of the Kalahari, Namib and Karoo Deserts, which occupy much of the western and central parts of the region south of the Zambezi River (Fig. 2.5). These deserts are of moderate size and have distinctive climates. True desert conditions with intense aridity are restricted to the coastal Namib Desert, which occurs as a narrow strip 80–150 km broad and 2000 km long along the Atlantic coast of South Africa, Namibia, and Angola, bounded to the east by the Great Escarpment (Lancaster 2002). The Kalahari and Karoo are inland dry zones. The Karoo is in western South Africa, and is contiguous with both the Namib and Kalahari.

Overall, the southern African deserts are relatively cool because the continent narrows as it projects into the southern oceans, exposing the interior to cool maritime air masses that are not greatly modified. Additionally, the altitudes are relatively high, with interior deserts mostly lying above 1000 m. In

summer, there is a considerable temperature gradient between the west coast and the interior, with the Namib in summer about 12°C cooler than the Kalahari (Warner 2004). During the winter period, this difference is reduced to 2°C. The interior position of the Kalahari results in a greater annual and diurnal range of temperatures.

In the interior of South Africa and Botswana, rainfall is dominantly summer convectional, associated with a monsoonal flow of warm moist air from the Indian Ocean. Annual rainfall decreases to the west. The coastal areas of the Namib Desert receive very little rain, with precipitation events occurring mainly in summer (January–April) (Lancaster 2002). Walvis Bay, for example, averages 22 mm of annual precipitation. The relative humidity ranges from 64 to 95% in all seasons. The interior of the Namib also receives scant but unreliable summer rain that increases eastwards and northwards. Windhoek, Namibia, averages 362 mm of rainfall, and has a less humid climate than the coast. The Karoo receives between 200 and 400 mm of rainfall per year, mostly during the summer months. The southernmost part of South Africa is influenced in winter by weather disturbances associated with the belt of westerly winds in the Southern Ocean.

The relatively dense vegetation of the Kalahari is generally savannah, with grasses, shrubs, and trees. The vegetation density increases to the north. The succulent Karoo Desert has the highest desert plant species density in the world, with over 3500 species of succulent. Overgrazing by domestic sheep and goats has altered much of the native vegetation. Annual grasses occur on gravel plains in the Namib, and riparian vegetation includes large ana trees (*Faidherbia albida*) and tamarisk (*Tamarix usneoides*) (Walter 1986).

The Namib Desert rises with a gradient of 1:100 from sea level to 1000 m. It is nearly rainless and, owing to the proximity of a cool ocean current, air temperatures are cool, with only minor fluctuations. The Namib is characterized by high relative humidities and heavy fogs, which dissipate inland (Fig. 1.3). The coastal town of Swakopmund experiences 100–120 days per year of fog. In the eastern Namib, the frequency decreases to 5–10 days per year. The fog deposits moisture and moderates annual and diurnal temperatures. Coastal deposition of moisture from fog is estimated to be equivalent to 15 cm annually (Nagel 1962). Rain occurs when the air aloft over the Namib contains moisture derived from the Indian Ocean, the normal inversion is not present, and convectional air movement develops. Thunderstorms that develop in association with these conditions account for almost all of the annual precipitation. Rivers carry water arising from the semiarid eastern escarpment, creating linear desert oases (Lancaster 2002). Minor river flows are usually dammed by the coastal dune fields that run perpendicular to their courses, but occasional large floods are capable of breaking through a dune field and flushing out the river course, allowing the small flood events in subsequent years to flow down the channel (Krapf et al. 2003). Dune fields are extensive, with the Namib Sand Sea extending between Luderitz and the Kuiseb River, covering approximately 34,000 km². Some of the tallest dunes in the world (300 m) occur here (Fig. 2.6). Dust storms in the southern deserts are principally associated with thunderstorms.

The Kalahari and Karoo Deserts (Fig. 2.5) are found within the Kalahari depression, which is both a tectonic and morphologic basin, filled with tens of meters of unconsolidated sandy deposits of Tertiary and Quaternary age. Beneath this lies Paleozoic to Mesozoic sedimentary rocks and basalt of the Karoo Supergroup (De Vries et al. 2000). Groundwater is found in the Karoo rocks at depths of 20–100 m beneath the surface.

The Kalahari depression appears as a sand-covered surface bordered by surrounding higher ridges and plateaus. The desert is a gently undulating sand plain, interrupted by inselbergs or minor chains of hills (Shaw 1997). Rainfall occurs during the summer period from September to April. The northeastern Kalahari, with a mean rainfall that exceeds 400 mm year^{-1}, is covered by inactive, pedogenically modified, vegetated, and degraded linear dunes. The southwestern Kalahari has a lower mean annual rainfall (150–200 m), and local aeolian activity is episodic and confined to dune crests. Aeolian activity in the past produced the Mega Kalahari, the world's most continuous area of aeolian sand sea (2.5 × 10^6 km²) (Stokes et al. 1997). Overall, the modern Kalahari is neither arid nor windy enough for widespread dune construction or reworking (Stokes et al. 1997). Perennial surface water is restricted to the north where the Okavango and Zambezi Rivers bring water from catchments in Angola and Zambia.

Fig. 2.6 Some of the tallest dunes in the world occur in the Namib Desert, with some reaching 300 m in height. This image shows chains of star dunes. Location: The Namib Sand Sea, 24.4°S, 15.6°E. Source: NASA/GSFC/ METI/ERSDAC/JAROS, and U.S./Japan ASTER Science Team.

2.2.2 MIDDLE EAST AND ARABIA

The Middle East (Fig. 2.7) has experienced at least 8000 years of anthropogenic landscape alteration, including deforestation, rangeland degradation, and watercourse alterations. Modern problems include rapid population growth, political conflict, and water scarcity.

Most of the Middle East has a Mediterranean climate, with wet winters and dry summers. However, there is a complex relationship between landscape and climate (Evans et al. 2004). Regional trends include a northward transition from desert, to steppe, to cool highland climate over a distance of about 400 km. Coastlines and mountain ranges locally modify the climate. Orographic precipitation is important in the Mediterranean coastal range including the hills of Lebanon, the Taurus Mountains of Turkey, the Zagros Mountains of Iran, and the Elburz Mountains near the Caspian Sea. The Taurus and Zagros Mountains supply the flow of the Euphrates and Tigris Rivers of Iraq. The highest terrain, near Lake Van and Mt Ararat, has snow every winter and runoff from the spring thaw smoothes the discharge hydrograph of rivers. The Red Sea and the Persian Gulf are sources of moisture for orographic precipitation, but trigger little local rainfall owing to the overwhelming influence of the descending branch of the Hadley Cell. The deserts of Syria, Iraq, Jordan, and Saudi Arabia are also made drier by the surrounding mountain ranges, which provide a rainshadow effect. Like other desert regions, there is a paucity of climatic data for this region (Evans et al. 2004).

The vegetation of the Middle East has a long history of human disturbance (Fall et al. 1998), with many modern species existing as only remnants of their former populations. Overall, the region is dominated by open steppe and desert vegetation. Elevation changes contribute to climatic and floristic variability, with temperatures ranging from 50°C in the Jordan Valley, to cool, highland climates in the mountains (Zohary 1973). The precipitation decrease from north to south and from west to east away from the Mediterranean Sea, with climatic, soil, and elevational differences, gives rise to a range of vegetation types that include forests, woodlands, steppe vegetation, and desert scrub (Davies & Fall 2001). The observed plant diversity also reflects the region's position between the three great landmasses of Europe, Asia, and Africa. As in other desert regions, annuals appear in abundance during wet years (Davies & Fall 2001).

2.2.2.1 Negev and Sinai Deserts

The deserts of the Negev and Sinai link arid zones of North Africa, the eastern Mediterranean, and Arabia (Fig. 2.7). The Negev Desert of Israel and the Sinai Desert of Egypt are considered one geographical desert, approximately 12,500 km² in area, characterized by many linear sand dunes. Their arbitrary separation resulted from a political border drawn in 1906 (Tsoar et al. 1995). The Negev region accounts for nearly two-thirds of Israel's land area, but only 9% of the population resides there. At present, there are 15 towns and cities, occupying less than 1% of the land area, but containing more than 90% of the region's population.

The Negev incorporates semiarid, arid, and hyperarid zones. Rainfall is difficult to characterize, as in some years the area may lack any precipitation, and in others up to 150 mm may fall. There is a general decrease in precipitation from north (around 200 mm year^{-1}) to south (50 mm year^{-1}) (Nativ et al. 1997). The rainy season is from October to April, with most of the events occurring between December and February. The maximum recorded temperature in the Negev is 44.8°C and the minimum −1.8°C. Annual pan evaporation averages 2300 mm (Nativ et al. 1997). In the desert areas of Israel (mean annual rainfall <200 mm), desert sands constitute 13% of the area. In the Sinai (mean annual rainfall <100 mm) they cover 21% of the area (Danin 1983).

Until 1948, only Bedouin pastoralists inhabited the hyperarid areas of the Negev. They used the sandy surface for growing certain crops, accompanied by goats, camels, and donkeys that grazed the dunes. Qualitative analysis of air photos from 1944 and 1945 shows sparse vegetation and no contrast in dune albedo between the Sinai and Negev. However, the creation of Israel in 1948 caused some of the Bedouin tribes to migrate from the Negev side into the Sinai owing to Israeli military occupation. Postmigration pressures led to the removal of shrubs for firewood, shelter, cultivation, and livestock grazing. Shrub density increased in the Negev and decreased in the Sinai. The opening of the border in 1967 allowed free movement of the Bedouin into the Negev and severe overgrazing occurred. In 1982, the political border was reestablished. Between 1968 and 1982, straight linear vegetated dunes in the Negev changed into linear-braided and sinuous seif (sharp-crested) dunes with no vegetation (Tsoar & Møller 1986; Tsoar 1989). Satellite imagery from the

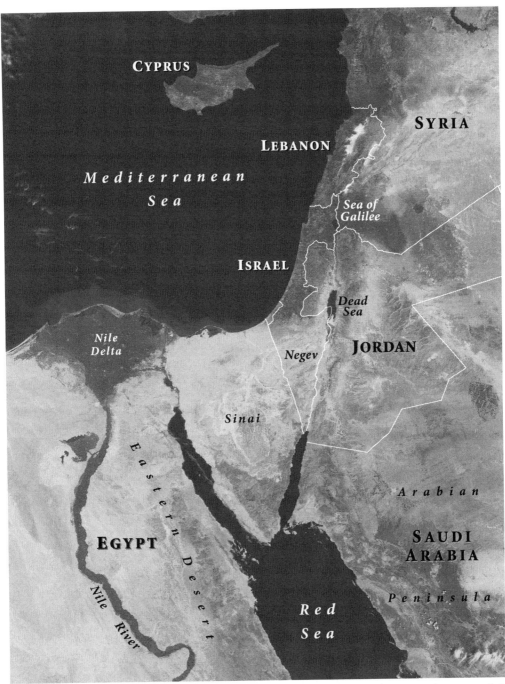

FIG. 2.7 View of the Middle East including the countries of Egypt, Israel, Jordan, Syria, Lebanon, and Saudi Arabia. The deserts of the Negev and Sinai link arid zones of North Africa, the eastern Mediterranean, and Arabia. The general decrease in rainfall from the north to the south in the Negev Desert is visible in the vegetation cover. The Arabian Peninsula is extremely arid, averaging less than 100 mm annually. The Dead Sea and Sea of Galilee occupy pull-apart basins formed along the Levant Fault. Source: MODIS Land Rapid Response Team, NASA/GSFC. See Plate 2.7 for a color version of this image.

1960s to the present highlights the contrast in albedo between the bright Sinai and the darker Negev (Fig. 2.7) (Otterman 1974; Otterman & Tucker 1985; Danin 1991), caused primarily by a biogenic crust cover that has been established on the sand surface of the Negev dunes, and is lacking in the Sinai (Tsoar et al. 1995).

The Negev has been an important site of research in fluvial geomorphology, including work on floods, sediment transport, and wadi network evolution (Yair et al. 1980). Runoff is derived from storms that are commonly high-magnitude and low-frequency (Yair & Lavee 1985). High erosion rates result from infiltration rates that rapidly decline following the onset of rainfall.

2.2.2.2 Deserts of Syria and Jordan

The Syrian Desert is located in the northern Arabian Peninsula (Fig. 2.4). It is a rock- and gravel-strewn plateau that extends through southern Syria, northeastern Jordan, and western Iraq. The topography of the area is relatively subdued and lacks high mountains. Elevations range from 1736 m ASL in southern Jordan to 400 m BSL in the Dead Sea. Sedimentary rocks dominate, although basalt plains of late Tertiary and early Quaternary origin cover extensive areas (Allison 1997). Desert pavements are widespread, covering both sedimentary and basalt rock outcrops.

2.2.2.3 The Arabian Peninsula

The Arabian Peninsula, covering 2.6 million km² in area, is arid or extremely arid, except in highland areas where orographic rainfall is promoted. It is bounded by the Red Sea on the west, the Gulf of Aden and the Arabian Sea on the south, and the Persian Gulf on the east (Fig. 2.4). There are two major geologic provinces: the Arabian Shield and the Arabian Shelf. The shield lies in the western peninsula and is an ancient landmass consisting of Precambrian crystalline and metamorphosed sedimentary rocks and volcanics. The shelf is formed of an exposed sequence of continental and shallow marine sedimentary rocks overlying the rocks of the Arabian Shield (Al-Juaidi et al. 2003).

Over much of the Arabian Peninsula, the annual precipitation is less than 100 mm (Glennie 1998). The Rub' al Khali, the largest continuous body of sand in the world, receives less than 50 mm year⁻¹. Winter is the rainy season and summers are hot (regularly >50°C) and dry. Along the southern coast,

the southwest monsoon brings rain in the late summer. Rainfall amounts are greatest and variability is least at the higher elevations. The mountains of southwestern Arabia have local rainfall totals that exceed 750 mm year⁻¹ (Glennie 1998). The lower elevations are often subject to periods of prolonged drought. Aquifers produce good-quality water and some springs provide sufficient water for irrigation.

During the winter months, eastern Arabia is dominated by the Shamal wind, which blows down the Arabian Gulf from the north-northwest. The average annual number of recorded dust-storm days ranges from 6 to 79 (Abd El Rahman 1986).

The Arabian Peninsula has been subject to significant climatic and environmental changes during the Quaternary (Edgell 1989; Glennie 1998; Al-Farraj & Harvey 2004). Fluvial deposition was most extensive in the Pleistocene and the Pliocene (Glennie 1998). Today, the Oman Mountains are flanked by alluvial fans that are seldom active and are covered at their northern end by dune sands that postdate fluvial activity. Cooler and wetter conditions during the late Pleistocene resulted in enhanced sediment supply to alluvial fans bordering the coast of the Arabian Gulf (United Arab Emirates, UAE), providing sediment that was driven by strong northwesterly winds to form dunes (Al-Farraj & Harvey 2004). About 7000–8000 years ago, during the Climatic Optimum, rainfall stabilized dunes and filled temporary lakes (Schulz & Whitney 1986; Glennie 1998). Over the past 5000 years, the climate has become increasingly arid. Dunes are more active, lakes have desiccated and been transformed to salt-covered playas (inland sabkhas, or sebkhas), and wadis have minimal flow.

Relief on the peninsula is generally low, seldom rising above 500 m. In southeast Arabia, the Hejaz Asir and Oman Mountains reach heights of 3000 m. Large ergs cover 795,000 km², with the main fields being the Rub' al Khali (or Empty Quarter), An Nafud, and Ad Dahna of Saudi Arabia, and the Wahiba Sands of the Sultanate of Oman. These sweep in a broad arc and are approximately aligned with the clockwise path of the Shamal wind across Arabia (Glennie 1998). The Rub' al Khali consists of stabilized sand and mobile dunes, with sand hills reaching heights of 150–250 m (Abd El Rahman 1986). It lacks rivers or sizeable springs. The northeast is dominated by crescentic dunes, the eastern and southern margins by star dunes, and the west by linear dunes, the latter with interdune corridors approximately 2 km in width.

The Wahiba Sands are the most extensively examined sand sea in the region and provide evidence of several periods of dune-building activity under different climatic regimes (Goudie et al. 1987; Allison 1988; Dutton 1988). These dunes are related to the southwest monsoon system. About 50% of the grains are quartz, 30–40% carbonate (from coastal areas), and approximately 10–20% are dark ophiolite grains derived from the Oman Mountains (Glennie 1998). Areas devoid of sand are covered by reg, and in places incised by wadis. Orographic rain in the highland areas gives rise to relatively lush vegetation (Allison 1997).

Along the coast of the UAE there is a broad zone of coastal sabkhas. These are saline flats above the mean high tide level, but subject to periodic inundation. In a landward direction, interdune sabkhas (akin to pans or playas) are found. Sites of ancient human habitation suggest that these once held shallow freshwater lakes (Glennie 1998).

2.2.2.4 Iran and Iraq

Much of Iran is a continental plateau interior of relatively low relief. There are two significant mountain ranges: the Elburz Mountains on the southern margins of the Caspian Sea, where elevations exceed 5000 m, and the Zagros Mountains to the northeast of the upper Persian Gulf, reaching to over 4000 m (Figs 2.4 and 2.8). Only the northern slopes of the

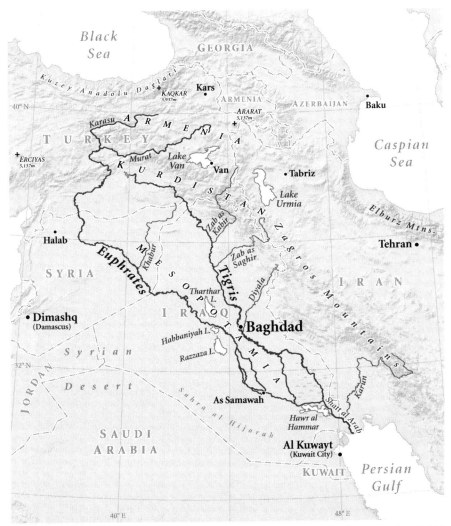

FIG. 2.8 The Tigris and Euphrates Rivers are fed from snow that falls in the high mountains (3500–4500 m ASL) of Turkey, Iraq, and Iran. The Euphrates River is the longest river (3000 km) in western Asia.

Elburz Mountains were glaciated during the Quaternary (Breckle 1983).

The annual temperature range in Iran is high, exceeding 50°C in summer and falling below freezing in winter. Strong local winds and low rainfall (usually less than 200 mm year^{-1}), and a lack of surface and ground water, result in a harsh environment (Allison 1997). The Dasht-e Kavir (Great Kavir) (Fig. 2.9) is characterized by salty conditions and flat pans and playas (Beaumont 1968). The Dasht-e Lut is a sandy desert, covered by loose sand and desert pavement. Extensive fields of yardangs in the Lut Desert are formed in playa sediments. Gullies on the flanks of these yardangs deepen during the winter rainy season, but aeolian processes dominate during the rest of the year (Krinsley 1970).

Iraq may be divided into four topographical areas (Fig. 2.8): (1) the northeastern highlands, which include the Zagros Mountains, (2) a largely arid region between the Tigris and Euphrates Rivers, (3) a marshland zone, the Shatt el-Arab (Shatt al-Arab), where the Tigris and Euphrates coalesce (discussed in detail in Section 4.7), and (4) an extensive desert region in the south and west which constitutes part of the Arabian and Syrian Deserts. Most of the rainfall occurs between December and March. The mean annual precipitation at Baghdad is about 140 mm. The prevailing winds are the Sharqi or Sirocco, a southeasterly dry, dust-laden wind, and the Shamal, a northwesterly dry cool wind. Shamal winds cause some of the most destructive dust storms in the Middle East.

2.2.3 EUROPE

There are no areas of extreme aridity in Europe. However, there are areas of semiarid climate along the Mediterranean rim, including areas of southern Italy, Sardinia and eastern Greece, and central and eastern Iberia (Shaw 1997). These regions have a long history of human settlement.

FIG. 2.9 The Dasht-e Kavir is a largely uninhabited area of Iran characterized by salty conditions, flat pans, shorelines, and playas. NASA Landsat image, USGS EROS Data Center.

Fluvial erosion has been the dominant geo-morphic process. Areas of the Mediterranean have become important for the study of alluvial fans (Harvey 1988) and piping, rills, and badland slope processes (Gutiérrez et al. 1988; Imeson & Verstraten 1988). Yardangs in the semiarid Ebro Depression of Spain attest to the power of wind erosion during more arid climatic phases (Gutiérrez-Elorza et al. 2002).

2.2.4 ASIA

The dry zones of central, east, and south Asia extend across vast areas (Fig. 2.4). They are dominantly deserts of continental interiors, far from the moderating influence of the ocean, and periods of cold temperatures occur each winter. Regional ecosystem studies have not advanced far beyond the descriptive stage. The deserts occur over a wide range of elevations and tectonic settings, from the Turpan Depression (just over 150 m BSL) to the eastern area of the Tibetan Plateau (elevations of more than 5000 m). The term Middle Asia has been used to describe the western part of the deserts, and Central Asia for the eastern regions (Walter & Box 1983d), with the two parts climatically and floristically different from one another. The boundary between the two is formed by Dzungaria and the mountain systems of the Tien Shan and the Pamiro-Alay.

In Middle Asia, the precipitation maximum is in winter, with springtime vegetation well developed. Precipitation decreases from west to east. The desert extends to the south toward Afghanistan and Iran. In Central Asia, frontal storms from East Asia move to the west in summer, and precipitation decreases from east to west, reaching a minimum in the Taklimakan Desert. Springtime vegetation is absent, and plant growth begins after the first summer rains (Walter & Box 1983d). To the south is the high plateau of Tibet and Pamir, a cold desert, with a very short summer and frost occurring throughout the year.

2.2.4.1 Middle Asian deserts

The Caspian Lowland is at the meeting point of the Middle East, Europe, and Asia (Fig. 2.4). It is a semi-desert that becomes increasingly arid from north to south and from west to east. On the border of the Caspian Sea, the world's largest lake, the annual precipitation is 156–177 mm (Walter & Box 1983a). In winter, the Siberian high-pressure system causes very cold temperatures, from −30°C on the coast to −40°C inland. Summers bring hot weather (maxima up to 40°C) and dust storms that cause extensive damage to cultivated land.

At times during the Pleistocene, the lowland area was under water, with a connection through to the Sea of Azov. The surface slopes at a low angle from north to south, falling to below sea level (28.5 m BSL) at the surface of the Caspian Sea. Twentieth-century fluctuations in lake area and volume have caused serious economic and environmental problems, adversely affecting fishing, agriculture, and oil and gas exploration (Leroy et al. 2007). The water surface area was 420,000 km^2 in 1969 and by the 1970s had diminished to 370,000 km^2 (Walter & Box 1983a). It began to rise again after 1978 (Clauer et al. 2000), flooding towns and engineering works. The present surface area is approximately 435,000 km^2. The causes of the historical changes in sea level are not yet fully understood, with climate, human impact, and tectonic influences appearing to interact to influence fluctuations (Leroy et al. 2007).

In the southern Caspian Lowland (Fig. 2.4), over-grazing by sheep, excessive wood use, and the cultivation of sandy areas in the mid-1800s destroyed the plant cover and steadily increased the area of shifting barchan dunes. An average of 1.4 million m^3 of sand was blown annually into the Volga, disrupting shipping (Walter & Box 1983a).

To the east of the Caspian Sea lie the arid areas of the Karakum Desert, part of the Turanian Lowland (Fig. 1.2). The Amu Dar'ya river system, fed by the snow and glaciers at elevations around 5000 m in the Pamir Mountains, has changed course continuously across the region, originally emptying into the Caspian Sea, but in the latter part of the Pleistocene, moving north to reach the Aral Sea. Regional dune forms include nebkhas (low sand hills anchored by vegetation), individual barchans that migrate up to 100 m per year, and barchan fields where dune migration is much slower (Walter & Box 1983e).

The Aral Sea, once the world's fourth largest lake, has shrunk in area by 60% since 1960 owing to diversion of water from its tributaries for irrigation (Micklin 1988). The exposure of the former lake bed and consequent dust and salt storms are discussed in more detail in Chapter 13. Other interferences in the natural environment are those connected with grazing by goats, camels, and sheep, whose destruction of native vegetation around watering holes has led to the formation of barchan fields (Walter & Box 1983e).

2.2.4.2 Deserts of India and Pakistan

The Great Indian Desert, or Thar Desert, occupies about 2.34 million km^2 in the northwestern sector of India and the adjoining area of Pakistan (Fig. 2.4). It lies at the easternmost reaches of the Sahara-Arabian desert tract and at the northwestern limit of the southwestern monsoon. About 85% of the desert lies in India, mainly in Rajasthan. To the west and northwest lie the enormous alluvial plain and delta of the perennial Indus River. Along the coast is the Great Rann of Kutch, a zone of siliclastic sabkhas. The foothills of the Himalayas and an attendant belt of alluvial fans border the northeast. Rainfall in the desert averages between 100 and 500 mm year^{-1}, 90% of which is received between July and September during the monsoon season. The northern reaches of the desert receive some winter rainfall and minimal summer monsoonal precipitation. Winds are relatively strong, averaging 4–8 m s^{-1} in June (Mann et al. 1984). Climatic change in the Thar Desert appears to be strongly influenced by variability in the southwest monsoon, which is affected by changing land–sea thermal contrast and solar insolation. During the LGM, eustatic lowering of sea level, combined with changes in climate and neotectonic factors, caused incision of river systems near the coast (Juyal et al. 2006).

River flows in the Thar Desert tend to be short-lived and occasionally catastrophic. The Luni River, which originates in the Aravalli Hills, is the only well-integrated river in the desert. The riverbed is normally dry, and measurable flows occur only in response to high-magnitude rainfall events during the monsoon season (June to September). The river flooded catastrophically in July 1979, widening from a pre-flood condition of 40–700 m to about 500–1360 m after the flood. The sediment loads of the Luni River are large, and post-flood deposits were as much as 20–60 cm in depth (Kale et al. 2000).

Sand dunes cover much of the Thar, with sediment supplied from the coastline, the alluvial plains, and the weathering of sandstones and granites. Close to the coast, the dunes may be highly calcareous and cemented, forming aeolianites (calcarenites), or lithified sand dunes. Inland, quartz sands dominate, although foraminiferal tests may still be found (Goudie & Sperling 1977). Elongate parabolic dunes are the dominant form in the Thar Desert (Breed et al. 1979), often with small dry lakes within their noses. Luminescence dates on relict linear and parabolic dunes to the south and east of the region suggest

that the last major phase of spatially extended aeolian accumulation of high dunes occurred about 13,000 years ago (Chawla et al. 1992).

Relative to other deserts, the Thar is heavily populated, reaching a density of 84 persons per square kilometer in the Rajasthan region (Derbyshire & Goudie 1997). In areas where the rainfall exceeds 250 mm year^{-1}, much of the land is under cultivation in a system of fallow farming. Farming on sandy, erodible soils has led to instability and accelerated sand movement. In other areas, livestock herding is the dominant land use, but overgrazing, combined with fuelwood collection, has resulted in land degradation (Mann et al. 1984). The natural basal cover of dune plants is probably 10–12%, but in areas of high grazing and fuelwood pressures it is less than 3%. Trampling by animals has ruptured the fragile inter-particle bonds and destroyed root systems, resulting in a significant increase in sand mobility over the past century (Kar et al. 1998). Sand is most mobile during the strong winds of March through July that precede the monsoon. Partially stabilized, but recently reactivated, parabolic dunes move at 0.5–9 m year^{-1} (Kar 1993). Barchans have developed in the immediate vicinity of settlements and there is secondary salinization in irrigated lands. Efforts to combat desertification have included the introduction of fast-growing exotic tree species, including eucalyptus and acacia; the stabilization of dunes by afforestation and barrier construction; the planting of shelterbelt plantations to reduce wind velocity; and ecological regeneration by the use of fencing and exclosures (Chauhan 2003).

Satellite imagery reveals that dust is mobilized from the Thar Desert. Loess deposits are extensive in nearby areas, and late Pleistocene ocean core sections in the Indian Ocean suggest that the wind was formerly more intense (Kolla & Biscaye 1977).

2.2.4.3 Deserts of China and Mongolia

Aeolian sediments of the Chinese loess plateau indicate that the onset of Asian aridity began at least 22 Ma in the early Miocene (Guo et al. 2002). During the Pliocene, Himalayan tectonic movements and the uplift of the Tibetan Plateau intensified desiccation (Songqiao & Xuncheng 1984). Quaternary fluctuations in climate in China and Mongolia were marked. High lake levels suggest much more humid conditions during Marine Isotope Stage (MIS) 3 (60–20 ka), although there appear to be significant regional variations in the timing of the more humid phases.

During the LGM (≈21–18 ka), sea level was 120–130 m lower than today and the Chinese coastline was located 600–1000 km east of its present-day position. The reduced sea-surface temperatures and enhanced continentality reinforced the cold, dry interior climate, depressing the mountain snowlines in western China by 300–1000 m, and causing glaciers to expand. The dry winter monsoon strengthened (Lu et al. 2000), deserts were more extensive than at present, and sand-dune areas were found eastwards and southwards of their present position. Wind-borne dust from the deserts helped to develop the Chinese Loess Plateau, an area of 500,000 km², with a thickness varying from 150 to 300 m (An 2000). The plateau lies between 30 and 40° north, with the thickest deposits in the western part of the region. During cold climatic periods, when the winter monsoon dominated, coarse particles were deposited and the sedimentation rate was higher than when the summer monsoon was strong. The expansion of the loess deposits was coeval with low levels of inland lakes (see Chapter 5). The strengthening of the summer monsoon brought more rainfall to the loess plateau, and during these periods dust particles were finer, pedogenic processes were intensified, and palaeosols developed (Lu et al. 2000). During the Holocene, the distribution of deserts and loess was reduced. In western China, high lake levels indicate wetter conditions during the early Holocene (Yang et al. 2003, 2004).

The modern arid and semiarid zones of northwest China constitute about one-quarter to one-third of China's total land area and include all of the Xinjiang Autonomous Region, the Hexi corridor in Gansu Province, and the area west of Helan Mountain in Inner Mongolia. The topography is rugged, with alternating high mountains and basins, ranging from the Turpan Depression (155 m BSL; Fig. 2.10) to the western Tibetan Plateau (≈5000 m ASL). Precipitation ranges from 40–150 mm year⁻¹ on the plains to 300–900 mm year⁻¹ in the mountains (Genxu & Guodong 1999). Meltwater from high-altitude snow and glaciers accounts for 23–43% of the total runoff and flows towards intermontane basins to form terminal lakes. The ongoing tectonic activity of the region is expressed in very high erosion potentials and rates of sediment production. High mountains and plateaus are juxtaposed next to deep basins, rivers, and lakes, providing a sedimentary sequence with the potential to yield high-resolution records of past environmental changes.

Fig. 2.10 Arid and semiarid zones of northwest China occur in areas of rugged topography. Meltwater and stormflow from the high-altitude Bogda Shan flow south towards the Turfan Depression (see also Fig. 8.10). Ongoing tectonism is expressed in high erosion and sediment-production rates. The enormous fans transport course, rounded debris, forming the gobi or gravel plains of the region. Wind activity is expressed in the dune field and in the presence of yardangs immediately to the west of the dunes. Source: Landsat-7 image, NASA/U.S. Geological Survey. See Plate 2.10 for a color version of this image.

For the most part, gravels or sands cover Chinese deserts, with an estimated 42% gobi, or gravel deserts, and the remaining 58% shamo, or sandy deserts. There is a general north–west to south–east transition from gobi, through shamo, to loess (Fullen & Mitchell 1991).

Deserts in China are affected by both the East Asian monsoons and the global westerlies, and a change in either of these systems influences regional climate. The interplay between summer and winter monsoonal circulation plays a critical role. The winter monsoon is governed by the Siberian high-pressure system, which affects western and central China and brings cold and dry continental air southward to about 20°N. The summer monsoon brings warm and moist air masses from the Indian Ocean to southern Tibet and southern China, and from the Pacific Ocean to eastern China. Changes in the intensity and range of these monsoons influence regional and temperature patterns across China. The eastern Tibetan Plateau is extremely dry as it is partly sheltered from the summer monsoon and is influenced by the cold and dry winds of the winter monsoon. Average annual rainfall in the summer is less than 100 mm and in the winter it is less than 10 mm. Northwestern China is also dry and lies beyond the limits of both the Pacific and Indian monsoons. Summer rainfall is less than 100 mm and winter rainfall, brought by the westerly winds to the northernmost part of Xinjiang, is generally less than 50 mm (Yu et al. 2001).

Biogeographically, the arid areas of northwestern China are affiliated with the eastern section of the Eurasian deserts and steppes. Conifer (*Picea*) forests are found at altitude in the Altay and Tien Shan Mountains. The modern snowline lies between 3700 and 4200 m in the Xinjiang region.

The ice mass over the mountainous regions of China provides a key source of water in the summer months. As an apparent response to global warming, about 95% of the glaciers in the Tianshan, Altay, and Pamir have shown a negative mass balance over the past 30 years, with the rate of loss accelerating in recent years. It is anticipated that this will have alarming consequences for arid areas of China, where glacial meltwaters form a major water resource. Glacial melt provides the principal dry-season water source for the 23% of Chinese that live in the arid western regions of the country. Although supplies of water are currently plentiful, if trends continue the glaciers will decline to the point that their contribution to the region's water supply will cease (Barnett et al. 2005).

Many of the desert areas of China appear to have spread over the past 25 years, and an area of about 176,000 km^2 has been described as desertified (Zhu & Liu 1988; Zhu 1989). Mitchell and Fullen (1994) describe a "blister-like" process in which desert-like conditions develop as small pockets within rangelands, away from the desert areas. These "blisters" enlarge, spreading and merging, and gradually increase the area of desert. The fragile desert margin has also been deforested within historical times and shrubs and trees continue to be removed. About 8% of desertified land has formed within historic times.

The largest deserts in China are the Taklimakan, Gurbantunggut, Badain Jaran, Qaidam Basin, Tengger, Ulan Buh, and Mu Us (Fig. 2.4). Of these, the most extensive (337,000 km^2) is the Taklimakan (Taklamakan or Taklemagan), located in the Tarim Basin. In the Uigur language, the name of the desert means "journey without coming back," alluding to the extensive dune fields and highly arid conditions that characterize the core of the desert. Mean annual precipitation is approximately 70 mm on the desert margin, decreasing to approximately 20 mm in the center. Mobile dunes cover about 85% of the Taklimakan and reach a height of over 100 m. There are numerous forms, including compound longitudinal dunes, compound barchans, dome-shaped dunes, and barchan dunes. The eastern Taklimakan has the strongest wind activity and lowest precipitation (Wang et al. 2004a). As the wind energy is generally low in other areas of the Taklimakan, it is likely that dune scales were determined in large part by the wind environment of the LGM (X. Wang et al. 2005).

The Gurbantunggut (Guerbantonggute) Desert in the Dzungarian or Zhungar (Junggar) Basin is the second largest desert in China (49,000 km^2; Fig. 2.4). A temperate desert, it receives more rainfall (70–150 mm year^{-1}) than the Taklimakan, and most of the dunes are stabilized by vegetation. Most rain falls in the spring. Vegetation consists of psammophyte (sand-loving) communities, salt desert plant communities, and ephemeral plants. In the southern part of the desert, microbiotic crusts cover as much as 70–80% of the surface (Chen et al. 2007). Moving from west to east across the basin, the landscape changes from desert steppe, to stable dunes, to barren gobi.

The Baidan Jaran Desert (Fig. 2.4) is the site of the highest dunes in China (400–500 m) (X. Wang et al. 2005), interdunal lakes (Yang et al. 2003), gobis in the piedmont zones, and extensive relict lake beds that are undergoing deflation. The Qaidam (Tsaidam, Chaidamu) Basin is the site of large yardangs that reach a height of 50 m, are several kilometers in length, and are aligned northwest–southeast, parallel to the predominant wind direction. The Tengger Desert consists largely of low mobile dunes. Lakes and swamps were widely distributed in this area during the late Pleistocene. The Tengger Desert transitions northward into the sandy Ulan Buh Desert (Fig. 2.4). East of the Helan Mountains lie several other small sandy deserts (such as the Mu Us Desert), in which the dunes are partly or totally stabilized by vegetation, including not only grasses and shrubs, but also trees. About 70% of the annual summer precipitation comes from the summer monsoon, and the average annual precipitation ranges from 150 to 500 mm (Yang et al. 2004). Mobile dunes in this area occur largely as a result of human activity. Deserts above 3400 m ASL on the Tibetan Plateau include dunes that are often located in areas of mountain permafrost.

Owing to their high latitude, altitude, and interior position, the arid areas of Mongolia are relatively cool. At Bayankhongor (1900 m), for example, the rainfall averages 198 mm year^{-1}, the January mean air temperature is $-18°C$, and the July mean only $16.3°C$. Most of the deserts are gravel deserts (gobi) and sand covers less than 3% of the area. Increases in the total number of livestock in the region have raised concerns about desertification (Yang et al. 2004).

2.2.5 SOUTH AMERICA

South America includes a great diversity of arid and semiarid regions and grasslands (Fig. 2.11). To the far northwest, in Venezuela and Colombia, is the Pericarribean Arid Belt, with relatively limited arid areas and more extensive semiarid habitats. In north-eastern Brazil is the semiarid Caatinga scrubland. Further south, the South American dry zones stretch in an almost continuous belt along the Andes from the hyperarid, subtropical Pacific coast (Atacama Desert), diagonally across the high-altitude (3800 m ASL) intermontane desert of the Altiplano-Puna, to the Patagonian Desert on the southeast side of the continent (Berger 1997).

Uplift has created a region of great relative relief, with Andean peaks up to 7000 m ASL. The tectonic setting ranges from the active subduction margin along the west coast, to the uplifted plateau of the Altiplano-Puna, to the continental craton and passive Atlantic margin. The Altiplano is a large (200,000 km^2) closed basin composed of endoreic basins formed by Tertiary and Quaternary tectonic and volcanic activity. It stretches for 1800 km along the central Andes and varies in width from 350 to 400 km.

Several factors account for the aridity of the subtropical and midlatitude deserts in South America. The subtropical deserts lie at the latitude of the southeast trade winds, with a prevailing easterly flow direction. The Andes are a barrier to storms and produce a pronounced rainshadow effect on the west coast, preventing humid air from the Amazon basin from reaching the zone of the Peruvian and Atacama Deserts. The Peru, or Humboldt, Current flows northward along the west coast and cold upwelling water suppresses evaporation and convection. An anticyclone, positioned off the coast of western South America, brings dry air to the region via subsidence from the subtropical jet. It also acts as a block, limiting the incursion of winter storms from the south (Latorre et al. 2005). The Andes increase the positional stability of this anticyclone, forming a barrier

against its eastern margin (Hartley 2003; Warner 2004). As a result, there is a strong temperature-inversion layer at approximately 1000 m in altitude, which brings about a thick stratus layer, high humidity, and low, stable temperatures. Further south, in the mid-latitudes, the prevailing wind flow is westerly, and the Andes produce a rainshadow to the east where the Patagonian and Monte Deserts are located.

The western margin of South America has a long history of aridity. Hartley et al. (2005) suggest that the Atacama is the oldest extant desert on the planet, with sedimentary successions indicating that an arid to semiarid climate prevailed from the late Jurassic (150 Ma) to the present. Extensive gravelly fluvial deposits (the Atacama Gravels) are incised, resulting in prominent late Miocene- to Pliocene-age fans and terraces. Fluvial incision and deposition ceased near the Plio-Pleistocene boundary, as rainfall declined (Ewing et al. 2006). Hyperaridity in the Atacama Desert appears to have developed between 4 and 3 Ma, when the central Andes were already close to their present elevation. The transition to hyperaridity was therefore not linked to an intensification of the rainshadow effect, but was more likely associated with general global cooling in the Cenozoic and the closure of the Central American Seaway between 3.5 and 3 Ma, which severed the connection between the equatorial oceans of the Pacific and Atlantic (Hartley 2003).

2.2.5.1 The west coast deserts: Peru–Chile, Atacama, and Sechura deserts

Along the west coast of South America, deserts stretch from southern Ecuador through Peru to northern Chile. The Andes run from north to south along the coast of South America in a number of parallel chains, cut by deeply incised transverse valleys. Dry conditions exist along the coast, on the lower levels of the western mountain slopes, and in the inter-Andean valleys (Rauh 1985).

The Peruvian–Chilean desert is one of the driest regions of the globe. It is extremely long and narrow, extending 3700 km along the coast from north of Chiclayao in northern Peru to the Santiago area of central Chile. Ranging in width from a few kilometers to 15 km, it is marked on its eastern edge by the Cordillera Domeyko. The most intense aridity occurs in the Atacama Desert of northern Chile, with less than 10 mm of annual precipitation: at Arica it averages less than 1 mm (Larrain et al. 2002). The

FIG. 2.11 Map of the South American deserts.

The following labels appear on the map:

PERICARRIBEAN ARID BELT

Caracas

Venezuela
Guyana
Suriname
French Guiana

10° N

Bogota

Colombia

G U I A N A   H I G H L A N D S

Quito

Ecuador

Equator

A M A Z O N

S O U T H

B A S I N

Chiclayo

Peru

CAATINGA

Recife

Callao
Lima

Brazil

10° S

SECHURA DESERT

La Paz

Bolivia

❶

B R A Z I L I A N

Arica

A M E R I C A

H I G H L A N D S

ATACAMA DESERT

❷

Paraguay

Sao Paulo

Antofagasta

Asuncion

Tropic of Capricorn

CHACO

CHILEAN MATORRAL

MONTE DESERT

Buenos Aires

30° S

Pacific Ocean

Santiago

PAMPAS

Uruguay

Argentina

Atlantic Ocean

Patagonia

Chile

PATAGONIAN DESERT

50° S

❶ Altiplano/Puna
❷ Rio Loa

Deserts
Arid Regions

Source: *Names and areas based on WWF Terrestrial Ecoregions of the World*

90° W   70° W   50° W   30° W

Peruvian and Atacama Deserts are separated on the basis of history and politics, but are really one geographic unit that will henceforth be referred to as the Atacama Desert. The Sechura Desert is a subregion in Peru that is covered with migrating sand dunes, including extensive areas of barchans (Gay 2005). It includes the lowest point in Peru, the Bayóvar Depression, which is 37 m BSL.

Owing to the presence of the cold upwind Peru Current, the coastal air temperatures are moderate and vary little along the full extent of the desert. This can be seen by comparing two coastal cities separated by 11° of latitude. The mean annual temperature at Callao, Peru (latitude 12°S) is 19.2°C, whereas that of Antofagasta, Chile (latitude 23°S) is less than 1°C cooler at 18.6°C (Warner 2004). Similarly, the temperature difference between the coolest and warmest months is only about 4°C. Further inland from the coast, the range of temperatures is greater owing to factors such as cold-air drainage, shading from surrounding mountains, and increases in elevation (Warner 2004). The trade winds in the region are strong but shallow.

Three types of fog – advection, orographic, and radiation – bring moisture to the desert. These fogs are the most important sources of water for biological soil crusts and plants in the Atacama (Cáceres et al. 2007). The South Pacific Anticyclone produces strong subsidence and a thermal inversion along the coast, allowing only stratus and stratocumulus clouds to form. These clouds form hundreds of kilometers from the coast in the southeastern Pacific Ocean and are advected inland, entering the continent through corridors in the coastal mountain range. They reach the coast at altitudes of between 400 and 1100 m, with the horizontal fog flux increasing with altitude (Cereceda et al. 2002). The distance that the fog penetrates inland depends on air temperature, humidity, and wind speed, and the fog persists until the heat of the land evaporates the cloud droplets. In Peru this fog is referred to as the *garúa*. It occurs south of latitude 8°S (Rauh 1985) and condenses on plants, flowing down stems to provide considerable annual soil water. The Chilean fog is locally called *camanchaca*. In localized areas along the coast, orographic fog forms *in situ*, related to oceanic upwelling, relief, and topography. Radiation fogs sometimes form inland on the Pampa del Tamarugal, with moisture possibly derived from groundwater inflow to the salt flats present in the area (Cereceda et al. 2002).

El Niño, an increase in the surface water temperatures, regularly disrupts the climate of Peru and brings heavy rains to the desert. During intense El Niño cycles, there can be catastrophic effects on the ecology and economy. However, in most events, there is a strong positive biological response (Jacsik 2001), marked by a tremendous increase in the number of ephemeral plants (see Chapter 11).

The desert of northern Chile may be the driest location on Earth. Oberlander (1997) refers to stagnant erosion or "erosional paralysis" in this region, a result of hyperaridity. The Rio Loa is the only through-going river extending from the coastal range to the sea. Two other rivers lose water to irrigation and do not reach the ocean. Numerous channels carry water only during the rare heavy precipitation event.

The geomorphology is characterized by the important role of salt, in salt tectonics, salt-crust development, salt heave and slope equiplanation, and salt weathering. The salts are the result of dry fallout from the atmosphere and fog-condensate over millions of years, volcanic activity, the breakdown of rock, and salt accumulations in geological deposits (Ericksen 1981; Berger 1997; Berger & Cooke 1997; Ewing et al. 2006). The soils of the Atacama are unique in that silicate dust and salts have accumulated over long time periods with only negligible losses from weathering. Nitrate-rich soils mantle the landscape, with the highest concentrations adjacent to or downwind of playas (salars) (Ewing et al. 2006). Where nitrate is concentrated (>5% by mass), it is commercially viable (Ericksen 1981).

Chains of salars, nitrate and sulfate salt flats, lie in depressions at altitudes of 700–1100 m ASL. The Salar de Atacama (Atacama Salt Flat) is the remnant of a once vast lake that has now largely evaporated. Laguna Chaxa, in the middle of the salt flat, is a shallow pool of brackish water inhabited by several flamingo species. The Salar de Uyuni covers more than 10,000 km^2 and is one of the world's largest salt flats.

The dry climate of the Atacama causes nutrient- and water-cycle curtailment, to the extent that the life cycles and physiological behavior of plant life are closely tied to the small-scale spatial and temporal variation in microclimate. One of the most notable characteristics of the Atacama Desert is the almost complete lack of vegetation from sea level to 2500 m ASL. A few localities support fog-zone desert communities between sea level and the top of the fog zone. These thick, seasonal growths of grasses and low plants are termed *lomas* and are home to numerous endemic plants. Lomas subhabitats with unique floristic qualities are defined by distance from the ocean and altitude with respect to the fog layer. Along the Chilean coast, the thick stratus layer banks against broad terraces and supports flora and fauna between approximately 300 and 800 m. In the interior desert, plant life is limited to areas of salt-water intrusion (aguadas) or to arroyos with concentrated runoff (Latorre et al. 2002). Moist valleys in the northern Atacama harbor some trees (*Prosopis*, *Salix*, *Schinus*, and *Acacia*) and shrubby and herbaceous plants. At about 3000 m, a modest cover of shrubs and grasses appear. There are also limited areas of irrigated agriculture. In the hyperarid core, primary production is principally limited to hypolithic cyanobacteria that survive under select translucent stones, such as quartz, existing in small spatially isolated islands amidst a microbially depauperate bare soil (Warren-Rhodes et al. 2006).

Along the coastal Atacama Desert, the lack of vegetation, strong onshore winds (the southeast trade winds), sand availability, and aridity allow the development of dunes. Sand is provided to the beaches from the Andes Mountains and rapid coastline emergence continuously exposes new expanses to the wind (Gay 2005). The sand moves inland to elevations of 500–2000 m ASL, where it accumulates in sand seas covering hundreds of square kilometers, the largest of which (900 km^2) lies just west of Ica.

2.2.5.2 Altiplano/Puna

The Altiplano is a semiarid, high-altitude (3800 m ASL), volcanic plateau situated between the eastern and western cordillera of the central Andes between 15 and 28°S. It is sometimes considered an extension of the Atacama Desert. The southern end belongs to the active Ojos del Salado volcanic region and is characterized by major volcanic structures such as calderas, strato-volcanoes, ignimbrites, and compound volcanoes. Fault-bounded, topographically closed basins contain siliciclastic, carbonate, and evaporite sediments (Valero-Garcés 2000).

The cold and dry climate of the Altiplano is characterized by fluctuations in daily temperature (up to 40°C) that are greater than seasonal temperature variations. Potential evaporation is 1500 mm or greater and exceeds precipitation by a factor of about six. The majority (60%) of the precipitation falls from intense convective storms during the summer (December–February), with pronounced interannual variations that range from 50 to 500 mm. Precipitation grades from the very dry and stable atmosphere

on the southern Peru–northern Chile desert border to the more humid northeastern sector which lies to the west of the Amazon basin. In the southern Altiplano, the dominant source of moisture is winter precipitation in the form of snow. Intense summertime storms trigger flash floods over the slopes of the Andes. Snowmelt and summer rains recharge the groundwater and control the regional hydrology.

El Niño Southern Oscillation (ENSO) conditions play a role in the variability of the rainfall received. By contrast to coastal Peru and Ecuador, which suffer heavy rains during El Niño events, the Altiplano often experiences drought conditions (Thompson et al. 1992; Fontugne et al. 1999), which inhibit the preservation of sedimentary records. The hypothesis that climatic variability is out of phase between the coastal regions and Altiplano is difficult to test, but preliminary research suggests that such conditions may have characterized the Holocene (Fontugne et al. 1999). In the Altiplano, the warming of the tropical troposphere and strengthening of the westerlies over the central Andes that occurs during the negative phase of ENSO produces a dry pattern. Lake Titicaca, which occupies the northernmost third of the plateau, experiences a decrease in water levels during El Niño events. During the cold ENSO phase (La Niña years), there is a tendency for wet conditions, with easterly flow encouraging summertime deep convection by transporting moist air from the Amazon Basin (Garreaud & Aceituno 2001). The southward displacement of the Bolivian High also allows stronger easterly winds to entrain Amazonian air and bring it into the Altiplano (Rowe et al. 2002). However, as in other areas of the world, the correlation between ENSO events and rainfall is not always strong and there are other factors that affect rainfall variability.

2.2.5.3 Monte Desert

East of the Andes, the Monte Desert lies in the subtropical continental interior of northwest Argentina, in an area of basin and range structure lying between latitude 27 and 44°S. Owing to its interior position, daily and annual temperature ranges are considerable. Summer rains dominate the northern part of the desert and winter rains the southern. Rainfall ranges from 80 to 250 mm annually (Mares 1999). Landforms include pediments, badlands, arroyos, active- and palaeodune fields, deflation basins, and salt deposits. The vegetation is relatively homogeneous and dominated by open thorn scrub, typified by the genus *Larrea*. Tall cardón cacti are found in the northern areas of the desert on rocky slopes and bajadas. Human land use includes deforestation for fuel wood, cattle and sheep grazing, agriculture, mining, and oil exploration. As a result, many elements of the biome, including endemic species, are endangered (Asner et al. 2003).

2.2.5.4 Patagonian Desert

The Patagonian Desert, the largest arid zone in the Americas (670,000 km^2), is located to the east of the Andes in southern Argentina. It lies at temperate latitudes, stretching between 39 and 50°S, and is thus a relatively cold desert, with an average temperature of about 16°C in the north and 8°C in the south. Owing to its extent and diversity, the climate is difficult to characterize. The annual rainfall ranges from 100 to 260 mm (Mares 1999), with snow falling in winter. Summers can be very hot. The vegetation is cold desert shrub steppe and is associated with an abundance of animal species. During the Cenozoic glaciations, enormous volumes of sediment were transported into the region by fluvial processes, and this helps account for the largely stony nature of the surface (Berger 1997).

2.2.6 NORTH AMERICA

North America includes a diverse number of small deserts stretching from southeastern California to western Texas, and from Nevada and Utah to the Mexican states of Sonora, Chihuahua, and Coahuila and much of the peninsula of Baja California. Beyond these areas, semidesert conditions extend north to eastern Washington, south onto the central Mexican plateau, and east to link with the steppes of the High Plains. In area, about 55% of the North American deserts are considered semiarid, 40% arid, and only 5% hyperarid. This section focuses on the core deserts, namely the Chihuahuan, Sonoran, Mojave, and Great Basin Deserts (Fig. 2.12).

North American deserts owe their aridity to a rainshadow effect caused by mountains blocking moisture of Pacific origin in the winter and the Gulf of Mexico in summer, the occurrence of a subtropical high-pressure cell, and cold west-coast currents. Rainshadow effects are greatest in the Great Basin and Mojave Deserts, the effect of high pressure is important in the Chihuahuan and Sonoran Deserts, and cold coastal currents play an important role in the deserts of Baja California.

From a climatic perspective, the deserts are classified according to their temperature and the seasonality of precipitation. "Cold" and "warm" deserts are distinguished based principally on winter temperatures, as all deserts experience hot summer temperatures. The Great Basin, characterized by its northern position, high altitude, the receipt of 60% of its winter moisture as snow, mean monthly temperatures below 0°C from December through February, and mean annual temperature of 9°C, is considered a "cold" desert (Laity 2002). The more southern warm deserts have a mean annual temperature of 20°C, and essentially all of the precipitation occurs as rain. The number of frost-free days averages 80–150 in the Great Basin and 210–365 in the Mojave, Sonoran, and Chihuahuan deserts (McMahon & Wagner 1985). The Great Basin, Mojave, and western Sonoran deserts receive winter/spring precipitation derived from Pacific frontal systems. Moving from west to east from the Mojave to Chihuahuan Desert, the ratio of winter to summer rainfall decreases. Winter precipitation, falling from November to February, makes up 44% of the annual total in the Mojave Desert, 38% in the Great Basin Desert, 28% in the Sonoran Desert, and 18% in the Chihuahuan Desert. The eastern Sonoran Desert mainly has a bimodal rainfall regime, with a strong primary maximum in July–August and a secondary maximum in February. The Chihuahuan Desert receives summer rainfall. Summer precipitation, falling from June to September, comprises 59% of the annual total in the Chihuahuan Desert, 53% in the Sonoran Desert, 27% in the Mojave Desert, and 25% in the Great Basin Desert (Smith et al. 1997).

The North American monsoon, known also as the Southwest United States, Mexican, or Arizona monsoon, is centered over northern Mexico and acts to increase summer rainfall over the southern deserts, but the effects tend to be both spatially and temporally variable. In addition to regional differences in precipitation, there are also significant elevational gradients, with lower altitudes receiving less and more episodic rainfall than moister, higher elevations. Precipitation approaches 500 mm year^{-1} in the higher mountains, but is less than 100 mm year^{-1} at low elevations near the Gulf of California and Death Valley.

Freezing weather and the seasonal and altitudinal differences in rainfall between the deserts are reflected in vegetation structure and floristic composition. The cold desert areas are primarily semiarid with steppe vegetation, dominated by the evergreen

Key:

❶ Great Salt Lake
❷ Pyramid Lake
❸ Mono Lake
❹ Owens Lake
❺ Death Valley
❻ Searles Basin
❼ Mojave River
❽ Bullhead City / Laughlin
❾ Salton Sea
❿ Borrego Badlands
⓫ Gila River

Deserts
Arid Regions

Source: Names and areas based on WWF Terrestrial Ecoregions of the World

FIG. 2.12 Map of the North American deserts.

Artemisia spp. (sagebrush) complex. The lower and mid-elevation bajadas of the Mojave Desert are dominated by a creosote bush/bursage (*Larrea tridentata/Ambrosia dumosa*) community. Areas of stabilized and active dunes support an insular flora that is approximately 95% indigenous (Pavlik 1985). The Sonoran Desert, with infrequent freezing temperatures and bimodal annual rainfall, contains subtropical arborescent flora, including leguminous trees and large columnar cacti, with a species diversity that in some areas exceeds eastern North American deciduous forests. The Chihuahuan Desert, owing to its higher elevations, cooler temperatures, and greater precipitation, is more dominated by perennial grasslands than the other deserts.

The desert ecology of North America has been altered by the introduction of exotic, invasive plants such as *Bromus tectorum* (cheatgrass), *Salsola* spp. (Russian thistle or tumbleweed), and the woody phreatophytes *Tamarix* spp. (saltcedar). These invasions were hastened by habitat disturbance, including overgrazing and river regulation and damming.

The North American deserts are located mainly within the Basin and Range Province, a region characterized by an alternating pattern of long, narrow subparallel mountain ranges separated by alluvial basins, trending north or northwest, the result of late Cenozoic crustal extension and faulting that began 12–15 Ma. The Basin and Range Province is subdivided into five sections: the Great Basin, the Mojave-Sonoran Desert, the Salton Trough, the Mexican Highlands, and the Sacramento Mountains (Fenneman 1931). The largest of these is the Great Basin, an enormous bulge with a central area of elevated basins flanked by significantly lower terrain. Drainage is largely internal, with more than 100 closed basins, many of which contained perennial lakes during wetter climatic intervals. The topography of the Mojave-Sonoran section is relatively subdued, drainage is internal, and the closed basins are largely undissected. The Sonoran Desert drainage is more integrated, forming the Gila, Salt, or Bill Williams River systems, and basin dissection is more continuous and widespread (Dohrenwend 1987). The Mojave-Sonoran section is bounded to the southwest by the Salton Trough, a structural depression containing as much as 6000 m of late Cenozoic sedimentary deposits. The Mexican Highlands, which trend north to northeast, are transitional between the Sonoran Desert and the Colorado Plateau and dominate northwest Mexico. The Sacramento Moun-

tains are a transitional zone between the Rio Grande rift zone and the Great Plains farther east. The Colorado Plateau lies to the east of the Basin and Range Province and is a semiarid region of high plateaus and isolated mountains, 90% of which are drained by the Colorado River.

Active sand dunes only cover small areas (<1%) of arid North America, but older windblown sediments, including both sand and dust repeatedly reworked during Quaternary times, mantle broader regions. Despite their small scale, North American dune fields are among the best studied in the world. Wind erosion has formed widespread ventifacts (wind-eroded rocks) and minor yardang fields (Ward & Greeley 1984). Aeolian dust is widespread and shows a large increase in soil-accumulation rates around the Pleistocene–Holocene boundary (Reheis et al. 1996). High early Holocene accumulation rates of silt, clay, and $CaCO_3$ were associated with the incorporation of dust deflated from desiccating playas. Modern dust levels in arid regions of Arizona, Texas, and California deserts have increased owing to human activity, including construction, off-road vehicle use, lake drainage, and agriculture. Dust emission poses numerous problems, including low visibility that triggers car accidents, loss of topsoil and soil nutrients, and health hazards associated with inhaling fine particles.

2.2.6.1 Chihuahuan Desert

The Chihuahuan is the largest desert in southwestern North America (518,000 km^2), with three-quarters in northern Mexico, and the remainder in western Texas and southern New Mexico. It ranges in altitude from 600 to 1500 m ASL and receives an average of 235 mm of rain, with a range of 150–400 mm (Schmidt 1979). Summer thunderstorms associated with monsoonal conditions supply most of the annual rainfall. Annual temperatures average 18.6°C, reduced by cool winter conditions (e.g. January <10°C) (R.H. Schmidt 1989). Summer temperatures range from 25 to 30°C, with temperature extremes higher than 50°C being rare.

Most runoff drains into interior basins, which are predominately covered in grassland. Desert scrub or arborescent woodlands dominate the upper bajadas. Characteristic plants are tarbush (*Flourensia cernua*), creosote bush (*L. tridentata*), whitethorn (*Acacia vernicosa*), marriola (*Parthenium incanum*), candelilla (*Euphorbia antisyphilitica*), guayule (*Parthenium argentatum*), and lechuguilla (*Agave lechuguilla*).

Dunes of the Chihuahuan Desert are largely active today and are notable in that many include or are formed of gypsum. White Sands National Monument in south-central New Mexico is the best studied of these (McKee 1966; Langford 2003).

2.2.6.2 Sonoran Desert

The Sonoran Desert covers about 275,000 km² around the northern two-thirds of the Gulf of California, mostly in the Mexican state of Sonora, but extending across the narrow peninsula of Baja California to the open Pacific coast, and including about half of Arizona and the southeastern corner of California. The topography is broken by mountains, with elevations ranging from below sea level in the Imperial Valley, California, to about 1500 m in mountain foothills.

The Sonoran Desert flora is unique for its diversity and the abundance of trees and columnar cacti (Fig. 2.13a). Characteristic plants include saguaro (*Carnegiea gigantea*), blue palo verde (*Cercidium floridum*),

(a)

(b)

Fig. 2.13 The North American deserts are small but diverse in their nature. These photographs contrast the Sonoran Desert (a) and its leguminous trees and large columnar cacti, with the Great Basin Desert (b), which is colder and dominated by the evergreen *Artemisia* spp. (sagebrush) complex.

ironwood (*Olneya tesota*), and creosote bush. During periods of low sea level in the Pleistocene, an even larger area of core desert for xerophytic desert species existed and lower elevations acted as a refugium for Mojave Desert plant species (Cole 1986). The modern Mojave-Sonoran Desert boundary was established about 8000–9000 years BP, when junipers and oaks disappeared from desert lowlands (Van Devender 1990).

The region surrounding the lower Colorado River Valley is termed the Colorado Desert (Yeager 1957), and is the hottest and driest part of the Sonoran Desert (Ezcurra & Rodrigues 1986). Precipitation averages 50–100 mm year^{-1} near the Gulf of California and the total vegetative cover is less than 6%. Microphyllous shrubs and ephemerals dominate the flora, with few trees or cacti (Shreve 1964; Cole 1986). To the east, the higher terrain of southwest Arizona is wetter and cooler, and is floristically more complex.

The Sonoran Desert is more subtropical than other North American deserts, rarely experiencing freezing temperatures. Daily summer high temperatures of 49°C or more are not uncommon in the western half of the desert. Rainfall is more evenly distributed throughout the year than in the Chihuahuan Desert, resulting in vegetation that appears more verdant. Median winter frontal precipitation ranges from 30 mm in southeastern California to 93 mm in southeastern Arizona (Woodhouse 1997). Quasi-periodic increases in winter rainfall are associated with ENSO events (Cayan & Peterson 1989). Little rain falls in late winter, spring, and early summer, when the Sonoran Desert is affected by the eastern edge of the Pacific high-pressure cell. During high summer, monsoonal circulation results in summer thunderstorms (Ezcurra & Rodrigues 1986; R.H. Schmidt 1989). Across the Sonoran Desert, there are significant regional and elevational differences in precipitation.

North America's largest field of active sand dunes (5700 km^2), the Gran Desierto sand sea of northwest Sonora, lies northeast of the Gulf of California. It consists of chains and clusters of star dunes, active and relict crescentic dune forms, reversing dunes, linear and parabolic dunes, and sand sheets (Lancaster 1995). Other notable Sonoran Desert dune fields include the Algodones Dunes in southeastern California, the Mohawk Dunes of Arizona, and dunes along the desert coasts of Sonora and Baja California in Mexico (Inman et al. 1966). Summer monsoons bring "haboobs" or dust storms to Arizona, carrying clay- and silt-size material to heights of 3000 m or more. The source of the fine material includes abandoned or fallow farm fields and construction sites (MacKinnon et al. 1990).

2.2.6.3 Mojave Desert

The Mojave Desert is the smallest North American desert. It is roughly triangular in shape, and covers 140,000 km^2 in southeastern California and southernmost Nevada. Elevations, which commonly lie above 1000 m and reach 2000 m in some isolated mountain ranges, distinguish the Mojave or "high desert" from the neighboring Sonora or "low desert" to the southeast. The impacts of late Quaternary climate change are conspicuous in the surficial stratigraphy of the area (Wells et al. 1987).

Rainfall ranges from 76 to 102 mm annually across the desert floor, increasing with rising elevation to about 279 mm. Most precipitation is derived from winter frontal systems, although strong, localized thunderstorms enter the easternmost part of the Mojave Desert in July through September, sometimes causing significant flooding and erosion. During the summer, temperatures often exceed 40°C for 100 consecutive days or more. Winters are colder than in the Sonoran Desert: in valley bottoms where cold air settles at night, temperatures may drop below −18°C in winter and may be close to freezing even in summer (Laity 2002).

The Mojave incorporates vegetational elements from the Great Basin Desert to the north and the Sonoran Desert to the southeast, as well as some endemic species (Rowlands et al. 1982). A creosote bush/bursage community dominates the lower and mid-elevation bajadas. Winter frosts exclude many of the subtropical, arborescent forms common to the Sonoran Desert. Characteristic Mojave Desert plants include white bursage (*Ambrosia dumosa*), creosote bush, Mojave yucca (*Yucca schidigera*), Joshua tree (*Yucca brevifolia*), and Mojave sage (*Salvia mohavensis*).

Drainage is largely internal and most closed basins are relatively undissected. During wetter intervals of the Pleistocene, the Mojave River formed a more integrated drainage system than today, flowing north and then east from the San Bernardino Mountains, to form Lakes Manix and Mojave before spilling northwards towards Death Valley where it was joined by waters from the Owens and Amargosa drainages to form Lake Manly. Evidence from dunes and ventifacts indicates that aeolian activity in the Mojave

Desert has been both prolonged and episodic, but more extensive and intense in the past than today (H.T.U. Smith 1967; R.S.U. Smith 1984; Tchakerian 1991; Laity 1992; Zimbelman et al. 1995). The principal active dune fields are in the eastern Mojave, and include the Kelso Dunes and the Devil's Playground area, the Ibex and Dumont Dunes, and the dunes of the Cadiz and Palen valleys. Field studies in the Kelso Dunes (Sharp 1966), the Ibex Dunes, and the Dumont Dunes (Nielson & Kocurek 1987) document little net movement of dunes ridges, despite short-term changes induced by seasonal wind reversals. Inactive dunes are extensive and include sand ramps and climbing and falling dunes, many of which are mantled by coarse talus and fluvial debris, and are fluvially dissected.

2.2.6.4 The Great Basin deserts

The Great Basin is the second largest of the North American deserts, covering about 400,000 km² of the northern Basin and Range Province. Much of Nevada and Utah are in the Great Basin Desert, as are parts of Oregon, Washington, Idaho, California, and Wyoming. The central area consists of elevated basins and ranges flanked on three sides by lower terrain (Dohrenwend 1987). The desert floor is broken by numerous north–south-trending mountain ranges, the highest (White Mountains) reaching 4340 m ASL. The lowest basin elevation (85 m BSL) is the floor of Death Valley, California. Most drainage is internal and many basins contain playa remnants of former Pleistocene lakes (see also Chapter 5). Perennial rivers include the Truckee, Carson, Walker, and Owens rivers flowing from the east flank of the Sierra Nevada, California, and the Humboldt River of Nevada.

Lying in the rainshadow of the Sierra Nevada-Cascade chain, the Great Basin receives from 100 to 300 mm of precipitation annually, a little less than half of which falls in the summer. It is a temperate desert, with extremes appropriate to its mid-latitude, interior location. Winters bring freezing temperatures and snow; the average temperature in January is −2°C. Nonetheless, summer temperatures in the southern Great Basin are very hot, with the highest temperature recorded in the USA (57°C) in Death Valley.

Great Basin ecosystems are a function of temperature, latitude, elevation, rainfall, and geology. Characteristic plants include big sagebrush (*Artemisia tridentata*), shadscale (*Artiplex confertifolia*), black-

brush (*Coleogyne ramosissima*), and greasewood (*Sarcobatus vermiculatus*) (Fig. 2.13b). Grazing has caused various changes in several plant communities, notably a reduction in grasses.

A thin veneer of sand, anchored by vegetation, mantles much of the Great Basin Desert. Locally extensive dune fields include star dunes up to 208 m in height in Eureka and Panamint Valleys, and dunes in central and southern Death Valley (R.S.U. Smith 1982). Their location suggests substantial topographic control of near-surface winds.

The Colorado Plateau is a region of quite different tectonic style and landscape, passing at higher elevations into open coniferous woodland. It is a roughly circular area of about 384,000 km² that consists of plateaus and isolated mountains encompassing parts of Utah, Colorado, New Mexico, and Arizona. Although the plateau lies mostly above 1500 m and is a definable tectonic unit, it shows considerable internal variation (Hunt 1974). Differential erosion on nearly horizontal or only moderately deformed strata of varying strength characterizes a landscape dominated by canyons, cuesta scarps, and plains stripped to bedrock. Volcanic fields form a discontinuous belt along the southern and western margins of the Colorado Plateau, with lava flows ranging from 9.9 Ma to late Holocene in age (Patton et al. 1991).

Investigations of the Colorado Plateau have given rise to many contributions to geomorphology, from the classic work of Gilbert (1877) and Bryan (1928b), to slope development and mass wasting (Ahnert 1960; Schumm & Chorley 1964, 1966; Howard 1995), fluvial erosion (Graf 1978, 1983b; Patton & Boison 1986; Hereford 2002), groundwater sapping (Laity & Malin 1985); volcanism, tectonic geomorphology (McGill & Stromquist 1974), and to a lesser degree, aeolian processes (Hack 1941; Reynolds et al. 2006). Large-scale mass wasting has modified many of the escarpments, canyons, and mountain flanks of the Colorado Plateau (Bradley 1963; Schumm & Chorley 1964).

Dune fields on the Colorado Plateau are not extensive, but include a variety of forms. Parabolic and longitudinal dunes, with isolated barchans, transverse dunes, and falling and climbing dunes occur over about 65,000 km² of northeastern Arizona, in a region of high plateaus and mesas. The wind regime is very active, in part owing to topographic effects of the San Francisco volcanic field, which cause funneling and severe turbulence. Aeolian deposits, consisting principally of sand sheets, parabolic dunes, barchan-barchanoid dune complexes, and linear dunes, cover a large expanse of the southeastern Colorado Plateau (Hack 1941; Wells et al. 1990). Sand sheets comprise approximately 88% of aeolian deposits and range up to 2 m in thickness. In Utah, the Coral Pink Dunes lie on the boundary between desert and woodland. Wind erosion features of the Colorado Plateau include deflation hollows, yardangs, wind-fluted cliffs, and blowouts. Yardangs are developed in Mesozoic claystones and siltstones in the Painted Desert, with the largest individuals about 50 m long.

2.2.7 Australia

In terms of area, the most arid continent is Australia, with three-quarters classified as desert or semidesert, including most of the central and western areas (Fig. 2.14). During the late Cretaceous, Australia separated from other continents and began drifting northward. Lacking major mountain building or volcanic activity, it differs in character from deserts of interior Asia, South America, or southwestern North America. Long periods of tectonic stability and erosion leveled the continent to create surfaces with low gradients, where shallow and infertile soils cover old bedrock. There has been little transport of sediments and no sustained inward flowing perennial rivers. Disorganized river systems negotiated dune fields and petered out in abundant sinks (Kotwicki & Allan 1998).

Quaternary climatic oscillations from wet to dry conditions profoundly affected the desert landscape. Conditions ranged from those favoring massive lake expansions to periods of extensive dune building and saline playa development. During the more humid phases, the arid belt may have disappeared completely. Throughout the last interglacial (MIS 5), greater effective precipitation across the Australian landmass resulted in synchronously high lake levels both in areas dominated by the monsoon and westerly low-pressure fronts (Stone 2006). During periods of peak aridity, the desert expanded on its southern margin to the northeast tip of Tasmania, which joined to mainland Australia owing to sea level lowering (Bowden 1978): stabilized linear dunes preserved here appear to have been active at, or immediately after, the LGM (Duller & Augustinus 2006). The principal cause of climatic variability appears to be ice-volume-forced glacial–interglacial cyclicity, which in Australia was expressed predominantly as oscilla-

tions in moisture availability. Glacial periods were drier than today with restricted distribution of mesic plant communities, shallow or ephemeral water bodies, and extensive aeolian activity. Superimposed on this cyclicity is a tendency towards drier and/or more variable climates over the past 350,000 years. Increased biomass burning, associated with the activities of indigenous people, may have resulted in more open and sclerophyllous vegetation and a further reduction in water availability (Kershaw et al. 2003).

In overview, the center of the Australian continent is very dry, and surrounded by an intermediate semiarid zone, which is transitional to zones on the rim of the continent that support forests. Average annual dryland rainfall ranges from 200 to 400 mm. The central part of the continent averages less than 150 mm annual precipitation, but no area receives less than 100 mm year^{-1}. The principal factor causing the dryness of modern Australia is its position astride the subtropical high-pressure belt. The moderate aridity of the continent arises because of its island setting, the absence of significant relief barriers to the penetration of moisture (particularly from tropical sources), and the lack of a definite inshore cool coastal current along the west coast (Mabbut 1977a). Evidence of the moderate nature of Australian aridity includes a general mantle of vegetation and the erosive action of running water on sloping ground (Mabbutt 1977a). Aeolian sand cover is largely stabilized under a cover of plants.

Much of the desert rainfall results from the tropical monsoon during the summer months. Australia is subject to considerable rainfall and streamflow variability, in a region where water resources are very limited. Periods of enhanced rainfall are linked to variations in the Southern Oscillation Index (SOI) associated with La Niña conditions and to the incursion of cyclones. Dry periods are associated with El Niño conditions, and the ability to use this information to forecast streamflow is invaluable to the management of land and water resources (Chiew et al. 1998).

The distribution of Australian desert plants strongly reflects the character of the soil and profile characteristics that determine moisture availability (Mabbut 1977a). The most extensive vegetation type is tall acacia shrublands, found on shield deserts with soils whose clay content increases with depth. Almost as extensive are drought-tolerant sclerophyllous hummock grasslands, dominated by spinifex, which occur in areas of sandy desert. Xerophytic tussock

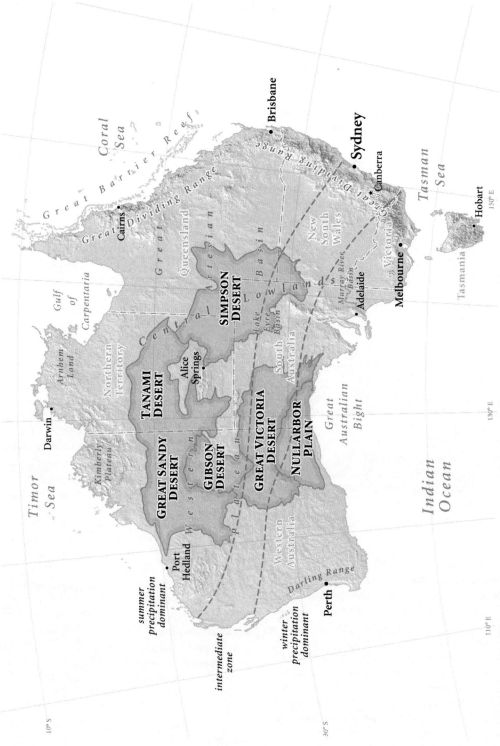

Fig. 2.14 Map of the Australian deserts. From Duller, G.A.T. and Augustinus, P.C. (2006) Reassessment of the record of linear dune activity in Tasmania using optical dating. *Quaternary Science Reviews* 25, 2608–18. By permission of Elsevier.

grasslands, characterized by perennial grasses, are found in association with clay plains in stony and riverine desert. Shrub steppe occurs on salt-affected clay loams. Large succulents such as cacti are absent from the Australian desert. Unlike Old World deserts, the Australian desert plants evolved in the absence of large herbivores. Today, because of the low rainfall, the main land use is the grazing of natural vegetation by domestic animals, and these have caused significant selective degradation.

True deserts with minimum vegetation cover are the Simpson Desert, Gibson Desert, Great Sandy Desert, Tanami Desert, and Great Victoria Desert. Along the south central rim of the continent is the Nullarbor Plain, a desert formed on a limestone substrate. Desert surface materials range from pebbles, to red soils, to windblown sand. Low-lying areas may contain small saline lakes.

Many Australian desert landforms were developed under different climatic conditions than those prevailing today. For example, extensive duricrust landscapes are a relict of more humid Tertiary conditions. Characteristic features of the Australian deserts include continental-scale longitudinal dunes, drainage-aligned systems of playas and clay pans, disorganized internal drainage patterns, and inverted relief features.

Dunes cover approximately 40% of the surface area of Australia, constituting 38% of the world's dune fields (Wasson et al. 1988). Most dunes are of simple longitudinal form, up to 300 m in length and 10–35 m in height, with an average interdune spacing of 160–2000 m. They form an anticlockwise whorl at the scale of the continent (Fig. 2.15), largely reflecting modern wind directions (Jennings 1968; Brookfield 1970; Wasson et al. 1988). The general

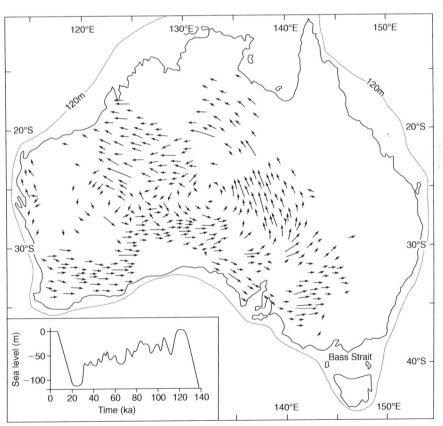

Fig. 2.15 The orientation of linear dunes in Australia, showing an anticlockwise whorl pattern. Linear dunes are also found in northeastern Tasmania. The 120-m contour shows the extent of Australia during the LGM, when sea level was considerably lower. The inset shows eustatic sea level changes over the past 140,000 years. After Wasson et al. (1988) and Duller and Augustinus (2006).

trend in dune orientation is thought to reflect the response of winds to the subtropical high-pressure system, which determines airflow in Australia (Brookfield 1970). Dunes of the Strzelecki and Simpson Deserts are markedly asymmetrical, with steeper eastern slopes, causing Rubin (1990) to suggest that they may be migrating slowly eastward over millenial timescales. However, Nanson et al. (1995) conclude that the dunes have not moved more than 100 m from their Pleistocene core. The color of the dune sand is pale adjacent to major rivers (Diamantina River and Cooper Creek) and, according to Wopfner and Twidale (1967), becomes redder and older downwind from sand sources. Wasson (1983a, 1983b) argues that the different sand colors are principally a result of provenance and are unrelated to dune age.

The source of the dune sand varies according to location, but is mainly derived from the sedimentary basins underlying the different regions of the dune field. Material was deflated from the floodplains of major rivers, salt lake systems, and weathered underlying sediments. The downwind transfer of material was not likely significant, as the grains lack surface features indicative of aeolian transport, and sand color, heavy mineral characteristics, and grain surface features change quickly along the dunes, suggesting localized sources (Pell et al. 2000).

Both the timing of the onset of dune activity and the age of the dominant dune forms remain controversial. At the present time, divergent palaeoenvironmental reconstructions based on luminescence ages are difficult to reconcile. Wopfner and Twidale (1988, 1992) conclude that the large dune fields of the Simpson and Strzelecki Deserts are entirely Holocene in origin, resting on a Pleistocene substrate, with older phases of dune activity only present on the desert's margins. Their OSL dates indicate that most dunes are younger than 2000 years and rest on a fluvial/lacustrine base of 36,000 years' age (Twidale et al. 2001). Then again, other studies suggest that aeolian activity was widespread within the Pleistocene, including thermoluminescence dates of dune material from approximately 100 ka (Nanson et al. 1995) and 243 and 167 ka (Gardner et al. 1987). Lomax et al. (2003) found fluvial-lacustrine sediments in interdune corridors that were deposited at least 210 and 160 ka (OSL ages), and later alluvial sedimentation at between 69 and 75 ka. Subsequent aeolian activity occurred between 60 and 50 ka and a palaeosol, indicating moister conditions with vegetation cover, formed atop this unit and is in turn overlain by another aeolian sequence (28,000–35,000 years old). Loose aeolian sands at the top of the dunes are less than 5000 years old, probably reflecting aeolian remobilization of the dune field in the late Holocene. It appears likely that dune sediments deposited during the LGM, a phase of maximum aridity and high aeolian activity (Bowler & Wasson 1984; Shulmeister et al. 2004; Duller & Augustinus 2006), were totally reworked during a late Holocene phase.

The dust record of Australia also supports the idea of enhanced Pleistocene aeolian activity. During the formation of the dune fields, large quantities of dust were entrained and transported as dust plumes far beyond the deserts. In Australia, the term parna has been used to describe desert dust-derived soils (Butler 1956), although Hesse and McTainsh (2003) suggest that such sediments bear a strong similarity to conventional loess deposits. Loess deposits are scattered, incomplete, and thin, and are largely found in the southeast part of the continent, where they occur as an extensive sheet over the Murray-Murrumbidgee riverine plain of New South Wales and Victoria (Butler 1956).

There are two general dust pathways in Australia, which direct dust towards the southeast and the northwest (Bowler 1976). The southeast pathway is stronger than the northwest and directs dust across the Tasman Sea to New Zealand and the southwest Pacific Ocean, with estimates of between 2 and 3.4 million t being moved in single events, and totals of between 10 and 20 million t during three events in May 1994 (Raupach et al. 1994; Knight et al. 1995). The overall significance of this loss is such that Knight et al. (1995) assert that more sediment is lost from the continent in the air than in the river systems. According to Hesse and McTainsh (2003), the most active global dust-source regions are the lower parts of inland drainage basins, where floodplains and dune fields merge. This is the case in Australia, where the principal modern dust-source areas are in the Lake Eyre Basin and the western sector of the Murray-Darling Basin.

The best records of Quaternary dust flux and deposition patterns are in the marine records, which show at least a threefold increase in dust flux relative to the Holocene during the LGM (Hesse & McTainsh 2003). Hesse and McTainsh (2003) attribute the higher LGM dust fluxes to a weaker hydrological cycle and drier westerlies, and Shulmeister et al. (2004) suggest that the effect of a reduced vegetation cover may have been critical. It is likely that the westerly winds were also stronger at that time (Shulmeister et al. 2004).

The arid areas of Australia are only sporadically penetrated by moisture-bearing weather systems and the streams are characterized by extreme variations in discharge. Three inland drainage systems in Australia have received the most attention: (1) the large anastomosing rivers of the Channel Country, (2) the "stony desert" rivers to the northwest of Lake Eyre, and (3) the rivers draining the Central Highlands of the Macdonnell and Musgrave ranges near Alice Springs (Croke 1997). The rivers of the Channel Country of east-central Australia, and in particular Cooper Creek, are the best studied. Cooper Creek has its headwaters on the southwestern side of the Great Dividing Range at an elevation of 230 m. It flows for 1523 km before terminating at Lake Eyre, the world's largest internally draining catchment with the lowest mean annual runoff of any major drainage system. Cooper Creek is notable for its distinctive anastomosing system of channels that constitute the main planimetric form of this river. The narrow and clearly defined channels are incised into cohesive floodplain muds (Knighton & Nanson 1997).

Australian deserts are also notable for the extent of duricrusted plains, composed mainly of silcrete and ferricrete. Silcretes were formed during wetter climatic periods and probably first developed on stable low-relief landscapes of the Jurassic, with the last phase of development in the late Pleistocene (Langford-Smith 1978). Angular silcrete fragments, known as "gibbers," constitute the surfaces of the "stony deserts." Ferricretes, distributed along the margins of the continent, are believed to have formed during warm, seasonal environments.

3

THE CLIMATIC FRAMEWORK

3.1 INTRODUCTION: CLASSIFICATION OF DESERTS BY TEMPERATURE

Climatologists have divided the world's deserts into a threefold division (Meigs 1953): hot deserts in tropical and subtropical latitudes, such as the Sahara, Sahel, Kalahari, Thar, Middle Eastern, Arabian, and central Australian deserts; temperate deserts in higher latitudes in continental interiors, including the Gobi and Kazakhstan deserts of central Asia, interior deserts of western North America, and the Patagonian region of Argentina, where temperatures are high during the summer, but winters are cold or very cold; and cold-water coast deserts found on the west coasts of continents in tropical latitudes, including the Namib and the Atacama and, to a lesser extent Baja California and the Atlantic coastal fringe of the Sahara.

High temperatures, and the persistence of high temperatures, characterize hot inland deserts. They have absolute maximum temperatures of 40°C and above, with maximum temperatures commonly between 45 and 49°C (Evenari 1985a). Mean annual temperatures in the Sahara exceed 30°C, with the global temperature record of 58°C set at El Azizia, Libya, in September 1922. Typical temperatures of Alice Springs in central Australia are 35–40°C in the warmer months (Mabbutt 1977b). Hot deserts may also experience a considerable diurnal range of temperatures (E. Smith 1986). Owing to clear skies, nighttime outgoing radiation causes large drops in temperatures and daily ranges of 17–22°C are normal (Fig. 3.1). In Death Valley, California, a maximum diurnal range of 41°C was recorded in August 1891, and the mean diurnal range of temperature for that month was 35°C. However, the fluctuations of the mean annual temperature from year to year are very small.

Temperate deserts are characterized by considerable seasonal variations in temperature, great diurnal ranges of temperature, and a dependable period of cold temperatures, when some precipitation occurs as snow and soil moisture is frozen. These deserts are at higher latitudes than hot deserts, and thus in the belt of the westerly winds. Mountain barriers deprive temperate deserts of moisture. Owing to the more interior position of the Eurasian deserts, indices of continentality are greater than in western North America or Patagonia (Fig. 1.2). The conditions of the Betpak-Dala Desert (annual precipitation approximately 100 mm), east of the Aral Sea in Kazakhstan, are extreme. Summer air temperatures are 42–44°C (ground temperatures 65–70°C), whereas winter temperatures drop to –40°C. Snow covers the ground for four and a half months during the winter, but is thin, not more than 15–20 cm thick (Walter & Box 1983b). Likewise, in the Caspian Lowlands, the winters are very cold, with minima from –30°C along the coast to –40°C inland. By contrast, July mean temperature is about 26°C, with maxima up to 40°C (Walter & Box 1983a).

Coastal deserts tend to have relatively low seasonal and diurnal ranges of temperature. The Benguela Current adjacent to the Namib Desert has a sea surface temperature of about 12°C. A cold layer of air up to 600 m high extends over the current and the adjacent land, forming a temperature inversion (Walter 1986). Sea breezes carry ocean fogs inland. Walvis Bay, Namibia, has a mean annual temperature range of around 6°C. Adjacent to the Humboldt or Peru Current, the Peru–Chile desert also experiences small mean annual temperature ranges, with that of Callao, Peru, around 5°C (21.6°C in March and 16.9°C in September) (Rauh 1985). Daily ranges also are low, generally around 11°C, or half that expected for the Sahara. Annual temperatures are generally moderate, with average temperatures of 17–19°C.

3.2 WEATHER DATA

Weather data for desert areas often lack essential spatial coverage; the records are short, inconsistent,

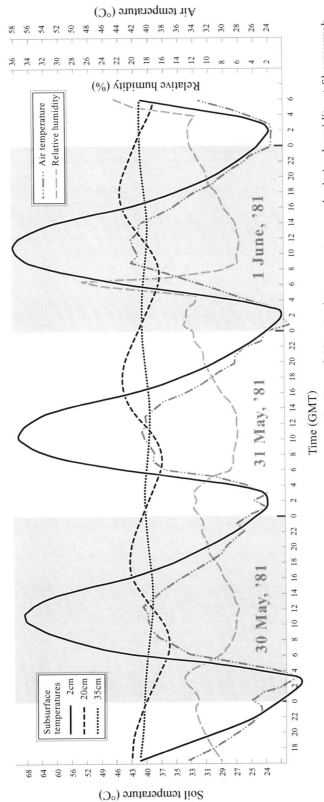

Fig. 3.1 Time series (15-minute sampling) of subsurface temperatures (2, 20, and 35 cm), air temperature, and relative humidity at Sharouwrah, Saudi Arabia, 30 May–1 June 1981. The amplitude of subsurface temperatures decreases with depth and is phase-shifted, so that temperatures are cool at depth when surface and air temperatures are high. Thus, animals retreating into burrows experience relief from the harsh surface environment. From Smith, E.A. (1986) The structure of the Arabian Heat Low. *Monthly Weather Review* **114**. By permission of the American Meteorological Society.

and sometimes inaccurate; and the number of variables measured may be low, consisting principally of temperature and precipitation. Long-term weather data are notably absent for many significant areas of desert, including, for example, the southern Rub' al Khali Desert (Empty Quarter) of the Arabian Peninsula (E. Smith 1986), North Africa (Shata 1992), and the central Kazakhstan area (Walter & Box 1983b). Nicholson (1985) noted the difficulty of obtaining meteorological data from Africa, where many stations have closed owing to political instability or economic woes, and some services, such as those in Chad, ceased to function for multiple years. Frequently, there is also a delay in data publication.

There are few instrumental records prior to the twentieth century, and the bulk of those since that time have been rainfall observations, with temperature records even less common and of shorter duration (Nicholson 2001). These problems compound the difficulties of regional analysis, and make it difficult to resolve issues concerning the degree to which the apparent movement of desert margins stems from human activity or natural rainfall fluctuations. Even many arid parts of the western USA lack long-term coverage, and stations are frequently discontinued or moved. For example, southwestern Arizona has almost no long-term historical climate data (Comrie & Broyles 2002). By contrast, some of the meteorological records for the Australian deserts are of surprising length, exceeding 100 years.

The lack of good spatial coverage in weather observations provides an inherent bias. Warner (2004) notes that most observations are made in inhabited areas, frequently at oases. These locations intrinsically have different climatic characteristics (for example, wind and humidity) than the surrounding desert. Observations made in large cities, such as Las Vegas, Nevada, may also suffer from heat-island effects.

Weather stations in remote areas are often subject to various disruptions, both from humans and animals. Shata (1992) reports the case of an automatic rain-gauge station protected by a local bedouin along the Mediterranean littoral zone. When the station was visited by a group of inspectors on a rainy night, it was discovered that the guard had taken the rain gauge to his tent and wrapped it in a blanket!

Weather data, particularly precipitation records, are essential for long-range hydrological planning and resource management. Water is the most essential resource determining the future of arid lands.

Large regions of North Africa lack even basic information on water resources (Shata 1992), leading to potentially large errors in management strategies.

3.3 ATMOSPHERIC CONTROLS: SURFACE BOUNDARY LAYER

Hot and temperate deserts are characterized by high values of solar radiation, low amounts of atmospheric water vapor, and clear skies that lack cloud cover. The climatic environment of noncoastal deserts is thermally extreme due to the high values of solar radiation available at the top of the atmosphere and to scarcity of water, both on the surface and in the atmosphere.

3.3.1 ATMOSPHERIC WATER VAPOR AND CLOUD COVER

Except during periods of dust storms, the desert atmosphere is very clear, clouds are a rare occurrence, and the water vapor content is low. The paucity of clouds has several important consequences. During the day, the incoming flux of solar energy approaches a maximum for the latitude and time of year. Approximately 80% of solar radiation at the top of the atmosphere reaches the surface (Oke 1978). Daytime surface temperatures are very high, despite the relatively high albedo of sandy soils, because of the large receipt of solar energy and the aridity of the soil (Otterman & Tucker 1985). A large proportion of the available energy is expended in sensible heat, whereas latent heat transfer to the atmosphere (through evaporation) is usually negligible, because of the dryness of the soil and lack of vegetation. At night, the air temperatures decrease rapidly, as there is little counterradiation from clouds. Another consequence of sparse cloud cover is that clear-sky conditions return rapidly following rainfall, so that water in the soil surface is rapidly evaporated. The mean number of sunshine hours per year reaches a maximum of 4000 (91% of the maximum possible) in the Atacama Desert and a minimum of 2300 in the Namib fog desert (Fitzpatrick 1979).

Desert air is commonly perceived as dry. Hot deserts typically have mean annual values of relative humidity between 40 and 50% (Evenari 1985a). In the Sahara, both relative and absolute humidity are very low. During the warmest part of the year, relative humidity averages about 20%, with values as low as 5% not infrequent. Daytime humidities on the southern Arabian Peninsula average 10–12% in June (E. Smith 1986), accounting for the lack of summertime cloudiness. Relative humidities of 100% are rare, occurring only where there is a local source of moisture, such as at an oasis, or in the Nile Valley and delta (G. Smith 1984).

The lowest humidity values occur during windy conditions. Within 4 hours, relative humidity can drop from 100% to less than 10% during periods of hot, dry *khamsin* winds in the Negev Desert (Evenari 1985a). In semiarid regions of southern California, hot, dry *Santa Ana* winds blow intermittently during the winter months, with their source the high desert interior of the American west. Relative to mean conditions, humidity drops by 5–10% in the deserts, and up to 25% along the coast, where descending air heats adiabatically (Conil & Hall 2006).

Humidity is inversely related to air temperature (Fig. 3.1). At Bullhead City, Arizona, in the southwestern USA, maximum relative humidity is at sunrise during both winter and summer, and minimum values coincide with the late afternoon peak in daily temperatures (Walsh & Hoffer 1991). At Bilma, Niger, mean monthly relative humidity in July is 39% at 0700 hours and 18% at 1300 hours.

Humidity may also vary seasonally. At Bullhead City, Arizona, the highest relative humidity values occur in the winter months (monthly mean of 35–60%), and the lowest values (monthly mean of 5–10%) in the summer (Walsh & Hoffer 1991). The number of days with relative humidity exceeding 90% is greatest in winter (Table 3.1) (Walsh & Hoffer 1991).

The relative humidity of deserts influenced by fog, such as the coastal parts of the Namib and the Peru–Chile deserts, can be on the order of 70–80%. Deserts under the influence of monsoon conditions, such as the Thar Desert, can show extreme short-term variations. Evenari (1985a) reports a range of 2–100% within the course of 1 day.

3.3.2 RADIATION

Incoming solar energy approaches a maximum in arid regions owing to the lack of cloud cover. Even in the absence of clouds, however, day-to-day differences in solar radiation of as much as 15 or 20% result because of changes in the concentrations of water vapor and dust (Fitzpatrick 1979). These concentrations are governed by synoptic patterns that affect atmospheric stability, wind strength, and turbulence.

In the central core arid areas, the seasonal distribution of radiation is characterized by a unimodal curve of medium to high amplitude, normally reaching a peak in June in the Northern Hemisphere and December in the Southern Hemisphere. Towards the equatorial margins of the arid zone, the amplitude of the annual curve diminishes, and the curve changes to become bimodal, owing to the double passage of the sun across the zenith within the tropics (Fitzpatrick 1979). Measurements by E. Smith (1986) for an interior site in the Saudi Arabian Empty Quarter illustrate the almost perfectly rhythmic daily nature of radiative properties of extremely dry desert regions. The processes show continuous periodicity and a virtually invariant amplitude of daytime peaks. The maximum value at noon in April in Tunisia is 951 W m^{-2} (see Table 3.2) (Wendler & Eaton 1983), and in June in the Saudi Arabian Empty Quarter it is 1015 W m^{-2} (E. Smith 1986).

Maps of the geographic distribution of the mean annual total of global solar radiation reveal that the largest annual receipts are at latitudes between about 20 and 25°, where about 200 kcal cm^{-2} (840 kJ cm^{-2})

TABLE 3.1 Number of days on which relative humidity exceeded 90%, Bullhead City, Arizona.

Season	No. of days on which relative humidity >90%
Winter	72 days
Spring	46 days
Summer	19 days
Fall	35 days

Source: Walsh and Hoffer (1991).

TABLE 3.2 Radiation budget of southern Tunisia in April.

Parameter	Value
Maxima values (noon)	951 W m^{-2}
Mean flux (24 hours)	481 W m^{-2}
Albedo	38%
Total incoming radiation (24 hours)	5581 W m^{-2}
Reflected radiation	2198 W m^{-2}
Shortwave balance	3383 W m^{-2}

Source: Wendler and Eaton (1983).

are received on average per year (Landsberg 1965). A broad band extending across the Sahara and Arabian deserts is the largest area to receive radiation of this magnitude. Similar values are also recorded for a small core area in central Australia. The values for fog deserts are less than those of hot deserts: from 100 to 140 kcal cm^{-2} (420 to 580 kJ cm^{-2}) per year (Evenari 1985a).

3.3.3 TEMPERATURE OF THE AIR, SURFACE, AND SUBSURFACE

Air temperatures are measured at about 2 m above the surface at shelter height. Profiles are typically logarithmic; with a strong temperature gradient in the lowest 1.5 m. Ground surface or subsurface temperatures are less commonly assessed.

3.3.3.1 Air temperature of hot deserts

Temperatures in hot deserts vary greatly between day and night, summer and winter. For organisms, it is the incidence of high temperatures liable to cause heat damage, and of low temperatures – which limit the growing season – that are of most significance. Temperature has a marked influence on the evaporation rate, and thus the availability of water to plants.

The dry climate, infrequent cloud cover, and lack of vegetation have a distinct effect on the diurnal temperature regime. Solar radiation is intense during the daytime, but is followed by strong radiative cooling at night. Fig. 3.1 illustrates a temperature profile for the Saudi Arabian Empty Quarter during June 1981: air temperature rose rapidly after sunrise, peaking at about 43°C an hour after local noon, then declining gradually until the subsequent sunrise, with a minimum temperature of about 27°C (E. Smith 1986). Other North African stations record similar large temperature ranges. During December, a daily range from −0.5 to 37.5°C was recorded at Bir Mighla, southern Tripolitania, and an annual range of shade temperature from −2.0 to 52.5°C was recorded at Wadi Halfa, Sudan (Cloudsley-Thompson 1984). Most places in the Sahara experience freezing temperatures at least once a year (G. Smith 1984).

Altitude may exert a moderating influence on temperature, resulting in lower mean daily maximum and extreme maximum values during the warmer months. The mean daily maximum at Tamanrasset (altitude 1405 m) on the slopes of the Ahaggar Mountains in the west central Sahara is 35°C,

whereas at Bilma (altitude 355 m) it is 43°C (G. Smith 1984).

Other aspects of topography may also significantly affect local temperatures, influencing cold-air drainage and thermal inversions. The flat valley floor and low mean elevation of the Colorado River valley emphasize the effects of daily heating. Additionally, pooling of dense cool air at this lower elevation tends to decrease the monthly mean minimum temperature (Walsh & Hoffer 1991).

Minimum temperatures have received less attention than maximum in desert areas. Low temperatures are principally recorded in inland deserts, such as in Asia and North America: regions that also record great ranges in temperature. Freezing weather plays an important role in the distribution of vegetation. Subfreezing temperatures for 36 hours or longer seriously impair the survival of some plant species, particularly the large columnar cacti (Shreve 1911; Nobel 1980; Mourelle & Ezcurra 1996).

Rain, wind, sand, and dust storms cause short-term fluctuations in temperature. Temperatures may drop rapidly for the duration of a rainstorm, changes of wind direction may cause rapid and irregular temperature oscillations, and during sand storms, frictional heat generated by the collisions of grains raises air temperature. In areas where dust storms are common, they may impact the overall temperature record. In Kuwait, up to 25% of summer days include dust storms that last from several hours to several days. These events lower the daily maximum temperature by as much as 5°C (Nasrallah et al. 2004).

3.3.3.2 Surface temperatures

The penetration of solar radiation into desert soils depends on the size, color, and mixture of particles. The interaction of the extinction of radiation by absorption with depth (commonly at less than 1 cm), upward reflection, and infrared radiation within and from the soil, result in maximum temperatures at about 0.3–0.5 cm below the soil surface (Larmuth 1984). In the Sahara, the rate of rise of surface soil temperatures is about 5°C per hour from 0900 to 1100. Temperatures at sunrise are about 20°C, climb to 60–70°C at midday, cool to 30–35° by early evening, and are 25°C at 2100 (Larmuth 1984).

Desert land maximum surface temperatures are the highest recorded on Earth. Maximum surface temperatures appear to lie in the range 55–70°C, with some records in the 80°C range. However, few maximum land surface temperatures have been recorded. Measurements made from an aircraft over the central Australian desert suggested maxima of about 62°C (Garratt 1992).

Weather conditions conducive to high surface temperatures include cloud-free conditions, low winds, and high air temperatures, which occur in summer with warm-air advection (Garratt 1992). Extreme temperatures are found at bare-soil surfaces (which have low aerodynamic roughness), where the soil is dry and has a very low albedo and thermal conductivity. For example, in a desert valley in the Great Basin, USA, the highest daytime surface and air temperatures occurred at a station with tall sagebrush (plants with the lowest reflectivity of those measured), and the lowest temperatures occurred at a station with short bushes, a lighter soil surface, and the highest albedo (Malek et al. 1997). Dry sandy soils have low thermal conductivity and are good insulators, protecting the subsurface from the extremes of temperature. As a consequence, sand surface temperatures are often very high, reaching 74°C in the Tengger Desert of China (Berndtsson et al. 1996).

Ground temperatures have an even greater diurnal and annual temperature range than air (see 2 cm values, Fig. 3.1). In extremely dry desert regions, unblocked shortwave radiation leads to intense daytime heating and strong longwave radiation results in efficient nighttime cooling (E. Smith 1986). In Alice Springs, Australia, ground temperatures are characteristically 20–25°C above air temperatures at midday, and between 5 and 10°C lower at night (Mabbutt 1977b). During the Sahelian nighttime, Wendler and Eaton (1983) found that temperatures at the surface were about 2°C less than at shelter height. Diurnal differences in surface temperature are as much as 50°C in June in Saudi Arabia (E. Smith 1986) (Fig. 3.1). Peel (1974) recorded a 40°C daytime temperature range for basalt surfaces in the Tibesti region of the central Sahara: from 40°C at 0600 hours to 80°C at 1600 hours. Daytime heating can be intense enough to form a super-adiabatic layer of air several centimeters in depth. This shallow air may be sufficiently unstable to raise a veil of sand from the desert floor (E. Smith 1986).

Temperatures drop rapidly at the soil surface during the night. Minimum winter temperatures may result in frost in some areas. However, there appears to be less difference between minimum surface and air temperatures than maximum surface and air temperatures. For example, in the Great Basin desert, annual minimum surface and 2 m air

temperatures were −29.6 and −28.2°C, respectively, whereas the annual maxima were 51.9 and 36.0°C (Malek et al. 1997).

Rain storms act to temporarily cool the soil, with temperatures dropping within seconds of the onset of rain. After the cessation of rain, energy for evaporation of soil moisture is mostly drawn from heat contained in the upper 5–6 cm of sands. Larmuth (1984) reported a temperature drop of 5°C within minutes of a rain's end.

The aspect (orientation) of a surface has a marked effect on its thermal characteristics. In the Negev desert, summer soil temperatures at 10 cm depth are 4–6°C cooler on northern slopes than southern ones (Evenari 1985a). For the Sahara Desert, Larmuth (1984) noted that north and west exposures were cooler than those facing east or south, with the difference enhanced by increasing slope. Wendler & Eaton (1983) observed that the surface temperature of dune sand was very sensitive to exposure. On sand dunes with a uniform grain size and an albedo of 47%, horizontal surfaces had a mean temperature of 39.6 ± 0.6°C, south slopes 41.1 ± 0.7°C, and north slopes 33.2 ± 0.6°C. The measurements were conducted at 1300 hours local time, and the dune slopes for both the north and south exposures were about 15°.

3.3.3.3 Subsurface temperatures

Temperatures decline very rapidly from the soil surface skin to depth (Wang & Mitsuta 1992). E. Smith (1986) placed soil thermistors at depths of 2, 20, and 35 cm below the surface during the month of June at Sharouwrah, Saudi Arabia. Fig. 3.1 illustrates the phase-shifted amplitude decrease with depth. At 2 cm depth, daily temperatures fluctuate by as much as 47°C: for example, on 31 May 1981, the daytime maximum was about 68°C and the nighttime minimum was about 22°C (a range of 46°C). For the same period the corresponding air temperature maximum was about 40°C and minimum was 20°C (range 20°C). At 20 cm depth, soil temperature averaged about 41°C, with a diurnal range of about 4°C, and at 35 cm depth there is essentially no difference in day/night temperatures, with the soil maintaining an essentially constant temperature of 38°C. Thus, at night the surface soil temperatures are considerably cooler than those at depth.

At night the surface is cooling, but the sand at a depth of 10–20 cm is still warm and transferring heat to layers above and below it. Deeper layers continue to gain heat, so that maximum temperatures at 30–50 cm occur when the surface is at its coolest, and minimum temperatures are reached when the surface is hottest (Larmuth 1984).

Measurements have been taken in animal burrows at depths of 70–100 cm, and records for the midday period indicate temperatures ranging from 25 to 31°C (Larmuth 1984). Inter-particle humidity increases with depth in sandy conditions (Wang & Mitsuta 1992), and humidity in burrows is higher than at the surface. Eissa et al. (1975) recorded a relative humidity of 72% in burrows during a period when the air above the ground was at 12%. Thus, animals retreating into burrows enjoy a higher relative humidity and a temperature at depth in soil or sand that is out of phase with that at the surface, and is normally near a minimum at the time when temperatures are highest at the surface.

3.3.4 ALBEDO

Interest in the reflectance of desert surfaces has increased since the proposal by Otterman (1974) that desertification along the southern border of the Sahara Desert was a result of albedo increases associated with overgrazing. Charney (1975) and Charney et al. (1975) used general circulation models (also known as GCMs) to model such effects, and found that albedo increases lessened the absorption of solar energy at the surface, causing desert surfaces to cool. The resultant stability of the lower atmosphere may reduce convective activity and rainfall.

In order to provide more realistic input data for climatic models, Wendler and Eaton (1983) carried out ground-based measurements of albedo in the northern Sahel of southern Tunisia (Table 3.3). The lowest albedo values are obtained for oases and the

TABLE 3.3 Albedo of surfaces in southern Tunisia.

Site	Albedo range (%)
Chott (salt lakes): in wet condition	31–47
Sand, loose (white to yellowish color)	32–51 (mean 47%)
Sand with lichen and bacteria	20–35
Desert vegetation	22–36
Oasis (vegetation)	10–23

Source: Wendler and Eaton (1983).

highest for salt lakes, although for each surface type there is considerable variation. For dune sands, the mean albedo is 47%, with a minimum of 32% and a maximum of 51% (Wendler & Eaton 1983).

Athough albedo is generally high for salt lakes, it varies greatly both across individual lakes and between different playas. At the Salar de Atacama, a 3000 km² playa in northern Chile with a broad range in crust types, the albedo ranges from 18 to 66% (Kampf et al. 2005). Salt crusts with the highest albedos (44–66%) tend to have low surface roughness and to be situated around the margin of the playa. Rugged, cavity-rich crusts in the center of the playa had lower albedos (18–25%) owing to the accumulation of dark sediments on the surface (Fig. 3.2). Daily variations in albedo are minor, with small increases resulting from decreasing solar elevation angles (Wendler & Eaton 1983).

The albedo of plants is higher in deserts than in other environments because the plants tend to be grayish in color rather than a dark green. This light color increases the plant reflectivity, thereby reducing the transpiration of plants, since less heat from the sun is absorbed (Wendler & Eaton 1983).

The temperature of the desert surface is linked to its albedo. On salt lakes, albedo and surface temperature are related, with surface temperature increasing as albedo decreases. Vegetated ground has a lower temperature than other surfaces, being up to 18°C cooler than sand. Unlike bare soil, however, lower plant albedos correspond to lower temperatures, as dark green vegetation has more water available for transpiration (Wendler & Eaton 1983).

3.3.5 PRECIPITATION

Precipitation is a major factor governing the hydrologic cycle, and its measurement is particularly important for planning purposes. An understanding of rain and snow amounts, the seasonal timing of precipitation, the intensities and durations of individual storms, the reliability of precipitation, and the accuracy and length of the available record are all essential in regional planning. Forecasting concerns include the choice of crops, dam design, irrigation and flood control, and floodplain development. Even so, there are no rain gauges whatsoever for many large areal expanses of deserts, including the Red Sea Hills between Egypt and Sudan, the Great Sand Sea Area between Egypt and Libya, and the Great Salt Desert of Iran (Shata 1992).

FIG. 3.2 Albedo of a saline playa, Death Valley. Salt crusts with low surface roughness and high albedo are located around playa margins. Rugged, cavity-rich crusts (in the foreground) are located in the playa center: these have lower albedos owing to the accumulation of dark surface sediments.

Most water that reaches desert surfaces is in the form of rain, but snow and dew are important in certain regions. However, both snowfall and dew are difficult to determine and are rarely recorded, and only rainfall is commonly gauged.

Deserts generally receive relatively low amounts of total annual precipitation, and this rainfall is of an unpredictable nature. For plants and animals, it is essential to develop strategies to cope with this very great quantitative, temporal, and spatial variability (see Chapters 11 and 12). Humans also have had to find means to deal with unreliable water resources, by using underground water, developing long-distance water transport systems, building storage mechanisms, or living along allochthonous rivers, such as the Nile and the Euphrates.

3.3.5.1 Storm types and seasonality of precipitation

Arid areas are characterized by a low frequency of occurrence and high variability of precipitation in space and time. Desert precipitation results from similar processes as in more humid regions, and can be local convective, frontal cyclonic, non-frontal cyclonic, and orographic, and from general convergence. In any given region, one or more of these systems may contribute rain. In arid Australia, for example, precipitation may result from monsoon rains, tropical cyclones, mid-latitude fronts, and/or anticyclonic fronts (Williams & Calaby 1985). The difference between arid and humid areas is the *probability* of rainfall rather than the rain type.

Local convective or thunderstorm rainfall is particularly prevalent in summer and autumn months, and is the major type of precipitation in the low-latitude sections of the subtropical deserts, such as the southern Sahara, northern Australia, and central Mexico (Bell 1979). Short bursts of high-intensity rain, lasting from a few minutes to about an hour, initially fall over an area of 1 or 2 km^2, but expand rapidly to cover the whole core area of the cell, which may be 5–10 km in diameter. Examples of such high-intensity events include 14 mm in 7 minutes in the Negev, 100–300 mm in 1 day in Australia, and 500 mm in 1 day in the Thar Desert (Evenari 1985a). Nonetheless, the majority of storms are of lower intensities. In the western Negev Desert, 60–80% of storms generated less than 5 mm of rain. Only 20% of storms yielded above runoff threshold intensities ($\geq$9 mm h^{-1}), and the duration of these medium- and high-intensity storms seldom exceeded 15 minutes. High convective intensities (>30 mm h^{-1}) lasted only for several consecutive minutes ($\leq$8 min), and totaled less than 6% of the precipitation (Kidron & Pick 2000).

Instability leading to thunderstorms results from a rapid decrease in air temperature with height, common on summer afternoons because of strong surface heating. Another cause of instability is rapid cooling of higher layers of the troposphere, either through movements of cold air aloft or by radiation from upper cloud surfaces. Convective rainfall is very rare in coastal deserts, where cold ocean currents inhibit the development of moist, unstable air masses. An exception is convective rainfall in late summer in the northern Peruvian Deserts associated with El Niño conditions. Convective precipitation is frequently "spotty" in nature, owing to the almost random spatial distribution of the cells. In Israel, a poor correlation of daily rainfalls has been shown for stations only 5 km apart ($r = 0.5$–0.6) (Sharon 1972).

Frontal cyclonic systems may bring precipitation to the poleward margins of the arid zone, particularly in winter. Areas affected include the northern Sahara, the Middle East, continental interiors of Eurasia and North America, and high-latitude deserts such as Patagonia (Bell 1979). Unlike more humid regions, the patterns of frontal cyclonic storms in arid regions are extremely variable, ranging from bursts similar to those of convective storms, to periods of widespread and relatively uniform rainfall (Fig. 3.3). The temporal and spatial pattern of rainfall depends on characteristics on the individual storm, including its strength, speed of movement, stage of development, and atmospheric moisture content. In general, however, the storms are depleted with respect to more humid regions, and rainfall or snowfall tends to be relatively light and patchy. Virga, precipitation that evaporates in the atmosphere before reaching the ground, is common (Fig. 3.4). Along the northern margins of arid Eurasia and North America, frontal cyclonic storms may be associated with outbursts of cold polar air from high latitudes, resulting in snowstorms. In areas such as Dzungaria and Turkestan, the major part of the total annual precipitation is in the form of winter snowfalls, and the duration of snow cover exceeds 3 months per year (Arakawa 1969).

Non-frontal cyclonic precipitation derives from upper cold troughs and "cut-off" lows, continental heat lows, and tropical cyclones. Cut-off lows may result in widespread inland rain falling with moderate intensity over periods ranging from 12 hours to 3 or 4 days. They are largely responsible for wet periods of 3 days or longer in central and southeastern arid Australia (Foley 1956), contributing 34% of the long-term total rainfall in Broken Hill, New South Wales (Bell 1979). They contribute an even greater proportion (80%) of the rainfall in the Egyptian and western Mediterranean coastal zone (Soliman 1953).

Heat lows develop in summer as a result of strong solar radiation producing extensive areas of warm buoyant air and general uplift. Subtropical desert heat lows develop in southwestern North America,

Fig. 3.3 Winter frontal precipitation falling in the Mojave River valley, near Barstow, California, USA. The pattern of precipitation in such systems is usually much more spotty and the rainfall more depleted than in more humid regions. Rain is falling in the area shown on the left side of the photo, but not the right.

Fig. 3.4 Virga in the Mojave Desert, California, USA. Much of the precipitation that falls above deserts evaporates in the dry air before reaching the ground. Photograph courtesy of Steve Adams.

North Africa, Arabia, Iraq and Pakistan, and Australia. The Arabian heat low is a dominant spring and summertime feature of the southern Arabian Peninsula and constitutes part of the southwest monsoon system, which controls the summertime weather patterns over the South Asian and Arabian subcontinent. It is one of a series of thermal lows embedded in a subtropical heat-low belt extending from the west African Sahara through Libya, Egypt, and the Arabian peninsula and across the deserts of Iran, Pakistan, and India (E. Smith 1986). The Saudi Arabian heat low is a shallow, cyclonic, warm, low-pressure area in the lowest kilometer of the atmosphere (Blake et al. 1983). At irregular intervals moist air may penetrate to normally dry areas, and periods of scattered summer convective storms may occur. If the smaller heat lows are particularly intense, rainfall may be more continuous and uniform. The southwest monsoon plays a critical role in rainfall receipt in the Thar Desert, with 89% of the total rainfall received during the months of June to September. The variability of this rainfall is quite high, however, and the total rainfall is made up of a few heavy falls (Gupta 1986).

The North American monsoon, known also as the Southwest United States monsoon, the Mexican monsoon, or the Arizona monsoon, is a pronounced increase in precipitation between July and mid-September over large areas of the southwestern USA and northwestern Mexico (Adams & Comrie 1997).

The monsoonal region is defined by sites that receive at least 50% of annual precipitation during the summer period. The precipitation region is centered over the Sierra Madre Occidental in the northwest Mexican states of Sinaloa, Durango, Sonora, and Chihuahua, where up to 70% of annual rainfall is related to monsoon activity (Fig. 3.5). It extends northward into the US states of Arizona, New Mexico, and Colorado, and frequently affects the lower Colorado River valley and the eastern Mojave Desert of California. The convective storms of the North American monsoon pose numerous dangers to the growing population of southwestern USA, including lightning, dust storms, and flash flooding.

North American monsoonal systems develop in response to atmospheric moisture supplied by nearby warm oceans (moisture advection from the Gulf of Mexico and the Gulf of California/tropical eastern Pacific) combined with thermal lows developed over warm land surfaces in lowlands during the summer months (particularly the lower Colorado River valley and nearby low desert areas, and the coastal lowlands of Sonora and Sinaloa along the Gulf of California) and elevated areas (Adams & Comrie 1997). Along Baja California, rain diminishes owing to the influence of cool Pacific waters and atmospheric subsidence under the Pacific subtropical high. The cold water of the California current and strong subsidence also stifle summer convection in most of California, and a winter precipitation regime dominates. As

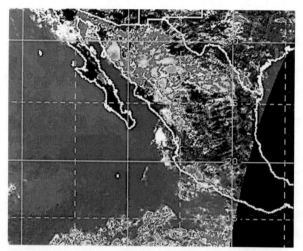

Fig. 3.5 In North America, a pronounced summer monsoon develops, centered over the Sierra Madre Occidental in northwest Mexico (shown here). It extends northward into the several southwestern states in the USA, providing convectional summer rainfall. Source: NOAA.

Orographic precipitation results from the forced uplift of moist air over elevated land. Precipitation is concentrated on the windward slopes and usually increases with altitude (Fig. 3.7). It is most pronounced where mountain ranges lie perpendicular to prevailing winds. Steppe climates are found on the Ahaggar and Tibesti highlands, which are surrounded by some of the driest areas of the Sahara Desert (Griffiths 1972). For the North American deserts, altitude is the most important influence affecting the total amount of precipitation. In southeastern California and southwestern Arizona, 88% of the variation in mean annual precipitation is associated with altitude (MacMahon & Wagner 1985). In the Death Valley area, rainfall increases semi-logarithmically from 0–4 cm year^{-1} on the valley floor to about 50–60 cm year^{-1} at elevations of about 2500 m, after which it levels off (Fig. 3.8) (Jayko 2005). If the uplift of air stimulates instability and convective cell activity, there will be randomly distributed centers of high-intensity rain within a broader matrix of low-intensity precipitation (Bell 1979).

3.3.5.2 Forms of precipitation other than rainfall: fog, dew, and snow

Precipitation that involves little or no atmospheric uplift includes light rain, drizzle, and snow from layer clouds (stratus, altostratus, and nimbostratus), fog drip, dew, and frost. These forms of precipitation may be important to living organisms in the desert, although quantitatively they appear relatively insignificant. Precipitation from layer clouds is most commonly experienced in middle to high latitudes and at high altitudes, occurring in Asia and North America during the winter months. Dew may result from the interception of fog droplets (as in the Peru–Chile deserts and the Namib Desert) or by radiational cooling of the air (as in the Negev Desert, Israel). It is not possible to distinguish between these two dew types.

Arizona, New Mexico, and Colorado are on the fringes of a principally Mexican phenomenon, precipitation is more variable and its distribution is more strongly influenced by topography, with higher elevations receiving most of the rainfall.

There is considerable interannual variability in the onset date and precipitation intensity and extent of the North American monsoon (Higgins et al. 1999). Studies underway relating ENSO and the monsoon are as yet inconclusive, but suggest that the impact of La Niña or El Niño events on the spatial extent and intensity of summer precipitation events varies according to region. In Arizona–New Mexico and northwest Mexico, drier-than-average monsoons are associated with La Niña (Higgins et al. 1999).

Tropical cyclones or hurricanes may penetrate the outer margins of some desert areas, bringing high winds and enormous amounts of rain for several days. Such incursions have been recorded around the Indian Ocean, especially northern and western Australia (Fig. 3.6) and the Pakistan–Persian Gulf region. Depressions that move slowly produce heavy and widespread rain, whereas faster-moving storms result in intense rain over restricted areas (Williams & Calaby 1985). Whim Creek, Western Australia, recorded 900 mm of precipitation over a period of 34 hours in 1898, three times the average yearly total (Hunt 1929).

Areas affected by fog drip include sections of the Peru–Chile deserts, Baja California, coastal districts of Namibia, and the coastal arid part of Madagascar. Although extremely low values of precipitation characterize coastal deserts, rainfall records underestimate the total moisture gain, as coastal fogs may distribute more water than actual precipitation. As a result, fog often has a greater biological importance than rain. Additionally, it contributes to rock-weathering processes. In Namibia, rainfall may account for only 9–27 mm of precipitation, whereas fog may

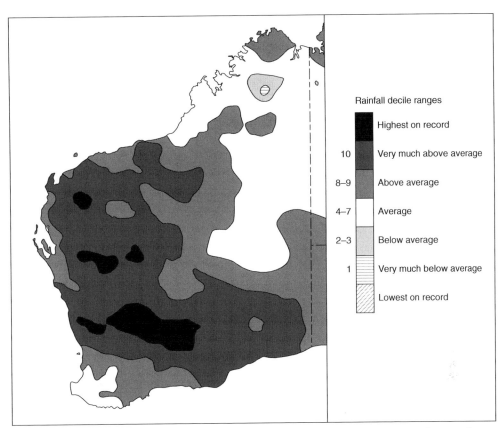

FIG. 3.6 The incursion of hurricanes into desert areas may result in heavy and widespread rain. The movement of Hurricane Vance into Western Australia on 22 and 23 March 1999 brought very high rainfall totals (up to 200–300 mm), including the highest on record for some areas. Source: Australian Bureau of Meteorology.

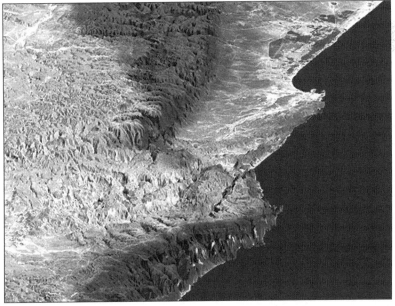

FIG. 3.7 Orographic influences on rainfall, Oman, southern Arabian Peninsula. The steep escarpment of the Qara Mountains creates an orographic effect. Summer monsoonal rainfall provides moisture for natural vegetation (dark areas along the mountain fronts and canyons). The city of Salalah, the second largest city in Oman, is located on the coastal plain. Irrigated crops shown as dark areas near the coast. Source: NASA/JPL/NIMA.

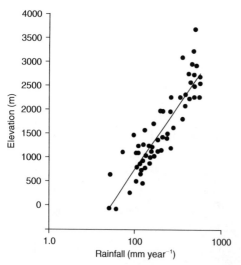

Fig. 3.8 Orographic precipitation, resulting from the forced uplift of moist air over elevated land, is an important source of moisture in arid and semiarid regions. Stations in the Death Valley area, California, USA, show rainfall increasing semi-logarithmically up to elevations of about 2500 m, after which it levels off. The length of the record is approximately 30 years. From Jayko, A.S. (2005) Late Quaternary denudation, Death and Panamint Valleys, eastern California. *Earth Science Reviews* **73**, 271–89. By permission of Elsevier.

occur on 200 days year^{-1} and extend as far as 110 km inland, distributing 34 mm year^{-1} at the coast, increasing to a maximum of 184 mm year^{-1} 35–60 km inland, and decreasing to 15 mm year^{-1} or less to the east of the desert (Lancaster et al. 1984; Walter 1986). A similar situation occurs in the Peruvian desert. Particularly in winter, the area south of 8°S has a high relative humidity owing to the *garua* fog (500–700 m in thickness), which moves 30–40 km inland into Andean valleys (Rauh 1985). Fogs occur in winter because of enhanced oceanic upwelling and colder land temperatures, and disappear during the summer months owing to heating of the air by the warm coastal land. The fogs wet the upper 1–2 cm of soil, and, on hills near the coast, a *loma* vegetation develops. In Baja California, fogs are a common occurrence, but they lack the regularity of fogs in other areas (Rundel 1978).

Dew formation by radiational cooling has received relatively little attention in deserts outside of Israel, where several studies have been conducted. This process requires moisture in the atmosphere and efficient nocturnal radiational cooling, associated with clear skies, light winds, and cold dry air overly-ing a shallow moist layer near the ground surface (Zangvil 1996). Owing to the proximity of the Negev Desert to the Mediterranean Sea, total yearly dewfall in Israel is about 30 mm (Slayter & Mabbutt 1964; Evenari et al. 1971; Evenari 1985a) and dew occurs on average 195 nights per year (Evenari 1985a). Like rainfall, the amount of dew precipitated increases with altitude. The largest values are recorded on low-angle slopes (0° and 30°) and values decline as slope angle increases, until at 90° they are only 25% of those of horizontal surfaces (Kidron 2005). Dew amount and duration are greater in winter than summer, because of longer nights and the increased moisture supply near the ground from rainfall (Zangvil 1996). Dew is uncommon in the Australian deserts, but may develop for several mornings following rainfall (Williams & Calaby 1985).

Dew is generally ineffective in changing the soil moisture content, as it evaporates quickly, but it is important for some plants, including lichens and algae, and animals like the isopod *Hemilepistus reaumuri* and Negev desert ants (Evenari 1985b), as it represents a relatively constant and stable water source. Dew affects plant stomatal opening, photosynthesis, and transpiration, and may assist in plant growth, flowering, and seedling development (Kidron 2005).

The susceptibility of a region to frost depends on topographic and ground-surface factors, altitude, and synoptic weather conditions. In the semiarid Puna of Argentina, high altitudes (3400–4500 m ASL) result in a very high frequency of frost: 176 days per year at La Quiaca. Patagonia, at a more southerly location, experiences frost on more than 265 days per year (Mares et al. 1985).

Mountain snow plays a significant role in water storage and release in deserts that are in proximity to high-altitude mountain ranges (Fig. 3.9). The snow provides a reliable source of runoff for agriculture and helps dampen the variability of the hydrograph. The Tigris and Euphrates Rivers, which supply the Mesopotamian region with needed water, are fed by orographic precipitation that falls in the Taurus and Zagros Mountains. Runoff fluctuates with the snow cover, which may vary considerably from year to year, from 180,000 km^2 in 1985 to 330,000 km^2 in 1993 (Evans et al. 2004).

3.3.5.3 *Variability in precipitation*

Rainfall is the most variable of meteorological measurements made in desert areas (Dolman et al. 1997). Research into its spatial and temporal distribution is

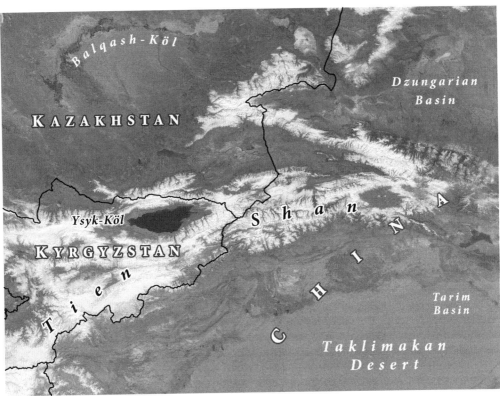

FIG. 3.9 Mountain snow plays an important hydrologic role in desert areas, providing a reliable source of runoff for agriculture and recharging the groundwater system. In this image, snow covers the high mountains of central Asia, most notably the Tien Shan Mountains on the border between northwestern China and Kyrgyzstan. To the south, lies the vast Taklimakan Desert, covered by dunes. Meltwater from the mountains penetrates into the margins of the desert areas. Source: NASA MODIS Rapid Response Team, NASA/GSFC. See Plate 3.9 for a color version of this image.

often limited by a lack of long-term weather records (Adams & Comrie 1997). Rainfall totals are not normally distributed, and the use of mean values for arid regions is often misleading.

The quantitative temporal uncertainty of rainfall can be expressed by variability coefficients, for example, the maximum annual rainfall divided by its minimum (M/m), with values of 6–20 in hot deserts (100 or more in extreme deserts); or the standard deviation as a percentage of the average annual rainfall, with values commonly between 30 and 70% (150% or more in extreme deserts) (Evenari 1985a). The year-to-year variability of precipitation is inversely correlated with the amount of precipitation. Yuma, Arizona (average annual precipitation 81 mm) has a coefficient of variation of 62%, whereas Tucson (average annual precipitation 275 mm) has a coefficient of 30% (Hastings & Turner 1965). Values

between 30 and 50% are typical of the arid parts of California, Nevada, and Arizona (Bailey 1981). Rainfall variability in the exceedingly arid North African deserts generally exceeds 60% (Djanet 99%, Timassinine 132%, Bilma 84%, Murzuq 151%, and Dakhla 225%), whereas the tropical regime in the southern Sahara results in lower variability (Tombouctou 29%, Gao 33%, and Tessalit 49%) (Le Houérou 1986). Rainfall variability may also differ by season. The coefficient of variation for summer precipitation in Tucson is 40%, but 54% for winter (Hastings & Turner 1965).

In extreme deserts, such as the Atacama, many years may pass without rain. Baghdad, in the Mojave Desert, California, went without measurable precipitation for over 2 years (767 days between October 1912 and November 1914) (Bailey 1981). In less extreme deserts, the number of days with more than

0.1 mm typically lies between 4 and 45, but with large variability (Evenari 1985a).

For deserts, the maximum rainfall in one location over short periods may approach or exceed the mean values for the region. Mean rainfall for Death Valley is about 45 mm, but monthly rainfall was greater than or equal to 38 mm on six occasions between 1911 and 1960 (Bailey 1981). In the summer of 1984, a weather station at Wildrose Ranger Station in Death Valley (elevation 1250 m and mean annual rainfall of 190 mm) recorded 305 mm of rain in a 24-hour period (Jayko 2005). Single, large-magnitude storms may have sufficient power to cause considerable modification of the landscape through erosion, incision, and hillslope failure.

Desert precipitation is more spatially variable than that of humid regions. It is often described as "spotty," with the area affected often limited by the radius of the clouds. The spatial variability in rainfall is difficult to assess because budgetary constraints limit the number of rain gauges in any given area. To examine precipitation variability at a fine spatial scale in the Sonoran Desert, Comrie and Broyles (2002) used precipitation data from 139 stations, including 49 low-cost, low-tech storage gauges (with a layer of mineral oil that floats on the accumulated water to halt evaporation) spaced 8–16 km apart. The region experiences a bimodal rainy season: from June to September localized and intense convective storms occur in association with the southwest monsoon, and between November and March the rainfall is more widespread and predominantly from frontal storms. Annual and seasonal precipitation was found to vary by orders of magnitude (<2.5 to >25 cm in sequential years at the same station). Even at a fine scale, there were strong spatial gradients in precipitation resulting from the effects of terrain, moisture advection, and the extreme patchiness of precipitation processes. Convective rainfall is much more localized and the patterns are more complex than those for frontal systems. Summer precipitation is spatially biased towards higher terrain, whereas winter precipitation is more evenly spread out. In the placement of rain gauges it is often assumed that the differences between rain gauges will be evened out over the course of a season, so that a single station will suffice to record meteorological data for a region. The fine-scale data (Comrie & Broyles 2002) show that this is not the case, as stations consistently showed spatial and temporal biases and patterns that were not eradicated over time. This patchiness of precipitation is an important factor when considering wildlife and grazing-management practices.

The seasonal and annual growth of vegetation are influenced by the length of the rainy season, the amount and timing of rainfall, and the date of the first and last rain. These factors can be of considerable significance to vegetation, depending on the thermo- and photoperiod that is affected. Germination of plants depends on the environmental conditions following rain: hot dry winds will deter germination by drying the soil, whereas, for the same amount of rainfall, cool and humid weather will be a trigger (Evenari 1985b). Seasonality of rain is also significant to plants, as a desert with winter rainfall will have more effective moisture than one with summer rainfall, owing to differences in evaporative losses.

The shortness of rainfall records and the extreme variability of rainfall raise difficulties in determining whether there are secular trends, and much attention has been given to the determination of rainfall cycles. This aspect of climatological research has received considerable attention with respect to the droughts and famines of the Sahel in Africa (see Section 3.4.2).

3.3.6 WIND

Wind plays many roles in the desert, influencing atmospheric transparency, temperature, evaporation, vegetation, and geomorphic processes. Strong winds (≥ 100 km h^{-1}) may severely damage plants, partial destroy fruits and flowers, or interfere with pollination. Wind erodes soil where the natural cover of vegetation has been removed.

Solar radiation and the daily cycle of afternoon instability and nighttime stability have a marked effect on wind speeds and direction. Peak daily wind speeds are often observed during the late afternoon hours and the lowest wind speeds during the night (Fig. 3.10) (Walsh & Hoffer 1991).

In areas of strong topographic control, wind directions are often aligned with valley axes. Although synoptic-scale flow in the American southwest is generally from west to east, valleys in the Basin and Range Province trend broadly southeast–northwest. The valley floors may be governed by locally generated circulation patterns, with air flowing up and down the valley axis, as seen in the Bristol-Danby trough and Owens Valley, California, and along the Colorado River (Laity 1987; Walsh & Hoffer 1991).

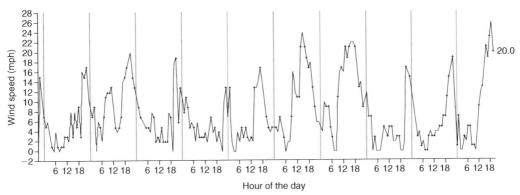

FIG. 3.10 Solar radiation and the daily cycle of afternoon instability and nighttime stability have a marked effect on wind speeds and direction. Peak daily winds often occur in the mid- to late-afternoon hours and the lowest wind speeds during at night and in the morning. Location: Mojave River sink, California, USA. California Department of Water Resources weather station.

FIG. 3.11 Air pollution is a growing problem in deserts. In China, the use of coal and lack of pollution controls is causing a decline in visibility in the western desert areas.

3.3.7 EFFECTS OF POPULATION GROWTH AND URBANIZATION ON DESERT CLIMATOLOGY

3.3.7.1 Air pollution

The global growth of human population has led to expansion of settlement into regions previously considered too arid. In the past, the desert atmosphere was characterized by excellent visibility and lack of particle and chemical pollution. Declines in the quality of the desert atmosphere are becoming increasingly evident. In China, the widespread use of coal for heating and industry has led to severe air pollution in industrialized eastern sectors and, although problems are less apparent, visibility has declined in desert areas owing to industrial expansion and lack of pollution controls (Fig. 3.11).

The desert areas of the southwestern USA are the fastest-growing regions in the nation and pollution stems largely from high power consumption (for example, air conditioner use) and from increases in vehicle miles driven. The population of Tucson, Arizona, increased from 265,000 in 1960 to about 522,000 in 2004 (with an additional 305,000 in unincorporated areas of Pima County) (Pima Association of Governments 2005). Owing to low-density development, the average vehicle miles traveled increased from approximately 1 million to over 20 million, resulting in motor vehicles emitting a majority of the area's ozone-precursor pollutants (Diem & Comrie 2001).

Ozone pollution is related both to locally produced pollutants and to regional pollutant transport, both within the USA and between the USA and Mexico. The highest concentrations of ozone occur in August, in part due to the North American monsoon. Transport of pollution from Phoenix, which lies to the north, may have an impact on elevated ozone concentrations in Tucson during April, May, and June, whereas the El Paso/Ciudad Juarez metropolitan areas, which lie to the east on the Texas/Mexico border, contribute to elevated ozone concentrations in August and September. Elevated ozone concentrations are promoted by hot, sunny, and calm conditions associated with anticyclonic conditions that prevail for significant periods of the year.

The air-quality trends over a 10-year period in Laughlin, Nevada/Bullhead City, Arizona, a small urban center surrounded by unspoiled desert terrain, are also related to regional air transport and local development trends (Walsh & Hoffer 1991). Air pollutants from the Los Angeles metropolitan area may be transported to the area and beyond by winds that blow from west to east. Locally however, the rapid rise in population (540% in 17 years) and the related economic and physical growth has introduced particulate matter (dust) into the atmosphere during construction, increased the emissions of halocarbons and other pollutants owing to demands on power and water utilities, and increased emissions of nitrogen oxides as automobile traffic has risen.

3.3.7.2 Heat islands

Urbanization may create heat islands, the relative warmth of a city compared with surrounding rural areas. These effects appear to be relatively small in deserts (Atwater 1977), but there has been little research on this topic. Several studies have been conducted in the southwestern USA. An examination of Tucson, Arizona, by Comrie (2000) indicated that the urban heat-island effect had caused a warming of approximately 3°C over the past century and more than 2°C in the past 30 years. These results were questioned by Peterson (2003), who found no statistically significant impact of urbanization in the contiguous USA. Daily mean temperatures at Bull-

head City, Arizona, showed an apparent increase over the 6.5-year data record (since 1981) associated with the peak tourist season (Walsh & Hoffer 1991). Similarly, Phoenix experienced an increase in maximum and minimum temperatures for the period 1948–84 (Cayan & Douglas 1984).

Local environmental effects probably are significant in heat-island development. Despite the rapid urbanization of Kuwait City, Kuwait, urban warming effects are negligible owing to the closeness of the city to the sea, low building heights, and use of local building materials (Nasrallah et al. 1990).

3.4 TEMPORAL AND SPATIAL VARIABILITY OF CLIMATIC INFLUENCES

3.4.1 ENSO FORCING OF DESERT CLIMATES

The cyclical ENSO is one of the most prominent large-scale climatological disturbances in the oceanographic/meteolorogic system. It affects hydrologic processes and primary productivity, an important determinant of community dynamics and structure in global ecosystems. El Niño was a term first applied by fishermen to a period of reduced fish catch accompanying a rise in sea surface temperature off the coast of Ecuador and northern Peru. Normally, off the Pacific coast of South America, cold nutrient-rich bottom water upwells to the surface owing to the influence of strong southeast trade winds that are maintained by a steep air-pressure gradient between the eastern and western Pacific. Dry, cool air sinks over the cold waters of the eastern Pacific and flows westward along the equator as part of the southeast trade system, driven by the difference between relatively high surface pressure over the eastern Pacific and relatively low pressure over the huge region of warm water in the western Pacific and around Indonesia. The air is warmed and moistened as it flows westward over the progressively warmer water and its moisture condenses and rises as towering cumulonimbus clouds over the western Pacific. The warmest part of the equatorial Pacific (water temperatures >29°C), termed the "Western Pacific Warm Pool," is found north of New Guinea and lies between 140 and 170°E. The Southern Oscillation refers to changes in the pressure differential between the South Pacific high-pressure region and the Indonesian low-pressure region. When this pressure differential reverses, it allows the trade winds to relax, and warm water from the western Pacific moves eastward to the central and eastern Pacific, initiating an El Niño event (Philander 1985).

The onset of a southward coastal current around the Christmas season (El Niño refers to the Christ child) signals the beginning of an annual warming of the surface waters along the Peruvian coast and marks the end of the fishing season. Not all annual warmings are the same, however. Every few years, an unusually large rise in the sea surface temperature of the eastern Pacific disrupts the marine ecosystem, changing it from a normal highly productive condition to one of greatly reduced productivity (Barber & Chavez 1983). A typical El Niño is short-lived, lasting only several months, and principally affects the waters off northernmost Peru, providing beneficial rains to the regional environment. Occasionally, however, an intense El Niño persists for a year or longer, spreading anomalously warm surface waters along the entire coast of Peru and over large areas of the east and central equatorial Pacific. In the strong El Niño event of 1982–3, abnormally heavy rainfall in Ecuador and northern Peru caused widespread flooding, loss of life, catastrophic property destruction, agricultural damage, the spread of diseases, and the loss of a generation of anchoveta fish (Waylen & Poveda 2002).

The opposite of El Niño is La Niña, a cold cycle in the eastern Pacific, which occurs when the area of warm surface waters contracts toward the western tropical Pacific, so that heavy rainfall and low surface pressures are confined to that region. The trade winds are intense during such periods, driving the warm surface waters westward and exposing cold water to the surface in the central and eastern tropical Pacific. El Niño and La Niña are two complementary phases of the Southern Oscillation, a continuous back-and-forth change in sea surface temperature, atmospheric pressure, and rainfall that occur on the order of every 3–6 years (Philander 1985).

The effects of the El Niño are felt well beyond the local area (Philander 1983; Chiew et al. 1998) and are well documented. However, the consequences for arid environments are often complex and remain poorly understood, largely owing to the lack of long-term records and the paucity of meteorological stations. The warm sea surface temperatures and the anomalous westerly low-level flow are associated with a major shift in tropical precipitation patterns and with heavier than normal precipitation along the equator. In Australia, El Niño events are associated with an increased probability of dry conditions and severe droughts may result (Gill & Rasmusson 1983;

Chiew et al. 1998). The seas around Australia cool, and slackened trade winds feed less moisture into the region. Crop yields, pasture growth, and the vitality of grazing animals are strongly tied to ENSO (Chiew et al. 1998). During the 1982–3 event, millions of sheep and cattle died and a \$2 billion loss in crops was suffered. The Pacific El Niño also has a close association with rainfall variability in many parts of Africa. ENSO events are associated with below-normal rainfall over much of southern Africa, particularly during the summer peak rainfall months of December through March. There is a weakening of tropical convergence over the subcontinent and a shift of the zone of convergence to the east and north (Mason 2001). As in other areas of the world, the effect is sometimes a weak one, as illustrated by the 1997–8 El Niño, which did not produce widespread drought over southern Africa. Continental-scale studies have indicated a lack of influence in Sahelian West Africa, although some studies suggest that rainfall may be reduced (Nicholson & Selato 2000). Whereas El Niño often brings rains to north-central Chile (Gutiérrez et al. 2000), research by Latorre et al. (2002) suggests that the effects are different on the Altiplano. On the Altiplano, El Niño years are characterized by high-altitude westerly winds that inhibit moisture advection from the eastern Cordillera and convection over the western edge of the Altiplano. On the other hand, La Niña phases are associated with enhanced easterly circulation that produces greater advection and increased precipitation (Latorre et al. 2002).

The effects of El Niño on terrrestrial desert ecosystems have only recently been explored. The most commonly noted is the spectacular short-term greening and flowering of deserts (Dillon & Rundel 1990; Holmgren et al. 2001). Barren, arid islands in the Gulf of California (mean annual rainfall of 59 mm) normally have a plant cover of 0–4%: this increases during rainy El Niño periods to 54–89%, largely caused by a profusion of annuals (Fig. 3.12) (Polis et al. 1997). During the 1997 El Niño event in semiarid north-central Chile, ephemeral plant cover increased by five times on a north-facing dry slope and by three times on a south-facing mesic slope (Gutiérrez et al. 2000). The seed bank in the soil is enhanced sixfold. Additionally, perennial herbs, shrubs, and trees increase their flower and fruit production (Polis et al. 1997), tree and shrub growth is impressive, the recruitment of shrubs and cacti increases (Barbour & Diaz 1973; Nicholls 1992), and plants that are established during such wet periods may persist through drier conditions.

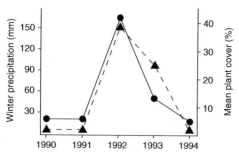

Fig. 3.12 During El Niño years, arid and relatively barren islands in the Gulf of California show a marked increase in the mean percentage of plant cover. The graph shows the winter precipitation (in mm; solid line) plotted against percentage of plant cover (dashed line for the 1992–3 El Niño event). From Polis, G.A. et al. (1997) El Niño effects on the dynamics and control of an island ecosystem in the Gulf of California. *Ecology* **78**. By permission of the Ecological Society of America.

Finally, some El Niño effects are subtle. The El Niño of 1997–8 caused an increase in air temperature and the water content of fog, which favored exuberant vegetation growth and an expansion of the range of some floristic elements of the fog oases of the northern Chilean coastal desert (Muñoz-Schick et al. 2001).

An anticipated result of the El Niño-driven expansion in primary productivity is a subsequent increase in herbivores, followed by carnivores. Several long-term studies have showed that such bottom-up effects do occur, although the picture is often complex (Holgrem et al. 2001; Jacsik 2001). On islands of the Gulf of California, herbivores dominate during wet years and detritivores and scavengers during dry periods. In Chile, there are documented rodent outbreaks, with population levels that may be 20 times higher than normal. Small rodents irrupt within months, but larger ones take a full year to increase. An increase in the abundance of mice generally followed within 6 months of an El Niño (Fig. 3.13) (Jacsik 2001). Predators respond to prey abundance with a 1-year lag in their increase in density (Jacsik 2001).

Some attempts have been made to reconstruct palaeo-El Niño events. Studies of the coastal region of southern Peru, on the northern edge of the Atacama Desert, suggest a lower intensity and frequency of occurrence of El Niño events in the mid-Holocene, during a period of important human settlement (Fontugne et al. 1999). Fine-grained deposits lie between two coarser debris-flow units, interpreted to be the result of El Niño events at about 9000 and 3380 years BP. The fine-grained and organic-rich deposits are thought to have been associated with a permanent water supply, possibly related to condensation from an increased frequency of intense coastal fogs (Fontugne et al. 1999).

During cold La Niña events, rainfall generally increases in Southern-Hemisphere regions such as Australia and southern Africa. It is generally expected that the Australian climate varies from excesses of wet to dry about every 5 years. The anomalously wet conditions of the 1970s were associated with La Niña

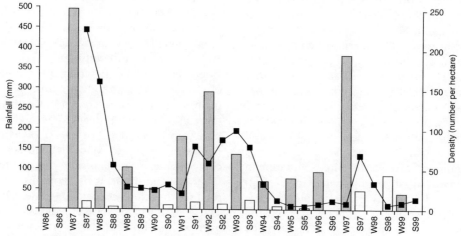

Fig. 3.13 A plot of rodent abundance (solid line) and seasonal rainfall (bars; W is autumn and winter; S is spring and summer) for northern Chile. Rodents reach their maximum abundance approximately 6 months after the peak of rainfall. Note the strong peak of rodent abundance after the 1987–8 El Niño year. From Jacsic, F.M. (2001) Ecological effects of El Niño on terrestrial ecosystems of western South America. *Ecography* **24**. By permission of Blackwell Publishing.

phases and stronger summer monsoon activity over northern Australia (Suppiah & Hennessy 1998). During the La Niña event of 1989, 133 meteorological records, some more than 100 years old, were broken in Australia. In Africa, Nicholson and Selato (2000) found a robust La Niña signal over much of the continent for the period 1948–79, particularly in those events that produced strong cold phases in the Atlantic and Indian Oceans. Increased rainfall was recorded in the first few months of the post-La Niña year in 14 of 17 events in Namibia and western South Africa. The rainfall relations appear to be principally manifestations of the influence of El Niño and La Niña on the Indian and Atlantic Oceans.

3.4.2 EXPANSION AND CONTRACTION OF THE SAHARA DESERT

The margins of the world's deserts are in a state of constant change (see Section 1.6). These transformations have taken place over long timescales (pluvials and interpluvials during the Pleistocene), medium timescales (for instance, Holocene wet environmental conditions in the Sahara which ended about 5500 years ago) (Foley et al. 2003), and short durations, such as the Sahelian droughts of the 1820s and 1830s, 1968–73 (Nicholson 1985; Goudie 1991), and 1982–3 (Todorov 1985). This section will focus on the position of the southern boundary of the Sahara where it merges into the semiarid Sahelian zone, a region noted for its considerable climatic variability, over both short and long temporal scales (Nicholson 2001). The consequences of this variability include subcontinental-level droughts, often lasting decades. The dryness of 1968 heralded a major 30-year drought in the Sahel, with as much as a 20–40% reduction in rainfall (Nicholson 2001). Such periods can have disastrous effects on human life, livestock, and agricultural productivity.

Considerable controversy surrounds the position of the Sahara's margins. In particular is the question of whether the desert is expanding or contracting in size, or simply experiencing natural boundary fluctuations associated with the variations in rainfall that typify arid regions (Tucker et al. 1991; Nicholson et al. 1998; Nicholson 2001; Oba et al. 2001). Several studies have examined whether feedback processes that occur between the surface and the atmosphere when the land becomes degraded have contributed to a decline in rainfall, thereby promoting desert expansion (Charney 1975; Foley et al. 2003). The potential influences of global-scale climatic variation (North Atlantic Oscillation (NAO) and ENSO) on rainfall and the desert's margins are also being examined (Oba et al. 2001). Of additional concern is the degree to which global warming may be influencing vegetation productivity and land (Hulme 1996).

The idea that the Saharan Desert may be expanding as a consequence of human activity has been considered since the early 1900s. Stebbing (1935) wrote of the decline of forests in parts of Mali, Niger, and Nigeria in an article entitled "The encroaching Sahara." Later in the century, satellite imagery of the Saharan Desert, coupled with photographs showing deep drought, starving humans, and land stripped of vegetation, suggested that the Saharan Desert was expanding into the zone of the Sahel. Desert expansion was attributed to "desertification," a largely human-driven, possibly irreversible, expansion of the world's deserts (see Chapter 13). In 1975, Charney speculated that albedo changes related to overgrazing caused a decline in regional rainfall, suggesting that the region was experiencing a process of desertification that may have exacerbated or caused the drought. A positive feedback was thought to develop, with albedo changes reducing rainfall, and thereby further altering the vegetation and soil, and promoting further loss. The processes operating along the Saharan border are probably very complex, involving both natural climatic variability and human-induced changes. Foley et al. (2003) believe that relatively rapid fluctuations from wetter to drier regimes in the Sahel are driven by strong feedbacks between the climate and vegetation cover, triggered by slow and subtle changes in the degree of land degradation, sea surface temperatures, or incoming solar radiation. Given the lack of long-term data, the issue of desertification has been difficult to resolve. However, ongoing examinations of the nature and causes of rainfall variability in northern Africa suggest that many of the changes originally attributed to desertification may be largely the result of natural fluctuations in Sahelian rainfall.

A critical question in the discussion of Saharan desertification is whether the desert is truly expanding in area. In recent decades, satellite imagery has helped to resolve this question by allowing a year-by-year examination of the position of the desert border. Tucker et al. (1991) and Nicholson et al. (1998) documented a seesaw-type movement of the boundary, employing the satellite-derived Normalized Difference Vegetation Index (NDVI), which is used to measure primary productivity (Fig. 3.14).

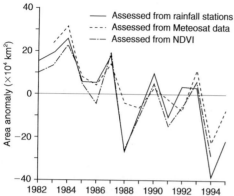

FIG. 3.14 The extent of the Sahara Desert (calculated as the area between the 200 mm isohyet and 25°N) varies over time, creating a seesaw-like pattern. The solid line represents the positions as assessed from rainfall stations; the dashed lines is from Meteosat data, and the dotted line from NDVI data. From Nicholson, S.E. et al. (1998) Desertification, drought and surface vegetation: an example from the West African Sahel. *Bulletin of the American Meteorological Society* **79**. By permission of the American Meteorological Society.

The NDVI correlates well with the percentage of vegetation cover, biomass, and biological productivity of arid and semiarid lands. It does not directly indicate land degradation, as this process implies a change in species composition, soil erosion, or other factors that are not measured (Nicholson et al. 1998).

The southern boundary of the Saharan Desert is in reality a transition zone, with a gradient from desert vegetation into Sahelian steppe vegetation, accompanied by an increase in mean annual precipitation of about 1 mm year^{-1} km^{-1}. Satellite imagery revealed that from 1981 to 1985, the border moved southward by 242 km, at a rate of about 60 km year^{-1}; from 1984 to 1986 the border moved northward again by 143 km, but in 1987 moved southward by 55 km; in 1988, it moved northward 100 km; and in 1989–90, it moved southward 77 km, such that the mean position in 1990 was about 130 km south of its position in 1980 (Tucker et al. 1991). Owing to variations in rainfall, the change in position was not latitudinally uniform, being greater in Mali through central Niger and from eastern Chad through western Sudan. There was less movement of the isoline in Mauritania and in the area from central Niger through western Chad. Nicholson et al. (1998) found similar back-and-forth movements of the border when they monitored the extent of the central and western Sahel for the period 1980–95. The boundary moved southward in 1990, northward in 1991 and

1992, southward in 1993, and then in 1994 moved to its northernmost position since 1980, as a result of an abrupt increase in rainfall. Together, the studies of Tucker et al. (1991) and Nicholson et al. (1998) suggest that the Sahara did not progressively expand during the 14-year period studied, but rather experienced interannual fluctuations in boundary position associated largely with rainfall. A study of many decades would be necessary to determine any long-term trends in the size of the Sahara.

One important aspect of the climate of the Sahelian region is that the variance tends to be on decadal timescales: that is, there is a strong year-to-year persistence of anomalous (either wet or dry) conditions. The Sahel region has been subject to many significant climate changes in the Pleistocene and Holocene. Semiarid climates probably set in about 2000 years ago (Lézine 1989). Since that time, both more humid and more arid periods have occurred. The ninth through the fourteenth centuries were humid and the Mali empire flourished (Nicholson 2001). Relatively humid conditions also prevailed from the sixteenth through eighteenth centuries. More arid conditions set in at the end of the eighteenth century and persisted through the first decades of the nineteenth century. The dry conditions of the 1820s and 1830s caused Lake Chad to dry up, a process which did not occur again until the 1980s (Nicholson 2001). Moist conditions at the end of the nineteenth century were manifested by the expansion of woodlands into the Saharan margins, the presence of lakes, and the growth of wheat in the Tombouctoo region. The twentieth century began with relatively dry conditions, and widespread and severe droughts in the 1910s; however, the rains returned within a decade. The more extreme fluctuations developed in the latter half of the twentieth century, culminating with a three-decade-long drought (Fig. 3.15). The decadal length of droughts in northern Africa implies that the controls on climate are likely to be different than in those regions, such as western South America, where the global-scale rainfall fluctuations are largely affected by the shorter-term influence of the ENSO.

The nature and causes of the large variations in Sahelian rainfall have received considerable recent attention (Nicholson et al. 1998; Zeng et al. 1999; Nicholson 2001, 2005). The droughts in the latter part of the twentieth century may now be considered in a broader context, and natural and anthropogenic factors potentially separated. The Sahel is characterized by a single, short rainy season associated with the northward movement of the ITCZ. Dry, dust-laden air from northeastern Africa is undercut by a

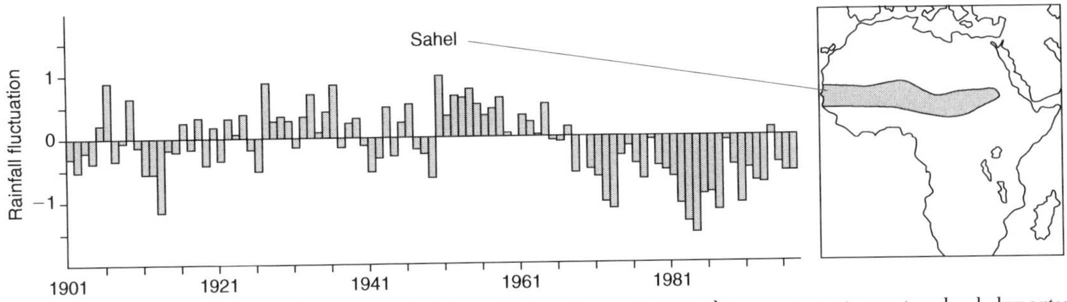

FIG. 3.15 Rainfall fluctuations in the Sahel of Africa (1901–97), expressed as a percentage standard departure from the long-term mean. From Nicholson, S.E. et al. (1998) Desertification, drought and surface vegetation: an example from the West African Sahel. *Bulletin of the American Meteorological Society* **79**. By permission of the American Meteorological Society.

wedge of warm, moist air from the Gulf of Guinea, giving rise to storms. Some storms are associated with organized easterly waves that propagate from east to west, and others are strictly local affairs. The African Easterly Jet (AEJ) helps to generate and maintain the wave disturbances (Nicholson 2001). At any one time, only 1–10% of the region is covered by storms, giving rise to highly variable rainfall in both space and time.

Rainfall in sub-Saharan Africa appears to be strongly influenced by the combined effects of the NAO and the ENSO (Nicholson 2001; Oba et al. 2001). The NAO is an alternation in atmospheric mass balance between pressure centers over the Azores and Iceland. It influences the speed and orientation of winds across the Atlantic Ocean from the Gulf of Mexico to northern Europe (Lamb & Peppler 1987), and induces temperature and rainfall variations on both sides of the Atlantic in the Northern Hemisphere. According to Oba et al. (2001), 75% of the interannual variation in the extent of the Sahara Desert is accounted for by the combined effects of the NAO and ENSO, with most of the variance the result of the NAO. Latitudinal changes in position of the 200 mm isoline were most strongly influenced by the NAO. The drier years in the Sahel tend to be associated with warm sea surface temperatures in the southern oceans and Indian Ocean, and anomalously cold sea surface temperatures to the west of the continent. During dry years, the AEJ is stronger but is displaced equatorward, the Tropical Easterly Jet is unusually weak (Nicholson 2001), and the ITCZ is weak and displaced southward. Thus, recent studies of African climate have increased our understanding of the factors governing interannual variability, and largely support the idea that the productivity of Sahelian rangelands is more a measure of natural climatic variability than anthropogenic disturbance of the land.

In summary, the rainfall conditions of the past 30 years are not unprecedented, with similar droughts having been recorded in the nineteenth century. The processes governing rainfall generation in the sub-Saharan region are now better understood, and recognized to correlate with sea surface temperatures. Changes in the land surface cover as a result of human action (vegetation cover, soil moisture, and surface albedo) probably influence climate, particularly at a local scale, but the decline in rainfall over the past few decades appears more closely tied to natural climatic variation. The process of desertification appears to be occurring at relatively small scales. Nonetheless, there is still a need for continued monitoring of the land surface to understand the degree of change and to determine the type and extent of recovery that occurs when wetter cycles return.

3.4.3 THE SAHEL: LAND-SURFACE–ATMOSPHERE INTERACTIONS

The interaction between the land surface and the atmosphere involves many processes and feedback mechanisms, all of which vary simultaneously. Global circulation models and mesoscale models have the potential to reproduce these interactions. As a result of the worldwide attention garnered by the drought conditions of the latter part of the nineteenth century, and the need to understand the land-surface interactions highlighted by Charney (1975) and others, an international land-surface–atmosphere observation program was launched. The HAPEX-Sahel (Hydrological and Atmospheric Pilot Experiment in the Sahel) project was initiated in western Niger, with an intensive observation period in the wet–dry period of August–October 1992 (Goutorbe et al. 1994). The aim of the project was to understand the role of the Sahel in the general atmospheric circulation, with particular emphasis on the large interannual land surface fluctuations. Additionally, the project sought to determine how the general circulation is related to the persistent droughts that have affected the Sahel.

The Sahara and Sahel zones are a major source of sensible heat for the atmosphere, and any expansion or reduction in the deserts associated with rainfall and vegetation changes will likely be manifested in strong feedbacks to the general circulation through the energy balance and other processes. Whereas large-scale effects, such as changes in Atlantic sea surface temperatures, are shown to correlate with fluctuations in Sahelian rainfall, the relative importance of external factors with respect to local ones (such as vegetation degradation) remains poorly understood. The interdisciplinary HAPEX-Sahel study involved over 200 scientists visiting and working in the field sites, supported by detailed boundary layer and aircraft hydrometeorological measurements.

The results of the HAPEX-Sahel research, as well as additional studies, provide evidence that there is a feedback between the surface and the atmosphere. Surface property changes affect climate by modifying the atmospheric energy and water budget. Taylor and Lebel (1998) found that rainfall was correlated with evaporation anomalies, as wet ground evaporates into the air and promotes future rainfall. This feedback effect enhances the patchy nature of the rainfall and creates pockets of wet and dry conditions within the overall field of precipitation. Modeling results from Zeng et al. (1999) also showed a high degree of correlation between surface moisture and precipitation, and allowing vegetation feedback to the atmosphere substantially enhanced the results. The spatial variation in rainfall and soil types makes conventional hydrologic modeling very difficult. At a local scale, surface runoff is very high from crusted surfaces, and infiltration is concentrated in non-crusted areas. The discontinuous surface tends to produce irregular flow patterns and finally pools.

Vegetation influences climate through its control of the evapotranspiration process and its modification of surface albedo. Vegetation growth responds principally to the interannual variations of water availability, and is less influenced by temperature and nutrient limitations. It modifies precipitation through a series of positive feedback loops: decreased rainfall reduces vegetation, causing higher surface albedo and reduced evapotranspiration, which in turn weakens the large-scale atmospheric circulation

by reducing the energy and water flux into the atmospheric column, further decreasing the local rainfall (Zeng et al. 1999). Such interactions amplify the interdecadal variation of rainfall. It is possible that desertification may account for any remaining differences between the model and observations. As the vegetation feedback amplifies Sahelian rainfall variability that originates with sea surface temperature variations, significant effects can occur with relatively small vegetation changes (Zeng et al. 1999).

Human modification of the land surface, which strongly affects vegetation and soil cover, thus has the potential to influence regional climates. Despite the decline in rainfall over the past 30 years, population growth in the Sahel has continued, subjecting the land to greater pressures. Agricultural crops replace natural savannah (Goutorbe et al. 1994) and grazing and fuelwood collection reduce native vegetation, leading to increases in surface albedo and soil degradation. Soil degradation causes more long-term effects than albedo changes (which may recover with increases in rainfall), as the soil is irreparable over human lifespans and stripping of surface layers strongly reduces its moisture-holding capacities. As land degradation continues, it is likely that there will be an increase in the amount of bare soil relative to the vegetation mass, which may change the evapotranspiration and aerodynamic characteristics of the surface. Models of human modification of the land surface will need to differentiate between bare soil and vegetation, as well as the carbon fixation pathways of plants. Owing to different pathways of carbon dioxide fixation, C_3 and C_4 plants in a canopy show different responses to water-use efficiencies. For example, as soil moisture declines, grasses decrease their transpiration losses, but the carbon dioxide flux remains at a relatively high level, measurably increasing carbon dioxide concentrations in the atmosphere (Dolman et al. 1997).

The extent to which the dry periods of the late 1900s are related to global warming is undetermined. Models depicting future scenarios suggest that Africa could be 2–6°C warmer in 100 years, but the direction and magnitude of regional rainfall changes are less clear, as present global climate models only ambiguously represent variability due to changes in sea surface temperature, dynamic land-cover–atmosphere interactions, and dust aerosols (Hulme et al. 2001). Such interactions await additional study to resolve their full role in Holocene climate variability and more recent Sahelian desiccation.

4

THE HYDROLOGIC FRAMEWORK

4.1 INTRODUCTION

The desert is a region of hydrologic extremes. Beset by flash floods, it is nonetheless an environment characterized by an overall paucity of water. Many arid areas have fluvial forms clearly imprinted on the landscape as a result of the action of flash floods, the presence of rivers with sources in more humid locales, and past pluvial conditions. Owing to sparse vegetation and soils with low infiltration capacities, geology and tectonism play an important role in runoff and groundwater processes.

Endogenic drainage is derived from precipitation within the desert, whereas allogenic drainage originates from rainfall or snowfall outside the desert margins (for example, the Nile River, Fig. 2.7). Desert surface runoff may drain to oceans (exoreic drainage) or, more commonly, flow into closed basins (endoreic drainage) from which no flow reaches an ocean. The runoff in endoreic basins may be considerable, but it is disposed of entirely by evaporation (Fig. 4.1). The Okavango River in central southern Africa is a classic example, terminating in an extensive inland delta. Tectonism and structural factors may contribute to endoreism, as seen in the Basin and Range geologic province of the western USA. Considerable runoff may occur, but the streams terminate in lakes, sinks, or playas, and the water leaves the basin by evaporation. Playas are dry except after exceptional rainfall events (Fig. 4.1).

The hydrology of arid environments differs from that of more humid regions in several important respects. The rainfall is irregular in both amount and intensity and displays considerable seasonal and interannual variability. Orographic effects are particularly pronounced and have a strong impact on flora and fauna. Thus, the physiography of a region – that is, its relative flatness or its relief – plays an important role in hydrologic processes. Interception rates are low owing to the sparse vegetation cover, whereas evaporation rates from exposed surfaces are high in warm deserts, particularly during the summer. Infiltration is largely controlled by characteristics of the surface, which show considerable spatial variation, ranging from impervious surfaces such as exposed bedrock, duricrusts, and well-developed pavements, to more porous surfaces such as recent alluvium, weakly developed pavements, and sandy surfaces. Overland flow is common during storm events, but the flow is ephemeral, lasting only hours or days (Fig. 4.2). Owing to the infrequency of surface flow, groundwater may play a very significant role in providing moisture for plant growth.

The runoff of major rivers in humid, temperate climates is the integration of different tributary inflows, each dependent on the climate and physiography of its drainage basin. In such systems, the discharge is augmented in a downstream direction. By contrast, in deserts it decreases, as water is lost by evapotranspiration and seepage into a permeable streambed. Dryland drainage systems are poorly integrated (Desconnets et al. 1997): flow may not be continuous along an entire link, or not all links may connect and provide tributary inflows. Total stream power, therefore, decreases downstream, reducing energy available for sediment transport and channel change. Thus, in contrast with the perennial flows of most humid-region rivers, semiarid and arid climatic patterns dictate a discontinuity in fluvial dynamics (Graf 1988; Desconnets et al. 1997).

4.2 THE WATER BALANCE IN DESERTS

The concept of the water balance is intrinsic in understanding the hydrology of deserts and may be expressed as follows:

$$P = ET + R + \Delta S$$

where P is precipitation, ET is evapotranspiration, R is runoff, and ΔS is the change in water stored in

Fig. 4.1 Desert lakes are dry except after exceptional rainfall events, and largely lose their water by evaporation. This photograph shows Silver Lake in the Mojave Desert. Relict shorelines, indicative of past wetter conditions, surround the lake. A veneer of windblown sand covers much of the basin and ventifacts have formed on the slopes.

Fig. 4.2 Infiltration characteristics of the surface strongly affect runoff processes. In rocky areas, shown here, much of the rainfall runs off the surface in ephemeral events that may last less than an hour. Location: Mojave Desert, USA.

lakes, snowfields, glaciers, and aquifers. Evapotranspiration is the water returned to the atmosphere by evaporation from ground and water surfaces and by plant transpiration. Storage occurs principally by water infiltrating deeply into the soil and moving slowly to the groundwater table. This water may be stored for long periods of time, often on the order of thousands of years, if not brought to the surface by faulting or withdrawn from wells. Snow and glacial storage may be critical to deserts fringed by high mountains, where seasonal melt recharges subsurface aquifers and provides an important source of surface water. Runoff is the residual part of precipitation after evapotranspiration and storage have been extracted.

As simple and fundamental as the water-balance equation may seem, its application is fraught with difficulties. A relatively simple error in determining either precipitation or evaporation may give a large error in runoff. As discussed in Chapter 3, precipitation is measured nowhere in drylands with sufficient accuracy and there is at least an equal problem with the measurement of evapotranspiration. For very large basins, such as Lake Eyre, Australia, additional complexity in calculating a water balance is added by the latitudinal extent of the drainage basin. Some areas lie in different rainfall regimes, and thus have dissimilar totals and seasonal distributions of rainfall, and markedly varying evaporation rates (Kotwicki & Allan 1998).

If only a small amount of precipitation falls on a basin, most will be lost by rapid evaporation and by infiltration, and little or no direct runoff occurs. Most of the infiltrated water only percolates into the upper meter or two of the soil, to be returned to the atmosphere by evapotranspiration, with little or none passing on to the groundwater table (Gee et al. 1994; Izbicki et al. 2000).

The character of the land influences the relative amounts of water in each of the water balance components. In some highly permeable sandy soils, direct runoff never occurs (Desconnets et al. 1997; Kidron & Pick 2000). In less permeable soils, little water reaches the water table; most being returned to the atmosphere by evapotranspiration, particularly if storm precipitation is small. Thus, the disposition of precipitation depends on the geologic, topographic, and vegetative characteristics of the drainage basin, and on the pattern of precipitation and temperature.

Climate affects the partitioning of runoff, chiefly through its strong influence on evapotranspiration.

The amount of evapotranspiration is not directly proportional to precipitation because it also depends on the available moisture, temperature, solar radiation, wind, and land cover. The seasonality of precipitation affects the bioavailability of water: rain that falls in the cooler winter months is less likely to be evaporated. The time distribution of precipitation can be a factor by altering the amount of time when moisture is available for evaporation. Short-duration, intense storms will experience less evapotranspiration than low-intensity storms of long duration. Owing to the spottiness of desert rainfall, the larger the basin, the less likely that extreme weather conditions will occur over the whole basin in one period.

Runoff approaches zero in basins where the potential evapotranspiration (the amount that would occur if moisture were always available) is higher than the annual precipitation. However, mean annual runoff is seldom zero because heavy rainfalls always exceed the losses over short periods. Arid regions of the southwestern USA and northern Mexico have values of between 5 and 30% for the ratio of mean annual runoff R to mean annual precipitation P (Hare 1980). In the Besor Basin of the Negev Desert, only 5–15% of the total precipitation flows out of the basin, the remainder infiltrating or evaporating (Nativ et al. 1997). Between runoff-producing storms, streambeds are dry. Thus, all but the large allogenic rivers are ephemeral, flowing only after major precipitation events.

Owing to the spectrum of different surface types in deserts, there is often considerable spatial variation in runoff coefficient values, even within a given region or basin. Only 1–2% of yearly precipitation reaches the Niger River from a 7000 km^2 sub-basin marked by degraded drainage channels that are often blocked by sand. By contrast, on the less-permeable laterite plateaus, runoff coefficients may be as high as 75% (Desconnets et al. 1997).

4.3 WATER BUDGETS

4.3.1 PRECIPITATION AND ITS ASSESSMENT: PROBLEMS IN GAUGING AND NETWORK DESIGN

The temporal and spatial variability of desert rainfall were discussed in Chapter 3. This section briefly addresses problems in assessing precipitation in desert environments that occur as a result of poor network design and an insufficient number of gauges. Arid regions suffer from a paucity of measuring sites, even in wealthier nations. As many of the world's deserts are located in less economically and technologically advanced regions, the assessment of precipitation is especially problematic (Nicholson 2001).

Few rain-gauge networks have a rational basis for their present form, most having developed where observers were available rather than because a rainfall record was required at a particular point. Rain gauges in eastern Egypt, for example, are located along the Nile River valley and the coastlines of the Red Sea and Gulf of Suez. None are positioned in the Red Sea hills, where precipitation is probably higher owing to orographic effects, and these rainstorms are largely undetected. As a consequence, studies of flooding and aquifer recharge rely on estimates of rainfall rather than actual measurements (Gheith & Sultan 2002), producing considerable uncertainty in the results.

Spatial precipitation patterns and area averages are the most difficult to ascertain from mountainous areas. Rainfall can be strongly enhanced over the windward side of mountains (Jayko 2005) and reduced to the lee (forming a rainshadow). Near the Gulf of California and in Death Valley, precipitation approaches 500 mm year^{-1} in the higher mountains, but is less than 100 mm year^{-1} at lower elevations (Laity 2002). Lower altitudes not only receive less rainfall than moister, higher elevations, but they also tend to have more episodic precipitation (Comrie & Broyles 2002). Heterogeneous rainfall patterns have strong influences on runoff and vegetation patterns.

Inherent rainfall patterns also contribute to problems in precipitation assessment. Most extreme rainfall events in deserts are associated with localized thunderstorms of restricted areal extent that slip between the sizeable gaps in the gauge network. Total precipitation is underestimated as few showers coincide well enough with the gauges to produce overestimation. This poses a particular problem for planning if the record is short. Rainfall averages have more meaning with cyclonic storms, but, as shown in Fig. 3.3, are also likely to be underestimated.

The final factor posing a problem in determining averages and extremes of precipitation in deserts is the length of record, which rarely exceeds 25 years. In general, short records tend to underestimate maximum precipitation. As a result of all of the factors discussed above, it is probable that the error in determining the mean rainfall for a desert will be appreciable.

4.3.2 INTERCEPTION

Interception is precipitation that is caught by vegetation or litter. It is evaporated back to the atmosphere from the canopy, stems, or litter during and after a storm, or channeled to the ground by stem flow or foliar drip. The process of interception changes the quantity, quality, and distribution of water that reaches the soil surface. In densely vegetated areas, little rain reaches the soil surface directly, but instead drips to the forest-floor litter or reaches the ground by running down plant stems. Owing to the paucity of vegetation in deserts and semiarid regions, interception is quantitatively less significant than in humid areas. At the local level, however, it plays an important ecohydrologic role, funneling water towards the soil and increasing the available moisture for plants.

Both weather conditions and plant structure affect interception losses. Weather factors include the number and spacing of precipitation events, the intensity of rain, wind speeds during and following precipitation, and the availability of solar energy to evaporate intercepted water. Structural factors are vegetation type, size, density, form, and age. Generally speaking, the denser the foliage, the greater is the interception storage. Bark roughness, branch form, and the number of stems or trunks per unit area control stem flow. As a whole, the details of the rainfall-interception process remain poorly understood, even in humid areas (Wood et al. 1998).

Interception is a minor component of evaporative loss in drylands because many plant species are small in size and the total cover of vegetation is often significantly less than 50%. However, the study of interception losses is still in its infancy, and there are few reliable estimates of community-level interception losses (Dunkerley 2000a). Studies of cool-desert shrubs (big sagebrush, *Artemesia tridentata*) in Idaho, USA, by Hull (1972) found that heavy brush intercepted about 30% of rain and 37% of snow, with potential interception of 1.1 mm. Similarly, a mean interception rate of 1.5 mm was found for individual plants in studies of big sagebrush and shadscale (*Atriplex confertifolia*) by West and Gifford (1976). For entire communities, with surface covers of 17–19%, interception was about 0.25 mm. Species differences were significant, particularly at higher intensities of rainfall. Seasonal differences (owing to the retention of leaves) were also very important.

The percentage of annual rainfall lost to interception is not well known. For cool-desert shrubs of the

American southwest, it amounts to approximately 4% of the annual rainfall (West & Gifford 1976). Interception losses appear to be greatest in arid areas covered by grasses. Dunkerley and Booth (1999) suggest that a grass community in western New South Wales, Australia, loses 30% of annual rainfall to interception.

Interception is greatest at the beginning of a storm, before the foliage is fully moistened. During this initial period, the soil beneath the plant remains largely dry. As time passes, the interception capacity of the vegetation is reached, and the area under the plant receives precipitation similar to open areas. Stemflow ensures that water is funneled to the soil beneath the plants in a more efficient manner than if interception had not occurred, increasing the value of water from light rainfalls (West & Gifford 1976; Dunkerley 2000a). Most evaporation takes place after cessation of the rainfall event (Wood et al. 1998).

The ability of litter to intercept and store rainfall depends largely upon its thickness and its water-holding characteristics. In desert areas, litter is discontinuous in its distribution, tending to accumulate around the base of shrubs and in gullies or surface depressions (Tivy 1993). Its significance in dryland interception remains to be assessed.

The subtraction of intercepted water from gross precipitation becomes insignificant during very large rainstorms and interception has little effect upon the development of major floods. It plays only a minor role in preventing soil erosion during intense rainstorms as most rainfall directly impacts rocks or soils.

Some arid coastal areas experience a type of reverse interception process, known as fog precipitation, by which plants increase the water supply reaching the ground. Fog wets the foliage, and water drips to the surface (Wood et al. 1998), contributing 50–300 mm year^{-1} to the annual precipitation along the coasts of southern California, South America, Namibia, and Madagascar (Tivy 1993). On the arid littoral of northwest Africa, the increased water supply from nighttime fog precipitation supports a 100-meter-wide strip of green vegetation in one of the world's driest areas.

4.3.3 EVAPOTRANSPIRATION

4.3.3.1 Introduction

Evapotranspiration is a collective term for the processes by which water in the liquid or solid phase becomes atmospheric water vapor. Evapotranspiration is a combination of two distinct processes: evaporation when there is a "free" water surface, such as water in lakes or the soil, and transpiration when water is removed from the interior of plant leaves through their stomata.

A quantitative understanding of evapotranspiration is of vital practical importance. As discussed previously, the amount of water leaving a drainage basin as runoff may be as little as 1% of the yearly precipitation in hot deserts with permeable substrates. The difference between rainfall and runoff is largely explained by evapotranspiration. Over the long term, evapotranspiration determines the water available for human use and management. Much of the food supply of arid regions is grown on irrigated land, and efficient water application requires knowledge of crop transpiration rates. Evaporation also influences the yield of water-supply reservoirs.

Evapotranspiration is controlled by four factors: (1) energy availability, (2) the humidity gradient away from the surface (vapor-pressure gradient), (3) the turbulent transfer function (represented largely by wind speed immediately above the surface), and (4) water availability (Houston 2006a). A steep humidity gradient is essential for rapid evaporation. In near-calm conditions, evaporation will lead to a build-up of humidity in the air layer immediately adjacent to the surface and a decrease in the evaporation rate. In the presence of wind, with its propensity for turbulent mixing, the low-level moist layer will be removed and replaced with drier air. Usually this replacement is associated with large-scale horizontal advection, which introduces a new, drier air mass to the surface. The net result is that the windy conditions and dry air characteristic of most deserts tend to maintain steep humidity gradients and high evaporation rates. High rates also depend on a continuous supply of water at the surface. For an open water surface, such as a lake or reservoir, this is no problem. However, for a land surface, water movement from depth, whether through the soil or through plants, requires a considerable time. There is an upper limit to the rate of plant transpiration. In midsummer, plants can wilt in the early afternoon because water vapor is removed from the leaves faster than it can be brought in from the roots. The surface soil layers become dry even though water remains at depth.

The action of the four factors controlling evapotranspiration, any of which can be limiting, has led to the development of two concepts of evapotranspiration for practical purposes. The first is *potential evapotranspiration* (PET), the rate that will occur from a well-watered, actively growing, short green crop completely covering the ground surface. This is essentially identical with the values obtained over a large open water surface. It represents the rate controlled entirely by atmospheric conditions and is the maximum possible in the prevailing meteorological conditions. The second is *actual evapotranspiration* (AET), the amount that is actually lost from the surface given the prevailing atmospheric and ground conditions. Actual evaporation rates are always considerably less than potential rates in deserts, as they are limited by the supply of water (Houston 2006a). Both concepts are important, since PET provides some measure of possible agricultural productivity if, for example, irrigation is initiated, while AET provides information vital for the determination of soil moisture conditions and the local water balance.

Not all of the energy absorbed at the surface is used to evaporate water, since the other energy fluxes are maintained. For many land surfaces the soil heat flux is small and can be neglected, so that the major partitioning of energy is between sensible Q_h and latent heat Q_e. The relative contributions of the two can be expressed by the Bowen ratio, $B = Q_h/Q_e$. In general, surfaces act to keep the ratio at a minimum. As a result, a moist surface will increase little in temperature while evaporation is occurring, but there will be a rapid temperature rise once it has dried out and the sensible heat flux takes over as the major energy transfer mechanism. In arid regions, sensible heat transfer dominates ($B > 1$). A map of the Bowen ratio for arid areas of western North America shows values ranging from about 1.5 to 3.0 (Hare 1980).

4.3.3.2 Evaporation

A combination of low relative humidity, high temperature, maximum solar radiation, and surface winds create conditions favorable for rapid evaporation in deserts. Evaporation occurs from open water surfaces (lakes and reservoirs) and from the soil. Whereas climatic conditions cause PET values to be very high, AET values are very low owing to the paucity of water.

The measurement of evapotranspiration, potential or actual, is difficult. The most common method of directly measuring PET is to use an evaporation pan, a container of a standard size (often a "Class A" pan or evaporimeter) containing water freely exposed

to the atmosphere (Singh 1992). The water depth is measured at the beginning and end of a time period, and the difference, after correction for any precipitation received, is the evaporation. Energy transfer through the sides of the pan makes it difficult to relate the results to natural open water surfaces. Because the pan is small, evaporation will be greater than from open water surfaces. For example, the annual open water evaporation from a filled Lake Eyre, Australia, is estimated to range from 1800 to 2000 mm, whereas the Class A evaporimeter loses 2400–3600 mm (Kotwicki & Allan 1998). Moreover, the difference between pan and lake varies throughout the year because of seasonal differences in radiation, air temperature and wind, and heat storage within the larger body of water.

In order to use pan measurements to estimate lake or reservoir evaporation, a coefficient or multiplying factor must be applied to the pan measurements. This pan coefficient varies geographically and seasonally. In areas where pan coefficients have not previously been derived experimentally, an average annual value of 0.70–0.75 is generally assumed. In studies of salars (saline lakes) in northern Chile, Risacher et al. (2003) applied a pan coefficient of 0.65. The pan coefficient for Lake Eyre is about 0.5–0.6 (Kotwicki & Allan 1998). If the lake is very small, such as a shallow stock pond, a coefficient of 0.90 or even higher is appropriate to account for the high evaporative loss.

In hot deserts, evaporation from a Class A evaporation pan is typically 2000–4000 mm year^{-1}. For example, potential evapotranspiration in Barstow, California, is 2121 mm year^{-1}, exceeding precipitation by a factor of 19. Evapotranspiration is highest in July (304 mm) and lowest (66 mm) in January (Laity 2003). Greater values (4262 mm year^{-1} in Death Valley, California) occur in extreme deserts. Variations in annual evaporation from year to year are small (Evenari 1985a). Evaporation is also high in temperate deserts. In the Caspian Lowlands, the potential evaporation, estimated at 1400 mm annually, exceeds the annual precipitation tenfold (Walter & Box 1983a). Further to the east, in the Turanian Lowland (the northern part of the Irano-Turanian deserts), the aridity is very pronounced, with potential evaporation over 2500 mm in the interior of the desert (Walter & Box 1983c). In the relatively cool Central Valley of northern Chile (mean temperature 18–20°C), precipitation is less than 10 mm year^{-1} and potential evaporation is 1500 mm year^{-1}, a factor of 150 or greater (Risacher et al. 2003).

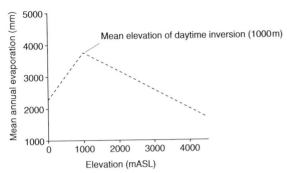

FIG. 4.3 Relationship between mean annual pan evaporation and elevation in the Atacama Desert. mASL, meters above sea level. From Houston, J. (2006a) Evaporation in the Atacama Desert: an empirical study of spatio-temporal variations and their causes. *Journal of Hydrology* **330**, 401–12. By permission of Elsevier.

Air temperature is a very strong control on pan evaporation, explaining 75% of the variance in studies in the Atacama Desert (Houston 2006a). As a result, pan evaporation values decrease with altitude, following the elevational control of temperature (Fig. 4.3). In the Atacama, they reach a maximum of 3800 mm year^{-1} at 1000 m altitude (above the thermal inversion) and drop to around 1700 mm year^{-1} at an altitude of 4200 m ASL (Houston 2006a). Evaporation rates are also strongly seasonal, following the annual cycles. These interannual variations are typically greater than the intra-annual variations. Intra-annual variations are weakly correlated with ENSO effects for the Atacama Desert, with La Niña years associated with higher evaporation rates (Houston 2006a).

Open water evaporation is vaporization from water surfaces free of vegetal cover, either emergent or overhanging. In arid or semiarid regions, there may be a net loss of water and high evaporation rates have a significant impact on water resources. The southwestern USA has the highest evaporation rate in the country (2180 mm) (Singh 1992), and much of the region's water is stored in a few large reservoirs. At Lake Mead, Arizona, 1 billion m^3 of water evaporate from the surface annually, equivalent to a depth of 2 m of water (Harbeck 1958). Lake Nasser, Egypt, loses between 10 and 16 billion m^3 year^{-1} from its total volume of 180 billion m^3 (Shaltout & El Housry 1997).

Several characteristics of the lake influence evaporation, including temperature, area, and salinity. Part of the energy received by a lake is used to raise the water temperature. In a deep lake, the increase in water temperature extends to considerable depth,

and involves large amounts of energy that would otherwise be used for evaporation. If a lake is large in area, the vapor-pressure deficit will decrease as moisture moves into the atmosphere. Therefore, there should be an inverse relationship between the evaporation rate and the size of the lake. Salinity reduces the vapor pressure of a water body (a 1% reduction in evaporation for every 1% increase in salt content), and thus there is a depression of evaporation rates as salinity increases (Singh 1992). This is an important factor in desert lakes, which are commonly saline.

Studies of evaporation from playa environments have found that evaporation rates are generally much reduced relative to free-surface open water values. For instance, evaporation from underlying groundwater through a sand-dominated playa surface was 0.4–0.5 mm per day, whereas measured Class A pan evaporation rates were 17–27 mm per day (Tyler et al. 2005). In many arid areas, playas represent the ultimate point of discharge for the basis's hydrologic system. The high salinity of groundwater limits capillary rise and the formation of salt crusts inhibits evaporation by forming an impermeable barrier. Moisture losses from the Salar de Atacama in northern Chile, for example, occur principally along the basin margins. Differences in crustal morphology can affect the evaporative response under the same climatic conditions, with thick and rough crusts exerting significant resistance to water-vapor transport (Kampf et al. 2005). Evaporation rates are difficult to determine in this environment and vary greatly according to surface type. Tyler et al. (1997) estimated average rates of 88–104 mm year^{-1} for Owens Lake. Brine-covered areas showed much higher annual evaporation rates (872 mm year^{-1}) than zones covered by a thin (3–10 cm) friable salt crust (104 mm year^{-1}) or sand-dominated surfaces (88 mm year^{-1}), and produced 50% of the annual evaporative loss. Ullman (1985) reported rates of 9–28 mm year^{-1} from crusted surfaces at Lake Eyre, Australia. Kampf et al. (2005) found that thick salt crusts are practically impermeable to evaporation, and block liquid and vapor transport, regardless of the depth to the water table.

Soil moisture values in deserts are generally quite low. Evaporation is driven by high rates of near-surface evaporation, as a result of the thermal gradient (Thornes 1994). Even in the absence of vegetation, soil moisture can slowly return to the surface and evaporate, with temperature and vapor-pressure gradients slowly moving the water towards the surface. This process may be aided by barometric pumping, caused by diurnal or irregular changes in barometric pressure. Dry air is forced downward into the soil when the pressure rises, and soil air escapes as it falls. The drier the soil, the lower the rate of evaporation, for as the soil surface desiccates, the remaining water is held more tightly by capillary and osmotic forces, and the evaporation rate declines (Houston 2006a). Water must be evaporated below the surface and the vapor must diffuse upward through the stagnant air in soil pores in order to reach the turbulent transfer regime in the atmosphere. This is a slow process and usually the rate declines to an insignificant value after 5–10 mm of water have been removed.

In hyperarid deserts, such as the Atacama, averaged evapotranspiration rates may be close to zero, as so little water exists that even microbial life cannot exist (McKay et al. 2003). When rain does fall, long-term storage can occur only if the wetting front penetrates more than 2 m below the surface, beyond the reach of evaporative energy potential. As this is unlikely at elevations below 2500–3000 m ASL, the majority of rain that falls on the soil surface evaporates (Houston 2006a).

Soil texture plays an important role in the evaporation processes (Thornes 1994). Coarse soils allow water to infiltrate more rapidly and to greater depths, and hold it at higher water potentials than fine-textured soils. Evaporation from sandy soils in the Heihe River Basin area of northwest China occurs to depths of 0.2 m. Below this depth, vapor values increase until the inter-particle humidity is almost saturated. Upward water vapor transport occurs through the sand above 0.2 m, causing water vapor to converge to the air/ground interface in the daytime (Wang & Mitsuta 1992). Fine-textured soils have a stronger capillary action, promoting the evaporation of subsurface soil water and bringing salts to the surface, suppressing the osmotically driven water uptake of plants (Key et al. 1984). Over long periods of time, the soils of arid regions have dried to depths of 15 m or so, creating giant desiccation fractures.

The Palmer Drought Severity Index (PDSI) is an index of meteorological drought, defined as "a period of prolonged and abnormal moisture deficiency" (Palmer 1965). The index is based on a water-balance approach to modeling soil moisture conditions that takes the water-holding capacities of particular soils into consideration. The PDSI value for a soil will vary depending on the accumulated balance of water added each month by precipitation and subtracted monthly by evapotranspiration, which varies with temperature. PDSI values represent departures from the long-term mean condition of a given location and do not provide a basis for comparing the absolute amounts of water in soils of different climatic regime. Thus, a soil with a PDSI indicating moderate drought in central Arizona may have more actual soil moisture than a soil in southeastern California with a PDSI indicating a wet period.

4.3.3.3 Transpiration

Transpiration is the loss of water from the stomatal openings in the leaves of plants. Water is vaporized within the leaf in the intercellular spaces and passes out by molecular diffusion. Anywhere from 50 to 500 stomatal openings may cover a square millimeter of leaf surface, with many more on the lower leaf surface than on the upper. Stomata tend to open and close as the plant responds to environmental conditions such as light and dark, or heat and cold (Singh 1992). They open in sunshine, allowing the diffusion of carbon dioxide into the leaf during photosynthesis, and when they are open, water vapor can diffuse from wet cells into the atmosphere. This transpired water is replaced by water taken into the roots of the plant from the soil. Through transpiration, plants control their temperature and perform other vital functions. A quantitative appreciation of the process is fundamental to understanding plant growth and distribution. On a very practical level, the measurement or calculation of water use forms a basis for choosing crops in areas of unreliable rainfall and calculating irrigation water requirements.

Plant tissue incorporates only about 1% of the water entering the roots. Most passes through the plant – from root, to leaf, to atmosphere – and the driving force is the total potential energy gradient from soil water to leaf surface. A measure of the potential energy available to pull water into plant roots from the surrounding soil is given by the osmotic suction within the roots (Singh 1992). This results from osmosis, which is the selective diffusion of molecules through a semipermeable membrane.

As the major subtraction of water from drainage basins, evapotranspiration dominates the water balance and controls such hydrologic phenomena as soil moisture content, groundwater recharge, and streamflow. In parts of Africa, the proportion of precipitation returned to the atmosphere by evaporation from plants and water surfaces exceeds 90%. Although evapotranspiration is necessary for the growth of plants, it is usually viewed as a "loss" from the water budget in that it reduces the amount of streamflow, lake storage, and groundwater available for direct human use. In the 1960s, attention was given to methods of reducing this withdrawal in the American southwest in order to increase streamflow (for example, removing vegetation along river channels) (Thornes 1994). As early as 1934, Conkling and Gleason suggested cutting down large cottonwood trees along the Mojave River, California, in order to set "water free for beneficial use," such as cultivating crops (Conkling & Gleason 1934). Replacement of one type of vegetation with another can change the transpirative loss and this is of particular concern in the southwestern USA where native riparian gallery forests have been largely replaced by the naturalized exotic *Tamarix ramosissima* (Busch et al. 1992; Sala et al. 1996).

Along river floodplains, phreatophytes, plants whose root systems draw water directly from groundwater, dominate the riparian habitat. Their depletion of groundwater flow significantly reduces water supplies and this is often of concern to water managers (Sala et al. 1996). As the economic value of such water grows, studies of the linkage between hydrogeologic factors and the riparian plant community are growing in importance (Busch et al. 1992). In the arid western USA, phreatophytes cover 6.5 million ha of valley floor and transpire over 30 billion m^3 of water per annum. This discharge can be equivalent to a runoff yield of 0.02 m^3 s^{-1} km^{-2} of drainage basin area, and may amount to 20% or more of the dry-weather yield of some streams.

Evapotranspiration reaches its maximum during the short, rainy periods in deserts, when soil moisture is not limited. During the longer dry periods, evapotranspiration is severely limited by the low soil-water availability. When shallow-rooted plants, such as grasses of the fallow savannah complex of the Sahel, experience soil moisture deficits, the evaporation flux declines rapidly (Dolman et al. 1997).

The best indices of consumptive use of water by plants are the total surface area of leaves and stems exposed to the air (Sala et al. 1996), and the total soil moisture accessible to the roots. These factors are very difficult to estimate. The Leaf Area Index (LAI) is used in several computer models to simulate evapotranspiration as a function of such variables as stand density, age, and type. The LAI (L^2/L^2) is the ratio of leaf to ground area, with values which range from 1 (young stands) to 10 (dense coniferous stands). Most major vegetation types in deserts have an incomplete canopy cover, with substantial areas of exposed bare soil. For tiger bush, a major vegeta-

tion type of the Sahel, the LAI for the vegetated sector is about 4, but owing to the exposed bare soil, the overall surface LAI is in the range of 1–1.25 (Dolman et al. 1997). Soil evaporation may reach 28% of the annual rainfall.

4.3.4 INFILTRATION AND SOIL WATER

Soil moisture values are low in most arid environments as rainfall infiltrates only to shallow depths. As a consequence, dissolved materials tend to be deposited in the upper soil horizons, promoting the formation of duricrusts. Spatially, there is considerable variability in infiltration, as water seeps into porous surfaces including soils around plants, stream beds or other alluvial surfaces, and sandy soils, but runs off from less permeable covers such as microbiotic crusts and duricrusts (Cornet et al. 1988; Abrahams & Parsons 1991; Dunkerley 2002; Belnap 2003; Buis & Veldkamp 2008).

Water in soils is continuously moving under the influence of one or more forces that determine its energy status or potential. Movement occurs where there is a difference in the total potential from place to place: movement is in the direction of decreasing total potential (positive to negative). The gravitational potential causes water to move in a downward direction. The matric potential (the matrix is the network of soil-particle surfaces to which water adheres) varies directly with soil moisture content. Finite values of the matric potential are always negative in sign: a value of zero is saturated. In dry soils, the matric potential is low (large negative values). It accounts for much of the movement of soil moisture: normally the movement is from wet to dry soil, and it can go in any direction: up, down, or horizontally. The osmotic (solute) potential reflects the attraction between dissolved ions and water: the higher the concentration of ions, the greater the magnitude of the forces. It is important in the movement of water through all cells of living organisms. Resolution of the gravitational, matric, and solute potential is used to determine the total potential.

Water deficits in soil layers immediately around rootlets increase resistance to absorption, but at the same time induce migration of water towards the root from soil deeper in the profile. When absorption is unusually rapid (as at noon on a hot day), the root/soil interface is the major impedance to absorption, translocation, and transpiration. During the night, moisture will migrate towards the roots, increasing the water availability near the root and favoring rapid absorption and rehydration of the plant before morning.

Measured matric-potential profiles show that vegetation type can affect the values, which are more negative under desert scrub and grasslands than woodland communities (Walvoord & Phillips 2004). Creosote bush (*Larrea tridentata*) has a more extended rooting depth and a longer growing season than grasses, favoring deeper drying depths in the soil. Both deep-rooted shrubs and relatively shallow-rooted grasses appear to be capable of extracting all of the water that has infiltrated into the soil over the past few thousand years. Soil moisture has a paradoxical role as both a cause and consequence of vegetation type (Rodriguez-Iturbe 2000; Walvoord & Phillips 2004): vegetation type is strongly influenced by soil moisture, but equally influences the soil moisture regime.

The salinity of the soil has an effect on water relations. Owing to the lack of deep percolation, desert soils typically have an accumulation of salts (mostly NaCl and some sulfates) at a depth of 10–100 cm below the surface. The water potential in these shallow zones is lower due to the addition of the osmostic potential to the matric potential. This reduces the water available to plants (Noy-Meir 1973).

The soil beneath most deserts is very dry except beneath stream channels. Volumetric water content (volume of water per volume of material) in the upper 10 m of alluvium in the western Mojave Desert ranges between 0.02 and 0.05. Water content increases with depth, but does not exceed 0.14 (Izbicki et al. 2000). Similarly, in the Chihuahuan Desert transition zone of western Texas, values range from 0.02 to 0.17 (Walvoord & Phillips 2004). Highly negative water potentials (a range of more than −1000 to −14,000 kPa) are found throughout the Mojave site, with the highest values at about 10 m below the land surface owing largely to the accumulation of soluble salts (giving a highly negative solute potential) (Izbicki et al. 2000). Water above this depth will move downward and water below this depth will move upward in response to total potential, largely via vapor transport.

The water content increases in alluvium beneath stream channels, owing to infiltration from runoff. Oro Grande Wash, Mojave Desert, has volumetric water contents as high as 0.27 and water potentials between −1.8 and −50 kPa. These values are lowest within the upper 3 m as a result of evapotranspiration. Water potentials are near zero at a depth of 12 m below the land surface (Izbicki et al. 2000).

The presence of even a small amount of clay (as little as 8%) strongly impedes infiltration and the water content increases at the contact between more permeable and less permeable materials. For example, Izbicki et al. (2000) found that water content rises at the contact between younger alluvium and older fan deposits.

Chloride profiles are used to evaluate moisture fluxes in the unsaturated zone. The chloride mass balance approach provides estimates of the downward moisture flux over long periods of time and the age of soil water (up to 50,000 years). Chloride is readily dissolved and moves with infiltrating water. If large amounts of chloride are present in the soil, it suggests that several thousand years may have passed since water has infiltrated. Water below the root zone (about 10 m) in the eastern Mojave Desert near Beatty, Nevada, and Ward Valley, California, is estimated to be 16,000 and 33,000 years old, respectively, corresponding to a cooler and wetter time period (Prudic 1994). In the western Mojave Desert, soil water at 10 m is about 13,000 years old (Izbicki et al. 2000). Walvoord and Phillips (2004) suggested chloride accumulation times of greater than 12,000 years in western Texas beneath areas of desert scrub. These results suggest that there has been little to no deep percolation for the majority of the Holocene.

There are several problems with assigning a precise age to desert soil water using chloride profiles. The calculations require knowledge of the chloride dry deposition rate, which is unlikely to have remained constant through recent geologic time. In the California deserts, it is probable that values increased at the end of the Pleistocene as lakes desiccated and became playas, exposing their surfaces to wind erosion. These playas are sources of chloride in present-day precipitation. The deposition rate in the western Mojave is probably greater than in the eastern owing to proximity to the Pacific Ocean and to anthropogenic effects (the collectors are located 130 km downwind from Los Angeles). Human activities, such as agriculture and construction, also disturb soils and increase chloride deposition rates.

Tritium, a naturally occurring radioactive isotope of hydrogen, is also used to trace water movement on timescales ranging from 10 to less than 100 years BP. Beginning in 1952, it was released to the atmosphere as a result of nuclear weapons testing, with values reaching a maximum in 1962 and declining after atmospheric testing ended (Michel 1976). Tritium in soils of the western Mojave Desert is detected only near the surface, at a depth of 2.5 m. This is a typical value for areas of low precipitation and high evapotranspiration, and thus shallow infiltration depths. Previous studies have consistently reported no percolation past about 1–2 m of the surface (Gee et al. 1994). By contrast, tritium is found at depths ranging from 16 to 29 m below stream beds, suggesting average downward movement of water of 0.5–0.7 m year^{-1}. If these rates were constant, it would take between 185 and 260 years for infiltrating water to reach the water table, which is presently located 130 m below the surface. However, clay layers tend to impede the downward flow and cause lateral spread of the water, increasing the recharge time (Izbicki et al. 2000).

In the Kalahari, infiltration rates are high and there is significant retention storage in the soils during the wet season. However, owing to high transpiration rates by the dense vegetation in the dry season, very little water passes the root zone and contributes to aquifer recharge (De Vries et al. 2000). In areas where rainfall is 400 mm or less, the average regional groundwater recharge is less than 1 mm year^{-1}, the remainder being disposed of by evapotranspiration. Losses from depths of tens of meters result from transpiration by deep-rooting acacias, and from depths of up to 20 m by capillary transport and water-vapor loss (De Vries et al. 2000). An average rainfall of 500 mm or greater would be required to recharge the groundwater table at a rate comparable to past pluvial periods.

The aforementioned studies suggest that water does not percolate to any great depth away from stream channels and that direct precipitation on the soil does not recharge the groundwater table under present climatic conditions. Although infiltration in small stream channels may provide groundwater recharge, more than 200 years may be required for water to reach the water table (Izbicki et al. 2000). This work has implications for rapidly developing desert areas, where groundwater may be the main water resource.

Soil water in dune sands responds principally to matric and gravitational potentials, as the solute potential can generally be disregarded as a result of the chemical non-reactivity of sand. Although dune sands display a general homogeneity in soil-particle size distribution, small-scale layering (cross-bedding) appears to play a role in the movement of water. Berndtsson et al. (1996) observed that the tops and bottoms of 15–20 m-high dunes in the Tengger Desert, China, had markedly higher volumetric water content (0.04–0.05) than slope sections (0.01–0.02). Water tended to move downward to the dune base along the zones of higher hydraulic conductivity associated with the sand layering. Infiltrating soil water was observed to follow paths that already had a high water content before rainfall.

4.3.5 GROUNDWATER, SUBSURFACE FLOW, AND SPRINGS

4.3.5.1 Role of groundwater in arid environments

Groundwater affects hydrologic, geomorphic, and biologic processes in deserts. Owing to the rarity of surface water flow, groundwater may be the only dependable water resource. The long-term availability of groundwater depends on a stable balance being achieved between water abstraction and recharge under present climatic conditions. However, quantitative estimates of recharge, whether direct or indirect, remain fraught with uncertainty (De Vries & Simmers 2002).

The geomorphic effects of groundwater are numerous, as it promotes the weathering of rocks and supports vegetation along stream courses. Rock weathering by sustained seepage influences the development of canyons and the retreat of cliffs. On the Colorado Plateau, groundwater sapping has maintained sandstone cliffs (Bryan 1928b; Ahnert 1960; Schumm & Chorley 1964; Howard & Selby 1994) and plays an important role in the headward retreat of deeply entrenched, theater-headed valleys (Laity & Malin 1985; Howard & Kochel 1988; Baker et al. 1990). Groundwater also plays a significant role in aeolian and fluvial systems of deserts. In channel systems, groundwater forced to the surface by faulting or bedrock may flow at the surface for short distances in zones marked by dense phreatophytic vegetation. Phreatophytes affect hydraulic roughness, depositional processes, and channel width. Vegetation that relies on shallow water tables also anchors dunes in some nebkha fields (Laity 2003).

Sustained seepage or high groundwater tables support streamflow in some of the world's more arid regions, such as the Darwin Falls area of the Argus Range, Panamint Valley (Fig. 4.4), and short sections of the Mojave River, both in California. In the Atacama Desert, outflow from aquifer system sustains perennial flow in the Rio Loa, allowing it to cross the desert and reach the sea (Houston 2006a).

Fig. 4.4 Sustained groundwater seepage supports minor streamflow in some of the world's more arid regions. This photograph shows the Darwin Falls area of the Argus Range, which forms the western boundary of the Panamint Valley, California, USA. Seepage supports important habitat for endemic plant and animal species. The Argus Range reaches a maximum elevation of 2694 m ASL. Annual rainfall at the valley floor averages 76–100 mm, and may be two to three times that at higher elevations.

Groundwater seepage also maintains wetlands, poorly drained areas of low relief in which the soil is seasonally or perennially saturated and which may include shallow ponds or swamps. Wetlands are relatively uncommon in deserts, but may occur along playa margins, particularly those bordered by faults. In the California deserts, wetlands border Deep Springs, Death, and Panamint Valleys, and Soda Lake. Additionally, many thousands of springs and seeps are scattered throughout xeric regions of the western USA, providing habitats for endemic plant

and animal species (Cole 1968). In the southern part of the Gran Desierto, Mexico, one of the driest areas of the Sonoran Desert, small artesian springs or *pozos* provide potable water at various points in the salt flats at Adair Bay, supporting a flora and fauna markedly different from the rest of the Sonoran Desert (Ezcurra et al. 1988).

4.3.5.2 Groundwater recharge

Groundwater recharge is strongly influenced by geology. Deep infiltration is greatly enhanced by jointing and by permeable cover materials (such as sand sheets). In the case of limestones, the capacity for recharge was enlarged by solution during wetter periods in the Pleistocene. The largest regional aquifers often occur in sandstones, whose porosity may reach 20–40% and whose permeability is enhanced by jointing. Sandstone has a relatively high yield of water (800–1000 $m^3 h^{-1}$). On the Colorado Plateau, USA, the Navajo Sandstone is the most transmissive of all lithologies, owing to its porosity (25–35%), permeability, and geometric configuration and continuity (Cooley et al. 1969; Laity & Malin 1985). Its recharge potential is high because of its widespread exposure at low dip angles, essentially uniform permeability, and pervasive fracturing. Interception and infiltration are promoted in areas where deposits of windblown sand and stabilized dunes mantle the surface. Seepage at the base of the Navajo Sandstone sustains springs and small streams throughout the region. In the Western Desert of Egypt, the Mesozoic Nubian Aquifer forms the basis of several oases, including Dakhla, Kharga, Siwa, and Farafra (Dabous & Osmond 2001). The thickness of the water-bearing sandstones ranges from 400 m near Kharga to over 2000 m at Siwa.

An artesian system is a pressurized groundwater system in which water rises above aquifer levels and sometimes to the ground surface from confined aquifers, wherein the water-saturated permeable layers are enclosed between aquicludes (impermeable rocks that do not transmit water). Artesian supplies usually occur at great depths and are held under pressure owing to the higher level of water uptake in distant uplands. Important artesian systems occur in large areas of Australia and North Africa, recharged by seasonal rainfall in bordering highlands. In Algeria, the artesian system is replenished by rain and snow in the Atlas Mountains.

The Great Artesian Basin (GAB), Australia, is the world's largest artesian system, with an estimated

8700 million ML of water. The GAB consists of alternating sandstone aquifers and impermeable siltstones and mudstones, which vary in thickness from less than 100 m to over 3000 m. Artesian bores range from 500 to 2000 m in depth. Much of the area overlying the GAB is arid or semiarid, and artesian water is the only reliable source of potable water. Prior to the 1870s, there were about 3000 flowing artesian springs ringing the basin (Habermehl 1982). The springs form local clusters (groups) that in turn form larger supergroups (Fensham & Fairfax 2003). Water temperatures vary from 20 to 45°C and are relatively high in dissolved salts. In some areas, carbonates precipitated from the water have combined with wind and waterborne sediments to create mounds up to 10 m in diameter and several meters high, giving rise to the popular name of mound springs.

The springs of the GAB supplied water to Aborigines as well as early explorers, workers, and pastoralists. Groundwater withdrawal associated with the drilling of thousands of bores led to the extinction of about one-third of the springs, and a reduction in flow for the remainder. Today, pastoralists use about 88% of the water, and as much as 90% of this is lost through evaporation and infiltration (Ponder 2002). Bore capping and control programs are slowly reducing this wastage. Local drawdown problems are associated with major mining ventures, which use increasingly large amounts of groundwater.

The springs of the GAB are home to many aquatic invertebrates and fishes; many of these are rare or indigenous with restricted distributions. Some are unique to given springs and each major spring group has a distinct fauna (Ponder 2002). Threats to these ecosystems include reductions in water flow, trampling by stock or feral animals, and the introduction of exotic plants and animals.

Crystalline rocks provide aquifers that are local rather than regional in nature. There is no continuous water table, but the supply is obtained from individual wells that tap joints in the system. Such localized aquifers are common in the Sudan, East Africa, and parts of India. The yield in volcanic materials depends very much on the nature of the deposit.

Shallow alluvial aquifers are one of the most important water resources in arid lands. These are replenished by infiltrating floodwaters (Izbicki et al. 1995, 2000; Lange 2005; Dahan et al. 2007). In the Mojave River basin, the shallow alluvial aquifer is in direct contact with the river and underlies the flood-

plain. It yields much of the groundwater supply to cities and agricultural interests in the basin (Laity 2003). Owing to the increasing demand for water from shallow alluvial aquifers, it is essential that they be managed in a sustainable manner. Imbalances between pumping and recharge have led to serious overdrafts in some regions.

4.3.5.3 Groundwater quality

In arid zones, groundwater is typically saline, regardless of its age.

The salinization results from high evaporation rates, low recharge rates (that limit flushing owing to low hydraulic gradients), and long groundwater residence times that increase water–rock interactions (Nativ et al. 1997). In some areas, saline sediments are present as a result of past marine transgressions.

Within a given aquifer, salinity values may show considerable spatial variation. In Egypt, low salinities (≤500 ppm) characterize the southern portion of the Nubian aquifer; in the northwest, near Siwa Oasis, the water is typically brackish (≈1000 ppm); and close to the Mediterranean Sea, the groundwater is highly saline (>100,000 ppm) as a probable result of marine water intrusion and evaporation (Dabous & Osmond 2001).

Groundwater salinity is derived from both atmospheric (dust and sea salt) and surface (rock type, surface salts, and locally dissolved soil salts) sources. Nativ et al. (1997) compared rainwater, surface water, and groundwater salinity in the Negev highland basins, where bare carbonate rocks constitute 60% of the land surface and additional salts can be derived from the Mediterranean Sea. Groundwater derived from the Eocene Avdat chalk is brackish (total dissolved solids (TDS) of 914–6180 mg L^{-1}), but used by local nomads to water their herds. Springs also sustain populations of wild animals. The TDS in rainwater averages 78.4 mg L^{-1}, a high concentration relative to humid regions, as a result of the dissolution of dust particles suspended in the lower atmosphere. Ca^{2+} and HCO_3^- are the dominant ions in both rainwater and dust, and reflect the chalky/limy nature of the rock outcrops. Na^+, Cl^-, and SO_4^{2-} are proportionally higher in rain than dust, suggesting a marine source (sea spray). The pH of the rain samples averages 7.18, a result of the limy dust and distance from major pollution centers (Nativ et al. 1997).

The major constraint on rock dissolution in the Negev is the short time that water is available. Floodwater runoff contains salts from rainwater, salts deposited on the surface as a result of evaporation of previous rains, and locally dissolved salts from the soil and rocks. Its salinity is significantly higher than that of precipitation, averaging 323 mg L^{-1}, and is also dominated by Ca^{2+} and HCO_3^- ions resulting from the interaction of surface flow with carbonate rocks. The concentration of all ions in the vadose zone of the soil reaches peak values at a depth of 1–3 m. Here, gypsum, halite, and other efflorescent salts are precipitated as a result of intense evaporation (Nativ et al. 1997).

In summary, abundant desert dust provides a steady flux of minerals that are dissolved in precipitation. As the falling raindrops evaporate, they are enriched, increasing their ion concentration relative to more humid regions. Rainwater in surface depression storage evaporates near or at the land surface, forming a salt pool that is rarely flushed. Little floodwater evaporates during flood events, owing to their short duration, and therefore salt enrichment occurs principally from water–rock interactions, including the dissolution of halite, gypsum, and calcite (Nativ et al. 1997). The enrichment of surface waters in salts, through precipitation and dissolution processes, ultimately contributes to the high salinity of soil and ground waters.

4.4 SURFACE RUNOFF AND FLOODS

Floods in deserts have hydrologic, geomorphic, and economic implications. They recharge groundwater systems, alter the form of channels, and provide life-sustaining water, but also cause substantial economic losses, endangering life and property. A flood along Wadi Qena in the eastern desert of Egypt, for instance, flowed at 50 km h^{-1}, destroying factories and railroads along its northern bank before discharging into the Nile River (Gheith & Sultan 2002).

Runoff in arid and semiarid regions is predominately initiated as Horton overland flow, a form of runoff that occurs when rainfall intensity exceeds the infiltration capacity of the soil or rock mantle (Buis & Veldkamp 2008). Changes in land use, including cultivation and grazing, further lessen the infiltration capacity of the soil, thereby increasing the amount of overland flow that occurs (Descroix et al. 2007). Interflow – flow between layers of soil – is usually insignificant owing to the thin nature of the soils. Saturation overland flow is rare, but Descroix et al. (2007) discuss its occurrence in areas of arid Mexico

during low-intensity, long-duration events (such as hurricanes or El Niño-related winter rain storms). Groundwater sustains minor baseflow in streams if springs or seeps are present, notably where faults, bedrock, or lithologic changes in sedimentary rocks bring groundwater to the surface. It also contributes to some larger rivers, such as the four perennial rivers that cross the Atacama Desert, buffering periods of low precipitation (Houston 2006b).

4.4.1 CONTROLS ON RUNOFF

Controls on runoff are climatic, geologic, and biotic. With respect to climatic controls, the most significant are rainfall intensity, duration, and amount. In low-intensity events, a high proportion of rainfall is absorbed by the ground and is lost by evaporation. Published data suggest that intensities of at least 1 mm min^{-1} and totals of about 10 mm at these intensities are normally required for channelized runoff to occur. However, these values are strongly influenced by the surface cover (Buis & Veldkamp 2008). Data from 1978–82 at the Sede Boqer research site, Israel, indicate that the threshold level of daily rainfall necessary to generate runoff varies from 2 mm for rocky regions, to 4 mm for stony colluvial surfaces, to 6 mm for material in the lower channel (Yair & Kossovsky 2002). The presence of antecedent moisture in the soil may lower threshold values. Likewise, thresholds are reduced if peak intensities occur late in the storm, after the surface has been wetted.

The principal geologic controls relate to permeability; that is, to rock and soil type, and the presence of fracturing (Buis & Veldkamp 2008). However, the great heterogeneity in soil types, and the very irregular patterns of precipitation in both time and space, create a complex hydrologic response on the land surface. In hydrology, this has led to the 'partial area contribution' concept, wherein it is recognized that certain portions of the watershed regularly contribute overland flow to streams, whereas others seldom or never do. In an arid area, cemented and compacted materials may lead to partial area contributions, whereas sandy surfaces tend to absorb most of the precipitation that falls on them, and thereby seldom generate runoff.

Runoff persistence depends strongly on rock type and soil permeability. Studies undertaken over a 10-year period in the Nahal Yael watershed indicated that only 20% of rainfall events initiated runoff, and that of these, only 5% of runoff events exited the basin (Schick 1988). Minor runoff on hillslopes may not result in streamflow in the main channel if water fails to reach the slope foot. Sandy soils commonly absorb all of the flow in minor runoff events. For example, in the Sanderao drainage basin, India, the main stream that emanates from the hills disappears into the deep alluvium and aeolian sand of the foothills, and thus does not reach the main trunk stream in the basin (Sharma & Murthy 1996). This represents a discontinuity of operation inherent in all desert drainage systems.

Over time, near-surface hydrologic conditions change as soils and surfaces evolve (Meadows et al. 2007). On the Cima volcanic field, California, infiltration into soils with desert pavements is reduced over time as the pavements become more mature. As the infiltration capacity lowers, runoff increases, and an age-dependent increase in drainage density occurs (Wells et al. 1985).

Owing to the scarcity of vegetation cover in deserts, plants probably exert less direct influence on overland flow than in humid environments. Riparian vegetation is an important influence along river courses and is discussed more fully in Chapter 11. When arid environments are dominated by shrub vegetation, the infiltration capacity of the soil is markedly patchy. Infiltration rates are commonly higher under shrubs owing to the concentration of litter, the reduction of raindrop impact, better soil crumb structure, and plant stems that funnel waters to the soil beneath their canopy (Schlesinger & Pilmanis 1998). At a much smaller scale, microbiotic soil crusts may enhance runoff relative to non-crusted sites. In the western Negev Desert, for example, dune areas covered by a 1–3 mm-thick microbiotic crust generate runoff with a threshold for initiation as low as 9–12 mm h^{-1} (the transition between light and medium rainfall intensities in the region), whereas no runoff occurs from uncrusted sand (Kidron & Pick 2000). However, Belnap et al. (2005) note that the hydrologic impact of biological soil crusts in the American southwest depends on their external morphology, with smooth forms enhancing runoff, regardless of precipitation event size, and rough crusts preventing runoff except in the most extreme monsoon events.

In some arid regions, runoff is increasing owing to changes in land use. In the West African Sahel, demographic pressure has led to a reduction in the area of bush and an increase in cultivated or grazed land. Subsequent erosion has decreased the soil water holding capacity. Thus, despite a decrease in rainfall of 20–30% from 1968 to 1995 (compared to 1950–67), the runoff coefficient rose (up to 100%) in experimental catchments and at the basin scale (Descroix et al. 2007).

4.4.2 RUNOFF FROM SLOPES

Runoff from slopes is dependent on the intensity of rainfall, the infiltration capacity of the soil, surface roughness, and slope angle. It is usually more dependent on rainfall intensity than duration, suggesting that the infiltration rate is a more important determinant than the degree of saturation. Infiltration rates in deserts are affected by sparse vegetation, lack of litter interception, and reduced detention storage. As in other areas, the infiltration rate decreases with time.

Some deserts are very rocky in character. Bedrock surfaces are essentially impermeable and produce runoff even at very low precipitation intensities (Yair & Kossovsky 2002). This runoff may be redistributed on the slopes, making some areas wetter than would be predicted by climatic factors alone, and fostering patches of denser vegetation, including some species more characeristic of wetter climates (Buis & Veldkamp 2008).

In the absence of soil and vegetation, rock debris and surface sealing exercise strong controls on runoff (Abrahams et al. 1994). Surface sealing results from swelling of soil aggregates, clogging of pores by the deposition of fine materials in ponded waters, or compaction of the surface by raindrop impact (Assouline 2004). A coarse mantle of debris tends to increase surface roughness and impede overland flow, thus enhancing infiltration. In some areas, water harvesting practices involve the removal of stones from a slope, which decreases infiltration and increases runoff by smoothing the surface, permitting crusting by rain impact, and creating a smoother downslope path.

Slope angle does not appear to be as significant a factor in runoff generation as roughness and infiltration. Greater runoff may develop from lower slopes that have smooth, less porous surfaces. The relationship between slope and runoff may be inverted because increases in slope angle strengthen the factors that control infiltration, such as roughness and coarse grain size.

The failure of runoff from slopes results from evaporation losses, absorption and depression storage in dry rock debris, and high rock porosity. For example, in the Borrego Badlands of the Anza Borrego Desert, California, light to moderate rainfall is easily absorbed by swelling clays, stabilized sand dunes, and fanglomerates.

4.4.3 RUNOFF IN CHANNELS

The majority of channels in drylands are ephemeral in nature, flowing only for short periods of time (usually hours) in response to rainfall events. The relative infrequency and short duration of flow, the remoteness of many channels, and the hazard and technical difficulty in measuring floods have made the study of channel runoff more complicated than in more humid regions. As a result, few direct runoff measurements have been made in desert uplands. Moreover, developing nations seldom have the financial resources to maintain an expensive gauging network or to carry out extensive hydrologic research (Lange 2005).

4.4.3.1 Ephemeral channels

Flow generation in ephemeral stream channels is largely controlled by rapid runoff from hillslopes. Many authors have shown that ephemeral runoff in small channels characteristically shows a very short time of concentration and rise to peak discharge. There is a typical hydrograph form with a nearly instantaneous rise and sharp peak, followed by a steep initial recession limb, and then a slower fall towards dry conditions. The duration of flow is usually short, perhaps 1–5 hours, with peak flow occurring within minutes or hours of the onset of rainfall (Fig. 4.5). Runoff normally occurs on only a few occasions a year, and bankfull discharge is unusual.

The "flash-flood" characteristics of arid-zone drainages are typified by a rapid wave-like advance of the flood. Two mechanisms contribute to the development of the wave front. First, infiltration rates are greatest at the wave front and diminish upstream: the effect is a steepening of the leading edge of the wave as it moves downstream. Second, the leading edge of the wave moves more slowly than the deeper part of the flood wave, such that over time the peak and front will coincide, forming a shock wave (Sharma & Murthy 1996).

The recession curve of the hydrograph shows that the duration of recession is typically much longer than the time to reach peak flow. Additionally, the recession limb of the hydrograph tends to become

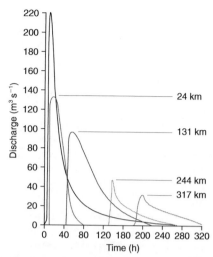

Fig. 4.5 Discharge hydrograph at downstream stations (with positions shown in kilometers) on the River Luni, India, during the period 19 July through 2 August 1981. In ephemeral channels, a rapid wave-like advance of the flood causes a steep rising limb on the hydrograph and a recession limb that is somewhat less steep. Transmission losses decrease the discharge in a downstream direction. Additionally, the recession limb tends to become less steep with distance downstream. From Sharma, K.D. and Murthy, J.S.R. (1996) Ephemeral flow modeling in arid regions. *Journal of Arid Environments* **33**, 161–78. By permission of Elsevier.

less steep with distance downstream (Fig. 4.5) (Sharma & Murthy 1996).

There are significant differences in the hydrograph characteristics of small and large channels. There are several reasons why small catchments and basin headwaters exhibit a rapid concentration of overland flow and short times to peak flood. First, slope lengths are generally short, slope angles are steeper than in larger catchments, and valley heads are zones of flow-path concentration. Second, soil cover and available soil moisture storage is low, and times to saturation are short. Third, it is easier for a small catchment to be covered by a single storm. In larger catchments only a part of the system may be functioning in a given storm. Fourth, larger channels may exhibit substantial storage of water in alluvial and slope deposits, which delays concentration in long channels and may even give rise to baseflow.

For these reasons, the storm hydrographs for larger catchments may display lower and flatter peaks and flatter hydrographs. Where there are steep headwaters, flashy peaks may be superimposed on delayed flow or baseflow characteristics. Generally, flow duration is longer. Rapid flash floods of small magnitude, but possibly catastrophic effect, are therefore more likely in small to medium streams. It should be noted that, in large catchments, snowmelt or frontal rainfall are likely to produce a second family of hydrographs with greater peaks and longer durations than those associated with localized storms.

4.4.3.2 Intermittent and perennial rivers

In most arid regions of the world, all but the major streams are ephemeral. Perennial streams flowing through drylands usually originate in mountainous regions, where the annual precipitation is higher (for example, the Colorado River), or have their headwaters in more humid climatic zones (for example, the Nile River). Very small streams may exhibit perennial flow when their source is groundwater springs (Fig. 4.4). The flow in such streams is relatively meager, but sustains important riparian habitat. Other streams are intermittent, flowing for several months of the year in response to seasonal sources of runoff, usually snowmelt (Fig. 4.6).

In temperate regions, the discharge of a stream often shows a cyclic pattern in response to the annual climatic cycle. The mean flows of large streams in arid regions are much more variable, although there may be seasonal patterns. The Tigris and Euphrates Rivers, for example, reach their highest discharges in the springtime, following snowmelt at higher elevations. Large rivers in arid areas are commonly heavily exploited and regulated, and this is commonly reflected in their discharge patterns, sediment yields, and biological systems. The construction of reservoirs, for example, regulates both the flow regime and sediment load of rivers, triggering changes in channel form, which may include degradation or aggradation, as well as channel metamorphosis (Ta et al. 2007).

Significant changes to channel morphology, as well as damage to property, may occur in flood events. Large dryland streams often have floods that are an order of magnitude or more larger than the mean annual flow. For example, the 50 year flood on the Gila River, Arizona, is 280 times the mean annual discharge (Graf 1988).

4.4.3.3 Low-flow events and the ecological effects of drought

The term drought is not easily defined, and represents a human judgment about the extent and nature of prolonged dry periods that has social, political, or economic implications. During drought conditions, water flow and availability fall to very low levels for extended periods of time. As a result, hydrologic connectivity is disrupted (McMahon & Finlayson 2003). Such disruptions range from flow reduction to complete loss of surface water and connectivity.

FIG. 4.6 Braided intermittent streamflow in Chaco Canyon, New Mexico. Snowfall is retained beneath the shaded slopes to the lower left of the photo.

In freshwater systems, droughts are of two types: seasonal and supra-seasonal (Lake 2003). Seasonal droughts are those related to the climate of a region and are generally predictable. Biota have evolved adaptations to survive these conditions, such as life-history scheduling. Supra-seasonal droughts are marked by an extended decline in rainfall beyond that normally expected, resulting from the failure of adequate precipitation over a number of seasons. These aseasonal droughts are less predictable and thus species have a more difficult time adapting to them (Lake 2003).

In stream systems, a lowering of water levels reduces the amount of habitat available for most aquatic biota. The stream may be reduced to a series of short-lived, fragmented pools (Stanley et al. 1997) in which aquatic species can become trapped and concentrated. For some species, such as fish, rising water temperatures may be lethal (Matthews 1998).

Low-flow conditions associated with drought impact population densities, species richness, age structure, and community composition and diversity. In addition to direct effects on ecosystems (loss of water, habitat, and stream connectivity), there are indirect effects such as the deterioration of water quality, alteration of food resources, and changes in the nature of interspecific interactions. However, natural ecosystems are adapted to the frequency and magnitude of dry periods. Organisms can resist droughts by the use of refugia, and isolated pools often harbor biotic assemblages with high densities of invertebrates and fishes (Lake 2003).

The regulation of rivers and continuous release of water from below dams mitigates the severity of low flows. This, in turn, may lead to new problems in an ecosystem that is adapted to naturally occurring low-flow conditions (McMahon & Finlayson 2003). In the Murray-Darling basin of Australia, irrigation needs drive river flows to be high in the summer and low in the winter, the opposite of natural conditions (Lake 2003). It is thought that these changes to the seasonal flow negatively affect biodiversity (Everard 1996), reducing fish recruitment and allowing populations of invaders to flourish.

4.4.4 TRANSMISSION LOSSES DURING FLOODS

An evaluation of runoff in ephemeral stream channels would be incomplete without consideration of the changes that take place in such channels during runoff events. Many channels contain deep and extensive coarse-grained alluvial deposits that are characterized by high infiltration capacities and high water-storage potential. These alluvial sediments promote the rapid infiltration of floodwater, which recharges local alluvial aquifers (Izbicki et al. 2000; Dahan et al. 2007).

When rivers flow from well-watered upland areas into the desert, the loss of discharge owing to progressive downstream transmission losses limits the distance of flow. The Guir-Saoura-Messaoud catchment is formed of rivers that flow from the High Atlas Mountains into northern Algeria. Transmission losses are significant in the sandy middle sections of large channels, whereas, in downstream reaches, water is also lost by lateral discharge into sump basins by distributaries. The limit of multiple annual floods is usually 640 km, but larger floods have extended as far as 950 km into the desert (Mabbutt 1977b). In a similar manner, large amounts of water are lost along the Mojave River, California, which flows from the San Bernardino Mountains, near Los Angeles, into the Mojave Desert. Almost all of the runoff is derived from the mountainous headwaters, but only large flood events are able to complete the 200 km journey to reach the terminal playa at Silver Lake (Enzel & Wells 1997).

The magnitude of transmission losses into alluvium during flood events is variable, and related to flow duration, channel length and width, antecedent moisture, prior streamflow events, peak discharge, the pattern of flood-wave sequences, temperature, and the physical properties of the alluvium and suspended sediment load (Gheith & Sultan 2002; Lange 2005). In the most favorable circumstances, it is possible for the entire available precipitation to be lost within a basin as a primary groundwater recharge source. The effect of transmission losses is to decrease flow downstream, to steepen the flood-wave front, to decrease the flood rise time, to increase relative sediment load, sorting, and deposition downstream, and to decrease stream width.

Transmission losses show pronounced spatial and temporal variability. Some studies suggest that most transmission losses take place in the early stages of a flood event. This is evidenced by small translatory waves, a few centimeters high, which drain rapidly as they pass downstream. By contrast, Lange (2005) found that along the Kuiseb River in Namibia, small to medium floods traveled considerable distances without substantial losses, whereas high-runoff peaks were associated with significant transmission losses. This apparent contradiction is explained by the presence of thin clay drapes on the channel surface, which form a seal, and are disrupted principally at high discharges, allowing infiltration (Dunkerley & Brown 1999; Lange 2005). During periods of high flow, overbank areas may also flood, causing large losses from evaporation and infiltration.

The few measurements of transmission that have been made show substantial losses over relatively short distances. In western New South Wales, Australia, surface flow was entirely lost over a distance of 7.6 km at a transmission loss rate of 13.2% km^{-1}. However, there was marked spatial variability in the loss pattern, as flow was abstracted into pools or other low points, and infiltration was affected by mud drapes that sealed the channel or by more porous channel materials (Dunkerley & Brown 1999). Transmission losses of 7300 m^3 km^{-1} in Nevada (Savard 1998) and 7200 m^3 km^{-1} in the Negev Desert have been recorded (Ben-Zvi 1996). These may be significantly reduced when the time interval between two successive floods is relatively short (Lange 2005).

The most marked spatial variability in transmission loss is between the steep headwaters of a catchment and the alluvial floor. The degree of loss affects stream velocity, channel morphology, and sediment transport. In mountainous and foothill regions, channels are narrow, and the area of contact between the channel bed and the runoff is limited, reducing transmission losses (Gheith & Sultan 2002). In general, velocity increases faster downstream within arid mountain catchments than in equivalent streams in more humid areas. Gradients are usually steeper and runoff more complete from slopes with thinner soils and vegetation, so there is less likelihood of infiltration or obstruction to flow. When the flow encounters an alluvial floor, however, the contact area with the channel bed enlarges and transmission losses increase, accompanied by changes in flow depth, velocity, discharge, and channel geometry.

4.5 THE CHEMICAL QUALITY OF SURFACE AND SOIL WATER

The movement of water on and through the soil and rocks dissolves inorganic materials, moves sediments, and provides a medium for development of biota. Consequently, the quality of surface water ranges widely according to the character of the land. To some extent, it is also influenced by constituents contributed from the atmosphere (Nativ et al. 1997).

The chemical quality of water is the composite of the concentrations of the many constituents. The simplest descriptor is the average concentration of TDS. The average concentration of the major chemical constituents is an alternative descriptor, but is less useful in comparing water quality among streams with different major constituents.

Problems in settling arid lands revolve around a lack of water and water that is too saline. The three principal reasons for the accumulation of salts in deserts are: (1) evaporation, (2) endoreic drainage, and (3) atmospheric recycling. In the hot, dry, and windy environment of the desert, the tendency is for any moisture in the regolith to be drawn upward so that salts are concentrated in the upper soil layers. In the vicinity of Newberry Springs, California, historically high water tables (0.6–1.0 m below the surface in the 1930s) result in strongly saline and alkaline soils (Thompson 1929; Storie & Trussel 1933). An alkali soil is one in which there is so high a degree of alkalinity (pH 8.5 or higher), or so high a percentage of exchangeable sodium (15% or more of the total exchangeable bases), or both, that plant growth is restricted. Near Newberry Springs, vegetation is absent or composed of salt-tolerant shrubs, grasses, and forbs. In some areas, there is a thin, patchy, white crust of salt on the surface. Cultivation is not possible on such soils. In regions of endoreic drainage, salts are not flushed out to the sea. In addition, salts are readily drawn to the surface owing to high water tables in areas of inland drainage. The atmosphere introduces salt into desert areas that might otherwise be free of salinity. The wind picks up salts from the surfaces of evaporative bodies such as the sea, salt lakes, sabkhas, and playas. In maritime environments, rain and fog may also precipitate salt particles.

The total volume of salts usually defines the hazard to cultivation or water consumption. Humans are not physiologically well adapted to the desert environment and water loss beyond 10% of body weight is debilitating. The maximum tolerated salinity for any length of time is around 3000 ppm, but the World Health Organization recommends 500–700 ppm. All domesticated animals of the desert drink relatively small quantities of water and tolerate high salinities, and because they feed on the natural vegetation, grazing is the basis of most economies in deserts.

Native plants of arid zones often have mechanisms to resist the toxic effects of saline soils and groundwater. In many cases, the accumulated salts are well dispersed within the soil, and the water table is at depth. Irrigation presents a problem when the water table begins to rise and salts precipitate at the surface. The increased salts in the soil can affect native plants, increasing tree senescence and mortality, and wildlife and waterfowl populations can be impacted (Froend et al. 1987).

Cultivated crops are quite sensitive to salinity and around 700 ppm is generally considered the norm for suitable irrigation supplies. As soil salinity increases, crop yields diminish. In addition to areas that have natural soil salinity, there are about 77 million ha of land that have been salinized as a consequence of human activities, in large measure (58%) a result of irrigation. In countries such as Egypt, Iran, and Argentina, more than 30% of irrigated lands are affected by salts (Ghassemi et al. 1995). It is anticipated that the salinization hazard will grow as populations and the demand for food in deserts increase. The economic costs include soil degradation and erosion, loss of farm property values, eutrophication of rivers and estuaries, and damage to infrastructure (roads and buildings) (Metternicht & Zinck 2003). Salinity problems may be ameliorated by controlled irrigation and flushing methods, and selective breeding of cultivated crops with a high degree of salt tolerance.

One problem with respect to the salinity of water resources is that over time many rivers become a drain, as well as a water source. The Murray River in southeastern Australia increases its salinity from 25 mg L^{-1} at its headwaters, to 480 mg L^{-1} downstream, with approximately 30% of the load human-induced. These values rise during periods of drought, causing major management problems, as the river provides a source of both drinking and irrigation water (Ghassemi et al. 1995). The USA and Mexico share a 3141 km border, marked in large measure by the Rio Grande (2019 km) and the Colorado River (38 km). This rapidly growing arid to semiarid region has a total population exceeding 10.5 million, and the rivers provide water for irrigated agriculture, manufacturing, mining, and urban growth. Irrigation has increased the salinity of river water and alluvial soil water due to the leaching of salts from agricultural fields. The Colorado River is highly regulated, with two main storage dams, Glen Canyon Dam and Hoover Dam. Water is diverted from the river via aqueducts to serve the cities of Los Angeles and San Diego, California, and Phoenix and Tucson in the Arizona desert. An aqueduct transports water to the cities of Mexicali, Tecate, Tijuana, and Ensenada

in Mexico (Bernal & Solis 2000). As early as 1961, Mexico complained that the introduction of highly saline drainage water from Arizona into the Colorado River compromised its utility as irrigation water. To address the salinity problem, the USA constructed a bypass conveyance channel for the discharge of saline waters to a point preferred by Mexico. Cooperative efforts continue to resolve problems associated with the distribution and quality of Colorado River waters (Bernal & Solis 2000).

4.6 WATER RESOURCES

Water controls the existence, distribution, and density of humans, animals, and vegetation. Sources of water in the desert are limited. Of these, the most significant are groundwater and regulated surface flow. Groundwater, obtained from springs or wells, is a dwindling resource in some areas. The utilization of surface waters depends to a large degree on the stream regime, controlled by meteorological conditions of the headwaters. During floods, water may be in excess of irrigation demands, whereas in early summer discharge may be too low for the requirements of cultivation. Hence, in many areas runoff is regulated by systems of dams and storage reservoirs. There are competing demands on this water for irrigation, power, recreation, and industry, leading to international disputes relative to water-resource development and management, such as for the waters of the Tigris-Euphrates Basin, shared by Turkey, Syria, and Iraq (Beaumont 1998; Cullen & deMenocal 2000; Altinbilek 2004). Additional means of obtaining fresh water include desalination, importation via long-distance transfer (canals and aqueducts), and rainmaking.

The need for water in arid regions constantly rises, due both to increases in population and standards of living. In Israel, for example, per-capita water consumption increased by 26.5% between 1991 and 1998, from 88 to 111 m^3. In absolute terms, the growth of population ($\approx$2.5% per annum) and per-capita consumption resulted in a change in total volume consumed from 445 million m^3 in 1991 to 672 million m^3 in 1999, an increase of more than 50%. Measures to conserve water, including the banishment of cotton plantations and restrictions on private swimming pools and lawn watering, have been largely offset by other factors, such as increased housing, private gardens, and public infrastructure (Portnov & Safriel 2004).

4.6.1 GROUNDWATER

For many areas of the world, the water supply is derived almost entirely from groundwater, and pumping has been increasing hand-in-hand with population growth. Under present-day climatic conditions, the quantity of recharge is small relative to the amount of water in storage and the quantity of water pumped from aquifers. Consequently, water levels in many wells are declining (Fig. 4.7).

The groundwater present in today's deserts was derived largely from rainfall in formerly wetter times. Research suggests that such water may be thousands to hundreds of thousands of years old. The absolute dating of groundwater is important to determine the sustainability of drinking and irrigation water sources. A number of radioactive tracers have been employed, according to the projected age of the groundwater reservoir. The measurement of cosmogenic ^{81}Kr was used by Collon et al. (2000) to assign ages of around 230,000–400,000 years for the mean residence time of groundwater in the southwestern GAB, Australia. In the western Mojave Desert, most of the recharge to the upper Mojave River basin occurred between 5000 and 20,000 years ago (Izbicki et al. 2000). The main period of aquifer recharge for the Chad Basin of northeast Nigeria occurred between 24 and 20 ka, prior to the LGM. This palaeowater is unsuitable for large-scale development as its location in the middle and lower confined aquifers is decoupled from the present environment and its exploitation represents mining of the resource without hope of recharge (Edmunds et al. 1999).

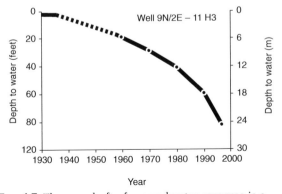

FIG. 4.7 The overdraft of groundwater reserves is a pressing environmental problem in arid regions. Near Barstow, California, USA, water levels were near the surface in 1930, but had dropped to 24 m below the surface by about 2000.

Some regions provide geomorphic evidence of formerly higher groundwater tables and wetter climates. Dry valley systems in the Kalahari, formed by groundwater seepage processes, may have been active several thousand years ago during wetter conditions. No springs outcrop under present climatic conditions and modern recharge would have to be between two and three times modern values for reactivation of these groundwater systems (De Vries et al. 2000).

There are several sources of modern groundwater recharge: (1) transmission losses along rivers, (2) deep artesian inflow, (3) seasonal rainfall in bordering highlands, and (4) artificial recharge from irrigation, flooding, reservoirs, or injection wells. Many basins have more than one source. For example, at Bahariya and Farafra Oases, Egypt, artesian water in the Nubian aquifer migrates from the south (Dabous & Osmond 2001) and planning is underway to inject excess Lake Nasser water into the aquifer. Additional water may infiltrate from proposed irrigation projects (Kim & Sultan 2002). Lake Nasser has been recharging the Nubian aquifer since its construction in 1965, and the groundwater table in the vicinity of the lake is significantly elevated. Although the water reserves of the Nubian aquifer are vast, it is probable that current pumping in the oases exceeds deep aquifer inflow by a factor of 10 (Thorweihe 1990). The water that is being withdrawn was principally recharged during past pluvial periods (Dabous & Osmond 2001).

The overdraft of groundwater reserves is a pressing environmental problem in many arid regions. When more water is withdrawn from an aquifer by pumping than can be returned by natural recharge, the system is considered to be in overdraft. The overuse of groundwater reserves is expressed in the declining yields of wells, the lowering of the water table, a loss of riparian vegetation and, sometimes, by land subsidence. In many cases the use of groundwater has been based on uncontrolled economic activity. Artesian groundwater located at depths of between 400 and 600 m in Quaternary sediments of the Chad Basin of northeastern Nigeria was extensively developed during the 1950s and 1960s, but subsequently underwent a significant decline in pressure, both due to abstraction and uncontrolled discharge from artesian wells (Edmunds et al. 1999). In the Gila River Basin and its tributaries in Arizona, the annual withdrawal in the 1950s was 4 billion m³ year⁻¹, approximately 30 times the rate of annual replenishment.

The Mojave River basin, California, has a mix of agricultural and urban demands for groundwater that have caused considerable overdraft. The river is a 200 km-long exogenic stream within a closed basin, originating in areas of high relief (up to 2585 m) in the San Bernardino Mountains, where over 90% of the basin precipitation (1067 mm year^{-1}) falls (Enzel 1990). Since the 1960s, the Mojave River basin, located in close proximity to the urban Los Angeles area, has experienced population growth in excess of 5% per year (Martin 1994). Water supplies are derived almost entirely from groundwater. The Mojave River is the main source of recharge, with about 80% of the input derived from floodflows along a 160 km reach from the headwaters to the Afton Canyon section of the channel (Martin 1994). Recharge decreases in a downstream direction, owing to transmission losses along the river course. An imbalance between pumping and recharge has resulted in serious overdraft, and in the subarea near Barstow, the annual deficit in groundwater storage during the 1990s exceeded 7000,000 m^3. Owing to dwindling water supplies, a pipeline is being constructed to carry water to a string of seven recharge basins designed to raise groundwater levels (Laity 2003).

Excessive groundwater withdrawal may generate land subsidence, with serious economic consequences. Continuous declines in groundwater levels in the Las Vegas Valley, Nevada, by as much as 90 m in some areas, have resulted in surface subsidence and fissuring, with damage to housing and other structures from differential settling (Bell et al. 1992; Morris et al. 1997). At Yucca dry lake, Nevada, parallel fissures as much as 2 km long and 500 m deep have developed. Similar problems have emerged at Rogers Lake, in the Antelope Valley of the western Mojave Desert, California. This dry lake lies within the confines of Edwards Air Force Base and has served as a runway for Space Shuttle operations. Groundwater in the Antelope Valley has been used for irrigation and domestic use since the 1880s, with as many as 500 wells in use by 1921 (Thompson 1929). Artesian water occurs in fluvial sands and gravels beneath impermeable lakebeds. By the 1950s, aquifer compaction following excessive groundwater withdrawal had caused surface subsidence and the development of sinkholes and giant desiccation cracks (greater than a meter in width and more than 5 m deep), which began to pose problems for military and aerospace operations (Ikehara & Phillips 1994; Orme 2004).

In order to maintain groundwater as a long-term water source in deserts, an assessment of the water budget is essential. In some cases, this may require international cooperation, as the aquifer does not stop at the border.

4.6.2 Dams and reservoirs

Dams and reservoirs are used extensively to store surplus water in arid lands. However, there are a number of disadvantages to reservoir storage in drylands relative to humid environments, including greater evaporation and transmission losses, higher amounts of sedimentation, and greater uncertainty of supply. It is sometimes necessary, as in Israel, to store as much as three times the amount of water actually needed. Full utilization of a fluctuating supply can rarely be achieved. In the southwestern USA stock owners have constructed large numbers of small dams that dot the landscape. Losses by evaporation and by seepage, often not recoverable as groundwater, represent a direct loss to the water economy. Water accumulates in reservoirs during the wet season and, when it is required for crops, much has already been lost by evaporation. Evaporation losses from reservoirs are a continuous challenge and have received considerable attention. The most viable technique is probably to control reservoir shape as evaporation accounts for losses of up to 50% in shallow reservoirs and only 20% in deep reservoirs.

4.6.3 Long-distance transfer: canals and aqueducts

Large-scale water transfer has physical, social, political, legal, and economic elements. Legal and administrative problems in conveying large quantities of water across state boundaries occur wherever doctrines, principles, and regulations vary. The large-scale and long-term environmental effects are often not considered.

Although regional water transfer now occurs at a very large scale, it has historical roots, having developed in early Babylonian times. Extensive irrigation of the Mesopotamian Valley occurred as early as 4000 BC. Today, water transfer in deserts influences regional growth patterns, allowing development at an unprecedented scale. Faced with overpopulation problems and the need for new agricultural land, Egypt has developed two national long-distance water-transfer projects: the El Salam canal, which diverts Nile river water to the Sinai Peninsula, and the Tushka Canal, which channels water from Lake Nasser to overflow lakes in the Baris Depression in the Western Desert (Fig. 4.8) (Gheith & Sultan 2002; Kim & Sultan 2002). These developments have resulted in part because Lake Nasser has approached its maximum storage capacity and it is essential to reduce possible downstream flooding damage. Under investigation is the construction of a network of canals to carry Nile water through several different oases in the Western Desert (Kim & Sultan 2002).

During long-range water transport, canals lose both water volume and quality by evaporation and evapotranspiration (Fig. 4.9). Despite attempts to conserve water, some loss is inevitable. Where phreatophytes line streams or irrigation canals, transpiration accounts for as much as 40–60% of the wastage. However, removal of native vegetation may cause sedimentation and/or flooding in downstream channels. Seepage also occurs along irrigation canals, which may be lined by compacted clays to reduce loss.

4.6.4 RAINMAKING

Weather modification, or cloud seeding, is sometimes seen as a viable means of augmenting the natural water supply of an arid region. Owing to the expense of these operations and their uncertain results, they principally take place in developed nations. In the USA, winter orographic clouds are modified to augment snowfall (Czys 1995). Large-scale seeding has been in operation in Utah since 1973 and studies indicate that such projects generally increase the winter precipitation by 14–20%, with economic benefits exceeding the costs. For the water year 2005 (1 October–30 September), Utah had four large-scale projects using silver iodide (Utah Division of Water Resources 2005).

Cloud-seeding experiments have been conducted in Israel since 1960 and appear to offer positive statistical evidence of precipitation enhancement (Cyzs 1995; Rosenfeld 1997). Seeding is conducted during the winter months (November–April) using aircraft to release silver iodide at cloud base along a line upwind of the targets (Czys 1995).

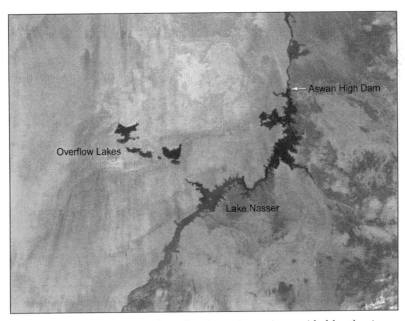

FIG. 4.8 Water storage and flood control along the Nile River, Egypt, is provided by the Aswan High Dam, constructed in the 1960s, which impounds Lake Nasser. Unusually high levels of water in 1998 caused flow to be diverted westward through the Tushka spillway into a series of overflow lakes, which were essential to reduce downstream flooding damage. Image from 10 October 2000 by Robert Simmon, Reto Stöckli, and Brian Montgomery, NASA GSFC.

Fig. 4.9 Regional water transfer by canals is a common solution to the aridity of deserts. Both water volume and quality diminish with distance, owing to evaporation. This photograph shows a canal near Dunhuang, China, with gobi in the foreground and dunes in the background.

Cloud seeding remains controversial, expensive, and relatively untested in arid regions. The development of a reliable and scientifically accepted cloud-seedling technology still lies in the future.

4.6.5 DESALINATION

All desalination involves large capital expenditures, large amounts of fuel or other energy sources, and a skilled, if small, labor force. It may be achieved through distillation, by freezing or chemical absorption, or electrodialysis (passing an electric current through water). Distillation is used in the Karakum, in Kuwait, and in Israel. Israel operates 30 desalination facilities, processing 9.8 million m^3 of water annually (Portnov & Safriel 2004).

4.6.6 FOG-WATER COLLECTION SYSTEMS

An innovative approach to securing water in coastal deserts is to use a fog-water collection system (FWCS), also termed a Standard Fog Collector (SFC). These have been implemented successfully along the west coasts of South Africa (Olivier & de Rautenbach 2002), Namibia (Shanyengana et al. 2002), and South America (Schemenauer et al. 1988; Cereceda et al. 2002; Larrain et al. 2002). The South African system consists of two 9 m × 4 m sections (36 m^2 each) of shade netting mounted vertically, and oriented perpendicular to the direction of fog-bearing winds. Similar systems are employed in South

America (Larrain et al. 2002). The screens intercept both fog and rain droplets, which flow downward to a gutter and thence through a pipe to a storage tank. The water yield from such systems depends on a high occurrence of fog, persistent fog, and the presence of fog-bearing winds. In South Africa, the ideal site is located within 5 km of the coastline, at elevations between 200 and 700 m ASL (Olivier & de Rautenbach 2002). Higher elevations in both Africa and South America yield more water owing to the increase in wind speed and fog moisture content with height.

Fog-derived water is more potable than groundwater, which is commonly contaminated with salts and heavy metals, or surface water that is polluted and contains parasites. It is low in TDS and major ion concentrations and there are no disease-causing bacteria present (Schemenauer et al. 1988; Shanyengana et al. 2002; Olivier 2004). Fog-derived water can be used to dilute saline ground water to make it more suitable for drinking.

4.7 CASE STUDY: THE WATERS OF THE TIGRIS-EUPHRATES BASIN AND THE IMPACT OF MODERN WATER MANAGEMENT

The Tigris and Euphrates Rivers have perhaps the longest history of development and water management of any arid zone river system. They served as the basis for the well-known empires in Mesopotamia (Iraq), including the Sumerians,

Acadians, Babylonians, and Assyrians. In "the land between two rivers," these civilizations maintained an irrigation and flood-control system that supported as many as 20 million inhabitants.

Water-related disputes in the Middle East and in the Tigris-Euphrates Basin extend back as far as 6000 years BP (Beaumont 1998; Altinbilek 2004). Throughout this period, water management only involved the manipulation of water in the lower part of the basin between April and July, when snow-fed meltwater waves moved down the river. Only a small proportion of the total water was used for irrigation or other human uses. However, these lands were progressively abandoned owing to waterlogging and salt accumulation.

The modern problems with respect to water utilization began with the fall of the Ottoman Empire, and the development of new political boundaries in the 1920s with the establishment of the states of Syria, Iraq, and Turkey. At present, the regional population of the Middle East is increasing at a rate of about 3.5% per year, and irrigation consumes about 80% of the available water (Cullen & deMenocal 2000). Much of the current focus of Middle Eastern water policy is on the environmental and socioeconomic impacts of the control structures that are in

place, or are planned, for the Tigris and Euphrates Rivers (Beaumont 1998; Cullen & deMenocal 2000).

The Tigris and Euphrates Rivers are fed from snow that falls on the high mountains (3500–4500 m ASL) of Turkey, Iraq, and Iran (Fig. 2.8). Annual precipitation at these altitudes may exceed 1000 mm (Beaumont 1998), and is largely confined to the cool period from October to April. The rivers exhibit strong snowmelt peaks during March, April, and May, when about half of the total annual discharge occurs. The Euphrates River, the longest river (3000 km) in western Asia, has a total discharge of about 32,000 million m^3. The discharge of the Tigris is greater, attaining an average value of about 50,000 million m^3. However, as in other dryland rivers, the discharge is highly variable from year to year (Fig. 4.10) (Beaumont 1998; Altinbilek 2004). The interannual variability is linked to the NAO, a large-scale meridional oscillation of atmospheric mass between a high-pressure cell located near the Azores and a subpolar low-pressure center near Iceland, which results in distinct, dipole-like climate-change patterns between western Greenland/Mediterranean and northern Europe/northeastern USA/Scandinavia. Research by Cullen and deMenocal (2000) suggests that its influence extends as far as the eastern Mediterranean,

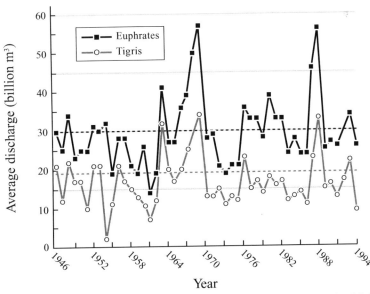

FIG. 4.10 The natural interannual flow variability of the Euphrates and Tigris Rivers is quite high and appears linked to the North Atlantic Oscillation (NAO). Discharge as recorded near the Turkish border, where the discharge of the Euphrates exceeds that of the Tigris. Downstream, near Al-Fallujah, the discharge of the Tigris exceeds the Euphrates owing to tributary inflows. From Altinbilek, D. (2004) Development and management of the Euphrates-Tigris Basin. *Water Resources Development* **20**. By permission of Taylor & Francis.

thus expressing itself in droughts or high-flow events in the Tigris-Euphrates Basin. The NAO is most pronounced during the winter months, because of the greater contrast in sea and air temperatures.

The major tributaries of the Euphrates are located in the upper part of the basin, with about 88% of the flow generated in Turkey and 12% in Syria. Thus, a single dam in the upper catchment, the Ataturk Dam in Turkey, achieves a high level of control. The Tigris River receives inflow from several tributaries and the overall water control has required construction of a series of major dam projects. Turkey contributes about 44% of the total flow of the Tigris. Water management in Turkey is both for the generation of hydroelectric power and to provide water for irrigation. The dams have resulted in a more uniform flow in the rivers, eliminating both major floods and extremely low-flow conditions (Fig. 4.11). Turkey plans additional controls of the water as part of the South-East Anatolia Project, with more storage reservoirs on the Euphrates River providing an effective stranglehold on the flow (Beaumont 1998). Its controls of the Tigris watershed are more limited, allowing Iraq more available water. Competing demands for water in the region have already resulted in tense relations between the watershed's neighbors and led to claims of decreased water supply by Syria and Iraq

(Cullen & deMenocal 2000). A combination of a low-flow year (from natural climatic variability) and the withholding of upstream flow have the potential to lead to more serious conflict.

To date, the water quality of the Tigris-Euphrates Rivers has been high, especially during the snowmelt period. As more water is diverted for irrigation, the potential exists for the return waters to carry a wide range of dissolved chemicals, including naturally occurring salts, as well as pesticides, herbicides, and petroleum products.

The effects of the attenuation of the snowmelt peaks will be largely felt in the wetlands of the Shatt el-Arab, which is the name given to the river, almost a kilometer wide and 190 km long, that results from the combined flow of the Tigris and Euphrates Rivers (Altinbilek 2004). Once the largest wetland ecosystem in the Middle East (covering 20,000 km²), the marshlands have been extensively drained and diminished and are now only a fraction of their former size (Fig. 4.12). The United Nations Environment Program reports that between 1970 and 2000 90% of the Mesopotamian marshlands were lost (Eden Again Project 2003). The causes of the loss included a diminution of inflow and the construction of drainage engineering works such as levees and canals by Iraq. The wetlands formed

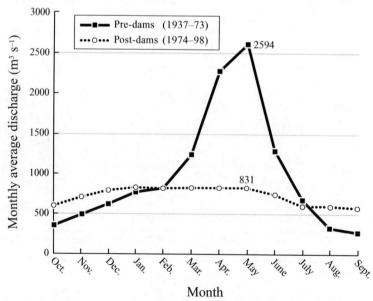

FIG. 4.11 Pre- and post-dam flow regime for the Euphrates River at Hit-Husabia, Iraq. Dams have resulted in a more uniform flow of the river, eliminating both major floods and extremely low flows. The spring snowmelt flood peaks have been considerably attenuated. From Altinbilek, D. (2004) Development and management of the Euphrates-Tigris Basin. *Water Resources Development* **20**. By permission of Taylor & Francis.

(a)   (b)

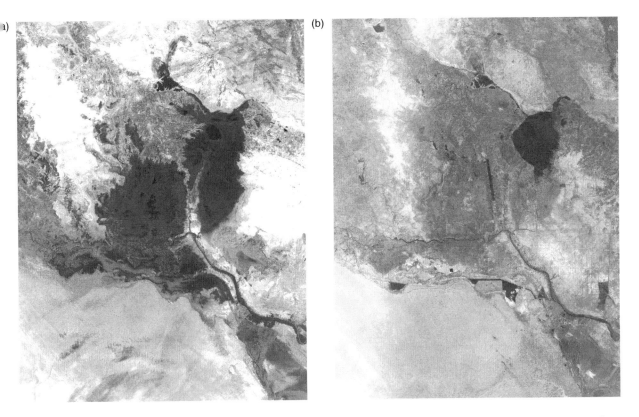

Fig. 4.12 The Shatt el-Arab is the name given to a river that results from the combined flow of the Tigris and Euphrates Rivers. It once formed the largest wetland ecosystem in the Middle East (20,000 km^2). Owing to diversion of flow and the elimination of spring floods, the marshland has been reduced by 85%. (a) Conditions in 1973–6, when dense marsh vegetation (mainly phragmites) appeared as dark patches; (b) the marsh in 2000, a largely arid, lifeless, and salt-encrusted surface. Source: NASA GSFC Science Visualization Studio, based on data from Landsat 7 and the USGS EROS Data Center.

part of an intercontinental flyway for migratory birds, sustained endangered species and freshwater fisheries, and for millennia provided a home for an indigenous human community, the Marsh Arabs. Present-day efforts to restore and remediate the region have resulted in some success. Uncontrolled releases of river waters after the 2003 war flooded 20% of the original wetland area and partially restored some areas of marsh in southern Iraq. In other areas, high salinities are causing problems with the restoration effort, but it is likely that flushing with high-quality water will eventually result in reestablishment of a productive wetland environment (Richardson et al. 2005).

5

LAKE SYSTEMS: PAST AND PRESENT

5.1 Introduction to desert lakes

Desert rivers are typically characterized by endoreic (internal) drainage or are "disorganized," that is, they lack integrated surface drainage. Seldom is there enough flow to allow water passage to the sea and this results in the formation of lakes. Three types of desert lake are found: perennial, ephemeral, and relict (palaeolakes). Only a few are perennial, and these are commonly highly saline: they include the Caspian Sea, the Dead Sea, Mono Lake, and the Aral Sea. These lakes have sources in more humid areas, such as surrounding highlands or more distant well-watered regions.

Most desert lakes are ephemeral, occupying their basins for short periods of time following localized flooding. Ephemeral lakes are often referred to as playas or pans, and are widespread throughout the world's arid lands, varying in scale from a few tens of square meters to more than 10,000 km². In total, they probably occupy about 1% of the desert landscape (Shaw & Thomas 1997), but their significance is greater than their area suggests, as they provide important temporary sources of water for both humans and animals. In some areas, they occur in very high numbers: in Mozambique densities of up to 200 pans km^{-2} are attained (Goudie & Wells 1995), and there are between 30,000 and 37,000 basins on the southern High Plains of Texas and adjoining New Mexico (Osterkamp & Wood 1987).

Relict lake surfaces are larger in area than modern playas, and represent the extent of water bodies during wetter periods, often referred to as "pluvials." Thus, the Great Salt Lake, Utah, occupies a small fraction of the land once filled by Pleistocene Lake Bonneville.

5.2 Types of lake

5.2.1 Perennial salt lakes

Perennial salt lakes never dry out, although their water levels may show secular fluctuations. Individ-

ual lakes vary little in their salinity over the long term, although there may be considerable differences between lakes. The Caspian Sea, for example, has maintained a relatively constant salinity of about 12 g L^{-1}, and the Aral and Dead Seas, before changes in their volume owing to human impact, had salinities that remained at 10 and 200 g L^{-1}, respectively, over several decades (Williams 2002). The composition and nature of the biota is different in saline lakes than fresh, as permanent or temporary adaptations to osmotic stress are needed. In general, with increasing salinity, biodiversity decreases.

The three most important climatic factors determining salt lake development are temperature, net evaporation, and precipitation. Changes in climate have profound impacts on the volume, and thus salinity and ecology, of salt lakes. Many human activities have adverse effects on salt lakes. These include surface inflow diversions, soil salinization, mining, pollution, biological disturbances, and anthropogenically altered climate. The most important of these changes is the diversion of freshwater inflows for agricultural purposes. The Aral Sea in Asia (Fig. 1.2), Mono and Pyramid Lakes in the USA, and the Dead Sea of Israel/Jordan are notable examples of lakes whose levels have fallen and salinities concomitantly increased. In addition to chemical and biological effects, falling lake levels have exposed saline lakebeds, causing increases in locally generated dust (Gill 1996). The environmental degradation of desert lakes is discussed more fully in Sections 13.4.4 and 13.5.

5.2.2 Ephemeral lakes: playas and pans

Playas occupy topographic lows in the landscape. The processes by which basins form are numerous and often act in combination. They include: (1) block faulting, (2) broad shallow warping, (3) crater-lake development in volcanic areas, (4) deflation, (5) biogenic activity, (6) meteorite impact, (7) interdunal development, and (8) groundwater dissolution and removal of material beneath a playa surface. In the latter, solutes and clastic material are transported laterally in the subsurface along solutional pipes, leading to gradual subsidence of the playa surface (Paine 1994). Karstic processes, such as those on the Moroccan coastal plain, may be significant in areas underlain by limestone (Goudie & Wells 1995). Small playas may form in interdune hollows, but the largest usually have some degree of structural or geologic control on form. Many basins, particularly those that are large in scale, have a polygenetic origin.

Most playas share several characteristics in common. They have relatively flat, vegetation-free surfaces and lack surface outflows. Where vegetation does occur, it is in distinct associations with a tendency for salt tolerance, most commonly along the periphery of the lake, clustered around zones of groundwater seepage (Fig. 5.1).

Playas exhibit a great range in morphology, hydrology, and sedimentology. Not only is the modern playa environment often complex and poorly understood, but most playas have a long history of geologic and environmental change which has proven difficult to decipher, particularly as the paucity of preserved organic material often precludes radiocarbon dating. Despite these difficulties, they have received a great deal of attention as long-term repositories of detailed climatic information and have been the focus of multidisciplinary scientific research.

The hydrologic budget of a playa is characterized by greater evaporation than water input, often with a ratio of greater than 20:1 (Bryant & Rainey 2002). Input is generally from streams, groundwater discharge, or precipitation directly on the playa surface. The general water balance of a closed lake can be expressed as follows (Bryant & Rainey 2002):

$$dV/dT = R - A(E - P) - D + G_i - G_o$$

where V is lake volume, T is time, R is runoff, A is the surface area of the lake, E is evaporation, P is precipitation, D is discharge, G_i is groundwater inflow, and G_o is groundwater outflow. Lakes respond to changes in the local water budget by fluctuating in depth or area. In *cascading* systems, as water elevations rise, lakes spill out over a sill, the lowest part of a divide that separates adjoining sub-basins, and the water in one sub-basin flows to another sub-basin (Benson et al. 1995). *Closed lakes* grow in size and depth as inflows increase, but do not overflow to interconnect with other lakes. Water loss is thus a function of evaporation rates and groundwater outflow/seepage.

Both closed-basin and cascading lake systems have been used to reconstruct past climate changes. Closed-basin lakes are considered one of the best sensors of past environmental changes because of their sensitivity to small fluctuations in the effective

FIG. 5.1 Most lake surfaces in deserts are vegetation-free. Salt-tolerant vegetation is most commonly found along the periphery of lakes, clustered around zones of groundwater seepage. This photograph shows groundwater seepage along a playa margin in Death Valley, California, USA.

moisture budget (Kotwicki & Allan 1998; Valero-Garcés 2000). Large basins more faithfully record major events than small playas, which register small hydrologic changes. Large playas may retain enormous thicknesses of sediment, and thus retain records spanning long time periods (greater than 100,000 years). By contrast, coastal playas or interdunal playas often have a short-lived record owing to the frequent migration of coastlines and dunes. For all types of lakes, the modern inundation regimes are largely unknown or unrecorded, particularly for time spans of several decades.

Crater lakes are a sub-type of closed lakes, with relatively uniform shapes, which tend to act as gigantic rain gauges, as most of the water that enters is from precipitation (Williams et al. 1993). They fill during periods of high rainfall and evaporate to shallow, saline conditions during drought intervals. Like other closed basins, however, they may have significant inflows and outflows of groundwater. At Lake Hora, a crater lake in Ethiopia, about 40% of total water inflow is derived from groundwater. Seepage accounts for 3% of water loss, the remainder being evaporated (Lamb 2001). In addition, crater lakes are affected by volcanic and hydrothermal activity, factors that make it more difficult to reconstruct past climate from the palaeolimnological data (Lamb 2001).

From a human perspective, playas are very important, serving as sources of fresh water and minerals, as surfaces for the development of airfields and racetracks and, in the case of Lop Nor (Lop Nur, Lop Nuur), China, for the testing of nuclear materials. Playas have been the sites of human occupancy, either by native peoples or, in later times, by travelers or armies. Soda Springs, an oasis on the western rim of Soda Lake playa in the eastern Mojave Desert, California, has served as a water source and resting area for native Americans, Spanish and Mexican explorers, early American explorers, surveyors, settlers, and explorers and a small contingent of the US Army. It has served as a civilian way station, as a religious colony, and as a health resort (Duffield-Stoll 1994). Today, it operates as the Desert Studies Center (in Zzyzx, California), run by the California State University system.

Owing to their global distribution, playas have a rich terminology, with some terms being widespread and others having a more local application. The term *playa* is perhaps the most commonly used for an arid or semiarid lacustrine basin. However, in some instances, the term *pan* is used to describe small basins and playa for a depression with a saline surface. Attempts to regiment language use have led to some confusion in the literature, as these schemes have not always received close adherence. Saline or salt lakes are referred to as *sebkhas* or *sabkhas* in North Africa and Arabia, as *salars* in South America, and *salt pans* in the USA. Playas with silt or clay surfaces are called *takirs* in central Asia and *claypans* in Australia. The large regional nomenclature is tabulated and reviewed in more detail in Rosen (1994) and Shaw and Thomas (1997, p. 295).

5.2.2.1 Wet (salt playas; discharge playas) and dry (recharge playas; claypans) systems

In general, there are two basic playa types that exist: "dry" systems that are fed primarily by surface flow, have a predominantly smooth clay floor, and for which the groundwater table is usually more than 5 m below the surface; and "wet" systems, for which groundwater is a significant contributor, with the water table typically less than 5 m below the surface, leading to a saline surface (Fig. 5.2) (Rosen 1994; Reynolds et al. 2007). When unoccupied by a lake, a dry playa (Fig. 5.3) is firm to the tread, can be driven upon, and occasionally serves as a landing site for planes. Rogers Dry Lake in California is a runway for the NASA Space Shuttle program. The second playa type is considered "wet" because beneath the apparently dry saline surface the sediments are commonly moist and sticky: walk on these surfaces at your peril (Fig. 5.4). Over the short term, these differences in moisture and salinity are more a function of topography and geology than climate. A third type of playa, much more restricted in nature, is the coastal sabkha (Glennie 1970), which is a saline flat that receives periodic marine incursions.

The features of desert lakes are attributable to their water, sediment, and salt budgets, which, in turn, reflect relationships with groundwater and salt systems (Gill 1996; Reynolds et al. 2007). Lakes which occupy structural basins act as terminals for both surface and groundwater movement and constitute evaporative cells in linked hydrologic systems. Groundwater imparts salts in varying degrees depending on the nature of inputs (atmospheric, volcanic, subsurface from ancient saline strata), the frequency of flooding, and the supply of sediment. The balance between surface water and groundwater inputs, as reflected by the position of the water table beneath the playa surface, is one of the most important determinants of the morphology and sedimentology of playas.

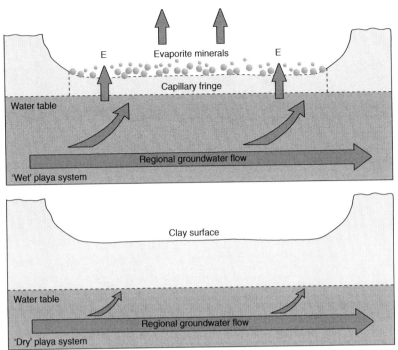

Fig. 5.2 Elements of wet and dry playa systems. The upper illustration shows a "wet" system, in which the water table lies close to the surface, and evaporation (E) through the sediments is associated with a wet, sticky surface and a veneer of evaporite-mineral crystals. The lower illustration is representative of a "dry" playa, wherein the water table lies at depth, such that capillary rise of water and dissolved salts do not reach the surface. From Reynolds, R.L. et al. (2007) Dust emission from wet and dry playas in the Mojave Desert, USA. *Earth Surface Processes and Landforms* **32**. Copyright John Wiley & Sons. Reproduced with permission.

Fig. 5.3 A "dry" playa has a smooth clay floor and is fed primarily by surface flow. East Cronese Lake is one of the terminal lakes of the Mojave River, California, USA. The alluvial fans to the north and east are veneered with windblown sand along their distal fringes.

Fig. 5.4 A "wet" or discharge playa has a water table close to or at the surface. Seasonal groundwater outflow may occur (shown in the upper right) and the surface is commonly covered with accumulations of evaporite deposits, which are replenished by capillary action. Beneath the apparently dry surface, the sediments are commonly moist and sticky muds. The photograph shows Soda Lake, California, USA.

Lakes on intermediate slopes may be bypassed by groundwater flow and therefore be largely dependent on surface flow so that silts and clays are more important contributions to the lake floor than salts. These playas, where the groundwater table lies at a depth such that the surface is above the zone of capillary rise, have minimum accumulations of salts on their surfaces and tend to be clay-floored. They are often referred to as recharge playas or claypans and the playa floor is permanently above the water table. They have firm level surfaces of clay that are commonly covered in fine polygonal cracks. Such floors may resist colonization by plants, although in some areas they are at least partly vegetated by salt-tolerant species.

Discharge playas are those where the water table is close to or at the surface. They experience considerable seasonal groundwater outflow and commonly are associated with accumulations of evaporite deposits, which are replenished by capillary action (Figs. 3.2, 5.4, and 5.5) (Crowley & Hook 1996; Reynolds et al. 2007). The depth to the water table varies according to seasonal and longer-term rainfall contributions, to human pumping of groundwater, and to regional climatic change.

The fluctuations in the water table over time lead to variations in the distribution of groundwater-fed and claypan playas. For example, Lake Eyre has been both a perennial lake up to 25 m in depth and a groundwater-controlled playa (Magee & Miller 1998). Troy Lake in the Mojave Desert of California was a discharge playa in the early 1900s (Thompson 1929), but by the late 1990s was a claypan owing to pumping of groundwater (Laity 2003). Water-table fluctuations also affect aeolian deflation processes and thereby the accumulation of aeolian sediments around the playa periphery (Jankowski & Jacobson 1989). Between the two extremes of salt lakes and claypans, many intermediate types occur.

Sediment inputs to lakes are invariably fine-grained and composed of silts and clays. Lakes often form the terminal sumps for streams, which, near their terminus, are only carrying suspended load. Similarly, where playas are adjacent to alluvial fans, the coarse material is deposited on the upper fan surface, leaving the suspended load to be deposited on the playa floor. Aeolian sediments may also aggrade on a playa and these, by their very nature, are fine-grained. Along the playa margins, lacustrine sediments may interfinger with coarser materials, such as those of alluvial fans, or with fine-grained aeolian deposits (Fig. 5.3; Nicoll 2004). Playas usually have flat, horizontal surfaces because fine-grained sediments or irregularities associated with evaporite growth are smoothed out by water movement and dissolution. If flooding is infrequent, sand dunes may develop on the surface, or evaporite growth may create uneven surfaces.

Fig. 5.5 In playas with thick salt crusts, the crust may crack into polygonal plates a meter or more in diameter. As crystallization continues, the plate margins develop raised rims that may overlap and buckle. Death Valley, California, USA.

Where saline groundwater is close to the surface, excess evaporation leads to a zonal concentration of salts in accordance with solubility (Houston 2006a). A crust of sodium chloride commonly develops in the lowest part of dry lakes, above a saturated mixture of mud and gypsum crystals. Crusts of gypsum-cemented sediments may occur on the higher parts of the floor. Calcium carbonate crusts are characteristic of the upper margins and deltas. Efflorescent salt crusts have been characterized using remote-sensing data. Studies in Death Valley, California, show that the crusts may reflect the chemistry of inflow waters: at Cottonball Basin, inflow waters are rich in sodium, and evaporite crusts contain abundant thenardite (sodium sulfate); in the Badwater Basin, calcic groundwaters emerge at springs and gypsum (calcium sulfate) crusts are common (Crowley & Hook 1996).

In playas with thick salt crusts, the crust may crack into polygonal plates up to a meter or more in diameter. Additional crystallization along the plate margins may cause raised rims and eventually the plates may overlap and buckle (Fig. 5.5). Renewed flooding may temporarily destroy a salt crust by solution. Salts may be impregnated by dust and become discolored (Fig. 3.2).

Dry lake beds may include a variety of surface features, including phreatophyte and spring mounds, mud volcanoes, solution depressions, and giant desiccation polygons. The polygons have been noted in numerous playas of the American southwest (Neal & Motts 1967), and appear to show considerable temporal and spatial variability owing to climatic fluctuations and changes associated with human activity (groundwater pumping). Mapping of North Panamint Playa, California, by Messina et al. (2005) revealed fissures of varying ages that ranged from 4 to 911 m in length, with some deeply incised (over 1 m deep) and others forming positive relief features where wind-blown sediment and plants had accumulated.

Many groundwater playas show a moist seepage zone around the margins. Some, including Owens Lake in California and Lake Eyre, Australia (Alley 1998), have marginal calcareous mound springs that attain heights of a few meters above the lake floor, indicating former artesian outflows that were possibly fault-aligned.

5.2.2.2 Playa degradation

In addition to aggradational processes that build up the surface of the playa, there are degradational processes that cause it to lower. These include deflation; solution, piping, and subsurface karstic collapse; and the excavations and trampling of animals. Erosion of playa sediments by the wind (deflation) is often an integral part of landscape development, with Space Shuttle photography showing that pans are important source regions for dust storms (Middleton et al. 1986).

Saline sediments, surface material, groundwater levels, and animal trampling exercise important controls on wind action (Goudie 1989; Reynolds et al. 2007). In some environments, salinity favors wind erosion by preventing vegetation growth and causing pelletization of normally cohesive clays. These transportable aggregates may accumulate downwind in crescentic lunette dunes. A major episode of deflation between 60 and 50 ka is believed to have excavated the present Lake Eyre basin and deposited a gypsum- and clay-rich aeolian phase around the lake (Magee & Miller 1998). In Mojave Desert playas, the production of evaporite minerals on playa surfaces results in several centimeters of "fluffy" sediments that are easily deflated. Reynolds et al. (2007) used automated digital photography to record the susceptibility of these surfaces to wind erosion. The regional geology also plays a role in playa formation. In some areas, pan initiation and growth appear to depend on materials that are susceptible to deflation, with depressions developing preferentially in poorly consolidated sediments, shales, and fine-grained sandstones.

The groundwater table sets a lower limit to deflation. During prolonged dry periods, the lowering of the water table may allow significant deflation to occur, marked in some areas by fields of yardangs. The rate of deflation in prehistoric to early historic times in Kharga, Egypt, paralleled the rate of water-table fall, with measurements suggesting that as much as 30 m of sediment was removed since the playas were last active (Haynes 1985). According to Haynes (2001) all playas of the Darb el Arba'in Desert of Egypt and Sudan are deflated to some degree, some so severely that no mud yardangs remain, with the surface scoured down to bedrock.

The trampling of animals exerts a feedback effect. As depressions become enlarged and retain water, they become more attractive to grazing animals, which further disrupt the surfaces, rendering them more prone to erosion (Goudie 1989).

5.2.3 PALAEOLAKE SYSTEMS: LAKES AS INDICATORS OF PAST CLIMATE CHANGES

In many arid areas of the world, a legacy of lake shorelines and sediment in now largely dry basins is testament to a once more effective water presence. Syntheses of lake status have become an important tool for reconstructing palaeoclimatic changes at a regional to global scale. The most studied lakes are those of the Great Basin of North America and the Lake Eyre basin in Australia (Nanson & Price 1998), although there are an increasing number of studies in China and Africa.

Lakes have existed in arid regions episodically throughout the Cenozoic. Numerous researchers have catalogued lake-level data and proposed lake-level chronologies for various lake systems. As the number of well-dated and studied sites increases, it will become increasingly possible to make syntheses of these data and better understand global climate change. At present, however, the accurate dating of palaeohydrological changes is hindered by the scarcity of adequate terrestrial organic matter in arid areas and by the contamination of old carbon (reservoir effect).

The naming of older (non-historic) lakes is distinguished from modern lakes. There are several conventions that are used. The deep-lake phase is commonly referred to as a megalake, as in Lake Mega-Chad or mega Lake Chad. The term palaeo-may also be used to name an older lake, as in the instance of Lake Palaeo-Baijan, China. In some cases, a flip-flop of the name occurs as in Owens Lake (modern) and Lake Owens (Pleistocene deep lake). Some lakes have entirely different names for their older counterparts – for example, Mono Lake (modern) and Lake Russell (Pleistocene deep lake), and the Dead Sea (modern) and Lake Lisan (older deep-lake phase). Various phases of the older lakes are referred to as *stands*.

Lake studies have used geomorphological (shorelines, beach ridges, geochemistry, stable isotope), archaeological, biostratigraphical (pollen; changes in the species assemblages of diatoms, ostracods, and mollusks; and aquatic plants), and historical records, and multi-proxy approaches to reconstruct palaeohydrological and palaeoenvironmental changes, wherever possible using radiometric methods to establish chronologies. Biostratigraphical and geochemical indicators result because changes in lake volume associated with a change in the climatic/hydrologic regime affect not only lake levels, but water chemistry and aquatic biota as well.

The most important source of information is lake cores. Changes in water salinity are indicators of relative water depth. Qualitative salinity estimates (fresh, brackish, saline) can be made based on sediment mineralogy and geochemistry, as well as from biotic assemblages. As the evaporative concentration of lake water increases, there is a shift in the mineralogy and geochemistry of the sediments. With increasing water salinity, the mineralogy may change in the following sequence: from carbonate, borate, mirabilite, and nitrate, through halite. This reflects the relative solubility of each salt (least through greatest).

Tectonically subsiding basins of the western USA provide a rich storehouse of lacustrine sediments. Alternating clastic sediments and evaporites of different chemical composition are sensitive records of aridity and changes in water inflow. Badwater Basin in Death Valley is underlain by up to 3000 m of unconsolidated material, including nearly 300 m of halite-bearing sediment. Analysis of a core extending 180 m in depth, with sediments up to 200,000 years in age, supports the contention that glacial periods in the American southwest were wetter than interglacials (Roberts et al. 1997).

Shorelines provide evidence of lake elevations and tectonic changes within a basin. Shoreline evidence includes wave-cut shorelines (shore terraces) (Fig. 1.5), tufa deposits, and barrier beach ridges (Fig. 5.6). These landforms are relatively common and have been intensively studied in the southwestern USA. Nonetheless, difficulties in dating such material, erosional modification of landforms, and regional tectonic deformation have led to numerous problems in the interpretation of lake extent, interconnections, and timing so that, despite a considerable body of research, many controversies persist (see, for example, Hooke 1999, 2002; Enzel et al. 2002). In areas of lower relief, such as Australia, the playas

have gently sloping margins and variations in lake volume cause fluctuations in areal extent. Rather than the shoreline terraces of the USA, evidence of lake stages largely takes the form of playa sediments and evaporites beyond the present limits of flooding (Kotwicki & Allan 1998).

The shorelines of large lakes are fringed by beach deposits that commonly show signs of reworking. Beach ridges, formed of rounded and sorted cobbles, not only indicate the presence of lakes, but the size and distribution of shoreline materials can be used to estimate palaeowind direction and velocity (Fig. 5.6). Wind transfers energy to the lake and waves are formed which entrain gravel, cobbles, and boulders. Orme & Orme (1991) and Adams (2003) inferred past wind velocities from barrier beach remnants preserved around the perimeters of pluvial Lake Owens and Lahontan, respectively. These studies suggest that the Pleistocene Great Basin was a windier place than at present. Such attempts to quantify palaeowind conditions at multiple locations may ultimately provide information on atmospheric pressure gradients, storm trajectory, and the vigor of atmospheric circulation. The lithology of the materials provides information on longshore currents. Older materials may be reworked into younger deposits, as shown during a recent highstand of the Great Salt Lake, Utah (Adams 2003).

Whereas most areas of the world show evidence of palaeolakes of much greater extent than modern

Fig. 5.6 Beach deposits may fringe the shorelines of desert lakes. This Pleistocene beach ridge, formed of rounded and sorted cobbles, is located on the eastern shore of Owens Lake, California, USA.

lakes, the highstands of these lakes were not necessarily synchronous from region to region. As a generalization, during the LGM, lakes in the intertropical zone were at a lowstand, probably because the atmosphere contains less moisture under colder conditions, whereas those of the southwestern North America were at a highstand. The cause of the highstand in North America has been the cause of some debate, centered around the degree to which lakes rose in response to depressed evaporation rates or to increases in precipitation (as oceanic moisture carried in by westerly flow was diverted southward around continental glaciers).

5.3 LAKES OF THE GLOBAL ARID ENVIRONMENT

The following sections address in more detail the distribution, extent, age, and unique characteristics of desert lakes, focusing on specific regions of the world.

5.3.1 WESTERN NORTH AMERICA

Western North America was the site of several hundred lakes of Pleistocene age, their presence favored by a basin and range topography (Fig. 5.7). The considerable scrutiny applied to these lakes suggests climatic oscillations on millennial timescales in the western USA during the last glaciation; however, as previously discussed, the correlation of lake histories with climatic oscillations is limited by uncertainties in age control and by complications imposed by tectonic effects. Russell (1885) and Gilbert (1890) suggested a close relationship between glacial moraines and high lake levels. Recent work suggests that this correlation may not be so simple, although many lake basins that dried out during the Holocene were occupied during the Pleistocene. The behavior of neighboring lake systems was often asynchronous owing to minor differences in the forcing factors, including the tracks of storm systems, the contributions from local glaciers and ground ice, and the connections between different lakes. The degree to which lakes were a function of increased cloudiness, decreased temperature, and decreased evaporation, rather than increased precipitation, remains an area of active inquiry (Menking et al. 2004). Antevs (1938) postulated that high pressure developed over the Laurentide ice sheet of North America and caused the jet stream and associated storm systems to be deflected

to the south. This idea is supported by general circulation models of the atmosphere (Kutzbach & Wright 1985; Manabe & Broccoli 1985; Kutzbach & Guetter 1986), which indicate that during the LGM ($\approx$21–18 ka), zonal flow occurred during glacial periods, bringing rainfall to the American southwest.

The largest Pleistocene lake in the Basin and Range Province was Lake Bonneville (Fig. 5.7) in northwest Utah, which preceded the modern Great Salt Lake. A 307 m core indicates that lacustrine events extend back over the past 3 million years, with intervals of deep lakes and desiccated surfaces. The basin has well-developed and pronounced shorelines, which have been the subject of longstanding research, and a complex clastic and evaporite stratigraphy. The lake was at moderate levels or was dry at 35,000 years BP, but rose in an oscillatory manner to achieve highstands between about 15,000 and 13,500 years BP (Benson et al. 1990). It reached an altitude of 1551 m ASL, was 372 m deep, and covered 51,700 km^2 (Currey et al. 1983), spilling to the Snake River and flowing to the Pacific Ocean. It maintained this level for a thousand years or more, cutting the Bonneville shoreline. Around 14.5 ka, the failure of weak deposits in the outlet channel caused 1 million m^3 s^{-1} of water to be catastrophically released into the Snake River, rapidly lowering the lake. This new level was maintained on the bedrock-controlled outlet at 1448 m, and over the next millennium formed the Provo shoreline. After 13 ka, Lake Bonneville declined in area as water evaporated. During the Younger Dryas interval, the lake rose again briefly to form the Gilbert shoreline. Subsequent to this, the lake fluctuated and lowered in elevation, reaching its lowest elevation (1278 m) in 1963 (Currey et al. 1983).

The coalescence of the water in seven basins in northwest Nevada formed Lake Lahontan (Fig. 5.7), which covered an area of 22,300 km^2 (Benson & Thompson 1987). At its highstand, the lake reached a maximum depth of 276 m at Pyramid Lake. The basins were linked by sills and fed by several rivers including the Carson, Truckee, and Walker from the Sierra Nevada, and the Humboldt, which emanated from northeast Nevada. Individual lakes, such as Pyramid Lake, formed in each sub-basin and these persisted even as the larger lake diminished in size. Lake Lahontan experienced numerous lacustral and interlacustral events during the Pleistocene, but did not spill over to the sea. As Lakes Bonneville and Lahontan lay at approximately the same latitudes (between about 38 and 42°N), they would be

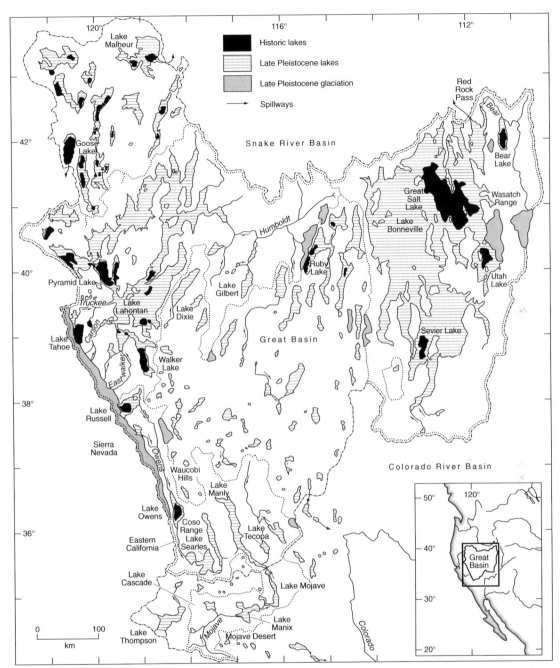

FIG. 5.7 Western North America was the site of several hundred Pleistocene lakes, including massive Lakes Bonneville and Lahontan. From Orme, A.R. (2002) The Pleistocene legacy: beyond the ice front. In: Orme, A.R. (ed.), *The Physical Geography of North America.* With permission from Oxford University Press. Developed from Snyder et al. 1964; Currey et al. 1983; Benson et al. 1990.

expected to show similar timings in the magnitude and nature of lake oscillations. Given the uncertainties of dating and the different elevation controls for both lakes, they show fairly good covariance in lake-level variation for several of the time frames, sug-

gesting that both responded nearly synchronously to the position of the polar jet stream (Benson et al. 1995). Lake Lahontan rose sharply about 26,000 years BP and was maintained at an elevation of 1265 m by overflow to the Carson Desert. It reached its highstand (1335 m) about 15,000 years BP, and began falling rapidly between 14,000 and 13,000 years BP. The recessions of both Lakes Lahontan and Bonneville were interrupted beginning around 11,500 years BP and temporarily stabilized for approximately 2000 years. These hydrologic responses, and associated glacial advances in the Sierra Nevada, were concurrent with the Allerød/Younger Dryas climatic intervals (Benson et al. 1992). By approximately 10 ka, Lake Lahontan existed as seven individual lakes, of which only Pyramid Lake persisted through the Holocene.

The formerly higher lake levels of Pyramid Lake, Nevada, and Mono and Searles Basins, California, are indicated by spectacular tufa deposits (Fig. 5.8). Tufa is composed of calcium carbonate that forms at the mouth of springs as calcium from spring water combines with carbonate dissolved in lake water to form mounds or pillars. The lake level limits their vertical growth and the elevations of many tufa mounds correspond to the elevations of sills. Radiocarbon ages suggest that most of the tufas at Pyramid Lake formed between 26,000 and 13,000 years ago (Benson 2004).

During the late Pleistocene, a chain of lakes in eastern California formed an interconnected cascading system along the Owens, Mojave, and Amargosa Rivers (Fig. 5.7). This is well documented, although many of the linkages have been disrupted by the active tectonism that characterizes the area. Whereas the basins associated with Lakes Bonneville and Lahontan once formed a single continuous lake, those of eastern California were connected like widely separated beads on a string. Each lake had to fill to a threshold before spilling downstream towards the next lower lake in the system. If lakes in the upper part of the chain did not fill, then the lower lakes responded only to local rainfall and runoff events. The Owens River system commenced at Lakes Russell (called Mono Lake today) and Adobe, and flowed southward into Lakes Long Valley, Owens, Searles, Panamint, and possibly Manly (Death Valley). Drainage from the San Bernardino Mountains (east of Los Angeles) flowed eastward into the Mojave River and linked Lakes Manix and Mojave with the Amargosa drainage through the Pahrump Valley and Tecopa Basin to Death Valley. Owing to the complex hydrologic conditions, lakes did not fill or desiccate simultaneously, and the lower lakes had shorter freshwater episodes and were characterized by greater evaporite deposition than higher lakes (Orme 2002).

There is abundant evidence for pre-Pleistocene and Pleistocene lakes in eastern California. A 323 m core from Owens Lake penetrated the Brunhes magnetic reversal and the 760,000 year-old Bishop ash

Fig. 5.8 Spectacular tufa towers in Searles Basin, California, USA, up to 36 m in height, indicate formerly high lake levels. Tufa forms at the mouths of springs as calcium from spring water combines with carbonate dissolved in lake water to form pillars. The towers can only grow as high as the lake is deep. Searle's Lake was fed by the Owens River system, and well-developed shorelines and tufa deposits around the lake margins also indicate its greater extent in the late Pleistocene.

(Smith & Bischoff 1997). The upper 693 m of a core from Searles Lake revealed climatic oscillations over the past 3.2 million years (G.I. Smith 1984). Flow along the Amargosa River from 3 Ma to 160 ka sustained a deep lake in the Tecopa Basin, which attracted elephant, horse, camel, llama, and flamingo (Morrison 1991). Lake Manix along the Mojave River attracted a similar rich mammalian fauna (Jefferson 1991).

The uplift of the Sierra Nevada subsequently increased the rainshadow effect to those areas lying to its east. Lake levels were relatively low between 120 and 30 ka, rose to attain a highstand between 22 and 18 ka, fell slightly between 18 and 14 ka, and oscillated between 14 and 10 ka, before declining in the Holocene. Owens Lake overflowed for most of the period between 52 and 15 ka (Benson et al. 1996). Downstream, the complex stratigraphy and tufas of different ages suggest that Searles Lake must have filled more than once from discharges from the Pleistocene Owens River. Lake Manly in Death Valley (Fig. 1.5) probably depended on local precipitation and runoff from the Amargosa River (Yang et al. 1999).

Today, Mono Lake still retains water, although at a much lower level than during the Pleistocene. Owens Lake contained water until 1912, when abstraction of the Owens River inflow by the city of Los Angeles caused it to dry out. Modern playas are dry except during exceptional rain events, such as the 2004–5 winter season which filled many lakes throughout eastern California and Nevada (Fig. 4.1).

The relationships between various climatic factors in the formation of Lake Estancia, New Mexico, which achieved a 1100 km^2 highstand, are controversial. Antevs (1935) concluded that a precipitation increase was required to form a lake of this size, but determined that the highstand was not synchronous with the LGM. Leopold (1951a) proposed that the lake rose at the LGM in response to a 50% increase in precipitation and a 9°C decrease in summer temperature; whereas Galloway (1970) called for a 14% reduction in precipitation and a 10–11°C annual temperature decrease; and Brakenridge (1978) suggested that the lake would be sustained at its maximum level without a precipitation change, but with a 7–8°C decrease in temperature. These studies examined relations among rainfall, runoff, temperature, and evaporation, but did not consider groundwater contributions, soil and vegetation properties, solar radiation, wind speed, or relative humidity.

Menking et al. (2004) incorporated these latter elements and used an array of models to suggest that precipitation during the LGM may have been double that of modern values. Their model suggests a substantial increase in surface runoff: an idea that is supported by the presence of a centripetal network of currently underfit stream channels. Modern analysis of dated shoreline deposits and stratigraphic units in basin-center deposits indicates that the lake reached maximum highstands at least five times during the LGM (20,000–15,000 years BP) and desiccated completely after 12,000 years BP.

5.3.2 SOUTH AMERICA

The drylands of South America are characterized by a wide diversity of aquatic environments. Lakes are found principally in the altiplano of the central Andes and the pampas of Argentina. They are affected by the north–south orientation of the topography, the considerable relief, volcanic activity, and very strong winds. In northern Chile, for example, there is intensive block faulting due to the subduction of the Nazca Plate beneath the South American Plate, creating a succession of north–south-trending basins and ranges that contain many saline lakes and salt crusts (Risacher et al. 2003).

South American dryland lakes have both economic and biological significance. In the Atacama Desert of northern Chile, playa groundwater and brines are a significant source of water for the mining industry. Playas provide an important habitat for flamingoes and other fauna (Kampf et al. 2005).

Most lacustrine research has concentrated on the largest of the playas, the Salar de Uyuni, in the Bolivian Altiplano, and the Salar de Atacama, Chile. The Salar de Uyuni (Fig. 5.9), located at an elevation of 3653 m, is 9000–10,000 km^2 in area, forming the largest salt flat in the world (Svendsen 2003).

The Altiplano of the central Andes of Bolivia, Chile, and Argentina contains a large number of closed basins within which lie saline lakes and salt crusts, collectively referred to as salars. Northern Chile has remained arid or semiarid since the Miocene, and saline lakes have occupied many basins since that time. During the Pleistocene, several large and deep saline lakes covered the Altiplano, and it is likely that neighboring Chilean basins also contained lakes. In the Salar de Uyuni, a 121 m-deep borehole revealed a succession of cyclic salt crusts and mud layers, representing periodic lowstands characterized

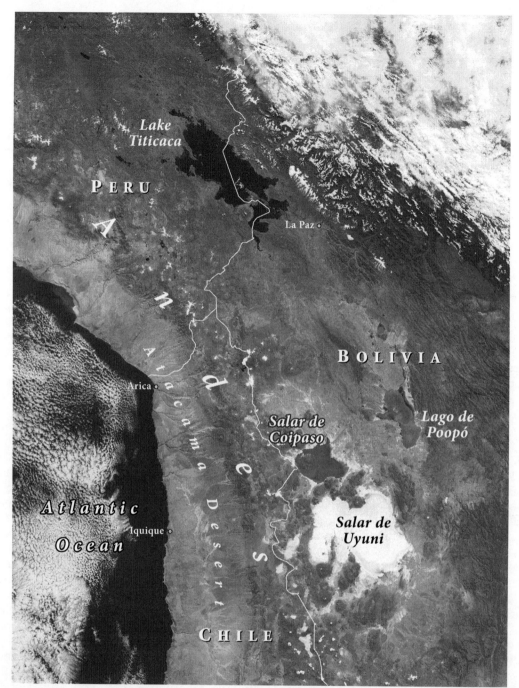

FIG. 5.9 Lake systems of the Altiplano of Peru and Bolivia, South America. Lake Titicaca (dark lake) is the largest high-altitude lake in the world. Part of a cascading system, it drains southward into shallow Lake Poopó (Lago de Poopó; greenish-colored salt-water lake; see Plate 5.9 for color) and the salars (salt-crusted lakes) of Coipaso (also known as Coipasa) and Uyuni. The Salar de Uyuni (white-toned surface) is the largest salt flat in the world. The Atacama Desert is located along the coastal plain to the west of the Altiplano. Source: MODIS Land Rapid Response Team at NASA/GSFC. See Plate 5.9 for a color version of this image.

by desiccation. A volcanic ash layer in mud at approximately 45 m depth has an $^{40}Ar/^{39}Ar$ age of $191,000 \pm 5000$ years BP (Fornari et al. 2001).

Lake levels in the Altiplano have responded over timescales of centuries to millennia to orbitally induced changes in solar insolation and, on an interannual scale, to ENSO variability that affects the moisture supply from the Atlantic-Amazon region (Rowe et al. 2002). Lake Titicaca, the largest high-altitude lake (8562 km^2, 3810 m ASL) in the world, is part of a cascading system, draining southward via the Rio Desaguadero into shallow Lake Poopó and the salars of Coipasa and Uyuni (Fig. 5.9). When the water drops below 3804 m ASL it becomes a closed basin. During the period 20,000–13,500 years BP, the lake level was somewhat higher than today and water was fresh. There was an oscillating regression between 13,000 and 7800 years BP, and by the mid-Holocene (7300–3600 years BP) the lake level was 85 m below the present level. More positive moisture conditions and higher lake levels (approaching the post-glacial maximum highstand) have characterized the late Holocene and the past 3600 years (Rowe et al. 2002).

There are several categories of lakes in the Altiplano: saline lakes, playas, and active and inactive salt crusts (Risacher et al. 2003). Saline lakes are permanent bodies of saline water, in which water occupies the entire basin. At maximum, they are only a few meters deep. Playas (in this region) are defined as shallow (<20 cm) and ephemeral saline pools that occupy localized areas in the central basin depression. There are two types of salt crusts found in the region: active (modern) and inactive (fossil). The capillary draw on subsurface aquifers causes a variety of salts to be precipitated within playa sediments, including gypsum, mirabilite, halite, and ulexite. Thin (<50 cm) active crusts of gypsum, mirabilite, or halite are forming at present from evaporating brines. Fossil salt crusts are found in the more arid areas and share a similar mineralogy to the active crusts, but are much thicker. The center of the Salar de Atacama contains a very rough efflorescent halite crust, with cavity-rich salt towers, partially darkened by windborne sediments. Roughness heights range from 15 cm to nearly 1 m. Such crusts form where the playa surface remains continually dry and is not subject to inundations. As the basin rainfall averages only 20 mm year^{-1}, the primary inflow to the salar occurs via groundwater flow, which in the central part of the basin is highly saline, with water densities exceeding 1.2 g cm^{-3}. The depth to groundwater in this area is 0.8–0.9 m (Kampf et al. 2005).

The halite crust of the Salar de Atacama is several hundred meters thick, but crusts on other salars range in depth from meters to tens of meters. For example, the crust of the Salar de Uyuni is about 10 m thick and covers a large body of subsurface brine (Svendsen 2003). Some crusts show dissolution pits, and tilted layers, evidence of their inactive status. Many salars combine both playas and salt crusts within one body. Water inflows to salars are from localized spring discharge, diffuse seeps that may extend laterally along tens or hundreds of meters of the shore, low-discharge streams (generally less than 500 L s^{-1}), and the discharge of subsurface aquifers in the central trough of the basin.

Salts within lake basins in the Altiplano have a complex origin and are affected by proximity to the ocean, volcanism, and groundwater flow. Risacher et al. (2003) examined the origin of salts and brines in 52 salars along a 1000 km north–south transect in the Andes Cordillera. They identified three main sources: (1) the westerly winds, which bring atmospheric inputs of sea salt and desert dust from the Pacific Ocean and Atacama Desert, (2) the weathering of volcanic rocks, which contributes the largest source of salts in inflow waters, and (3) brine recycling – the mixing of dilute meteoric waters with present lake brines – which is the main source of salts in most salars. In fractured volcanic material, brines may be transported at depth to adjacent basins or even further.

Aeolian deflation appears to play a role in the formation of the playas of the Altiplano, as Space Shuttle astronauts were able to photograph point-source dust storms rising from pan surfaces. The plumes appear to rise preferentially from evaporite zones on the perimeter of the playas, suggesting that salt weathering plays an important role in deflation. Goudie & Wells (1995) propose that the playa floors of several salars may have been lowered by tens of meters by this process. At the altitude of the Altiplano, wind velocities exceed 25 m s^{-1} and are locally accelerated by stratovolcanoes and changes in slope. The Potrerillos Depression (291 km^2 and 800–900 m deep) may have had as much as 160 km^3 of material removed by wind erosion. The rim of the depression is marked by wind-fluted surfaces and yardangs, indicative of the role of aeolian processes (Goudie & Wells 1995).

On the pampas, Argentina has a large number of pans developed principally on unconsolidated

Quaternary sediments with densities of up to 90 depressions per 100 km². They include small basins that lack apparent orientation, larger pans that form in a line and are probably related to the deflation of river channel sediments, and even larger basins (diameters of several kilometers) that are characterized by a "pork chop" morphology, with a crenulate windward side, a bulbous lee side, and a long-axis orientation that is transverse to the dominant winds. Lunettes to the lee of the pans are formed of clay pellets and indicate the role of deflation in basin formation. The largest of these dunes reaches a height of 43 m (Goudie & Wells 1995). Some of the pans are found in areas that are no longer arid, and it is likely that the basins were excavated during dry glacial conditions or during dry periods within the Holocene.

5.3.3 AUSTRALIA

By contrast to the lakes of South America and southwestern North America, many of which are confined to distinct intermontane structures, the lakes of Australia are considered a manifestation of the "disorganization" of desert drainage under aridity in a landscape of shields and structural platforms, with streams terminating in inland basins or dying out at intermediate points on catchment slopes (Mabbutt 1977a). Nonetheless, structure and relief play a role in the distribution of the lakes. In the Yilgarn Plateau of Western Australia, a shield desert, "river lakes" formed as sectors are isolated along broad trunk valleys. The larger lake basins of the interior lowlands, such as Lake Eyre (Figs 5.10 and 5.11) and Lake Torrens (Fig. 5.11), occupy structural depressions with centripetal drainages. Small claypans form in interdunal depressions in the Great Sandy, Simpson, and Strzelecki Deserts (Goudie & Wells 1995). Lakes form part of an integrated desert lacustrine landscape, which also includes aeolian features spawned from lake sediments, and palaeolake features that extend beyond the present lake. Australia is the type location for lunette dunes, which form to the lee of lakes, and are discussed in more detail in Chapter 8.

Lakes play a vital role in Australia's arid ecosystems, with thousands of temporary wetlands ranging in area from a few square meters to thousands of square kilometers. They include: (1) terminal water bodies filled by major drainage systems (for example, Lake Eyre), (2) lakes filled by local drainage (e.g.

FIG. 5.10 Lake Eyre is a 9600 km² playa at the lowest point (12–16 m BSL) in Australia. The low-gradient basin surface includes stoney (gibber) plains and continental dune fields (see upper right of image). False-color composite image acquired by Landsat 7 on 29 July 1999. Image from the Goddard Space Flight Center's Landsat Team and the Australian ground receiving station teams. See Plate 5.10 for a color version of this image.

Lakes Gairdner and Torrens), and (3) overflow lakes and floodplains associated with flood events, as along Darling and Cooper Creeks (Roshier et al. 2002). Many of these lakes are inundated on a highly infrequent basis (Roshier et al. 2002). Lake Torrens, the second largest salina in the world, is only known to have completely filled in 1878 and 1989 (Kotwicki & Allan 1998). Events on this scale are almost always associated with cyclone activity.

Lake Eyre is the largest playa (9600 km²) and lowest point (12–15 m BSL) in Australia and forms the depocenter of a 1.14×10^6 km² basin whose low-gradient surface incorporates the continental dune fields of the Simpson and Strzelecki Deserts, gibber plains, and bedrock (Fig. 5.10) (Nanson & Price 1998). The basin covers one-seventh of the continent and spans 14° of latitude, with the northern part extending into the zone of tropical summer rainfall and the southern into the temperate winter rainfall zone (Figs 2.14 and 5.11). The basin falls within the zone of Australia's lowest rainfall (≈ 125 mm year^{-1}) and its highest annual evaporation (≈ 3800 mm) (Nanson & Price 1998). In the north, high evaporation rates and warm temperatures offset the higher rainfall contributions: conversely, the south has lower rainfall, cooler temperatures, and less evaporation.

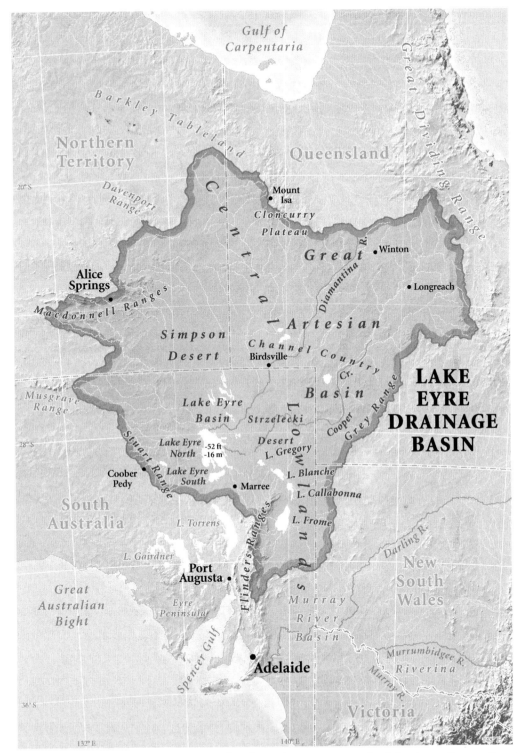

Fig. 5.11 The Lake Eyre basin occupies 1.14×10^6 km², or about one-seventh of the Australian continent. Lake Eyre fills a few times each century when flow from the more humid regions of Queensland to the northeast moves through a maze of channels (the channel country) to reach this terminal basin.

The majority of runoff is derived from the northern Australia monsoon, flowing southward along the Georgina and Diamantina Rivers and Warbuton and Cooper Creeks (total mean annual water yield of 2.40 km³). These rivers suffer enormous transmission losses into the floodplains and aeolian dunes. Rivers from the west and south of the lake (Macumba, Neales, and Frome Rivers, and Margaret and Warriner Creeks) provide a smaller, but nonetheless substantial, water yield (0.72 km³). Many of the drainage patterns were formed under past pluvial regimes and today are disconnected relics.

The Lake Eyre basin has the lowest mean annual runoff of any major drainage basin in the world, 3.5 mm in depth (Kotwicki & Allan 1998), and sediment transport is correspondingly low. The Diamantina River and Cooper Creek have low floodwater flow velocities and long flood transmission times. A flood event from either Cooper Creek, the Diamantina River, or the western tributaries may cause a minor lake filling, but a combination of two or all three is required for an extensive filling (Kotwicki & Allan 1998). The relative contribution of each sub-basin appears to change over the short term, and the long-term inputs of each are poorly understood. Many flash floods are not be detected except by passing pilots, as there is little ground hydrometeorologic equipment in the basin and the roads are inaccessible in flood periods. The first recorded filling of Lake Eyre was in 1949: up to this time, the existence of water in this very arid region was reported, but the evidence was often dismissed. Thus, the historic palaeohydrologic record is short and incomplete.

Flood events appear to be strongly related to ENSO phenomena, with most major events associated with La Niña phases (Kotwicki & Allan 1998). During such periods there is enhanced austral summer monsoon activity and incursions of the monsoonal system into central Australia (Allan 1988). Lake Eyre appears to flood about once every 8 years (80% lake coverage), with major events occurring every 20 years. The 1974 event reached a level of −9.5 m Australian Height Datum (AHD) and was considered a one in 100 year event (Kotwicki & Allan 1998). Four major floods occurred in this basin between 1986 and 1997, with as much as 1.8 million ha of surface water present in the upper basin in March 1991. This quickly declined to 160,000 ha within 6 months, and then continued to decline over the next 5 years (Roshier et al. 2002).

Pleistocene Lake Eyre is referred to as Lake Dieri. Aspects of its palaeohydrology remain controversial.

While studies by Nanson et al. (1998) and Magee and Miller (1998) both agree that the lake was full at times during Marine Oxygen Isotope Stage 5 (≈130–120 ka), the former argue that during the LGM the lake level was high and fed by tropical rainfall, and the latter that there was pronounced aridity. Stranded high beaches tens of meters above the present floor of the playa were thermoluminescence dated at around 22,000 and 26,000 years BP, near the beginning of the LGM (Nanson et al. 1998), suggesting high water levels at that time. However, this picture is at odds with the general view of glacial age aridity in Australia, a period during which dunes were built under colder, drier, and windier conditions. By contrast to North American lakes, whose shorelines are difficult to interpret owing to recent tectonic displacements, the Lake Eyre basin is believed to have remained relatively stable throughout the Quaternary, with coeval beaches as old as Stage 5 distributed widely about the basin at the same topographic level (Nanson et al. 1998).

Temporary wetland features, including both lakes and inundated floodplains, are critical for mobile waterbirds that move frequently to seek feeding and breeding habitat. Indeed, arid Australia supports extraordinary numbers of waterbirds as a result of these temporary habitats. Paradoxically, one of the driest areas in the continent, the Lake Eyre Basin, has the highest habitat availability for bird species with large dispersal capabilities for which the general aridity of inland Australia is not a barrier to movement. The wetlands of this basin are interconnected and linked by broad pathways to wetter parts of southeastern Australia. Along dryland river courses, the primary productivity increases by two orders of magnitude following floodplain inundation (Roshier et al. 2002). Waterbirds such as the grey teal (*Anas gracilis*) take advantage of the changing spatial distribution of water, moving 180 km day⁻¹ and covering up to 3200 km of straight-line distance in a few weeks.

5.3.4 AFRICA

Dry lakes are found in both the northern and southern arid belts of Africa.

In southern Africa, pans are most frequently found in areas that receive less than 500 mm of rain annually and have an average net evaporation loss of more than 1000 mm. Within the general arid area, there are certain areas with notable concentrations of pans. Dry lakes are found in southern Africa, Namibia, central Mozambique, Zimbabwe, and western Zambia. Their locations and densities are reviewed in detail in Goudie and Wells (1995).

The largest of the southern African pans is the Etosha Pan (Figs 1.3 and 5.12). A massive saline pan 130 km long by about 50 km wide, it is located in the semiarid north of Namibia. Inflow is largely from an extensive network of ephemeral channels located to the north of the pan. Water tables are generally close to the pan surface and springs found along the southeastern pan margin provide a water source for wildlife. Rainfall varies from 400 to 450 mm, with 2500 mm year^{-1} potential evaporation. Satellite data suggest that during periods of inundation, atmospheric dust loadings from this area are reduced (Mahowald et al. 2003). These temporary fillings attract thousands of water birds, including large flocks of flamingos.

In northern Africa, depressions are found on the Moroccan coastal plain and as large depressions in the Western Desert of Egypt. The Zone of Chotts (Fig. 5.13) is a large region of ephemeral lakes stretching from the lowlands of southern Tunisia to the Atlas Mountains of northern Algeria. *Chott* is an Arabic term for a dry, salt depression. Most of the chotts are terminal discharge systems, and groundwater tables are near to the surface for much of the year for those playas that are at or below sea level, and below the surface for those at higher altitudes. Seepage into the playa floors probably recharges the Bas Sahara Basin, a large aquifer system in the northern Sahara. Mean annual rainfall for the region is 90–120 mm, with most of the rain falling from November to February. Mean annual evaporation is 2500–3100 mm (Mahowald et al. 2003). The flooding ratio is the percentage of time that a playa is inundated in any one year, which may be the result of multiple events or one single large event. Chotts studied by Bryant and Rainey (2002) had ratios that ranged from around 10 to 41%. Recorded inundations all occurred within the winter period. Important periods of lake fillings appear to coincide with periods of strong positive winter NAO values.

The Fezzan region of southwest Libya, in the hyperarid Sahara Desert, is the site of three types of palaeolakes: (1) large open bodies of fresh water, (2) interdune lakes, and (3) playas. Shoreline features

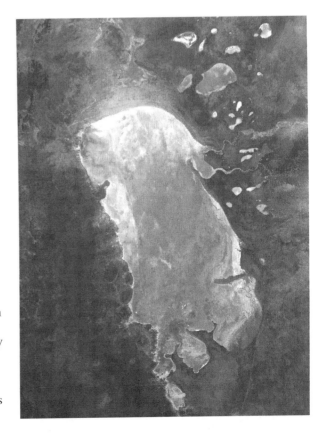

FIG. 5.12 The Etosha Pan is the largest of the southern African pans. Located in northern Namibia, this saline playa is 130 km long and 50 km wide. Inflow is largely from summer storms that feed a series of ephemeral channels that can be seen to the north of the playa. During the summer months, the water attracts considerable wildlife. During the dry season, the pan is a major source of dust. Source: NASA MODIS image.

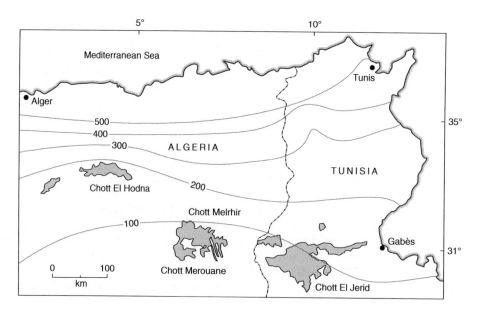

Fig. 5.13 The Zone of Chotts in northern Africa. The contours indicate rainfall in millimeters. From Bryant, R.G. and Rainey, M.P. (2002) Investigation of flood inundation on playas within the Zone of Chotts, using a time-series of AVHRR. *Remote Sensing of Environment* **82**, 360–75. By permission of Elsevier.

and sequences of lake sediments, which include dark layers of organic material and mollusk shells, attest to large palaeolakes. The shells were uranium-thorium dated to around 85,000 years ago (White et al. 2000). Other organic layers situated at higher elevations indicate that an earlier, deeper, and more extensive lake existed, although the timing of this lake is not yet known. Interdunal lakes of the Ubari Sand Sea are rich in lithics such as hand axes, pottery, grinding stones, and ostrich eggs, indicative of Pleistocene and Holocene human occupation of the area. The claypan playas, whose surfaces vary in the degree of surface roughness and salt deposition, probably represent the last phase of drying out of large Holocene water bodies, with final desiccation occurring around 3000 years ago.

Lake Chad is a region of internal drainage in the Sahel region of north central Africa, bordering Chad, Cameroon, Niger, and Nigeria. The present lake varies in elevation between 275 and 280 m ASL (Schuster et al. 2003). The climate is semiarid to the south of the lake, where monsoonal rain falls during June, July, and August, and arid to the north. The lake is divided into two sub-basins: the northern one is currently dry, and the southern one is fed by the Logone and Chari Rivers, which emanate from wetter regions to the south. These rivers provide approximately 90% of the lake's inflows (Birkett 2000). The northern edge of the Lake Chad basin is dotted by small perennial interdune lakes and playas,

some of which are saline. This zone provides some net recharge to the groundwater table, with flow towards Lake Chad (Edmunds et al. 1999). Water losses from the lake are via groundwater seepage, which accounts for as much as 20% of the loss, and evaporation, which accounts for the remainder. Evaporation rates are high owing to the low humidity, high temperatures, and shallow lake depths (a mean of 5 m). The surface area of the lake has fluctuated dramatically, both in the short and long terms. In the middle Holocene (about 9 ka), Lake Mega-Chad was as large as the present day Caspian Sea (350,000 km²).

During historic times, Lake Chad has occupied less than 1% of its drainage area. Over the past 40 years, lake levels and surface area have declined as a result of changes in the regional precipitation pattern and human impact. After a period of high inflows in the 1960s, the lake occupied an area of 23,500 km². By the 1990s, approximately half of the historic volume of flow entered the lake from the Chari and Logone Rivers and its surface area had shrunk to about 1350 km², so that the lake only occupied a small area in the south of its original basin (Walling 1996; Coe & Foley 2001). Upstream irrigation abstractions accounted for much of the loss of inflow (Birkett 2000). Coe and Foley (2001) suggest that climate variability still controls the interannual fluctuations of the water inflow, but that human water use accounts for roughly 50% of the observed decrease in lake area since the 1960s and 1970s. In late 1999, Lake Chad expanded in size again, with negative consequences for some lakeshore communities. The inherent high variability in rainfall and lake level engenders adaptive livelihood strategies, including multiple sources of income (from farming, fishing, or trade) and mobility, both seasonally and over the longer term (Evans & Mohieldeen 2002). Satellites provide valuable data with which to monitor the dynamics of the lake and allow for water management. During the 1990s, accurate variations in the lake surface level and area were derived using data from the TOPEX/POSEIDON satellite, and in the future similar assessments may allow improved flood and drought forecasting (Birkett 2000).

The genesis of the noteworthy depressions of the Western Desert of Egypt (Siwa, Qattara, Baharia, Farafra, Dakhla, and Khargha) has been subject to several interpretations (Figs 2.3 and 5.14). Several of these basins have floors that lie below sea level (e.g. Khargha −18 m and Qattara −143 m). Depressions occur along the boundaries of northward-dipping strata and are bounded to the north by escarpments. To the south, the valley floors rise gently. The depressions were initially cited as textbook examples of deflation, with the water table forming a base level (Ball 1927). Knetsch and Yallouze (1955) invoked tectonic action in their models of depression formation. Said (1960) excluded the possibility of a tectonic origin, noting the absence of faults. The role of wind action is suggested by the conformity of depression locations with areas of thinner and more easily breached limestone capping. However, the volume of the Qattara Depression (20,000 km^3) poses problems for the idea of wind action acting alone. Albritton et al. (1990) proposed that it was originally excavated as a stream valley, subsequently modified by karstic activity, and further deepened and extended by mass wasting, deflation, and fluviatile processes. Salt may also play a role in basin evolution, weathering, and preparing materials for deflation. This is a self-enhancing process, for the wetting of low areas allows new salts to form upon drying, whose crystallization weakens sedimentary cements, preparing grains for deflation (Haynes 1982).

5.3.5 ASIA

The best-described Asian lacustrine environments are in China and India, and these provide important evidence of regional climatic fluctuations (Fig. 5.15). Groups of pans occur to the lee of the Ural Moun-

tains and to the north of the Caspian Sea, developed in areas where the rainfall is 450 mm or less. Some have a preferred orientation with their long axes perpendicular to the dominant winds (Goudie & Wells 1995). Deflation basins are also found to the east of the Caspian Sea. The Aral Sea is discussed in detail as a case study in Section 13.5.

Arid conditions and low lake levels characterize the modern climate of a broad extent of Asia, from Central Asia to the Arabian Peninsula. Central Asia, surrounded by high mountains and with a continental climate, is one of the driest areas in the world. Winters are cold and dry, and summers are warm and dry. In northwest Asia and the Middle East, the subtropical anticyclone dominates summer conditions, and cold and dry air flows from the Siberian anticyclone in winter, resulting in an extremely arid climate, with moisture brought in by the westerlies. In southern Asia, including the Arabian Peninsula, the Indian subcontinent, the Mongolian Plateau, and the majority of China, a monsoon climate prevails (Qin & Yu 1998).

The most well understood changes in Asian lake levels are those of the LGM and the mid-Holocene (≈6000 years BP). These two periods show strongly contrasting climatic conditions: the LGM is noteworthy for the extent of the ice sheets in North America and Eurasia, and the development of loess and deserts and low lake stands in East and South Asia, and higher stands in northwest Asia and the Mediterranean; whereas for most areas, the warm conditions of the mid-Holocene were associated with dramatically higher lake stands, and a flourishing coniferous and deciduous broad-leafed mixed forest (Qin & Yu 1998). Holocene lakes in East and South Asia and Central Asia expanded, while those in northern West Asia registered some decline. The high lake levels at 6 ka are attributed to enhanced summer monsoonal activity, and a possible weakening and northward displacement of the winter Asiatic anticyclone. The interpretation of LGM lake levels is complicated because lake levels responded not only to temperature and precipitation, but were the recipients of glacial meltwaters. Rainfall was influenced by the southward shift of the westerlies and a weakening of the summer monsoon (Qin & Yu 1998).

5.3.5.1 China and Mongolia

There are about 276 saline and hypersaline lakes in western China, occurring in regions where evaporation is greater than precipitation and the mean

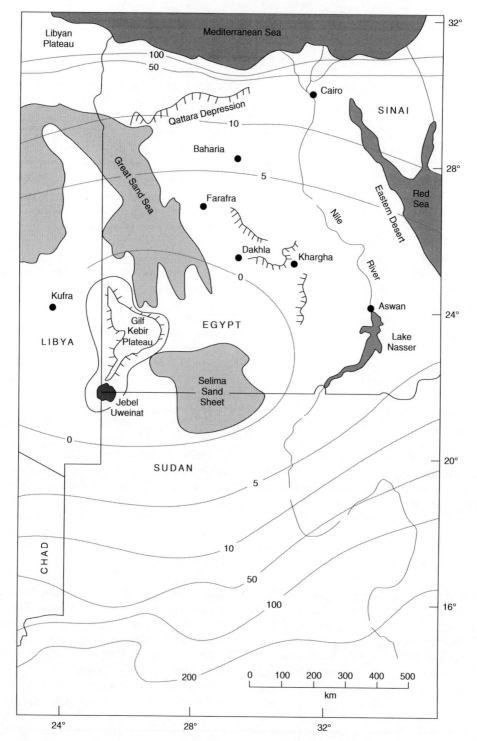

Fig. 5.14 Map of Egypt and northern Sudan showing locations noted in the text. The contours indicate rainfall in millimeters. After Haynes (2001).

FIG. 5.15 Dagze Co (Lake) is an inland lake on the Tibetan Plateau. Fossil shorelines that surround the lake attest to much higher lake levels in the past. Shorelines cross large fan deltas developed to the north and south of the lake. Source: NASA/GSFC/METI/ERSDAC/JAROS, and U.S./Japan ASTER Science Team.

annual rainfall is less than 600 mm (Yu et al. 2001). These saline lakes are found in areas relatively uninfluenced by the Pacific and/or Indian summer monsoons. A compilation of the work of Chinese scientists is presented in Yu et al. (2001). Most of the closed lakes appear to have desiccated in the past 3000 years and the lacustrine sediments have been exposed to aeolian erosion. Associated sand-dune encroachment is thought responsible for the abandonment or burial of many ancient cities or oases, such as Loulan on the western shores of Lop Nor in the Tarim Basin. As relatively little is known of Holocene environmental and climatic change, the degree to which oasis desertification and lake desiccation were the result of human impact remains unresolved. Lake core records provide a key to understanding these changes.

Millennial-scale changes in the strength of the summer and winter monsoons are recorded by a number of palaeoenvironmental records, including lakes. The interval between about 35,000 and 45,000 years BP was characterized by the existence of several mega lakes, including Palaeo-Baijian Lake (16,200 km^2) and Palaeo-Chaerhan Salt Lake (25,000 km^2). After this period, the lakes were considerably reduced in size and often split into hydrologically independent sub-basins (Yu et al. 2001).

Fresh-mesohaline palaeolakes were widespread in the Tengger Desert, developing under warm-humid conditions around 42 ka and reaching a highstand between 35 and 22 ka. Thereafter they experienced a fluctuating decline and disappeared completely by 18 ka (Zhang et al. 2002).

Most lake levels were high during the Holocene optimum, as illustrated by Lake Yiema, a closed lake in the Minqin Basin, which is filled with up to 200 m of Quaternary lacustrine sediment (Chen et al. 1999). Lake Yiema is surrounded by active dunes of the Badain Jaran Desert to the west and the Tengger Desert to the east. The presence of aeolian materials in lake cores suggests that the lake was dry before 14,000 years BP. The lake fluctuated in level between 14 and 10 ka, with silty-clay lacustrine sediments being succeeded by carbonate-enriched sediments and aeolian deposition at about 10 ka, possibly corresponding to the Younger Dryas event. Between 10 and 4.2 ka, the moisture level of the area increased and higher levels of runoff and sedimentation were recorded in the basin. During this period, however, a drought episode with sand deposition occurred at about 7.6 ka. Sand deposition is recorded in the core about once every 400 years, and Chen et al. (1999) believe that the sand storms that characterize present-day western China have been a recurrent feature throughout the Holocene. After 4.2 ka the lake desiccated in a seemingly abrupt manner as the deserts of northwestern China expanded. Within this general desiccating trend there were moister periods at 2.5 ka and within the past 1000 years (undated event). Archaeological findings also support a more humid period at approximately 2000 years ago, including large groups of settlement ruins at Loulan, Niya, and Keriya in the Tarim Basin (Yang et al. 2004). The hydrometeorological changes that are recorded in Lake Yiema appear to mirror those in Africa and Tibet, suggesting a larger regional response.

For the past 2500 years, the size of Lake Yiema has been closely correlated with the human population, suggesting that anthropogenic impact has accelerated the lake desiccation associated with natural climate change (Chen et al. 1999). Human activity in the Minqin basin can be traced back at least 5000 years. The conversion of economic activity in the area from pastureland to irrigated agriculture around 100 BC reduced the inflow of the Shiyang River into the lake. Historical documents show that since that time the total lake and swamp area has fluctuated in inverse proportional to that of population.

Population increases were associated with Han Chinese occupation and irrigated agriculture, and decreases with nomadic (including Mongolian) occupation. Similar conditions of lake decline and human activity are observed in the Hei River basin of the Hexi Corridor, Gansu Province, where 80% of the total available surface water has been diverted to farmland, threatening the existence of terminal lakes. West Juyan Lake had a surface area of 213 km^2 before the 1960s, but today it is a dry playa (Genxu & Guodong 1999).

Lop Nor, in the eastern Tarim Basin, has gained notoriety as the world's sole operational nuclear test site. It is a desiccated salt lake, 50 km in diameter, with a salt crust 30–100 cm thick occupying 35% of the surface. On orbital photography, its form appears as a giant ear, with concentric rings created by shorelines. Lop Nor is the terminal basin of the Tarim River. Precipitation in the immediate area of the lake is less than 20 mm and pollen from a Lop Nor core suggests a desert environment for the entire period of the Holocene (Yang et al. 2004). Deflation of sediments by northeast winds has given rise to a large field of yardangs. Hedin (1903) estimated that 6 m of sediment had been deflated in 1600 years, with an annual erosion rate of 4 mm year^{-1}; whereas Xia (1987) inferred that 5.3 m of surface was eroded between 1919 and 1959, for a rate of 133 mm year^{-1}.

Mongolia has two areas in which lakes are found. The Valley of Big Lakes (Ikh Nuuruud Basin) is located in western Mongolia between the Mongolian Altai and the Khangai Mountains, at an average elevation of between 1000 and 1500 m ASL. Several lakes are over 1000 km^2 in area, including the Uvs Nuur (3350 km^2). Higher shorelines of Uvs Nuur are probably middle Pleistocene in age, although strong tectonic activity complicates the interpretation of lake evolution. The Valley of Gobi Lakes lies to the southeast, and includes smaller salty lakes. A palaeolake of more than 90,000 km^2 occupied this area in the middle Pleistocene (between 280 and 177 ka), probably as a result of enhanced precipitation, meltwater inflows, and decreased evaporation (Yang et al. 2004). Following the LGM, during which many Asian lakes dried up, there was a lacustrine pulse. Several lakes provide evidence for relatively high lake levels as early as 13–12 ka and at 8.5 ka, although this pattern does not appear to be universal. During the late Holocene, around 2–3 ka, there was another more humid period during which lake levels showed small oscillations.

5.3.5.2 India and Pakistan

Closed lake basins in the Great Indian Desert or Thar include *dhands*, deflational features blown out of the noses of parabolic dunes and saline closed basins (*ranns*), an important landform in the region. The origin of the ranns is not entirely clear, with several potential formative mechanisms, including the blockage of streams by dunes, and the formation of depressions associated with strike-slip faulting. The source of salts in the region is the source of some debate. The proposal that the saline lakes are remnants of the Tethys Sea (that existed before the collision of the Indian and Eurasian plates 70 Ma ago) has been discarded on geochemical grounds (Ramesh et al. 1993). Another potential source of lake salts is aeolian transport from the Rann of Kutch, a salt marsh located to the southwest of the desert. Ramesh et al. (1993) suggest that the lake water is of meteoric origin, and the salt derived by rock weathering and transported from the northern part of the desert by palaeorivers that drained to the lakes. At present, the closed-basin lakes receive water during the short monsoon season (June–September) and evaporate during the remainder of the year.

The lakes have helped to shed light on the regional palaeoclimatology. The Indian summer monsoon appears to have been weaker during the glacial maximum and stronger during interglacial conditions. A wet phase was recorded at Lake Didwana during Oxygen Isotope Stage 5 (about 130–120 ka, during the last interglacial). During the LGM the western Rajasthan region was covered by steppe vegetation and hypersaline lake conditions at Didwana suggest a weakened summer monsoon and relatively high winter precipitation (Kajale & Deotare 1997). Other records from the Thar suggest that summer precipitation was higher than today between 10.8 and 10 ka, that a middle-Holocene lake persisted between 7.2 and 5.6 ka, and that desiccation occurred around 5.6–4.8 ka (Enzel et al. 1999; Jain & Tandon 2003).

5.3.6 MIDDLE EAST

The most well known body of water in the Middle East is the Dead Sea, which is the lowest body of water on Earth (400 m BSL). It is fed by the Jordan River, which flows into it from the north. The waters of the Dead Sea are extremely saline, thereby excluding all forms of life except bacteria. Discontinuous plant life is found along the shoreline, consisting

FIG. 5.16 During the Pleistocene, a larger body of water, Lake Lisan, occupied the Dead Sea rift. This photograph shows the former shorelines of Lake Lisan and a sub-lacustrine fan/delta. Photograph courtesy of Tony and Amalie Orme.

principally of halophytes. The Dead Sea and the Sea of Galilee occupy pull-apart basins that were formed by left-lateral slip on major faults.

During the Pleistocene, larger bodies of water occupied the Dead Sea basin (Lake Lisan) (Fig. 5.16) and the basin in which the Sea of Galilee is presently found (Lake Kinneret). A topographic sill at Wadi Malih (altitude of 280 m BSL) separated these two lakes during most of the past 70,000 years. Throughout this period, the two lakes have undergone essentially contemporaneous and synchronous changes in level (Hazan et al. 2005). During a highstand of 164 m BSL at about 26,000–24,000 years BP, the two lakes coalesced into a single body of water. The relatively fast and intense lake level rise at this time must have been associated with a significant climate change, as the lake crossed the sill, which required a substantial increase in the total water input to the lake (Bartov et al. 2002). Lake Lisan began its retreat

about 17–15 ka, and reached its minimum stand about 13–12 ka. During most of the Holocene the level of the Dead Sea fluctuated around 400 m BSL, reaching a minimum stand of less than 420 m BSL between 3900 and 3400 years ago.

Research concerning palaeolakes in other areas of the Middle East is scant. Closed basins are developed to the southwest of the Euphrates plain in Iraq and in northern Saudi Arabia, formed in Pliocene–Cretaceous sedimentary rocks (Goudie & Wells 1995). Two major lake periods have been recorded for the sand sea of the An Nafud area of Saudi Arabia. Between 34,000 and 24,000 years BP, the area was occupied by freshwater lakes extending several square kilometers in area, with depths of about 10 m. Interdunal lakes developed during the Holocene (8400–5400 years BP): these were basically small swamps, fluctuating with the rise and fall of the water table (Schulz & Whitney 1986).

6

WEATHERING PROCESSES AND HILLSLOPE SYSTEMS

6.1 INTRODUCTION

This chapter introduces some of the most contentious topics in arid-land geomorphology, including the efficacy of insolation weathering, the process of salt weathering, the origin of cavernous weathering, the causes of rock varnish and its potential to yield age data for exposed surfaces, and the origin of pediments. Despite studies that stretch back over a century, the nature of weathering and slope formation in arid lands remains poorly understood (Walther 1893; Blackwelder 1925; Hume 1925).

Early weathering studies were influenced by travelers' accounts of heat, cold, and dryness in deserts that, in turn, influenced scientific views on temperature and moisture regimes. Subsequently, some of these views were amended, but even today the results of weathering studies are often equivocal and, in some cases, controversial. Process-driven studies have helped us better understand weathering, but there are several limiting elements to this research. The majority of *in situ* observations are of short duration and areally restricted (Smith 1994). Attention is naturally drawn to those forms that will exhibit measurable change over only a few years. Additionally, many processes, such as insolation weathering, salt crystallization, or freeze–thaw weathering are examined in isolation, often in laboratory-based simulation studies. One pitfall of this type of work is that environmental conditions may be unrealistically simplified and synergistic weathering conditions not considered (Smith et al. 2004). Such short-term microscale studies are valuable, but it is a much larger step to assess the cumulative impact on the landscape over long time periods.

Similar problems and pitfalls plague studies of slope processes. The number of field-based studies remains relatively small and areally restricted and, as such, provides only limited insight into the potential mechanisms of slope formation, the timing of forma-tive events such as rockfalls, and the effect of climate change. As is true for many geomorphic surfaces in deserts, the lack of reliable dating techniques means that the rate of hillslope modification is poorly understood.

In order to develop a more generalized approach to slope formation, Howard (1995) identified three main processes of scarp backwasting – uniform backwasting, fluvial incision, and groundwater sapping – and attempted to infer processes from statistical characteristics of the planimetric shape. Simulation models suggest that scarp backwasting is characterized by broad, shallow reentrants and sharply pointed headlands; fluvial erosion by dendritic drainage patterns with sharp terminations and canyons that gradually widen downstream; and groundwater sapping by canyons of nearly constant width with rounded headwalls. Diagnostic planform features allow a preliminary interpretation of landforms in remote desert areas.

6.2 WEATHERING

The weathering of rocks in deserts is well exposed owing to the sparsity of vegetation and soil. Weathering processes include flaking, spalling, splitting, and granular disintegration. The resultant forms include tafoni, gnammas, and rillenkarren. Weathering processes also strongly influence desert pavement formation and are important in generating sediment that is later transported by fluvial and aeolian processes.

The variables influencing weathering processes in deserts include insolation, salt, dust, and moisture. The relative significance of each of these is uncertain, but the presence of salt has been shown to be a potent force propagating cracks and thereby breaking up both natural rocks and building materials. At higher altitudes, such as in the mountains rimming the Taklimakan Desert, frost weathering is an important source of sediment that is subsequently transported into desert basins.

6.2.1 INSOLATION WEATHERING

The term insolation weathering refers to rock breakdown in response to heating and cooling, which is thought to set up a series of stresses that ultimately results in rupture. The concept of insolation weathering arose from travelers to the desert who reported sounds like pistol shots as rocks cooled in the evening hours. The strong diurnal heating and cooling cycles cause rocks to oscillate between periods of expansion and contraction, resulting in compressive and tensile stresses. Such stresses may cause rock fatigue, particularly on rocks with a westerly or equatorial exposure (Halsey et al. 1998). For any given rock, there may be differences in the rate of thermal expansion between minerals that enhance weathering. The availability of moisture may alter the mechanisms by which insolation weathering operates.

In order to test the hypothesis of insolation weathering, early experimental research subjected rocks to multiple heating and cooling events over many cycles. The dramatic rock breakdown observed in deserts was not replicated (Blackwelder 1925; Griggs 1936), leading many researchers to conclude that insolation weathering is not a viable process. Many of the rocks used in insolation weathering or combined insolation weathering/salt weathering experiments were subject to unnaturally high temperatures (Smith 1994). Roth (1965) extended measurements to the field, gauging temperature and humidity conditions within a quartz monzonite boulder in the Mojave Desert, California, and concluded that chemical alteration was probably more significant than insolation changes in causing rock disintegration.

As discussed in Chapter 3, the daily range of temperatures at a rock's surface is greater than that of the air. The most intense and significant thermal changes occur at the outer skin of the rock, with the interior of the rock undergoing only modest temperature change. Furthermore, the heating of rocks is asymmetric in nature, with the side facing the sun being warmer than the opposite face. In the American southwest, a significant percentage of large clasts (intermediate diameter >5 cm) have cracks with a nonrandom, moderately strong north–south orientation (McFadden et al. 2005). These steeply dipping cracks may be related genetically to thermal stresses caused by the differential heating and cooling of the rock associated with the daily movement of the sun across the sky. During summer, the greatest difference in temperature is between the east and west sides of the rocks, with temperatures in the mid-morning showing the greatest difference, on the order of 10–26°C. Surface parallel cracks, or spalls, may also be related to differential heating and are correlated with rock type, being most common in quartzite (86% showed spalling) and granite clasts (49%). Crack morphology and width vary with the age of the geomorphic surface: in general, cracks are wider on older surfaces. Small clasts appear less likely to develop cracks by differential heating, as there is less of a temperature gradient (McFadden et al. 2005).

Conditions of intense heating, such as fires, may also be associated with the splitting of rocks, but fires are probably not significant in sparsely vegetated true desert environments. It is possible, of course, that fires were more common during wetter, more heavily vegetated climatic intervals, and that significant splitting or spalling occurred at that time.

In summary, laboratory observations have been inconclusive regarding the role of insolation weathering in rock splitting or disintegration in deserts. Recent fieldwork in the American southwest suggests that cracks may be caused by tensile stresses that develop in the interior of rocks due to radial gradients in temperature associated with diurnal pattern of solar heating. Fire may be an infrequent cause of rock breakdown in semiarid environments. Regardless of the cause of incipient cracks in rocks, they provide an entryway for dust and salts that produce additional internal stresses ultimately leading to rock fracture.

6.2.2 SALT WEATHERING

Salt weathering refers to the breakdown of rock by principally physical changes produced by salt crystallization, salt hydration, or thermal expansion (Fig. 6.1). This process occurs in a variety of environments, including coastal, polar, and urban areas, but is notable in deserts owing to the abundance of salt sources.

The evidence for the power of salt weathering includes laboratory simulations, field observations and monitoring, and engineering studies of both modern and ancient structures (Winkler 1994). The action of salt is evident even in a hyperarid

Fig. 6.1 Salt weathering. Intense salt weathering has disintegrated rocks at the toes of alluvial fans on the eastern side of Death Valley.

environment. From a geomorphic perspective, salt weathering is important in generating finer debris that may subsequently be removed by wind and water erosion (Beaumont 1968; Goudie & Day 1980). It is also associated with the formation of tafoni (Huinink et al. 2004), and with the process of groundwater sapping (Laity 1983). Understanding salt weathering is also critical to preserving structures of economic, historical, and cultural value (Goudie & Viles 1997). In Egypt, the annual inundations of the Nile Valley prior to the completion of the Aswan High Dam in 1965 allowed for leaching of accumulated salts in the soil. By 1977, new salt deposits were discovered at Karnak, where they continue to deteriorate the foundations of ancient monuments (Burns et al. 1990).

There are three mechanisms by which salt may break down rocks: (1) salt crystal growth, (2) hydration, and (3) the thermal expansion of salt. The first results from salt crystal growth from solutions in rock pores and cracks. Crystal growth within a confined space sets up pressures that may cause disruption by wedging. The relative power of different salts (for example, sodium sulfate, sodium nitrate, calcium chloride, or magnesium sulfate) depends on such factors as crystal habit, growth rate, and salt solubility (Zehnder & Arnold 1989; Goudie 1997). The crystallization pattern may be strongly affected by the relative humidity (Rodriguez-Navarro & Doehne 1999). The second main mechanism of salt weathering involves phase changes between hydrated and

dehydrated forms in response to changes in temperature and humidity. Changes in volume generate hydration pressures against pore walls. The largest pressures occur when anhydrite changes into gypsum (Winkler & Wilhem 1970). The third mechanism involves the differential thermal expansion of salts, with salts having a greater coefficient of expansion than rock minerals. This process is thought to be less effective than salt crystal growth or hydration.

Salts may crystallize on the surface of a porous material (efflorescence) or within the porous system (subefflorescence). The latter typically creates the most extensive damage to stone, although the former is more visually apparent (Rodriguez-Navarro & Doehne 1999). Additionally, the wetting caused by the presence of a liquid (the salt solution) greatly increases the rate of crack propagation by reducing the interfacial tension between rock minerals. Thus, the stone may be damaged, even if little pressure is exerted by the salts. The porosity of the stone also plays a role in evaluating its susceptibility to salt weathering. Rock with a high proportion of micropores is more susceptible to decay than material with larger pores, probably because slow solution flow results in greater salt crystal growth beneath the surface (subefflorescence) (Rodriguez-Navarro & Doehne 1999).

Within an arid environment, there are numerous sources of salts, including seawater (for coastal deserts), ancient evaporitic beds, internal saline playas, volcanic activity, and rock breakdown. Salts may be brought to the surface by localized rock weathering, volcanic emissions, coastal fog, rain and snow fall, aeolian transportation, wash processes, and groundwater and capillary migration (Ericksen 1983; Berger & Cooke 1997; Nativ et al. 1997). In hyperarid regions such as the Atacama Desert, long-term aridity helps to preserve sources of salts (Berger & Cooke 1997). Groundwater dissolves minerals within rocks and transports salts to the surface, where they may actively weather rocks at sites of seepage (Laity 1983).

A wide range of salts is present in deserts, including mirabilite, natron, thenardite, hexahydrite, and gypsum (Goudie 1997). The nature and character of the salts is often unique to a region, owing to localized environmental differences in salt input and distribution. In the Atacama Desert, for example, the principal source of calcium sulfate is Andean volcanism, and this salt is widely distributed by a combination of groundwater flow and aeolian activity (Berger & Cooke 1997).

Some of the fastest rates of rock breakdown occur in coastal salt-pan environments, owing to the high levels of salt and the frequent wetting and drying associated with coastal fog conditions (Goudie et al. 1997). In a 2-year exposure trial, Viles and Goudie (2007) placed marble and granite blocks in and around a Namibian coastal pan and on a desert pavement surface. They observed breakdown of the marble blocks, but the granite blocks showed no visible signs of damage. Blocks placed further inland showed much lower rates of weathering, even where salinities within the sediments reached similar levels to the coastal zone. This implies that the frequency of wetting and drying is a major control on the breakdown rates. The intensity of coastal weathering has led to a flattening of relief along the coastal strip and may contribute to dust production (Viles & Goudie 2007).

Where alluvial fans abut saline playas, clast breakdown by salt weathering is often much more evident on the lower (distal) slopes of the fan than on the upper (Fig. 6.1). This results from the higher concentration of salts on the distal portion, explained less by wash processes from the upper fan than by: (1) capillary migration from groundwater, (2) past interaction of fan deposits with lake-level fluctuations, and (3) aeolian deposition of saline dusts. Such relationships have been observed in areas as widely separated as the Atacama Desert (Berger & Cooke 1997) and Death Valley, California (Goudie & Day 1980).

6.2.3 FROST WEATHERING

Frost weathering is not often considered in deserts, but is probably an important process in cold deserts with winter rainfall regimes (Goudie 1997). Similar to salt weathering, it may result in rock splitting and granular disintegration. Freezing may also act in combination with other processes, such as salt weathering. Freezing conditions may trigger the breakdown of rocks already weakened by salt weathering, or may enhance salt weathering by causing microfractures in the rock, particularly in deserts with a marked cold season (Smith et al. 2004). On the southern Colorado Plateau, there are 100–140 days annually where the freeze–thaw boundary is passed. Freeze–thaw cycles are one of several processes responsible for releasing sediment into the channel systems (Hereford 2002).

6.2.4 BIOLOGICAL WEATHERING

Biological weathering results principally from microorganisms and plants such as lichens and fungi. They mobilize some elements and contribute to the formation of some desert varnishes (Dorn & Oberlander 1982). Rock coatings may be destroyed owing to the secretion of organic acids or by removal of elements from the substrate (Dragovich 1993).

6.2.5 SILT INFILTRATION

Desert dust is nearly ubiquitous in rock fractures and extends deep within rocks. Even fractures that are very fine, which require considerable force to pry apart, will yield fines at depth (Fig. 6.2). To date, there has been little study of silt infiltration, despite its potential significance. Fines aid in the weathering of rocks by expanding and contracting as humidity levels change in the rock. Additionally, the material may contain salts (Amit et al. 1993).

There are two potential sources of fines in rock fractures: (1) *in situ* rock weathering and (2) aeolian dust (Condé-Gaussen et al. 1984). Villa et al. (1995) collected samples from the western USA and examined the texture and chemistry of the fines. In Death Valley, they found abundant elements in the fines that were not present in the host quartzite, suggesting an aeolian origin. However, there was also evidence that quartzite weathering was contributing silica to the fissures. Thus, it appears that fine

Fig. 6.2 Silt infiltration. Desert dust is nearly ubiquitous in rock fractures and penetrates deep within rocks. The fine material appears to be derived both from aeolian dust and weathering. Fines aid rock weathering by expanding and contracting as moisture levels change.

materials are derived from both aeolian dust and weathering, with the relative contribution of each varying according to the study site.

In summary, with respect to weathering, rocks in deserts often have a complex history of stresses that relate to the geologic history of the material, episodes of transport and deposition, climate history, and local environmental conditions. This history, combined with the size, confining stress, and mineralogy of the rock, determine in large measure the rock's susceptibility to further weathering. Conditions in nature are much more complicated than those that can be simulated in the laboratory, involving interactions between many processes, including insolation weathering, chemical weathering, salt weathering, silt infiltration, and freeze–thaw cycles. There remains much to learn about these important processes in desert environments.

6.3 WEATHERING FORMS

6.3.1 CAVERNOUS WEATHERING/TAFONI

Cavernous or honeycomb weathering is found throughout the world, from Antarctic valleys to temperate, arid, and tropical environments. Weathering hollows form in vertical rock faces, often in large numbers, to create a honeycomb-like form (Fig. 6.3).

Although many scientists have attempted to explain their origin, most publications largely describe their physical appearance. Unfortunately for students of geomorphology, the nomenclature has not been standardized, and this type of erosion has been referred to as alveolar weathering, stone lattice, stone lace, fretting, cavernous weathering, and tafoni. There is a general distinction into large features, greater than 1 m in diameter, referred to as tafoni or cavernous weathering; and small features (cavities several centimeters in diameter), called honeycomb weathering or alveolar weathering.

Cavernous weathering has been reported in many arid regions, including Jordan, Egypt, Australia, Chile, the southwestern USA, and various other desert locations. In the USA, Blackwelder (1929) observed honeycomb structures in outcrops of igneous rocks, sandstone, sandy shale, and conglomerate, and noted its absence in limestone, slate, and quartzite. Most examples of honeycomb weathering occur in homogeneous rocks, particularly sandstone, volcanic tuffs, and granite.

Numerous hypotheses have been advanced to explain the formation of honeycomb structures, but the processes of differential erosion that give rise to them have been a matter of great controversy. A major limitation of most studies has been the reliance on visual observation. In most cases, surface appearance provides too little information to explain

FIG. 6.3 Cavernous weathering near Urumqi, China.

physical and chemical changes that occur during weathering. Early reports (1890s) of honeycomb weathering from granitic rocks of Corsica assumed the cavities resulted from wind erosion. In other studies, the cause of honeycomb weathering could not be ascertained, and it was concluded that erosion must have occurred in the past. Blackwelder (1929) suggested that cavities develop through a process of exfoliation related to hydration reactions of feldspars in moist sites on the rock surface, with removal of resulting debris by the action of wind, rain, and animal movements.

The concept that honeycomb weathering results from the physical action of salt crystallization was first advanced by Hume (1925), who observed masses of fibrous salt crystals associated with honeycomb structures. Salt weathering has since become the most popular hypothesis for explaining honeycomb weathering in coastal environments. The action of evaporating salt solutions has also been cited for inland desert settings (Mustoe 1983). The evidence includes salt accumulation in hollows, highly soluble cation contents in the spall detritus, and the presence of the mineral gypsum. Salts may be introduced as migrating fluids or the salts may be contained within the original sediment.

Several studies conducted in quartz sandstones of the Colorado Plateau suggest the importance of salt weathering in the development of cavernous weathering and alcoves (Laity 1983; Mustoe 1983). According to Mustoe (1983), moisture enters the cliffs either

as precipitation or by the inward motion of condensed water. The source of the dissolved salts is the extraction of soluble minerals within the rock, and salts supplied from the surrounding alkali-rich soils in the form of wind-blown dust. The areas subject to salt weathering are determined by rock permeability (water migration along permeable strata dissolves salts before emerging at the surface), localized microclimatic conditions that control evaporation rates, water flow across the surface of the rock from precipitation (removes salts brought to the surface), shading, rock type, and upward "wicking" of salts from underlying soils or shales. Cavernous weathering is more common in lower, shaded elevations where surface salts go through repeated cycles of dissolution and recrystallization that favor the breakdown of feldspar and other unstable minerals. Weathering primarily results from physical disaggregation of the grains rather than from chemical decomposition. In a given locality, cavities tend to develop along bedding or joint planes, or other areas of structural or compositional weakness. However, well-developed honeycomb weathering has been observed to occur in apparently homogeneous sediments and massive crystalline rocks.

The degree of tafoni development has been used as a relative dating tool (Hereford & Huntoon 1990). The rate of tafoni growth is not linear, however, and invariably there are differences in growth associated with minor lithologic variations and the degree of slope on the rock face. The rate of enlargement of desert tafoni decreases over tens of thousands of years (Norwick & Dexter 2002).

6.3.2 GNAMMAS

Gnammas are shallow, largely circular, basins formed on horizontal rock surfaces (Fig. 6.4). Like cavernous weathering features, they are not limited to arid environments. They are often found on the top of domed inselbergs (Twidale & Corbin 1963) and are most common on igneous rocks and sandstones. The term gnamma is of Australian Aboriginal origin, used by Western Desert people to describe a rock hole, particularly one likely to contain water (Bayly 1999). They are quite common on the horizontal surfaces

FIG. 6.4 Gnammas, or weathering pits, are common on horizontal surfaces of sandstones on the Colorado Plateau. During winter, water freezes in these small ponds. The location is Muley Point, Cedar Mesa, Utah, USA.

of sandstones on the Colorado Plateau where, during winter, water freezes into miniature ponds (Fig. 6.4). Chemical weathering, and freeze–thaw activity in colder deserts, probably contribute to their formation.

6.4 DURICRUSTS

6.4.1 TERMINOLOGY

Duricrusts (or duricretes) are general terms used for hardened crusts formed at or near the Earth's surface. They are associated with rock weathering and soil formation, usually in tropical or arid regions. They attest to the mobilization and precipitation of minerals in the presence of water. In many cases, they provide useful environmental indicators. Lamplugh (1902, 1907) originally coined the terms calcrete, silcrete, and ferricrete for rocks cemented by calcium carbonate, silica, and iron. Calcareous-rich deposits are particularly widespread in arid regions and many local names are used. The term caliche comes from the Latin term *calx*, meaning lime or limestone. Gypcrete is used for gypsum crusts. Salcretes are halite crusts, composed of sodium chloride, which are usually thin and ephemeral and subject to rapid dissolution by water. Intermediate forms may also occur, such as siliceous calcretes or calcareous silcretes (Nash & Shaw 1998).

Massive crusts, such as those formed by calcrete, may be tens of meters thick, and mantle extensive areas of desert terrain. They protect the underlying materials from weathering and erosion and, by retarding the denudation of desert landscapes, act to preserve relict landforms. Whereas the occurrence of desert crusts in the geologic record has the potential to provide valuable information on palaeoclimates, the precise environmental conditions under which they develop are not always well understood.

In an extensive review of the duricrust literature, Dixon (1994b) summarized key aspects of our present knowledge. Calcretes, silcretes, and gypcretes are widely distributed in Australia, Africa, the Middle East, and North America. Although duricrusts have received considerable scientific attention, and in some cases detailed study regarding their chemistry, mineralogy, and environmental conditions of formation, there are significant geographic gaps in our understanding. The American southwest, for example, has received a great deal of geomorphic study, but duricrusts are incompletely documented. There is very little literature concerning potential Asian occurrences.

In the formation of duricrusts, it has been shown that there is a fundamental division between pedogenic (formed within the soil) and nonpedogenic (deposition by groundwater and in lacustrine or fluvial environments) categories (Dixon 1994b). Within each duricrust type, there is considerable variety in morphology, micromorphology, chemistry, and mineralogy, reflecting the diverse environments under which the duricrusts formed, as well as the effects of climate change.

6.4.2 SILCRETE

There is very little reference to silcrete in the standard earth science texts, probably because it is a substance not particularly conspicuous in North America and Europe. Silcretes are most frequently described from Australia, southern Africa, and parts of the Sahara. In Australia, silcrete duricrusts are prominent elements of the landscape, capping plateaux and mesas, and thereby preserving conspicuous remnants of a planation surface that once covered much of the continent. The most widespread and characteristic distribution of silcretes in Australia occurs in the arid interior, particularly in association with the drainage system of Lake Eyre. Eroded and reworked silcrete clasts provide the cobbles and pebbles of the gibber plains of stony deserts (Wopfner 1978).

Silcrete is a lithological term for a "very brittle, intensely indurated rock composed mainly of quartz clasts cemented by a matrix which may be well-crystallized quartz, cryptocrystalline quartz, or amorphous (opaline) silica" (Langford-Smith 1978, p. 3). A cryptocrystalline substance is crystalline, but so fine-grained that the individual components cannot be seen with a magnifying lens. Amorphous means "without form," and is applied to rocks and minerals having no definite crystalline structure. The texture of the silcrete reflects that of the host rock, and clasts may range in size from very fine sand grains to boulders. A claystone that has silicified to the extent that its clay minerals have been replaced by silica is often called a porcellanite, especially when it is contiguous with, and grades into, a more normal silcrete with a coarser texture.

A silcrete is produced by an absolute accumulation of silica. As most silcretes contain over 90% silica (many have more than 98%), silcrete formation requires the removal of most elements other than silicon that were present in the host material (Summerfield 1983; Dixon 1994b). A characteristic of most silcretes is the relatively high concentrations of the resistate elements titanium, zirconium, and niobium.

In the field, fresh, unweathered silcrete is usually grey. However, there is considerable color variation and the outcrop can also appear whitish, red, brown, or yellow. Fresh silcrete, when struck with a hammer, gives a characteristic ring, and emits a pungent odor. The rock is extremely brittle, and shatters into sharp, angular pieces with a conchoidal fracture and a semi-vitreous sheen. Beds of silcrete are usually 1–2 m thick, but may exceed 5 m or more (20 m occurs). Vertical jointing is common, and may give rise to a columnar structure (Langford-Smith 1978).

Many silcrete duricrusts are of great antiquity. Some Namibian silcretes may be as old as 50 million years (Watson & Nash 1997). Australia has extensive land areas that have been subject to subaerial denudation for hundreds of millions of years. Landform longevity and slow rates of geomorphic change characterize much of the continent's history. However, chronologies of landform evolution have been difficult to establish and therefore the evidence of duricrust materials is crucial. Three distinct periods of Australian silcrete formation have been recognized: late Jurassic, early Cenozoic, and late Cenozoic. Late-Jurassic silcretes are the oldest to have been recorded with any certainty. They are unusual in that they include abundant casts of plant fossils. Early Cenozoic silcretes are by far the most widespread and are developed on a quasi-planar land surface of enormous dimensions. They also overlie a deeply weathered, kaolinized profile, for which the silcrete/profile relationships have not been established. Late Cenozoic silcretes are more areally restricted. They are not associated with a distinctive, deep kaolinized profile. Their probable age is Pliocene, extending into the Pleistocene. Most of Australia's precious opal fields are associated with silcretes of this age (Langford-Smith 1978).

Although there is a general agreement that the development of both silcrete and laterite requires a warm to hot climate that is moderately humid, various views have been expressed as to the amount of rainfall required and the nature of its occurrence. It is likely that silcrete requires a drier climate than laterite. Factors other than climate that probably governed silcrete development include lithology and specific chemical environments influenced by groundwater conditions and drainage (Roberts 2003). The geographic and geomorphic settings of silcretes and geochemical considerations suggest that in Australia silcretes formed during warm, humid, subtropical to tropical climates. In the early Tertiary, the depositional basins of Australia likely were bordered by rainforests. Many silcretes display evidence of more than one period of likely genesis (Wopfner 1978; Roberts 2003).

Environments where silcretes formed included floodplains, where runoff is sluggish and drainage markedly impeded, areas of high groundwater concentration at the base of hillslopes and, in southern Africa, marine-planed surfaces (Roberts 2003). Although much remains to be investigated concerning the geochemistry of silcrete formation, it appears clear that silcretes developed under environmental conditions markedly different from those that now characterize the arid zones where they are presently preserved.

6.4.3 CALCRETE/CALICHE

Calcretes have a very widespread distribution and are thought to cover as much as 13% of the Earth's land surface (Watson & Nash 1997). According to Dixon (1994b), the term calcrete is generally preferred to caliche, although both are commonly employed. Calcium carbonate accumulations range from powerdery soil carbonate, through isolated nodules (Fig. 6.5), to massive, indurated carbonate (Fig. 6.6) (Goudie 1983a). Caliche is characteristically considered to be a product of semiarid or arid zones, where massive profiles are best exposed. In Africa, for example, calcretes characterize the more arid parts of the continent, occurring in areas where the mean annual rainfall is less than 850 mm (Goudie 1996). Vast areas of the Sahara, South Africa, the western USA, and Australia exhibit exceptional development. Calcretes are also known from Tanzania, Angola, the Kalahari Desert (Namibia and Botswana), Paraguay, Mexico, Argentina, parts of Spain, southern France, and Italy. Caliche occurs throughout the Middle East, but its distribution is not well known. In India, calcretes are found mainly in the Thar Desert and in other areas of Asia they have been reported from Armenia, the Karakum, Kazakhstan, and the Kyzylkum areas. Reliable maps for most of the world are not available and distribution is best known, of course, for inhabited areas.

Carbonate-rich horizons are common in soils of semiarid and arid regions. They may occur at some depth below the surface or may extend to the surface. As discussed in Section 7.3, there are three master horizons in soils of arid and semiarid climates; the A, B, and C horizons, in order of increasing depth. The modifier k, as in the Bk horizon, refers to horizons

FIG. 6.5 Carbonate nodules, isolated from one another, suggest Stage II carbonate development in stabilized aeolian sands south of Death Valley, California, USA.

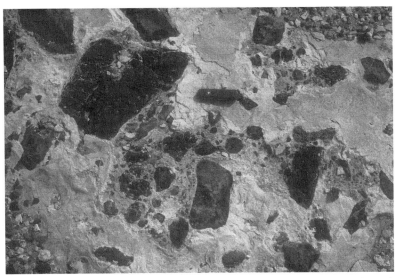

FIG. 6.6 Dense, indurated carbonate (calcrete) in the Cadiz Valley, California, USA. Such surfaces have a high runoff potential. A penknife is shown for scale.

with secondary carbonate (Machette 1985), whereas the K horizon has a prominent layer of carbonate accumulation (Gile et al. 1965). Both calcium and magnesium carbonates are present in the soil, with the former dominant (Birkeland 1984). Specifically, a Bk horizon is an illuvial accumulation of alkaline earth carbonates, mainly calcium carbonate, with properties that do not meet those for the K horizon. Illuviation refers to the deposition of colloids, soluble salts, and small mineral particles that have leached out of the overlying (A) horizon. A K horizon is a "subsurface horizon so impregnated with carbonate that its morphology is determined by the carbonate. Authigenic carbonate coats or engulfs all primary grains in a continuous medium to make up 50 percent or more by volume of the horizon" (Birkeland 1984, p. 138). The cemented uppermost part of the horizon is commonly laminated and corresponds to some caliches and calcretes.

The origin of carbonate horizons involves carbonate/bicarbonate and carbon dioxide/soil air equilibria. Carbon dioxide partial pressures in soil air are on the order of 10–100 times greater than in the atmosphere. Root and micro-organism respiration and organic matter decomposition cause the high carbon dioxide partial pressures. The highest pressure is found in the A horizon, with values diminishing down to the base of the root zone. As water moves vertically through the soil, the Ca^{2+} (which may be present or derived from weathering of calcium-

bearing minerals) and HCO_3^- content (which is formed from the disassociation of carbonic acid, H_2CO_3) may increase to the point of saturation after which further dissolution of calcium carbonate is not possible. The basic reactions involved in these equilibria are as follows:

$$CO_{2(g)} + H_2O_{(1)} \Leftrightarrow H_2CO_{3(aq)}$$

$$CaCO_{3(S)} + H_2CO_{3(aq)} \Leftrightarrow Ca^{2+}_{(aq)} + 2HCO_3^-{}_{(aq)}$$

Precipitation as a calcium carbonate-rich horizon causes a lowering of carbon dioxide pressure below the zone of rooting and major biological activity, and a progressive increase in the concentration of Ca^{2+} and HCO_3^- in the soil solution. This occurs as the water percolates downward and is lost by evapotranspiration near the surface. The depth of leaching directly affects the position of the calcium carbonate-bearing horizon and is related to the climate. Temperature affects calcium carbonate concentration. Precipitation of calcium carbonate occurs as temperature rises, because carbon dioxide is less soluble in warm water than in cold water (forming less carbonic acid). This may be significant in considering the effects of climate change and in comparing the depths to the tops of Bk or K horizons between regions of contrasting temperature.

An important aspect of carbonate formation is the origin of the calcium carbonate. The most important source of Ca^{2+} and calcium carbonate appears to be

TABLE 6.1 Stages of calcium carbonate morphology.

Stage	Diagnostic carbonate morphology
I	Filaments or faint coatings; thin, discontinuous coatings on underside of clasts
II	0.5–4 cm nodules, isolated from one another; some interstitial accumulations of carbonate in matrix; continuous, thin to thick coatings on tops and bottoms of clasts
III	Coalesced nodules; massive accumulations between clasts; moderately to firmly cemented matrix
IV	Matrix is indurated, platy, and massive, with 0.2–1.0 cm-thick laminae in upper surface, which may drape over fractured surfaces; relict nodules may be visible: K horizon is 0.5–1 m thick
V	Dense, indurated laminae greater than 1 cm thick; platy to tabular structure; incipient pisoliths in laminar zone; K horizon is 1–2 m thick
VI	Massive K horizon, commonly greater than 2 m thick with multiple generations of laminae, pisoliths, and breccia; recemented, indurated and dense; case hardening common

After Machette (1985).

the atmosphere. The carbonate is derived from aeolian dust and from Ca^{2+} dissolved in rainwater. A locally important source for the Ca^{2+} in carbonate is atmospheric gypsum, for on dissolution, calcium is released. In nonpedogenic calcretes, there is a strong association with river valleys, including both modern and palaeodrainage systems. In these calcretes, the calcium carbonate is derived from groundwater sources and reprecipitated in channels (Dixon 1994b).

There are two basic types of carbonate: pedogenic and groundwater (phreatic). Pedogenic carbonate forms within the soil by the precipitation of carbonates originally present in rainwater, or leached by infiltrating rainwater from aeolian fines deposited in upper soil horizons (Dohrenwend 1987). The accumulation of carbonate follows a number of stages (Table 6.1) that proceed from discontinuous clast coatings, to segmented filaments and nodules, and finally to dense, indurated accumulations that gradually plug the calcic horizon (Machette 1985). For soils with high amounts of carbonate (K horizons), pedogenic carbonate is mainly micrite and the original grains have been forced apart. Dense pedogenic carbonates containing more than 50% calcium carbonate may have developed over hundreds of thousands to millions of years (Machette 1985).

Nonpedogenic accumulations include gully-bed cementation, laminar layers on interfluves and hillslopes, and case-hardened surfaces, associated with lateral surface or subsurface water flow (Lattman 1973). The process involves the evaporative concentration of Ca^{2+} to supersaturated levels. Calcium carbonate is subsequently precipitated in areas of intermittent surface flow (e.g. sheet wash and gully

flow), temporary surface ponding, or where subsurface flow either discharges or reaches a near-surface position (Lattman 1973; Machette 1985). Groundwater carbonate tends to be more coarsely grained than pedogenic carbonate, fills only the original pore spaces, and, because of overburden pressure, has not forced the original grains apart.

In either pedogenic or nonpedogenic carbonate accumulations, the horizon gradually becomes impregnated by carbonate deposition until the voids become plugged and water percolation is restricted. Thereafter, water tends to collect over the plugged horizon and the resulting solution and reprecipitation produces a laminated upper K horizon. The horizon builds upward and therefore is younger closest to the surface. During the process of K horizon development, the volume of pedogenic carbonate eventually exceeds the original volume of the pores. The carbonate, upon crystallization, forces the silicate grains and gravel apart or, in some cases, the carbonate replaces silicate minerals.

Calcretes may occur on most rock types and sediments. Morphological features that may be present in indurated K horizons (stages IV, V, and VI) are ooids and pisoliths. Ooids are sand-sized round particles in which a nucleus is surrounded by one or more concentric layers of fine-grained calcite grains (micrite). Micrite is aggregates of calcite crystallites less than about 100 μm in diameter. Pisoliths are subangular to spherical bodies, 0.5 to more than 10 cm across, surrounded by multiple thin layers of micrite. Pisolith nuclei include rock fragments or pieces of broken and rotated indurated K-horizon material that may display internal laminar layers and pisoliths from previous pedogenesis (Birkeland

1984). Both of these features usually are set into massive carbonate cement.

Within the Basin and Range Province of the southwestern USA, the rates of carbonate accumulation vary considerably and are influenced by climatic, lithologic, and topographic factors. The accumulation of carbonate on noncalcareous fans is related to the proximity of external sources, such as playas, and to aeolian-borne carbonate dust and silt (Dohrenwend 1987). Calcretes are commonly associated with the presence of pans and playas in the USA, Australia, and southern Africa (Watson & Nash 1997). The roughness of fan surfaces promotes aeolian deposition, and the high surface temperatures occurring on basalt and andesite clasts induce calcium carbonate precipitation.

Carbonate accumulation depths have been used to suggest climatic change in some semiarid and arid environments. In southern California, the Holocene soils have a shallow accumulation of carbonate, whereas in Pleistocene soils it is three to four times as deep. Soils often have two carbonate accumulation maxima with depth: a shallow one related to the present Holocene climate, and a deeper one related to moister conditions in the late Pleistocene. It is possible that there are variations in accumulation rate with climatic change. For southern New Mexico, there is evidence that the rate of carbonate accumulation during interpluvials could be twice that of the pluvials (Machette 1985). This may be related to vegetation type and percentage cover.

Isotopic studies of pedogenic carbonate coatings provide both chronological and palaeoenvironmental information on alluvial deposits. Carbon-14 ages of pedogenic carbon suggest that older geomorphic surfaces in alluvial piedmont deposits along the western flank of the Providence Mountains, California, are of late Pleistocene age (around 47–17 ka) and that younger surfaces formed during the Holocene (around 11–4 ka) (Wang et al. 1996). Both the carbon and oxygen isotopic composition of soil suggest that the climate became warmer and drier during the Holocene (Wang et al. 1996).

Certain morphological features of K horizons owe their origin to climatic change. Solution pits on the upper surface, for example, indicate a change towards a wetter climate. Other calcretes show laminated K horizons, brecciation of these horizons, and subsequent cementation of the breccia. Such materials provide evidence of a complex history and climatic change, but the processes responsible are not clearly understood.

Aeolian and fluvial erosion are affected by the accumulation of carbonate material, which influences surface cohesion and permeability. Even small amounts of powdery carbonate in the soil reduce infiltration and promote surface flow. For example, on rock-mantled falling dunes in southernmost Death Valley, carbonate accumulation in dune sands reduces aeolian erosion, but promotes gullying of the lower dune surfaces. Drainage basins with significant calcrete development have a high runoff potential, and stream hydrographs tend to be of a flashy nature.

6.4.4 GYPCRETE

Gypsum crusts (gypcretes) are less widespread than other crusts and are found in some of the Earth's most arid regions. As defined by Watson (1985), gypsum crusts are "accumulations at or within 10 m of the land surface from 0.10 m to 5.0 m thick containing more than 15% by weight gypsum . . . and at least 5.0% by weight more gypsum than the underlying bedrock."

Most gypsum crusts are found in areas where mean annual rainfall is less than 200–250 mm (Watson 1983). In North Africa, there appears to be a transition from calcretes to gypsum crusts as mean annual rainfall drops below this amount. In the central Namib Desert, there is a gradual transition from gypsum crusts to halite crusts as mean annual rainfall drops below about 25 mm.

The main areas of widespread gypsum crusts are in North Africa, particularly central Algeria and Tunisia; the central Namib Desert (Watson 1988); parts of south and western Australia; and central Asia. They are also found throughout the Middle East, including Egypt (Ali & West 1983), Israel (Amit & Gerson 1986), and Iraq (Tucker 1978). Occurrences in the southwestern USA appear to be limited in distribution (Dixon 1994b).

6.4.5 SALCRETE: HALITE CRUSTS

Halite (sodium chloride) deposits in deserts are relatively ephemeral in nature, being easily dissolved. In the geologic record they are almost always interpreted as lagoonal or lacustrine evaporites. Halite occurs most commonly in sabkhas (desert basins or littoral flats where the water table lies just below the land surface) or in the basins of ephemeral lakes. On lake beds, crusts of almost pure halite form horizontal beds on the surface. Periodic influxes of fresh rainwater or runoff often result in the complex dissolution and subsequent reprecipitation of the deposits as, for example, in Lake Eyre, Australia.

6.5 DESERT VARNISH

Desert varnish is a thin coating, rarely exceeding 200 μm in thickness, developed on subaerially exposed pebble and rock surfaces (Fig. 6.7). It accumulates at rates ranging from less than 1 to 40 μm per 1000 years, making it the slowest known accumulating terrestrial sedimentary deposit (Liu & Broecker 2000). Although found in virtually all terrestrial environments, it is best developed in arid and semiarid regions.

Varnish is chemically, structurally, and morphologically different from its underlying substrates. It forms micrometer-thick laminations that parallel the topography of the rock substrate. In color, it is orange, grey, brown, or black and frequently has a lustrous appearance. It is principally composed of clay minerals (70%), with the remainder being oxides and hydroxides of iron and manganese admixed with detrital silica and calcium carbonate (Dorn & Oberlander 1982; Thiagarajan & Lee 2004). The darker varnishes contain more manganese (20% MnO_2 by weight), whereas those with an orange hue contain more iron (10% FeO_2 by weight and less than 3% MnO_2). Major elements in varnish include manganese, iron, silicon, aluminum, calcium, potassium, barium, sodium, strontium, copper, and titanium, while minor, locally significant elements are vanadium, phosphorus, cobalt, cadmium, nickel, lead, zinc, lanthanum, yttrium, zirconium, and boron (Lakin et al. 1963). Trace element concentrations are often considerably enriched relative to the host rock. Varnish is found on a wide variety of rock types, but is generally absent from rocks that are susceptible to rapid weathering or dissolution, such as carbonates.

The origin of varnish has been the subject of broad debate. Early research suggested that the varnish was derived from an internal source, that is, from the host rock itself. In this scenario, coatings develop as moisture is drawn out of rocks. Following this "sweating" process, minerals are precipitated on the rock surface (Merrill 1898; Blake 1905). Varnish has also been attributed to surficial chemical weathering and the accumulation of a residue rich in iron and manganese (Engel & Sharp 1958; Smith & Whalley 1988). Organic weathering by lichen (Laudermilk 1931) or by micro-organisms (fungi, algae, and bacteria) has also been proposed (Krumbein & Jens 1981).

The most widespread modern theory invokes the development of varnish from an external source of

FIG. 6.7 Desert varnish is a thin coating developed on exposed pebble and rock surfaces. Chips on this heavily varnished rock reveal the true (medium grey) color of the underlying rock.

manganese, iron, and clay minerals. The virtue of this argument is that it explains the presence of trace minerals that are absent from the host rock. Varnishes occur on nearly all rock types, including quartzites, which have little or no iron, manganese, or clay minerals. Furthermore, the contact between the varnish and the underlying rock is texturally abrupt.

There are numerous indicators that the coatings have an atmospheric origin as they: (1) contain small amounts of sulfate with a uniquely atmospheric oxygen isotopic signature, (2) incorporate a diversity of clay minerals that could not be generated *in situ*, and (3) include ^{210}Pb, a shortlived radionuclide that exists in the atmosphere (Thiagarajan & Lee 2004). Deposits of airborne dust are thought to be fixed by chemical or micro-organic processes.

Numerous studies of global deserts have suggested a principally biological means of manganese and iron fixation (Dorn & Oberlander 1982; Dorn & Dragovich 1990; Adams et al. 1992; Drake et al. 1993). The debate lies principally in the specific types of microorganisms involved. Several researchers have cultured bacteria from the surface of desert varnish, suggesting that they may play a role in varnish formation. Perry et al. (2003) collected desert varnish scrapings from the Mojave and Sonoran Deserts and found 13 amino acids, indicating a biogenic component. Although the presence of these biomarkers does not prove that bacteria produce the varnish, their presence is consistent with a bacterial role in varnish formation. It is well documented within the microbiological literature that bacteria induce the precipitation of manganese and iron in deep-sea manganese nodules, Frutexites (marine carbonates), and so-called black shrubs of hot-water travertines, and can concentrate manganese in mineral precipitates from waters with nearly undetectable manganese concentrations (Chafetz et al. 1998).

To explain the trace-element abundances, Thiagarajan and Lee (2004) invoke direct aqueous atmospheric deposition of varnish. According to their theory, iron, manganese, and trace metals in varnish are derived from leaching of dust particles that are entrained in rain and fog droplets. When the droplets contact the rock, iron–manganese oxyhydroxides and particle-reactive trace minerals are co-precipitated, with the residual fluid and dust particles washed off or removed by the wind.

Rock varnish constitutes a long-term, microscale, sedimentary archive of past environmental change in drylands. There have been many attempts to use it as a relative or absolute indicator of geomorphic age. To date, none have been completely successful. For instance, it might appear that varnish thickness would correlate with the age of geomorphic features, but data acquired by Liu and Broecker (2000) and Broecker and Liu (2001) indicate that varnish accumulation rates vary greatly from sample to sample, even at a given site. Earlier, attempts were made to absolutely date varnish using a cation-ratio method (Dorn 1983) and radiocarbon techniques (Dorn et al. 1992), but these innovative approaches were abandoned following critical reviews (Bierman & Gillespie 1991; Reneau & Raymond 1991; Beck et al. 1998). Thus, at present it is not possible to directly date geomorphic surfaces using varnish, although visual examination tends to show a progression from lightly varnished fresh surfaces to more heavily varnished stabilized surfaces, a trend particularly evident on alluvial fans. Meanwhile, a growing body of research examines the climate record through analysis of the microstratigraphy of varnish, using the microlaminations as a correlative dating technique (Dorn 1988; Broecker & Liu 2001; Liu 2003; Liu & Broecker 2008).

Varnish microstratigraphy has the potential to establish a climatic history in deserts stretching back from the Holocene, through the last glacial, and into the previous interglacial, a time period of 70,000 years or more. The climate record is obtained through the analysis of variations in color and chemical composition of the varnish. For example, varnish from Death Valley, California, shows a range in manganese dioxide concentrations from 4% in the Holocene to as much as 40% during parts of the last glacial. The color of the varnish varies according to manganese dioxide content: for low values, it is yellow in color; for intermediate values, orange; and high content varnish is black (Perry & Adams 1978). For lava flows in the Mojave Desert, ultra-thin sections show layers of yellow, orange, and black varnish (Liu 2003). Several researchers have observed a correlation between layers of high manganese content during wet periods, and low manganese content during dry periods (Zhou et al. 2000; Broecker & Liu 2001; Liu 2003). The yellow (low-manganese) layers occur at the surfaces of all of the varnish samples examined in the southwestern USA, potentially indicating the dry conditions of the Holocene. Beneath these lie alternating black and orange layers, which have been correlated with the ice-rafting episodes recorded in deep-sea sediments of the North Atlantic (Broecker & Liu 2001; Liu 2003).

A stratigraphy may be built by examining varnish formed on substrates of known age, such as raised shorelines or basalt flows. From this, it may be possible to use varnish microstratigraphy to estimate surface exposure ages of desert landforms, where datable surfaces are relatively rare. Varnish micro-lamination dating is a correlative age-determination technique that can be applied to such surfaces as alluvial fans, desert pavements, and colluvial boulder deposits (Liu & Broecker 2008). It is used to establish minimum surface-exposure ages using the assumption that the climatic signals recorded in the varnish are regionally contemporaneous. Few varnish records appear to extend back to the penultimate glaciation, largely because varnish tends to peel once it reaches a thickness of about 200 µm. In some cases, aeolian abrasion also removes material.

Chemical fingerprints indicate that varnish is still accreting in the modern environment. During the past century, a large number of anthropogenic substances have been added to the atmosphere, including lead from automobiles. Lead shows a tenfold increase in concentration in the outer 1 µm of varnish relative to its ambient Holcene concentration (Broecker & Liu 2001).

The widespread nature of varnish on exposed rock surfaces in deserts has made it the subject of intense research and debate for over 100 years. This thin deposit has received more scrutiny than almost any other geomorphological feature, and the interest shows little signs of abating owing to its potential to yield information on past climates and the age of underlying surfaces.

6.6 HILLSLOPE PROCESSES

Hillslopes in arid regions differ from those in other climatic zones owing to their lack of vegetation and soil, which leaves rock surfaces clearly exposed. Slopes today are largely weathering-limited, meaning that the rate at which soil and debris can be removed by erosion exceeds the rate at which it is produced by weathering. As a result, weathering determines the rate of ground loss and the hillslope form is largely controlled by the relative resistance of the rock masses.

Unlike aeolian dunes or fluvial channels, which show clear and distinct changes within a human lifetime, rock slopes in deserts are only sporadically active in both space and time. Thus, with few exceptions, slope processes are rarely witnessed (Schumm

& Chorley 1964). Most process studies have been conducted in badlands, where softer sediments respond rapidly to rainfall events. In some cases, artificially generated runoff has been used to understand the generation of overland flow on desert slopes (Abrahams et al. 1986).

An added complication to understanding slopes is that it is widely agreed that many of them are largely relict of earlier, wetter climates. It is therefore difficult to fully understand their formative processes based solely on observations made in today's more xeric conditions. For example, the central Sahara has a very high concentration of fossil landslides, which are probably of early Pleistocene age. It is inferred that the climate must have been considerably wetter than semiarid to form these landslides, all of which appear to be developed in association with clays of high swelling potential (Busche 2001).

Finally, each rock slope is a "law unto itself" (Oberlander 1997, p. 135). By this, it is meant that there are many unique slope forms, each dependent on a host of petrologic and structural variables, interacting with different microclimates and biochemical conditions. Nonetheless, geomorphologists have classified three basic types of desert slope: (1) rock slopes, including (a) hillslopes formed on bodies of massive rock and (b) scarps and cuestas in layered rocks dominated by outcropping-resistant rock layers; (2) rock-mantled slopes; and (3) badland slopes (Howard & Selby 1994). Many slopes incorporate elements of all three types.

6.6.1 ROCK SLOPES

6.6.1.1 Hillslopes in massive rocks

Massive rocks are those with few joints and high intact strength. They are well exposed in desert regions and have been the subject of numerous studies on the application of rock mechanics to hillslope development. Influences on the nature of such slopes include the intact rock strength, the degree of weathering, joint spacing, orientation, width and continuity, and groundwater outflow. Sheeting, the development of parallel shells of rock separated from a massive parent rock, is common on granites and sandstones. In deserts, domed inselbergs are commonly called *bornhardts*, after the German geologist Wilhelm Bornhardt (1900), who described distinctive granitic hills with a bare rock surface, a domed summit, and steep sides. Hills with similar forms include the famous Uluru (Ayers Rock) (Fig. 6.8),

FIG. 6.8 Massive rocks with high intact strength are well exposed in deserts. In deserts, domed inselbergs are commonly called bornhardts. Hills with such a form include Uluru (Ayers Rock) in central Australia. Domed hills tend to shed water rapidly and lack soil development. Photo: Lloyd Laity.

formed of arkosic sandstone, and Kata Tjuta (The Olgas) of central Australia, composed of a massive conglomerate (Twidale 1978a). Although few large or continuous joints pass through the body of the bornhardt, sheeting is a common feature.

Domed hills are not unique to an arid environment, but are well exposed here, and are not subject to vigorous erosional processes. Weathering is limited by the form of the bornhardts, which tends to shed any water that is precipitated. Tafoni, pits, and other features are largely superficial. The presence of bornhardts in arid regions is related to their position on the surfaces of cratons that have been deeply weathered, with the resistant domal form having survived erosional stripping of the regolith (Howard & Selby 1994).

6.6.1.2 Scarp and cuesta forms

The erosion of sedimentary sequences of flat-lying or folded rocks gives rise to a landscape of scarps or cuestas, which are capped by more resistant rocks (Fig. 6.9). These are underlain by sequences of more erodible sedimentary rocks, sometimes interbedded with tabular intrusive or extrusive volcanic rocks. In arid environments, the resistant rocks are usually sandstones, limestones, or volcanic flows.

The cuesta form evolves to permit rocks of different erosional resistance to be eroded at roughly

equivalent rates. The most obvious features are the scarp profile and the scarp planform. As discussed in the introduction to this chapter, simulation experiments by Howard (1995) demonstrate that the planform of an escarpment correlates with the processes that form it. The profile (Fig. 6.9) consists of a backslope, rim, face, and rampart (also known as debris-covered slope (Cooke & Warren 1973), substrate ramp (Oberlander 1997), or footslope (Ahnert 1960)).

Scarp retreat proceeds as weaker underlying materials (often shales or weak porous sandstones) are eroded, undermining the massive caprock (often sandstone, limestone, or calcrete), leading to the development of cliffs, and backwasting of the scarp. The scarp is eroded by rockfall, block-by-block undermining, and slumping (Howard & Selby 1994), according to the thickness, jointing, and resistance of the caprock. Of these processes, rockfall is the most common. The blocks typically travel a short distance downslope, and form a protective layer on the shales, thereby reducing additional erosion (Fig. 6.10). In order for further retreat to occur, the debris must weather and be removed. Thus, the retreat of the scarp front is an inherently episodic process. Large blankets of debris offer considerable protection for the underlying weaker rocks, and such areas often stand out in relief above surrounding slopes. The amount of debris on the rampart varies considerably, being generally greater in reentrants and less on

Fig. 6.9 Cuesta on the Colorado Plateau. The profile consists of a backslope, rim, face, and rampart (or debris-covered slope). The face is nearly vertical and is principally formed of caprock, but may include some of the less-resistant underlying units. The rampart extends from the base of the face to the surrounding plain, and is commonly debris-mantled. See Plate 6.9 for a color version of this image.

projections (Howard & Selby 1994). In some cases, the talus becomes isolated from the scarp face, as erosion of the caprock and scarp retreat continue: this leads to talus flatirons (Koons 1955; Schmidt 1989a), steep triangular cliff facets which look like the base of an old-fashioned flat iron.

The processes by which scarps retreat have been most extensively studied on the Colorado Plateau, USA. Many of the prominent scarps within this region are formed of massive sandstones that are underlain by shales or other weaker materials. The retreat, or backwasting, of the scarp results from a number of processes, often acting in combination (Fig. 6.11). These include processes that affect the caprock itself, such as jointing, freeze–thaw activity in winter, groundwater sapping, and minor amounts of creep; and processes that erode the underlying layer which provides basal support. A high density

of fractures in caprock materials tends to make them much weaker, even if the material is inherently strong. Erosion of the ramparts involves both disintegration and removal of the talus that covers them and erosion of the weaker shales. Weathering processes affecting the talus include frost shattering, salt weathering, granular disintegration, and solution. Sandstone talus tends to disintegrate quickly, either by initial impact, or by subsequent weathering (Laity & Malin 1985), but conglomerates and limestones are more resistant and generate more conspicuous talus (Oberlander 1997). Rill erosion and gullying on unprotected material erode shales. Accumulated talus that lies near the base of the caprock acts to protect it from further backwasting (Koons 1955). If the rate of talus production exceeds that of destruction, the slope becomes moribund, as it is smothered in its own debris (Schumm & Chorley 1966).

Fig. 6.10 Talus blocks frequently travel short distances downslope and form a protective layer. In this photo, a sandstone block shelters the underlying shale from erosion, with as much as a meter of erosion on the surrounding slopes.

Fig. 6.11 Rotating block in heavily jointed sandstone on a cliff at Muley Point, Cedar Mesa, on the Colorado Plateau. In winter, rainfall freezes within the joints and frost wedging probably plays an important role in initiating rock fall processes. This block is in the same vicinity as the weathering pit shown in Fig. 6.4.

Fig. 6.12 Exposed rock slopes that lack a soil cover are often referred to as slickrock slopes. Weathering products are rapidly removed from such steep slopes by water, gravity, or wind erosion. Jointing strongly influences the erosion pattern and helps to emphasize the "bee-hive" topography. On the Colorado Plateau, such erosion is notable on the Navajo Sandstone (shown here). Note also the development of cavernous weathering.

If the caprock is very thick, it too is subject to erosional processes. Aspects of landform development on the backslope may be similar to that discussed for massive rocks. On the Colorado Plateau, such erosion is particularly notable on the Navajo Sandstone, a thick massive sandstone that is well exposed at Zion National Park, Arches National Park, and in the vicinity of Lake Powell. The term slickrock is often applied to these exposed rock slopes that lack a soil cover (Fig. 6.12). Weathering products are rapidly removed from these steep slopes by either water or wind erosion. In many cases, the bedding is emphasized by differential granular disintegration. Jointing strongly controls the erosion pattern. The fracture pattern allows the infiltration of water and aeolian sediment, and creates an environment in which plants can take root. During winter months, ice forms within the cracks and acts to wedge them apart. Thus, physical, chemical, and biological weathering processes act on the joints and further emphasize the "bee-hive" topography of the slickrock, so-called because of the convex slopes that develop in response to exfoliation or sheeting fractures (Bradley 1963).

The degree to which climate change has influenced the processes or rate of backwasting of escarpments on the Colorado Plateau is not known. Schumm and Chorley (1964, 1966) cite numerous

examples of rockfalls, which suggest that the process is still active. Schmidt (1989b) inferred long-term rates of about 3 m 1000 years^{-1} for scarp retreat. However, it is likely that caprock erosion today is less than during the Pleistocene (Reiche 1937; Ahnert 1960), when cooler and wetter conditions probably accelerated freeze–thaw activity and rill erosion, and promoted more areally extensive groundwater flow.

The role of groundwater sapping in cliff formation remains an area of active investigation, both on Earth and on Mars. Sapping results from groundwater outflow, which causes the disintegration of rocks and undermines the base of support for the caprock. It is often marked by zones of seepage, or by alcove development. Peel (1941) suggested that scarp retreat via groundwater sapping in the Gilf Kebir region of the Western Desert of Egypt must have accelerated during wetter phases of the Pleistocene. The area is presently hyperarid, with mean annual precipitation less than 1 mm. Over large areas of the Gilf, resistant beds of silicified sandstone outcrop at the surface. These beds are permeable to water infiltrating the plateau, and numerous caves at the cliff base are suggestive of spring activity.

The geomorphic significance of sapping processes in the maintenance of sandstone cliff form in the American southwest has long been recognized by individuals working in this region. Bryan (1928b, 1954) described the development of niches and alcoves in the walls of the Cliff House Sandstone, Chaco Canyon, New Mexico. The sandstone is broken up by a combination of cement dissolution, frost action, and salt crystallization. Schumm and Chorley (1964) noted that block movement is seasonal, increasing during the winter owing to frost action and wetting of the shale by snowmelt. Ahnert (1960) considered that basal sapping played an important role in the maintenance of cliff form in several sandstones. Laity and Malin (1985) reiterated the significance of groundwater to cliff development in canyon walls. This topic is discussed more fully in Chapter 7.

Although many escarpments on the Colorado Plateau develop by a combination of the processes outlined above, other less common mechanisms account for some features. One unique landform suite is the grabens of Canyonlands National Park, southeastern Utah, developed near the confluence of the Colorado and Green Rivers. This is a spectacular series of more-or-less systematically spaced grabens that average 150–200 m in width and 25–75 m in depth. They occur in a 460 m-thick plate of bedded brittle Pennsylvanian and Permian rocks that overlie ductile Pennsylvanian evaporites. The rocks dip 4° northwest toward Cataract Canyon of the Colorado River in a region where the river has eroded through the brittle plate down into the evaporites. The free face of the canyon permits the evaporites to flow down the slope, causing extension and faulting in the brittle plate. The graben faults are initiated at, or close to, the contact between the brittle and ductile rocks, and are propagated both upward and laterally (McGill & Stromquist 1974).

6.6.2 GRAVITY-RELATED ACTIVITY: TALUS AND SCREE SLOPES AND RELATED FORMS

Scree slope development in arid lands has received little attention relative to arctic and alpine environments, where freeze–thaw processes are dominant. However, studies of desert colluvial slopes adjacent to freestanding cliffs suggest that there is a continuum of forms and processes related to lithology, stratigraphy, structure, climate, and time. Slopes attain many different configurations: their form is a function of varying source materials, rates of formation, weathering processes, and gravity-driven contributions to the surfaces. Once in place, the colluvial

slopes may be modified by wind or water action, or protected by resistant clasts.

Talus and scree are accumulations of boulders that lie beneath an exposed free face or cliff. The principal cause of deposition is rock fall. In a review of gravitational processes acting in deserts, Abrahams et al. (1994) note that creep processes are relatively insignificant in deserts owing to the paucity of fine materials (with the possible exception of badlands; see Section 6.6.3); that landslides and other deep-seated failures are uncommon; and that mass movement processes are dominated by positional changes of surficial layers in processes ranging from debris flows to talus falls. Talus slope materials include those beneath cuesta scarps, discussed earlier, as well as more localized accumulations on mountain flanks and beneath vertical rock faces. The debris may be emplaced by processes that range from individual rock falls to debris avalanches.

The degree to which talus is being produced under present climatic conditions is poorly understood. For the Colorado Plateau, Ahnert (1960) commented on the limited amount of talus at cliff bases, and concluded that collapse must be limited under present climatic conditions. However, according to Schumm and Chorley (1966), talus is not common because it is largely destroyed during rock falls and further disintegrates into loose sand by weathering. Laity and Malin (1985) observed that the mechanically weak Navajo Sandstone shatters readily upon impact, and that boulders are commonly friable, rounded *in situ*, and surrounded by aprons of loose sand. At Lake Powell, Utah, the interiors of talus cones are exposed by slump failure, and the exposed cores reveal a very high percentage of fine-grained material, suggesting that boulders break down in place.

The relationship between the rate of debris supply and rate of disintegration on the footslope is critical in determining the nature of talus slopes. If rock wall collapse is infrequent and talus weathering rates are high, there will be only minor scree accumulation at a cliff base. The wetter conditions of many deserts in the recent past have probably had a significant influence on talus production and weathering, but these connections are poorly understood. Turner and Makhlouf (2002) suggest that for very arid areas of Jordan (<50 mm annual precipitation), increased rainfall inhibits the contribution of sandfalls and sand grainflows onto colluvial slopes, leaving rockfalls as the dominant, or perhaps the only, active disintegration process acting on cliffs. In high-altitude or interior deserts (such as the Colorado

Plateau), colder climatic conditions in the past may have exacerbated freeze–thaw processes.

In addition to talus, sand may be an important component of colluvial slopes in deserts. In eastern Jordan, colluvial fans are built up by rockfalls, rock-fall-derived debris flows, dry sandfalls, and sandy grainflows (Turner & Makhlouf 2002). The sediment is derived from the adjacent weakly cemented sandstone. Sandfalls are dry sandflows that cascade down the escarpment face. Grainflows are dry, cohesionless sand flows that evolve from sandfalls and move down narrow bedrock gullies. Rockfalls produce sandstone blocks that break up rapidly upon impact, to produce smaller clasts and volumes of dry sand. Some rockfalls accelerate downslope as cohesionless debris flows, with larger boulders floating at the surface, to produce a crude, upward-coarsening sequence.

A unique variation of the talus slope is a ramp slope (Mignon et al. 2005). This is a low-angle (<10°) slope, formed largely of fine debris, that lies beneath vertical cliffs and merges into adjacent wadi beds or playas. In southwest Jordan, they are developed in the massive Ishrin sandstone (part of the Nubian Sandstones), a poorly cemented and mechanically weak rock. The joints and bedding planes are separated by several meters and near-vertical cliff faces are common. The outcrops show extensive case-hardening (up to 5 cm thick), the presence of an algal layer, and extensive tafoni development. The rock slope/footslope junction appears to be the most aggressive weathering environment, probably as a result of groundwater sapping at the cliff base. When the cliff fails (rockslides), the impact of the debris produces large quantities of fine material. The remaining sandstone talus is rapidly disintegrated by salt weathering; fine material is distributed by wind and water down the footslope; and the ramp extends. Resistant ferruginous clasts mantle parts of the ramp and protect it from further erosion.

The development of gently sloping ramps appears to require a material that disintegrates very rapidly to fine fractions. In addition to the ramp slopes of Jordan, similar landforms probably develop in the presence of the Nubian Sandstone in North Africa and Arabia, where massive, iron-rich, cliff-forming sandstones surrounded by smooth depositional forms are common (Mignon et al. 2005).

6.6.3 BADLANDS

Landscapes with highly dissected, thick exposures of readily weathered and eroded rock units, usually shales or poorly cemented sandstones or alluvium, are referred to as badlands (Fig. 6.13). The term was

FIG. 6.13 Badlands are regions of very high drainage density that form where sediments are readily weathered and eroded. The Borrego Badlands are located near the Salton Sea (in the distance) and are formed in Tertiary and Quaternary conglomerates, fanglomerates, sandstones, and lacustrine claystones laid down in ancient Lake Coahuila.

probably derived from the expression "mauvaise terres à traverser" or "bad lands to cross," referring to the terrain in North American drylands, including Badlands National Monument, South Dakota (Campbell 1997). Owing to their very high dissection, large expanses of badlands appear maze-like and are disorienting to the traveler, leading to the alternate (nongeomorphic) definition of a badland as "a hell of a place to lose a cow."

Badlands are not limited to arid lands, but are perhaps more common there, although they comprise only a small percentage of the landscape (for example, 2% of the Sahara) (Clements et al. 1957). Erosion is relatively rapid which, coupled with the relatively infertile nature of the substrate, means that there is little vegetation growth. This feedback effect enhances erosion rates. Badlands with high percentages of swelling clay minerals are characterized by desiccation cracks, and there is often an associated development of subsurface flow in complex networks of pipes and tunnels, fed by sinkholes. Such forms are sometimes referred to as pseudokarst, to distinguish them from similar karst features that are the result of chemical solution.

Badlands are combinations of both slopes and drainage networks. In many cases, research has treated discrete portions of such landscapes; for example, slope development and shape (Davis 1892; Gilbert 1909; Moseley 1973; Yair et al. 1980), small-scale processes (Schumm 1956; Schumm & Lusby 1963; Campbell 1974), channel morphology (Schumm 1961; Cherkauer 1972), and piping (Harvey 1982; Jones 1990; Parker et al. 1990; Campbell 1997).

As badland slopes largely lack vegetation, the forms of the hillslopes and channels are largely a function of the properties and behavior of the underlying lithologies. The reaction of different soils to rainfall is closely linked to material properties, including texture, aggregation, and chemistry (Kuhn et al. 2004). Badland materials are inherently erodible by nature, and include swelling clays, silts, sands, and larger material, affected by surface water erosion, subsurface water flow (piping), rainsplash, creep, and small-scale slumping.

Surface water erosion on badland slopes constitutes falling or flowing water that is not concentrated in permanent channels. It includes rainsplash transport, unconcentrated surface wash, and rill erosion. The effects of these processes are moderated by the infiltration capacity of the soil. Raindrop impact is responsible for dislodging soil particles and transport-

ing them short distances by surface cratering, or longer distances by their inclusion in droplets. On sloping surfaces, the net movement of particles will be downslope. Soil particles loosened or detached by raindrops are the primary means by which thin sheets of unconcentrated surface wash can erode clay or silt slopes. When slope materials also include sands or larger particles, this erosive action by a thin water film is reduced, as the dislodgement of the finer particles by rainsplash tends to leave a relatively hard surface crust.

The development of convex divides in arid regions has been attributed to the unimpeded action of raindrops falling on a vegetationless surface (Gilbert 1909), the collecting area of the upper hillslope being as yet too small to allow overland flow. While the process of creep (Schumm 1956) and overland flow (Yair 1973) are also recognized as contributing to the development of convexities in some badland environments, Moseley (1973) demonstrated that in granular non-cohesive materials, convexities can develop solely under the action of rainsplash. If such activity was wholly responsible for the formation of slope profiles, the form might be convex throughout. However, in most cases rainsplash is also accompanied by soil wash and overland flow on the lower slopes with the resultant development of straight or slightly concave profiles in these areas. Convexities at the slope's base develop when the stream channel at the base of the slope moves laterally towards the hillslope, or downcuts more rapidly than the slope can adjust.

Owing to the absence of vegetation, the nature and rates of infiltration are directly related to rainfall characteristics and to type of material. In particular, if swelling clays and shales are present in a basin, their unique three-dimensional swelling properties will strongly affect infiltration behavior and the hydrologic environment. Swelling clays are materials of low inherent permeability, the initially high velocities of infiltration being dependent mainly on such natural defects as soil aggregates, desiccation cracks, fissures, root holes, and animal holes. The high initial permeabilities attributable to these defects are a transient phenomenon, declining in value as swelling and filling of the voids and defects proceed.

Desiccation cracks are responsible for high rates of infiltration in many badland clays. They develop initially in response to internal tensile stress that occurs during the drying phase. In natural crack systems, the forms are not ideal polygons, owing to the random nature of initiation and stress or fabric anisotropies resulting in specific planes of weakness. The cracks provide surface water access by gravity and capillarity to the lower layers of the soil. In swelling clay field experiments in Australia, Quirk and Blackmore (1955) found that an average of 70% of the total water intake during a 4 hour period actually infiltrated during the first 10 minutes. The availability of void space in the underlying soil mass is the main limiting factor under maximum crack entry conditions. In most cases, however, water entering the soil mass will be sufficient to close the surface cracks before total profile saturation occurs.

Unconcentrated surface wash in a thin uniform sheet is only possible on smooth surfaces. In badlands, most water is diverted into shallow rills, where flow concentrations are too local and flow durations too brief to produce permanent channels. Surface flow commences if precipitation is sufficient to produce soil aggregate expansion which closes the cracks, thereby sealing the surface and preventing further infiltration. Runoff will tend to follow the downslope component of the old shrinkage cracks, forming rills.

Creep processes are significant on swelling clay hillslopes, but are of minor importance relative to surface water erosion on nonswelling materials. Creep is primarily a surface process, and decreases markedly with depth. It is prompted by the swelling and shrinking of the surface aggregates which, when on a slope, expand or contract primarily on the downhill side of weight distribution (Schumm 1956).

Badlands are notable for their very high rates of erosion. Chen (1983) reported sediment yields in excess of 38,000 t km^{-2} year^{-1} from the Loess Plateau badlands of China. Campbell (1992) showed that a badland area that comprises only 2% of the Red Deer River basin, Alberta, Canada, nonetheless accounted for almost 80% of the river's mean annual suspended sediment load.

6.6.3.1 Case study: Borrego Badlands, California

The Borrego Badlands are formed in Tertiary and Quaternary continental deposits laid down in ancient Lake Coahuila, which formerly covered a large part of the Imperial and Coachella Valleys of southern California (Fig. 6.14). Much of the area is composed of moderately to poorly consolidated conglomerates, fanglomerates, sandstones, and lacustrine claystones.

FIG. 6.14 Small basin (known as Moth Basin) formed in lacustrine claystones, Borrego Badlands, California, USA. The hillslopes are convex in form. Processes operating in this basin include rainsplash, slope failure, overland flow, and piping. Small, ephemeral rills cover the surface of the basin, but subsurface pipes carry most of the runoff to the outlet. Dessication cracks cover the surface of the slopes and are responsible for high initial rates of infiltration.

The response to rainfall is highly nonuniform and strongly affected by lithology, as shown by an investigation of two basins (J. Laity, unpublished results). The first, called Moth Basin, is composed predominantly of lacustrine claystones. The thinly bedded shales dip 22°NE and are weathered to an average depth of 33 cm. This is typical of badlands, which typically have very thin regoliths, with maximum values of about 30 cm (Howard 1994). The alternate swelling and shrinking of the brown clays have resulted in a complete surface cover of desiccation cracks and a "popcorn-like" surface texture (Fig. 6.14). A second basin, Eagle Basin, includes brown swelling clays, as well as sandy clays grading to coarser sands and gravels. Piping is a prominent feature of both basins, and these tunnels appear to instigate both headward erosion and down-cutting in the stream channels. Drainage density values for Moth and Eagle Basins are high, 166 and 201 km km^{-2} respectively, as are the relief ratios (0.260 and 0.331; the ratio between total basin relief and the longest dimension of the basin parallel to the principal drainage divide). The effect of lithology on drainage density was also considered: values are 145 km km^{-2} for the swelling clays and 298 km km^{-2} for the pebbly sands. The difference in these values is related to the higher initial infiltration capacity of the swelling clays, and the proportional increases in throughflow and piping flow relative to surface runoff.

The maximum slope angles in the basins are very steep, with a mean for Eagle Basin of about 45–50°. As a result, soil slips are a common occurrence, occurring principally on hillslopes exceeding 34°, and along the steep slopes adjacent to main channels. Aggregate movement appears most rapid on the steep lower slopes, with excess basal material removed by adjacent channel runoff.

Rills are transient features, subject to position changes produced by the continual upheavals of clay under varying moisture conditions. Field observations show that 2 weeks after a rain, crack propagation has eliminated many of the new rills. As long as the rills remain fairly shallow, they tend to change position frequently. Dense networks of rills are less evident on sandy or pebbly surfaces, because water depth and velocity are insufficient to erode and transport the coarser debris. Instead, water flows in a relatively unconcentrated form over the short interchannel distances. However, there is a greater density of permanent channels than for basins formed of swelling clays alone.

6.7 COMPOSITE SURFACES (PEDIMENTS)

Pediments are found in many areas of the world, but are most clearly associated with arid and semiarid climates, where they have received the most scrutiny. Pediments of late Tertiary and Quaternary landscapes of the Basin and Range Province of the western USA are widespread and have been the most intensively studied (Hadley 1967; Oberlander 1972; Cooke & Warren 1973; Moss 1977; Dohrenwend 1987).

Despite geomorphic research spanning many decades and an extensive literature, these landforms remain poorly defined and their formative processes obscure. There are many definitions of pediments, each differing somewhat in detail (Moss 1977; Dohrenwend 1994). According to Oberlander (1997, p. 135), a pediment is a "gently inclined slope of transportation and/or erosion that truncates rock and connects eroding slopes or scarps to areas of sediment deposition at lower levels." In his view, the term has been applied to two geometrically similar geomorphic elements that have very different genetic origins.

The first of these two forms is an erosional surface beveling the weaker substrate of a more resistant caprock escarpment (Gilbert 1877): forms which are often referred to as *glacis d'erosion*. The beveled lower slopes develop from hydraulic planation of the underlying weaker material and they are commonly veneered by alluvial gravels.

The second, and more controversial form, is the rock pediment, composed of the same or similar material throughout (from the upland to the erosion surface), commonly (but not always) a granite, granodiorite, or quartz monzonite. These are the pediments discussed in this section. Pediments generally lack a gravel veneer, and the surface cuts both sound and decomposed rock. At the base of the relatively straight slope, there may be onlapping alluvium. The pediment expands in an upslope direction as the upland hillslopes erode back at a constant angle (Fig. 6.15). On the pediment slope, there may be isolated erosional residuals, termed inselbergs. Pediments lack terrace development.

One of the most notable aspects of pediment morphology is its generally planar and featureless surface. The longitudinal slope ranges between 0.5 and 11° (Tator 1952; Cooke & Warren 1973). In the southwestern USA, slopes typically range between 2 and 4° and seldom exceed 6° (Dohrenwend 1994). Pediments terminate against a mountain front whose angle is generally greater than 20° (Moss 1977). The

Fig. 6.15 Rock pediments are beveled surfaces that generally lack a gravel veneer. The granite pediments of the Mojave Desert, California, USA, have received considerable attention, but their formation is not fully understood. Photograph location: Sheephole Mountains, California.

long profile is, for the most part, concave upward, similar to that of a stream channel. Thus, pediments are generally considered to be graded surfaces of fluvial transport, with the slope adjusted to the discharge and sediment load. The transverse surface profile is more complex in form, owing to the tendency for water to channelize. The intersection of channels and erosion processes result in several general forms, including convex-upward, concave-upward, or rectilinear shapes (Dohrenwend 1994). Inselbergs may stand above the surrounding erosional plain. They vary greatly in abundance, size, and form.

The bedrock pediments of the southwestern USA have not been systematically mapped with respect to their tectonic environment, but appear to be more abundant in quasi-stable geomorphic environments where erosion and deposition have been essentially in balance for long periods of time (Dohrenwend 1994). With respect to the Basin and Range Province, alluvial fans are prominent where range height, relief, length, and volume are large, and pediments best developed where these values are low (Lustig 1969).

The influence of climate on pediment development has not been clearly established. Although there are similarities in form between pediments in tropical climates and arid lands, it is not clear whether these morphologic similarities also imply the same

processes of mass transport. In general, it is thought that pediments have formed over a very long period of time, suggesting that several climatic cycles may be involved (Lawson 1915; Bryan 1922; Howard 1942). The granite pediments of the Mojave Desert may have been fully developed by late Miocene time (Oberlander 1974), and since then largely modified by erosion of the regolith by slope wash, rill erosion, and channelized flow, and by chemical breakdown of the regolith along a subsurface weathering front. Radiometric dating of packrat middens in Joshua Tree National Monument, California, indicate no appreciable scarp retreat in quartz monzonite over the past 10,000 years (Oberlander 1974).

The formation of pediments remains contentious and has spawned a very large body of desert literature. There are several theories concerning pediment formation. The most commonly proffered is the headward growth of erosional pediments. In this model, the dominant process is backwearing of the mountain front accompanied by concurrent development of the beveled plane extending outward from its base. The beveled surface may develop by lateral planation (Gilbert 1877), sheetflooding (McGee 1897), streamflooding and sheetflooding (Davis 1938), rill cutting, or backwearing of the escarpment by moisture concentration (Twidale 1967). Moss (1977) emphasized the importance of downwasting by weathering to explain the origin of granite pediments in Arizona, where as much as 20 m of residual grus mantles the irregular pediment surfaces. Decomposition of the granite principally occurred during times of wetter climate and was followed by erosion. Modern processes largely modify a landform produced in the past. An entirely different theory was proposed by Twidale (1978b) on the basis of studies in Australia. In his view, granitic landscapes are structurally controlled and the landscape lowering is controlled by petrography and joint density. Given these different theories, it is possible that pediments may develop from more than one set of processes, with a history and origin than varies considerably according to the environment. Despite the long history of inquiry, pediments remain an enigmatic and controversial landform.

7

DESERT SOILS AND GEOMORPHIC SURFACES

7.1 INTRODUCTION

The desert surface is a mosaic of different soil types and geomorphic covers, some only recently formed, but others relics of the Pleistocene or even earlier. This chapter examines the nature of desert soils and several widespread surfaces, including desert (stone) pavements and inorganic and biological crusts. Stone pavements warrant special consideration, as they cover 50% of desert surfaces, more than any other surface type. Other surface types discussed in this chapter include extensive rocky surfaces (hamadas) and patterned ground (gilgai).

7.2 THE NATURE OF SOILS IN ARID AND SEMIARID REGIONS

Soils are an important component of the geomorphic environment of deserts, although they are perhaps more difficult to recognize than their counterparts in more humid regions. They are an essential substrate for plants, which depend on them for nutrients, water, and physical support. Desert soils often vary greatly from place to place, being influenced by factors and processes that have substantial spatial variation, over both large and short distances. A soil on the floodplain of a major river, such as the Tigris, for example, will differ greatly from one in an elevated or sandy location.

Key factors influencing soil development include vegetation, faunal activity, climate (precipitation and temperature), aeolian inputs, and time. Vegetation determines such important soil qualities as the organic content, and affects the biotic environment of the soil. The very strong interrelationships between geomorphology, soil development, and vegetation patterns in arid and semiarid environments are being increasingly recognized (Buxbaum & Vanderbilt 2007). Faunal activity, including burrowing, is often

a function of soil age and structure (Shafer et al. 2007). Such activity helps to redistribute minerals and nutrients, and increase aeration and infiltration. Precipitation affects weathering and the downward movement of nutrients and chemical compounds, and temperature affects the rate of decay of organic matter (Fierer et al. 2005). Aeolian inputs of dust and salts are important contributors to soil development, with loess soils relatively common. Time is also an important factor, as the characteristic properties of a soil require an extended period to develop (often many thousands of years), and may change dramatically during that period (Meadows et al. 2007).

The field study of soils is referred to as *pedology*, and *pedogenesis* is the process of soil formation. There are several ways to define a soil. From a geomorphic perspective, the definition of Birkeland is useful:

> . . . a natural body consisting of layers or horizons of mineral and/or organic constituents of variable thicknesses, which differ from the parent material in their morphological, physical, chemical, and mineralogical properties and their biological characteristics; at least some of these properties are pedogenic. Soil horizons generally are unconsolidated, but some contain sufficient amounts of silica, carbonates, or iron oxides to be cemented. (Birkeland 1984, p. 3)

Soils develop on a parent material, which consists of all forms of mineral matter suitable for transformation into soil. The parent material may be residual in nature, such as a layer of regolith above bedrock, or may consist of mineral particles transported into a region, by the action of streams, waves and water currents, or winds.

How do the soils of deserts differ from other regions? In general, it can be said that they are either thin or non-existent and their distribution tends to be spatially patchy. Soils are somewhat deeper and better developed in semiarid areas than true deserts (Yair & Kossovsky 2002). True desert soils are found principally on low-angle, stable surfaces, such as pediments or alluvial fans (Cooke & Warren 1973). They are commonly absent from steep slopes, bare rock surfaces, fresh lava flows, recent alluvium, and moving sand dunes. In deserts, high temperatures and low precipitation limit soil moisture and depths of leaching are shallow (Dohrenwend 1987). In the past, the leaching depths may have been greater, leaving an imprint on the soil.

Desert soils are often considered to be poorly developed: relative to other environments, there is less organic material and less horizon development. The low plant productivity restricts the soil-building properties of micro-organisms that convert organic matter into humus. As a result, dryland soils have a low water-holding capacity, and tend to shed much of the rain that falls on them (Dunkerley 2002).

Desert soils are commonly more saline than those in humid regions. Salts are deposited by the wind or are wicked towards the surface by capillary action where the water table lies close to the surface. As discussed in the section on desert pavements (Section 7.8.2), many desert soils represent accumulations of windblown materials, including sand, silt, and salts, that have been entrained from other surfaces such as playas, and deposited downwind (Reheis et al. 1995). Windblown materials provide allogenic minerals, salts, and clay in quantities that are more abundant than weathering in an arid environment would produce *in situ*. This is in contrast to humid regions, where weathering of parent materials produces much of the soil material. The accumulation of windborne and waterborne material in desert soils is very slow. As a result, climatic and environmental changes leave an imprint on soil development. In Australia, for example, episodes of enhanced swelling and shrinking in clay soils in the past have affected the development of stone sorting and gilgai (hummocky surface development) microrelief (Dunkerley & Brown 1997).

Soil degradation is an important environmental concern in arid regions, affecting the sustainability and productive capacity of the land. A degraded soil is associated with long-term changes in ecosystem function, physical structure and chemical components, nutrient content, and biodiversity (Melegy 2005). The conversion of semiarid grasslands to grazing or agricultural land has had significant impacts on soil quality (see Section 13.3.1). In some areas, a pattern of land use, exploitation, and abandonment develops, resulting in a mosaic of different soil types with varying degrees of degradation (Wang et al. 2007). Livestock trampling and the compressional and shear forces generated by vehicles in the desert, including military vehicles and various types of all-terrain vehicle, exacerbate soil loss. The loss of soil fines by wind and water reduces the productivity of the soil, as nutrients are lost and the water-holding capacity of the soil depleted.

7.3 SOIL DESCRIPTION AND CLASSIFICATION

A description of soils usually considers the soil profile, a vertical arrangement of soil horizons that extends to the parent material. In soil horizon nomenclature, capital letters are used to denote master horizons (in deserts, surface or near-surface horizons are commonly designated as A horizons; the B horizon lies beneath the surface horizon; and the slightly weathered C horizon lies beneath the B; with unweathered bedrock beneath the C). Lower-case letters refer to a specific characteristic or subdivision of the master horizon and Arabic numerals indicate a further subdivision of the horizon (see Birkeland 1984 for additional details). Not all soil horizon nomenclature is used in all climatic regions, owing to the different environmental characteristics of each.

Soils are classified into major types and subtypes that are recognized in their distribution over the Earth's surface. The top level of this classification scheme consists of soil orders; a second level consists of suborders. The orders and suborders are often recognized by the presence of a diagnostic horizon that has a unique combination of physical properties, such as color, texture, and structure, or chemical properties (the presence or absence of certain minerals) (Birkeland 1984). Many desert soils fall into a group with poorly developed horizons or no horizons. These include the entisols, soils that lack horizons, usually because their parent material has accumulated only recently, and the inceptisols, soils that have weakly developed horizons, and have minerals capable of further alteration by weathering processes. Entisols support plants but lack distinct horizons for one of two reasons: (1) horizons do not readily form in the parent material, often sand, or (2) there has been insufficient time for soils to form (for example, on recent alluvial deposits). The weakly developed profiles of inceptisols are often accounted for by the youth of the deposit. Other desert soils have more well-developed horizons, or have fully weathered minerals, as a result of a long adjustment to prevailing soil temperature and water conditions:

these are the aridisols, soils of dry climates that are low in organic matter and often have subsurface horizons of accumulation with carbonate minerals or soluble salts. Humus is lacking owing to the sparse vegetation; the soil color is often pale gray to pale red; soil horizons are weakly developed; and there are often important subsurface horizons of accumulated calcium carbonate or soluble salts. The salts may give the soil a high degree of alkalinity, often restricting plant growth. Although aridisols are traditionally used for nomadic grazing, they can be highly productive under irrigation, in areas such as the Nile Valley of Egypt, the Indus Valley of Pakistan, or the Imperial Valley of the USA. However, important impediments to agriculture must be overcome, including waterlogging and salt build-up. Vertisols are clay soils that crack deeply, with shrink–swell characteristics common (Birkeland 1984). Alternating periods of expansion during wet weather and contraction and deep cracking during hot, dry periods may give rise to terrains of low relief characterized by the presence of hollows, rims, and mounds that denote gilgai (Kariuki et al. 2004). This topography has been widely mapped in swelling soils of Australia, but is also found in other arid environments. Swelling and shrinking soils pose a problem for cities in arid environments, as they can cause widespread damage to infrastructure (Kariuki et al. 2004).

Deserts soils are less studied and understood than those of more humid environments. The taxonomic classification of desert soils is difficult, in that there is very little horizon differentiation. The soils often show accumulations of soluble salts and minerals, which are rare in more humid areas. Other features, such as pavement development, microbiotic crusts, vesicularity in the A horizon, and surface sealing are of relatively little use taxonomically (Dunkerley & Brown 1997), but are important geomorphologically. Furthermore, as desert soils show such a high degree of spatial variation over short distances – often forming two-phase systems of water-shedding, organically poor, unvegetated soils, and water-absorbing, richer, and vegetated soils; each with different moisture characteristics, salinity, and leaching characteristics – the taxonomic classification of soils is highly problematic. Changes in moisture along a transect, particularly associated with increased elevation and effective moisture, may also result in a range of different soil classifications (Chadwick et al. 1995). As a result of these complicating factors, taxonomic classification often receives little attention in reviews of desert soil.

7.4 SOIL CHARACTERISTICS OF ARID REGIONS

Soil characteristics vary widely in arid regions, over both large and small scales. The formation of soil relates to the mass balance between inputs to the initial parent rock or sediment, such as water, organic matter, and atmospheric salts and dust; and losses, including dissolved losses from chemical weathering and transport of solutes out of the weathering zone, or physical removal by water or wind erosion. These gains and losses are integrated over geological timescales. With decreasing rainfall, dissolved losses are reduced, and the slow gain of atmospheric inputs leads to gains in dust (McFadden et al. 1987; Reheis et al. 1995; Anderson et al. 2002; Young et al. 2004) and salt (Ewing et al. 2006). This section summarizes some of the common features of desert soils.

7.4.1 PHYSICAL CHARACTERISTICS

One of the more obvious characteristics of a soil is its color. Although soil color may be inherited from the mineral parent material, it is more commonly a result of soil-forming processes. In deserts, reddening of the soil may occur (Quade et al. 2007), notably in the argillic B horizons of Pleistocene soils (Dohrenwend 1987).

Soluble salts in the unsaturated zone of soils are widespread in arid lands. Associated with salinity are alkaline soils, which have a pH value greater than 7.3. In general, soluble salt concentrations are highest near and immediately below the surface and decrease with depth. However, the ion content of the soil may vary considerably from site to site. In northwest China, for example, soil pH values at Ejina Oasis, along the lower reaches of the Heihe River, range from 8.2 to 9.7 (Zhou et al. 2006), and stabilized dune soils in the Gurbantonggut Desert have pH values from 8.43 to 8.66 (Chen et al. 2007). Where the salt concentrations are very high, the soil may be cemented. In the Atacama Desert, anhydrite, halite, nitratite and other salts cause pedogenic mineralization in the absolute desert, cementing deserts to many meters in depth (Ericksen 1981, 1983).

Episodic aeolian additions have provided fine-grained silicate and carbonate minerals to many desert soils. In southern Nevada and California, the sub-2 mm fractions of the A and B soil horizons are very similar to modern dust in a number of factors, including grain size, calcium carbonate content, major oxides, and clay mineralogy, with modern dust accumulation rates similar to soil accumulation rates (Reheis et al. 1995). This dust is derived principally from dry lake beds (Harden 1990; Chadwick et al. 1995; Reheis et al. 1995) during interpluvial periods, when sediments are exposed to deflation. These climatically controlled inputs of loess affect the depth of leaching, water retention properties, and the type of mineral weathering that occurs.

The highest rates of dust deposition are in semiarid areas. As a result, soil properties may change along a climatic gradient, with areas covered by loess more extensive in semiarid than arid areas, where sandy and rock surfaces are more prevalent (Yair & Kossovsky 2002). In some areas of the world, such as the Sahel and southern Israel, the addition of dust has formed deep, silty soils of loess. In the Halluqim catchment in the Northern Highlands of the Negev Desert of Israel, for example, 56% of the surface is covered by loess soils less than or equal to 0.05 m deep, 31% by soils 0.05–0.20 m deep, and 13% by deeper soils (Buis & Veldkamp 2008). Soil depth influences hydrologic processes. Where soils are shallow, the infiltration capacity is limited, and runoff occurs even for small rainfall events. This water may be redistributed to the deeper soils of the slope.

At the transition from arid to hyperarid conditions, there is a shift from a biotic to a largely abiotic environment. A pronounced decrease in soil organic C is noted with decreasing rainfall. Other processes linked to water, such as solute leaching, also decline, and more soluble salts are retained (Ewing et al. 2006). The geochemistry of the soils principally reflects the inputs of atmospheric dust and salts, which physically expand the landscape as it undergoes long-term mass gain and volumetric expansion.

7.4.2 THE ORGANIC CONTENT OF SOILS AND NUTRIENT AVAILABILITY

The organic content in soils depends on litterfall. As vegetation is relatively scarce in arid environments, the percentage of organic matter is commonly very low and few desert soils show darkening of the surface A horizons. In central Asia, the organic content is commonly less than 2%, and often less than 1% (Cooke & Warren 1973). Organic matter in stabilized dune sands in the Gurbantonggut Desert, China, is very low, ranging from 0.078 to 0.158% (Chen et al. 2007). Even lower values are found in the deserts of the Atacama, where the percentage of

organic matter ranges from less than 0.03% in the absolute desert, to 0.05–0.06% on the absolute desert fringes. Values increase to 5–12% in the coastal lomas zone (Quade et al. 2007).

At a local scale, the organic content of soils is related to differences between unvegetated and vegetated patches, varying over distances of between 10 and 100 m (Dunkerley & Brown 1997). In the Mexican Chihuahuan Desert, observed organic matter levels were 2.6% in vegetated zones, and 0.8% in the intervening bare zones (Montaña et al. 1988).

The nutrient availability in soils is also low, reducing primary productivity. This is partially the result of the limited time in which soils are damp, restricting microbial activity. Microbial processing of organic matter to produce plant-available nutrients can be activated by even small amounts of precipitation. Activity rates also depend on temperature, light, and substrate (Belnap et al. 2005). Biological soil crusts can increase soil fertility, enhancing the biomass of vascular plants that grow in them (Belnap 2003). Rain falling on a biological crust will cause a pulsed increase in C and N fixation.

Desert dust provides an important source of soil nutrients in areas where the slow weathering of bedrock would otherwise produce a relatively impoverished soil. The input of dust, and thus nutrients, varies over time. On the Colorado Plateau, for example, dust deposition was enhanced in the late Pleistocene to early Holocene, when lake levels throughout western North America were low, exposing surfaces that emitted dust. As a result, past processes affect modern ecosystem function, with today's plants utilizing nutrients deposited around 15–12 ka (Reynolds et al. 2006).

7.4.3 ROLE OF THE PAST

Soils have proved valuable in dating deposits (on the basis of soil development) and reconstructing palaeoenvironments in deserts (Shlemon 1978; Dohrenwend 1987). As many desert soils formed during earlier climatic regimes, particularly the Quaternary, the study of environmental change is essential to adequately understand their formation. Soils experienced profound fluctuations in vegetation cover, moisture, and aeolian dust and salt influx, that affected both the processes and rates of soil development (Chadwick et al. 1995), leaving a polygenetic imprint on the soil. A polygenetic soil is one that records multiple mineralogical, morphological, and chemical imprints as climate changes over time.

Large climatic excursions, such as a change from glacial to interglacial conditions, with accompanying hydrologic and temperature changes, have strongly influenced the pedologic environment. Areas of the desert that have more deeply weathered mantles, for example, are considered reflective of more humid phases in the past. Soil profiles undergo constant evolution and modification in the context of changing effective moisture conditions which influence: (1) depth of leaching, (2) episodic additions of dust, (3) mineral weathering rates, which are slower during dry periods, (4) biomass, (5) organic carbon, (6) base saturation and exchange acidity, and (7) changes in the rates of types of pedogenic mineral synthesis. For example, carbonate, gypsum, and smectite may be synthesized during interpluvials, and kaolinite and vermiculite in pluvials (Chadwick et al. 1995).

Heterogeneous soil conditions caused by past climatic fluctuations can be the source of modern vegetation patterns that differ over short distances. Buxbaum and Vanderbilt (2007) examined a level surface (<1% slope) in central New Mexico and observed that plant species varied according to abrupt changes in soils, associated with a pattern of buried channels incised in a petrocalcic horizon (caliche) formed in a 0.5–1.2 million-year-old palaeosol beneath the current soil surface. Blue grama (*Bouteloua gracilis*) dominates in deep soils with thick argillic B horizons; black grama (*Bouoteloua eriopoda*) is abundant where the buried petrocalcic horizon lies less than a meter below the surface; and creosote bush (*Larrea tridentata*) dominates where the petrocalcic horizon is either exposed or lies close to the surface.

Stone pavements developed on soils have been used as the basis for subdividing and correlating Quaternary alluvial fans, and for studying climate change and neotectonics (Bull 1991; McFadden et al. 1998; McDonald et al. 2003). Pavements become more strongly developed on older surfaces, and their hydrologic properties change, as do plant size and vegetative cover (Young et al. 2004).

7.4.4 ROLE OF RELIEF AND ALTITUDE

The steepness of a slope affects the amount of runoff, infiltration, and erosion. In general there is a break point on slopes, above which, on steeper slopes, soils

are absent; and below which, on shallower slopes, soils may develop (Cooke et al. 1993). Soils may also be younger on the steeper slopes and older on the more level ones.

In mountainous areas, there may be considerable changes in soil properties with altitude. Chadwick et al. (1995) demonstrated that organic carbon in the soil increases with altitude, rainfall, and biomass, until cooler temperatures and a short growing season reduce biomass. Leaching depths may also increase with altitude owing to greater effective moisture.

7.5 INORGANIC AND BIOLOGICAL SOIL CRUSTS

Desert soil surfaces may be covered with a number of different surface types. This section examines inorganic and biological soil crusts, which may coexist within certain areas. Table 7.1 contrasts the eco-hydrologic properties of these two types of surface cover.

7.5.1 INORGANIC SOIL CRUSTS

Inorganic soil crusts commonly form in the absence of vegetation where rain beats down on an exposed soil. The term *surface sealing* is sometimes applied to the initial or wetting phase of development, and *crusting* to the hardening of the seal as the surface dries (Slattery & Bryan 1992). Inorganic crusts can be subdivided into structural crusts, related to raindrop impact, and depositional or sedimentary crusts, where seals develop when fine particles carried in suspension are deposited in depressions (Assouline 2004). This section will examine structural crusts. The complex development of crusts has been studied using experimental investigations and simulation models, and several models of formation proposed. The process, particularly in desert environments, remains incompletely understood.

Crusts commonly show a microstratigraphy in which the surface is covered in a very thin veneer of clay, usually about 0.1 mm thick (Epstein & Grant 1973; Chen et al. 1980). The crust thicknesses reported under field conditions (up to 20 mm) are generally much thicker than those noted from laboratory experiments (Assouline 2004). The uppermost clay layer is devoid of pores, with the percentage volume of pores increasing to 10–25% at a depth of 2–6 mm beneath the surface (Epstein & Grant 1973). As a result of compaction, and the infiltration of fines

TABLE 7.1 Although soil crusts and biological crusts may coexist within a region, they vary greatly in their ecohydrologic role.

Soil crusts	Biological crusts
Reduce water infiltration	Enhance water infiltration when rough; reduce infiltration when smooth
Inhibit plant establishment	May enhance establishment of native plants
Protect the surface from wind erosion when dry	Contribute to soil cohesion and stability, therefore reducing wind and water erosion
Dissolve when wet and are prone to erosion	
Are organically poor and lack fertility	Increase the fertility of soils; trap dust, augmenting plant-essential nutrients, including nitrogen, phosphorus, and potassium
Smooth surfaces	Rough surfaces in temperate deserts; smooth in hyperarid regions
High albedo	Lower albedo; absorb energy and increase soil temperature
High rates of evaporation from soil	Rates of evaporation are retarded relative to soil crusts

Sources: Belnap (2003) and Belnap et al. (2005).

into voids in undisturbed soil beneath the surface, there is an increase in soil bulk density towards the surface.

Factors leading to the development of crusts include those characterizing the storm (raindrop size and rain intensity) and those of the soil (water content, bulk density, organic matter content, and exchangeable sodium percentage, or ESP). Physical crusts are best developed in fine-textured soils (with clays and silts) and where salt is present. Raindrop impact is believed to be the principal cause of soil crust formation. Direct raindrop impact on unprotected soils causes rapid disruption of surface aggregates, compaction, slaking, segregation of particles, and the filling of pores by wash-in of fine material. Both McIntyre (1958a, 1958b) and Epstein and Grant (1973) consider compaction of the soil by raindrops to be important in seal formation. The raindrop size in desert environments may be somewhat greater than in more humid environments, as the smallest drops are lost to evaporation in the dry air (Dunkerley & Brown 1997). In addition to drop size, dispersion is affected by the chemistry of the soil water system, so that both chemical and physical dispersion occur (Agassi et al. 1981). Aggregate breakdown by raindrops is more probably more efficient after they have been wetted by rainfall. The uppermost crust or skin is likely to be the result of the deposition of fine particles in suspension after rainfall ceases, explained by the lower settling velocities of clays relative to coarser particles.

The development of a crust has numerous environmental effects, reducing the infiltration rate, the availability of water to the roots of plants, and crop yields, and increasing the amount of runoff and soil erosion. It can therefore enhance the desertification process through a feedback mechanism. As crusts form, the vegetation is reduced, overgrazing occurs on the remaining vegetation, and the bare exposed soil is exposed to more rainfall, incurring more soil surface sealing (Assouline 2004).

7.5.2 BIOLOGICAL/CRYPTOBIOTIC SURFACE CRUSTS

Biological soil crusts, also known as cryptobiotic (microphytic, microbiotic, cryptogamic, biogenic, or microfloral) crusts, are found globally on exposed sandy or silty soils in open shrub and grassland communities in arid and semiarid environments. The crusts are composed of nonvascular plants including algae, fungi, lichens, and bryophytes. They affect soil surface stability, water-infiltration capacity, and plant succession (Eldridge 1993; Belnap 2003). The distribution, ecology, and role of nonvascular plants in arid and semiarid environments are not fully known, but are rapidly gaining scientific attention. In North America, they are common throughout the Great Basin, Colorado Plateau (Fig. 7.1), and the Columbia Basin. They have also been described in arid areas across Asia, from the Middle East to China.

(a)

(b)

FIG. 7.1 Cryptobiotic soils on the Colorado Plateau. (a) General overview of a crust on sandy soil. (b) Close view of crusted soils, illustrating several centimeters of relief across rough surfaces. Ponding takes place in the depressions, increasing infiltration. Photographs courtesy of Robert Howard.

Cryptobiotic crusts are frequently overlooked, yet may be seen in a variety of geomorphic settings, including small, stone-free patches on alluvial fans. It is likely that future research will further emphasize the role of such crusts on the ecology, hydrology, and geomorphology of deserts.

Cyanobacteria, or blue-green algae, play a major role in the development and stability of cryptobiotic soils. A slimy mucoid substance is exuded by sheaths surrounding living filaments of blue-green algae. Sheaths and filaments become enlarged and swollen when wet, forming a net-like structure that binds and aggregates particles to enhance soil stability. When the soil is dry, the filaments are completely encased in the sheaths, but the sheath still holds firmly to soil particles. The three-dimensional network of filaments also promotes a mechanical binding of the soil (Danin 1991).

The composition of biological soil crusts depends upon the desert conditions. In cool deserts, such as the Colorado Plateau, a rich lichen–moss flora dominates. By contrast, crusts dominated by cyanobacteria largely cover hot deserts, such as the Sonoran Desert (Belnap et al. 2005). Furthermore, cyanobacteria are most significant in disturbed surfaces.

Cryptobiotic crusts develop on sandy soils that are augmented by the infiltration of airborne silt and dust. The substrate is relatively stable in the sense that deflation and accumulation of sands is minimal. Cyanobacteria are found when silt accumulation in the soil reaches 1.5–2% (Danin et al. 1989), and they increase in quantity with time as fine-grained material is added to the soil. The process involves feedback mechanisms, as the cyanobacteria increase sand stability and promote deposition of silt and clay, and the improved moisture regime resulting from higher proportions of finer materials allows the growth of still more cyanobacteria. On stabilized dunes, the high wind speeds near dune crests limit the establishment of crusts, whereas the low wind speeds and gentle topography of interdune areas improve the environmental conditions (including moisture retention and seepage), so that crust coverage, thickness, and species change along a gradient. In the Gurbantonggut Desert, for example, animalcule (minute, usually microscopic, organisms) communities occupy the top of dunes, algal crusts the upper and middle parts, lichen crusts the middle and lower parts, and moss crusts the lower dune slopes and interdune lowlands (Chen et al. 2007). Average crust thickness is 0.05–0.1 cm at the dune crests, and increases to 1.5–5.0 cm in the interdune areas.

Cyanobacteria in young crusts appear green in color and occur beneath a thin protective layer of sand. With time, they grow above the surface and become darker in color owing to pigments that protect the cells from photo-oxidative death (Abeliovich & Shilo 1972). Crusts in the early stage of development are not readily visible to the observer, and a microscope may be required to detect the presence of algae. Once established, microphytes are able to persist through extended dry periods and extremely high or low temperatures (Lange 1990). Spatial coverage of cryptobiotic crusts can be greater than that of vascular plants, and constitutes between 40 and 100% of the ground cover in some arid and semiarid communities (St. Clair et al. 1993). Cryptobiotic soil crusts provide up to 70% of the living cover on the Colorado Plateau (Belnap & Gardner 1993; Belnap et al. 2005) and as much as 70–80% in the southern Gurbantonggut Desert (Chen et al. 2007).

Microphytic crusts play an important role in protecting the soil from water (Booth 1941) and wind erosion. When crusts are well developed, winds are seldom able to move any soil particles. J. D. Williams et al. (1995b) used a portable wind tunnel to test the strength of the sheathing mechanism in 48 plots of cryptobiotic soil at Capital Reef National Park in Utah. Sixteen control plots were left undisturbed, an additional 16 were chemically treated to kill the cyanobacteria, but allow the filament structures to remain intact, and the remaining 16 plots were physically scalped to remove all vestiges of the cryptogams. Wind speed was increased until particle movement was observed. Little difference was noted in the threshold friction velocities of the control and chemically treated plots. The threshold friction velocity of the scalped plots was half that of the others and almost five times as much soil was dislodged. Comparable results were obtained on coastal dune blowouts in the Netherlands (Van den Ancker et al. 1985). Thus, soil aggregation by biogenic crusts is a factor in reducing erosion. On the Big Sand Mound in Iowa, Schulten (1985) demonstrated that beneath cyanobacteria-covered soil, 68% of the particles were aggregated in clumps greater than 2000 μm in diameter. For bare soil, only 4% of the particles fell in this size class.

Cryptobiotic crusts influence several hydrologic processes including infiltration rates, the depth of water penetration, runoff, and sediment production. Overall, crusted soils show increased depth of water penetration and less soil movement than non-crusted soils. However, the ability of crusted soils to absorb and hold more water is related to their biomass and microtopography (Brotherson & Rushforth 1983; Belnap et al. 2005). In hot deserts of the USA, where soils do not freeze, biological soil crusts smooth soil surfaces and tend to increase water runoff (Belnap et al. 2005). In cooler deserts, where soils freeze, a well-developed crust shows 2–3 cm of relief across rough surfaces (Fig. 7.1). Ponding takes place in the small depressions, increasing the time for infiltration to occur, while decreasing runoff. Sediment entrainment is reduced and the water percolates deeply. The ponding may also be aided by the hydrophobicity of cryptobiotic soils (J.D. Williams et al. 1995a). Because the sheath swells immediately upon wetting, a layer of tightly bound particles at the surface blocks infiltration. Water is shed from the higher crusted surface towards the depressions. Additionally, crusted soils trap wind-blown silt (Tsoar et al. 1995), tending to seal the surface.

Cryptobiotic soils have been observed to regenerate on soils. At Shapotou, China, crusts provide part of an overall program to reduce shifting sands in a desertified region. After stabilization of the sands by means of straw checkerboards, soil algae form the initial successional stage, developing a dark surface crust typically 1–3 mm thick and colonized by olive-gray bryophytes. Crust formation is enhanced by entrapment of fine silty particles that are deposited from desert dust storms in the spring. Over time, newly developed crusts form micromounds 15 mm in height, interspersed with microhollows (Mitchell & Fullen 1994).

Similarly, dust deposition affects crust formation in arid southwestern Tajikistan, an area used for sheep and goat grazing approximately 10 km north of the Afghanistan border, where 100–120 mm of rain or snow falls annually (Gillette & Dobrowolski 1993). In 1965, a sand-stabilization program of shrub establishment was initiated to protect cropland threatened by blowing sand and shifting dunes. Vascular plants reduced the rate of soil erosion to a level where cryptogamic crusting was possible. Over time, cryptogams (including algae, lichens, and mosses) and rainwater incorporated aeolian dust into a stable soil crust, furthering the growth of microphytes, which prefer silty soils. Lichen cover is associated with shrub interspaces, and mosses with shrub coppice dunes.

There is some evidence that cryptobiotic soils play a role in plant succession. However, contradictory and inconclusive studies make it unclear whether cryptograms increase or decrease vascular plant cover (Eldridge 1993; Belnap 2003). In addition to binding soil, promoting the entrapment of airborne dust, and improving soil water-holding capacities, cyanobacterial soil crusts raise the availability of nutrients (Danin 1991). Levels of nitrogen, potassium, iron, calcium, and magnesium are higher in annual grasses growing on crusted soils than non-crusted soils (Belnap & Gardner 1993). Geochemical analyses of developing crusts at Shapotou, China, showed an increase in Fe, Mn, Ca, Mg, Cl, P, and S during the period 1956–81, the nutrients mostly concentrated within a layer less than 40 mm thick, and decreasing with depth (Mitchell & Fullen 1994). The relatively thin layer of enhanced nutrients restricts natural plant colonization. Danin (1991) and Eldridge (1993) reported that as the cyanobacterial crust becomes more stable, mosses and lichens gradually increase in dominance at the expense of algae. Vascular plants, such as perennial grasses, may

arrive later in the order of succession. Belnap (2003) notes that field studies show that biological soil crusts either increase or have little effect on the germination and survival of native plants, but may slow the germination of exotic annual grasses.

Trampling by cattle and off-road vehicles can severely damage cryptobiotic crusts. Grazing near Navajo National Monument had a more pronounced effect on cryptobiotic cover and diversity than on vascular plants (Brotherson & Rushforth 1983). A single disturbance by hooves of the crust may not be significant, but multiple disturbances eliminate the crust (Gillette & Dobrowolski 1993).

7.6 SPATIAL HETEROGENEITY IN SOIL PROPERTIES AND THE ECOHYDROLOGY OF PATTERNED VEGETATION ZONES

An understanding of the relationship between plant cover, soils and surface characteristics, and soil hydrology is essential for understanding landscape processes in arid lands. Ecohydrology is an interdisciplinary study in which ecologists and hydrologists collaborate in order to understand soil water fluxes. It provides a valuable framework to help address such pressing environmental problems as invasive species, woody-plant encroachment, water quality, and desertification. Ecohydrology considers both vertical water fluxes in the soil (evaporation, infiltration, percolation, recharge, and transpiration) and horizontal fluxes (runoff and processes of soil erosion) as they relate to vegetation.

In arid and semiarid lands, there are often very striking differences in soil properties over short distances, creating a mosaic landscape. Although the soils may have the same taxonomic classification, they have a very different function in the landscape. It is not uncommon to have a binary or two-phase mosaic of soils, in which one component supports a relatively dense vascular plant cover of grasses, shrubs, or small trees (patches, groves, or bands), and the other component is composed of bare ground. The bare ground, which is commonly crusted and underlain by vesicular horizons, is less permeable, and sheds water rapidly downslope to the vegetated bands. These are mechanically strong soils. Australian studies of intergrove soils show that the bulk density and compressive soil strength increase towards the lowermost zones, probably reflecting the transportation of clays and salts downslope. By contrast, soils in the groves are soft and friable, and show features of shrink–swell phenomena (Dunkerley and Brown 2002). As a result of these differing soil properties, the vegetated patches may receive water amounts close to twice the climatological rainfall (Cornet et al. 1988), whereas the unvegetated zones are more arid than rainfall totals suggest. The result is heterogeneous landscapes of patterned vegetation, which include bands, stripes, arcs, and patches of vegetation with intervening bare patches of soil, widely described for gentle slopes (Cornet et al. 1988; Tongway & Ludwig 1990; Seghieri & Galle 1999; Dunkerley & Brown 2002; and others). It is thought that the banding or patchiness of vegetated and non-vegetated soils helps to conserve scarce water and nutrients within a part of the landscape.

Within the field of ecohydrologic study, the trigger-transfer-reserve-pulse (TTRP) conceptual framework is a model to describe how runoff is redistributed to create ecological patterns (Ludwig & Tongway 1997; Ludwig et al. 2005). Temporal (trigger) events, such as rainstorms, are spatially transferred (through runoff) to reserve (landscape patch) processes, to cause pulse events, notably plant growth. The trigger events transfer soil, microbes, seeds, plant litter, and water to downslope areas (Belnap et al. 2005). The transfer (runoff) processes are affected by the characteristics of the soil, with fundamental differences between the largely bare soil of the interpatch and the vegetated patch surface. In the interpatch, rainfall usually exceeds the infiltration capacity of the soil, resulting in runoff (R): any water that is retained is largely lost by evaporation. Runoff may be trapped in a reserve (a nearby vegetation patch) as runon (RN), and infiltrates into the soil adding to the storage of soil water (ΔS). Within the vegetated soil, water and organic matter promote biological activity by organisms such as invertebrates, further enhancing the infiltration capacity of the soil as soil aggregates and macropores are formed (Abrahams & Parsons 1991; Ludwig et al. 2005). Following runon, the replenishment of soil moisture causes a pulse of vegetation growth, replenishing the seed supply and returning litter as organic matter and nutrients for recycling. The pulse of growth increases the architecture of the vegetation (stems, logs, etc.); more efficiently obstructing future overland flows of water. Furthermore, vegetation patches effectively trap windborne materials and nutrients. In addition to standing vegetation, surface obstructions, such as ant and termite mounds, rocks, and fallen dead stems or logs, trap water- and wind-borne sediments and litter, further increasing inputs

of water and nutrients to the soil (Ludwig et al. 2005). In theory, the greater amounts of water available to the patch on the upslope edge may lead to colonization by pioneer species, and the gradual upslope migration of the entire patch. However, Dunkerley (2002) was unable to observe this effect in a comparison of aerial photographs 24 years apart. Individuals at the bottom edge may die out, as less water is available (Thiéry et al. 1995; Klausmeier 1999).

The TTRP process is characterized by feedback effects (Fig. 7.2). For example, positive feedback occurs when runoff fosters vegetation growth, enhancing the obstructive properties of the patch, increasing water storage, and allowing more plant growth. Associated changes in the type and abundance of soil fauna, nutrients, and organic matter occur. A negative feedback would occur when the increased plant biomass transpires more water, and decreases soil moisture and microbial activity (Belnap et al. 2005).

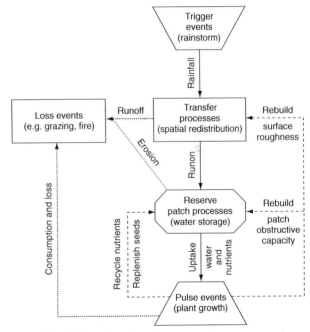

FIG. 7.2 Linkages in the trigger-transfer-reserve-pulse (TTRP) framework. Rainstorm inputs (trigger events) link as solid lines through spatial transfer (runon) and reserve water storage (reserve) processes, to plant growth (pulse events). Dashed or dotted arrows indicate feedbacks and flows out of the system. From Ludwig, J.A. et al. (2005) Vegetation patches and runoff-erosion as interacting ecohydrological processes in semiarid landscapes. *Ecology* **86**. By permission of the Ecological Society of America.

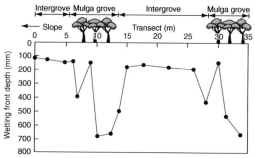

Fig. 7.3 The depth of soil wetting is typically much greater under groves than intergroves. Runoff moves from the crusted surfaces of intergrove areas towards groves, where vegetation, litter, and biota aid in water storage. From Ludwig, J.A. et al. (2005) Vegetation patches and runoff-erosion as interacting ecohydrological processes in semiarid landscapes. *Ecology* **86**. By permission of the Ecological Society of America.

Several hydrologic conditions characterize the general conditions of the vegetation patch. Relative to intergrove areas, infiltration rates are higher (Dunkerley 2002; Ludwig et al. 2005), the wetting front is deeper (Ludwig et al. 2005), and groves may garner up to 200% of the intergrove rainfall due to runon (Cornet et al. 1988) (Fig. 7.3). Infiltration rates in intergrove areas are low, perhaps 10–15 mm h^{-1} (Dunkerley 2002), and any excess is shed downslope. In Australian mulga groves, the highest infiltration rate was 292 mm h^{-1} in mineral soil that was porous, friable, and enriched in organic matter.

However, the specific conditions at a given location are more complex. Even within a given grove or vegetation patch, infiltration rates may vary widely, with the open areas between shrubs having rates similar to those of intergrove soils (Dunkerley 2002). Around bluebush shrubs (*Maireana pyramidata*) and in groved mulga (*Acacia aneura*) woodlands in arid central Australia, infiltration rates are highest close to the plant stem. The elevated infiltration rates extended to beyond the position of the tree canopy, but showed a power-function decline from the stem (Dunkerley 2000b; Dunkerley 2002). The architecture of the plant may help to increase stemflow to the site immediately beneath the plant, as observed in the upright leaf posture and inward-sloping stem form of mulga trees (Slatyer 1965). Thus, the plants are not solely dependent on the increased runon from intergrove areas for survival, but can also rely on the internal processes of canopy interception and stemflow, and the characteristics of the soil. Although

the groves showed a highly heterogeneous pattern of soil moisture, there was a clear trend for the soil to become drier downslope, suggesting that runon delivers the most water to the upslope sections of the groves (Dunkerley 2002). In the Sahel, it has been proposed that some of the excess water delivered to the upslope areas of banded vegetation may escape to deep drainage and be lost to plants (Seghieri & Galle 1999).

These concepts, depicting links and interactions between ecological and hydrological events and processes, used to describe the natural landscape, can also be applied to understanding degradation when, for example, grazing reduces the cover and size of vegetation patches. Grazing interrupts the cycle, by reducing the amount of resources trapped and retained, therefore increasing runoff and erosion.

7.7 SURFACE VOLUME CHANGES

7.7.1 THE PROPERTIES AND NATURE OF SWELLING CLAY SOILS

Many landforms in arid regions are affected by the swelling and shrinking properties of clay. Exposed clay slopes commonly lack vegetation and are easily eroded to form badlands, by a combination of surface-water erosion and piping processes (see Chapter 6). Wetting and drying phenomena influence the formation of desert pavement and strongly affect the development of gilgai, a form of patterned ground found in arid environments. The wetting and drying of clays involves swelling and contraction in both vertical and horizontal dimensions. The vertical relief changes are related to the swelling capacity of the clay and the sodium saturation of the exchange complex (sodium ions adsorbed to the clays produce larger more resistant clods, hence, a greater amplitude of undulations).

7.7.2 PATTERNED GROUND OR GILGAI

Patterned ground is a widespread phenomenon in warm deserts, often associated with stony deserts and playas. "Gilgai" is an Australian Aboriginal term, applied to a wide variety of patterned ground features, and referring to small-scale surface undulations related to wetting and drying of the soil. Australia is best known for the diversity and widespread distribution of its gilgai, but patterned ground has also been described in the Middle East, central Sahara, eastern and central Africa (Kariuki et al. 2004), Death Valley, California, and the Atacama Desert (Dixon 1994a; Oberlander 1994). The formation of gilgai and elements of the form are reviewed in Cooke & Warren (1973) and Cooke et al. (1993).

Gilgai form in regions with soils of high swelling potential and strong textural contrasts, and where the climate is characterized by strong seasonal contrasts in rainfall. Patterned ground in arid regions is generally attributed to two different types of process: (1) wetting and drying in fine sediments or soils with a high proportion of swelling clays, such as montmorillonite, and (2) solution and recrystallization of salts in zones of salt concentration such as the margins of ephemeral water bodies, along drainage channels, and within the capillary zone of water movement in soils (Cooke et al. 1993).

Gilgai form may include puffs or mounds, depressions or channels, and shelf areas between them. Their morphology and dimensions are extremely variable. Six principal types of gilgai morphology have been recognized: (1) normal gilgai, with randomly oriented mounds and shelves, (2) melon-hole gilgai with large mounds and depressions, (3) stony gilgai, with wide and flat stone-covered mounds, resembling patterned ground at high altitudes and latitudes, (4) lattice gilgai, occurring on gently sloping ground, with complex morphology, including those with the microtopography oriented more-or-less parallel to slope contours, (5) linear or wavy gilgai, developed on very low-angle ground with mounds perpendicular to the contour, and (6) tank gilgai, large-scale gilgai, usually rectangular in plan (Dixon 1994a). The type of gilgai that develops is related to the depth of the layer of maximum expansion and to the variations in expansion between different layers. Gilgai formation is most likely in regions with a marked alternation of wet and dry seasons or conditions (Twidale 1972). Uneven wetting also plays a role. Light showers will wet the surface, whereas heavier rains will penetrate deep cracks. It is probable that the production of isolated pressure zones, and the subsequent development of gilgai, may be closely related to the prior formation of cracks and crack patterns.

Australian soils are noted for the prominence and variety of soil micro-relief patterns of mounds and hollows, which affect the distribution and growth of indigenous plant species. In the Lake Eyre lowlands of Australia, landscapes with a well-developed mantle of stones or "gibbers" are found (Twidale 1972). The

boulder pavements on mesas and tablelands are locally patterned by circular networks of stony gilgai, formed by the heaving of underlying soils with expansive montmorillonitic clays. Southeast of the Lake Eyre lowlands are plains and undulating surfaces with dense stone mantles above solonetzic soils. Solonetz soils are black alkali soils; that is, soils with sodium carbonate (Na_2CO_3) present in the upper horizons, that tend to be black owing to the solubility of organic matter in alkaline solutions. These lowlands exhibit patterned ground on a remarkable scale, strongly modified by topography, with sorted stone steps following the contours, separated by vegetated flats with crabhole depressions (Mabbutt 1984).

7.8 SURFACE TYPES: HAMADA AND STONE PAVEMENTS

Stone mantles in deserts are of two basic types, pavements and hamada. Surfaces are termed pavements (also called reg, serir, gobi, and gibber plains) where the stones, ranging in size from boulders to fine gravels, are closely packed on flat or moderately inclined surfaces. Hamada is an Arabic word denoting difficult bouldery terrain that is "unfruitful" to cross.

7.8.1 HAMADA

Hamadas (hammadas) are surfaces consisting of coarse-grained, extremely angular, and unworked rocks. Many have a residual *in situ* origin; that is, they have not been transported. There are two types of surface to which the term hamada applies. The first is outcrop or rock hamada: hamada having the connotation of a structural tableland through much of the Sahara, generally in association with flat-bedded rocks or horizontal weathered crusts (Fig. 7.4). The second is boulder hamada, for which a residual origin is shown by the lack of sorting and angularity of the pavements and an affinity with the rock below (Mabbutt 1977b). The development and geomorphic significance of hamada have received little study.

7.8.2 STONE PAVEMENTS

7.8.2.1 Introduction

A reg or gobi is a stony desert where the surface consists of sheets of gravel with little or no vegetation (Fig. 7.5) (Evenari et al. 1974). Such features are

FIG. 7.4 Rock hamada in the Dead Sea rift area. Boulders are approximately 20 cm in diameter. Photograph courtesy Tony and Amalie Orme.

FIG. 7.5 The desert surface of China is covered by approximately 40% gobi, or gravel-covered surfaces. Vegetation is absent from such surfaces. In the background are the snow-covered Tien Shan Mountains, which provide an important supply of runoff.

known by a variety of names throughout the world, including gibber plains or stony mantles in Australia and gobi and saï in central Asia. Reg is an Arabic term meaning "becoming smaller," and surfaces of finer gravel produce trafficable desert pavements in Algeria, Israel, and areas of the Sahara. Serir replaces reg in the central Sahara (Libya and Egypt).

Where the stones are highly concentrated, the surfaces are referred to as stone or desert pavements, particularly in North America. Stone pavements generally consist of closely packed gravel in a surface layer a few centimeters thick (one- to two-particle-thick layer), above a soil that may contain a few stones dispersed through it, or may be relatively stone-free (Fig. 7.6). The gravel may be angular to

(a)

(b)

FIG. 7.6 Stone pavements are one or two stones thick and underlain by a fine-grained substrate that is generally stone free. (a) Cowhole Mountains, Mojave Desert, California, USA. The pavement is underlain by at least 30 cm of stone-free silt. (b) Heavily varnished, vegetation-free pavement in western China, Gansu Province, south of Liuyuan.

subrounded in shape. Pavements occur in a variety of relatively stable geomorphic environments where vegetation is sparse, but they are particularly prominent and abundant in hot desert regions. Well-developed stone pavements are found on abandoned alluvial surfaces such as fans and terraces, stabilized sand sheets and sand ramps, pluvial lake beach ridges, and associated with aeolian mantles on lava flows. The stability of pavements results from their gentle slopes, the resistance of coarse gravel to erosion by local wash, and the negligible input of runoff from distant surfaces (Amit & Gerson 1986). Pavements and their underlying soils are among the most intensely studied geomorphic features of deserts (Springer 1958; Jessup 1960; Mabbutt 1965; Cooke 1970; Evenari et al. 1974; Goudie 1974; Shlemon 1978; Dan et al. 1982; Wells et al. 1985; Amit & Gerson 1986; McFadden et al. 1987; Wells et al. 1994; Williams & Zimbelman 1994; Haff & Werner 1996; Wainwright et al. 1999; Al-Farraj & Harvey 2000; Quade 2001; Anderson et al. 2002; Valentine & Harrington 2006; Meadows et al. 2007; Pelletier et al. 2007).

Pavements develop in both extremely arid (less than 80 mm year^{-1} annual precipitation) to moderately arid (150–200 mm year^{-1}) climates (Amit & Gerson 1986). Rainfall can be either from winter frontal rainstorms alone, as in the northern Sahara, Middle Eastern deserts, and deserts of southeastern California, or from both winter frontal storms and summer convective storms, as in arid areas of Arizona.

Pavements are of two basic types. The first includes those of an alluvial character, characterized by gravel of mixed composition and distant origin, including pavements formed on alluvial fans, stream terraces, and beach ridges. The second are those of a residual character, such as angular flakes of local bedrock or less weatherable residue, and including pavements formed on basalt flows.

Pavement characteristics are used to map Quaternary surficial deposits, including alluvial fans (Shlemon 1978; Christenson & Purcell 1985; Al-Farraj & Harvey 2000), stream terraces, and lava flows (Wells et al. 1985; McFadden et al. 1987; Williams & Zimbelman 1994). Changes in the chemical composition of varnish and the presence of organic carbon in varnish coating gravel of pavements have been employed to estimate the age of the underlying materials on which the pavement formed (see Section 6.5) (McFadden et al. 1987; Dorn 1988; Liu & Broecker 2008). Owing to their

tendency to become better developed with age, desert pavements are widely used as a relative-age dating tool (Pelletier et al. 2007).

7.8.2.2 Description of stone pavements

Mature stone pavements are characterized by nearly level stony expanses that tend to be vegetation-free. As pavements form, there is a general reduction of the original surface relief, smoothing flow structures on lava beds or depositional bars on alluvial fans. The angular to subrounded pavement stones form a tightly knit mosaic, often coated with some degree of desert varnish (Figs 7.6 and 7.7). Heavily varnished, tightly knit, stone pavements are considered characteristics of surfaces of late Pleistocene age (Bull 1991; Al-Farraj & Harvey 2000), and generally indicate long-term surface stability. Amit and Gerson (1986) suggest that at least 100,000 years is required for the development of a mature smooth pavement surface consisting of fully shattered small clasts. For the Mojave Desert, however, Quade (2001) argued that all pavements above 400 m in altitude are younger than the latest Pleistocene due to the presence of pavement-disrupting plants at those elevations during the LGM. These findings were contradicted by Valentine & Harrington (2006), who found clear differences between pavements developed on an 80,000–75,000 year-old and an approximately 1 million-year-old volcanic pavement. They suggest that the older pavement was probably vegetation-free since its initial formation. Pelletier et al. (2007) concluded that although vegetation, animals, and other disturbances periodically influence pavement dynamics, the pavement surfaces are nonetheless good indicators of relative age on Pleistocene timescales. Varnish microlamination dating by Liu and Broecker (2008) confirm that once a pavement forms, it may survive for 74,000–85,000 years or longer without significant disturbance.

The stones of a pavement rest on a fine-grained substrate composed of clay, silt, and fine sand that is mostly stone-free (Table 7.2). Soils are quartz-rich, well-sorted sandy silts that have been transported largely as windblown suspended load (McFadden et al. 1987; Reheis et al. 1995). Between major periods of silt deposition, a lower aeolian flux rate permits development of soils in the deposits. Soil formation proceeds beneath the surface concurrently with pavement development. Over time, soil depth increases, horizons become more differentiated, and the B horizon becomes more argillic and red in color.

Fig. 7.7 At the surface, stone pavements are densely packed and composed of angular to subrounded fragments. (a) Angular fragments of mature pavement on an alluvial fan, Panamint Valley. (b) Rounded to subrounded beach gravels on a beach ridge, Beatty Junction area, Death Valley, California, USA.

Table 7.2 The depth of stone-free soil underlying desert pavements.

Location	Depth of underlying stone-free soil	Study
Pisgah basalt flow, California	0 cm (young surface)	Williams and Zimbelman (1994)
Panamint Valley, California	5–10 cm	Haff and Werner (1996)
Oman	Youngest surface: 10–15 cm	Al-Farraj and Harvey (2000)
	Middle-age surface: 20–28 cm	
	Oldest surface: 26–50 cm	
Cima volcanic field, California	100–270 cm	Wells et al. (1985)

FIG. 7.8 Immediately beneath the stone veneer of a desert pavement lies the A horizon with vesicular pores. Note the vesicles on the fragment to the left of the scale.

Calcium carbonate accumulates in the soil and soluble salts may be precipitated.

Pavement soils are defined by several horizons. At the surface is the gravelly desert pavement and inter-gravel crust (Ao horizon). Beneath this is an ochric A horizon with vesicular pores (Av horizon, where v stands for vesicular) (Fig. 7.8). The Av horizon, observed in many desert soils, is thought to have formed largely by incorporation of dust and contains significant quantities of silt, clay, and other constituents such as calcium carbonate, gypsum, or other soluble salts (McFadden et al. 1998). The Av horizon differs from the parent material in both mineralogy and grain size and is the horizon most similar in composition to atmospheric dust (Reheis et al. 1995). The spherical vesicles, with walls strengthened by calcium carbonate cementation, probably result from soil air that cannot escape freely after a rainfall event owing to the surface cover of stones (Evenari et al. 1974; McFadden et al. 1998). Vesicular horizons may show a pronounced columnar structure, attributable to the alternating shrinking and swelling of clays in the increasingly clay-enriched Av horizon, and a platy soil structure (McFadden et al. 1987; Anderson et al. 2002; Young et al. 2004). Silt, fine sand, and solutes are readily transported below the surface through cracks by rainsplash, surface wash, and translocation, and the walls of the peds are coated with loose silt. The dust may be transported horizontally along platy boundaries to ped interiors

(Anderson et al. 2002). Underlying the Av horizon is a cambic (Bw) or argillic (Bt) horizon enriched in silt and clay and commonly containing pedogenic calcium carbonate (Bk horizon) and soluble salts (By or Bz horizon) (Reheis et al. 1995). These deeper horizons represent dilutions of the original parent material by dust and may contain products of *in situ* weathering. The B horizon is virtually gravel-free.

7.8.2.3 Formation of pavements

Several theories have been proposed to explain the development of stone pavements. Until recently, they were thought to be lags or veneers caused by the deflation of fines, a process which continued until the stones were sufficiently closely spaced to act as a desert armor. However, studies show that wind erosion is virtually ineffective when the stone covering reaches 50% and the deflation mechanism cannot account for the stone-free zone immediately beneath most pavements. Unfortunately, although this theory has been largely eclipsed by recent research, it remains the most oft-cited mechanism in introductory textbooks.

It has also been proposed that stones are concentrated by surface wash, with rainbeat and erosion by overland flow more effective than wind in scouring fine-textured or sloping desert pavements. Studies by Sharon (1962) in the Negev Desert showed that wash accounted for most of the lowering. However, like the deflation theory, removal of fines by water cannot account for the stone-free zone, and though it may help to reestablish disturbed pavement surfaces (Wainwright et al. 1999), it cannot be solely responsible for pavement formation. Creep processes resulting in reduction of original depositional relief have also been invoked (Denny 1965, 1967; Hunt et al. 1966; Hooke 1972).

Another theory proposed for the concentration of stones at the surface is the upward migration of gravel through an increasingly clay-rich, gravel-depleted B horizon via alternating shrinking and swelling associated with wetting and drying and freezing and thawing (Springer 1958; Jessup 1960; Cooke 1970; Mabbutt 1977b; Dan et al. 1982). The process is believed to be of major importance in the stony tablelands of Australia. Soils that exhibit this phenomenon contain expansive clays and are subject to swelling and heaving on wetting, and to shrinkage and deep cracking on drying. According to Springer (1958), stones are lifted slightly as the soil swells. During desiccation, cracks are produced, but these are too narrow for the stone to fall into, although finer particles may wash or fall into the cracks. Over many cycles of wetting and drying the stone is displaced upward to the surface. This mechanism is not effective in all areas, however. For example in the Mojave Desert, California, rainfall infiltration is limited by arid conditions and the climate does not cause extensive freezing. In addition, this mechanism implies that stones at the surface should be of different ages, whereas varnish studies and cosmogenic dating indicate that pavement stones are of similar age.

In summary, traditional views of desert pavement formation favor the surface concentration of stones resulting from aeolian removal of fines, sheetflood removal of fines, the upward migration of stones through clay-rich soil, and creep processes. By contrast, the most recent research favors aeolian aggradation rather than deflation as the principal formative mechanism. This is a major shift in perspective, in which desert pavements are no longer considered zones of erosion, but rather zones of deposition.

7.8.2.4 The aeolian aggradation theory of pavement development

According to this model, salt- and carbonate-rich aeolian dust is brought in from surrounding plains and trapped among the pavement stones. The fine dust accelerates mechanical weathering of the surface rocks and accumulates within the pore spaces between clasts, promoting soil development and displacing these stones upwards as the aeolian layer grows in thickness. Field experiments in Israel indicate that rocky surfaces trap several tens of times more dust than adjacent pebble-free surfaces, and that entrapment rates increase as the rock fragments become smaller, more flattened and elongated, and the total cover density increases (Goossens 1995). Thus, desert pavements are part of an accretionary mantle, in which the clasts undergo syndepositional lifting with the surface of the accreting soil layer (Wells et al. 1985, 1987). This research suggests that many pavements are "born at the land surface," with pavement gravel never deeply buried in the underlying soil (McFadden et al. 1987). Cosmogenic ray surface exposure dating of pavements in the Cima volcanic field, California (Wells et al. 1994), showed that the ages of surface clasts were similar to those of their source bedrock, suggesting that the pavements had never been buried. The varnish microlamination dating of Liu and Broecker (2008) also supports the hypothesis that pavement clasts have been exposed continuously since their formation.

Most stone pavements become established in relatively stable areas where the processes of stone sorting, surface creep, and clast disintegration are dominant over concentrated wash processes. They are characterized by relatively little relief, as lower areas are gradually infilled, and the entire surface rises by infiltration of aeolian fines. On alluvial fan surfaces, for example, the original bar and swale microtopography is reduced until a smooth vegetation-free surface has developed. The surface clast size becomes progressively smaller with time owing to disintegration of surface clasts by physical weathering, particularly salt weathering. Salt-rich aeolian fines accumulate in fractures of the clasts, and wetting and drying of the fines result in volumetric changes related to crystal growth or shrinking and swelling of clay. This volumetric change enhances clast fracturing and causes vertical and lateral displacement. As stones are displaced, additional aeolian fines and salts are deposited between the clasts, further enhancing separation of clasts from the underlying bedrock.

Physical weathering is sensitive to climatically induced variations in the flux of aeolian fines. Aerosolic salts are a significant component of the soils beneath most stone pavements, particularly in the lower piedmont areas where their abundance is indicated by high soil electroconductivities and the presence of secondary gypsum. Weathering is conspicuous near the margins of saline playas and along the late Pleistocene shorelines of pluvial lakes. Medium- to coarse-grained plutonic and metamorphic rocks are particularly susceptible to this process (McFadden et al. 1987).

Several studies have examined aeolian deposition and pavement formation on lava flows in the Mojave Desert of California (Wells et al. 1985; McFadden et al. 1987; Williams & Zimbelman 1994). Pavement on lava flow surfaces of the Cima volcanic field developed as locally derived basaltic rubble, consisting of mechanically weathered blocks of flow rock derived from constructional highs, which moved into topographic lows by mass wasting and other slope processes. Aeolian accumulation in the topographic lows resulted from the high initial surface roughness, which caused local reductions in near-surface wind velocity. The extremely high permeability of flows inhibited fluvial erosion of the trapped silt. Accretionary mantles as much as 1.0–2.7 m thick have developed (Wells et al. 1985). Over time, the surface roughness of the flow was reduced and the trapping effect dampened. The probable sources for salts and aeolian materials on the volcanic flows were large playas and distal piedmont areas. Aeolian accumulation appeared not to be continuous, but episodic, with occasional rapid influxes of material.

In contrast to the mature pavements of the Cima volcanic field, young, well-developed stone mosaics on the Pisgah basalt flow in the Mojave Desert lack underlying salt and dust. Williams and Zimbelman (1994) propose that sheetfloods play a dominant role in the incipient growth of these small pavement areas (about 1 m^2). Aeolian processes do not appear to affect these early stages of development. Such pavements may represent an initial substage of accretionary mantle development, wherein the mosaic is formed prior to soil development, rather than contemporaneously with mantle accretion.

Mature pavements remain dynamic surfaces that show recovery from disturbance. Haff and Werner (1996) cleared small patches of stones and documented pavement recovery by repeat photography. Lateral displacement of stones by animal activity was a major component of the resurfacing process. The cross-surface movement of stones is one means by which topographic highs are lowered and depressions filled, contributing to the nearly flat surface characteristic of mature pavements. Wainwright et al. (1999) mixed the surface stones with the fine

substrate to investigate the effect of raindrop erosion processes on regenerating stone pavements following disturbance. Significant accumulations of coarse particles were observed within five 5 minute artificial rainfall events. A disturbed plot with an initial stone cover of 41% increased to 71% cover following the experiment: close to the 77% cover of the pre-disturbance pavement. About 10 runoff events are annually recorded at Walnut Gulch, Arizona, the site of the study. Thus, it could be anticipated that pavements in the area would be able to regenerate on an annual cycle following disturbance.

7.8.2.5 Pavement development as a relative-age dating tool

Pavements and their underlying soils have been used as a relative-age dating tool owing to the pronounced changes that characterize their development over time. In general, it is thought that fan surfaces of early Holocene age exhibit partial pavement development, and late Pleistocene surfaces are characterized by smooth, well-armored stone pavements (Christenson & Purcell 1985). There are exceptions, however, with some early- to mid-Holocene alluvial surfaces exhibiting well-developed pavements (Quade 2001). Varnish microlamination dates indicate that desert pavements in California can form in less than 10,000 years (Liu & Broecker 2008). As the development of pavement is strongly influenced by dust deposition rates, which are higher near source areas such as playas, it is likely that pavement-development characteristics vary regionally (Pelletier et al. 2007).

Pavement characteristics change as the surface matures and becomes older. In early pavement formation, vesicular A horizons constitute the initial soil horizon; weakly developed color B horizons are present below the vesicular A horizon of middle to early Holocene soils; but a weak to moderately strong and usually nongravelly argillic horizon is present below the vesicular A horizon of late Pleistocene and older fan soils. As surfaces become older, there are reductions in clast sizes, better particle sorting, increased angularity, and an increase in the surface area occupied by well-interlocked smooth pavement. Soil development proceeds concurrently, and involves increasing soil thickness, an increase in fines (silt plus clay) content, progressive horizon development, increasing redness of the B horizon, and increasing carbonate accumulation (Al-Farraj & Harvey 2000; Young et al. 2004). With time, the topography also changes, with a systematic decrease in bar-and-swale microtopography, and a rounding of gully and terrace edges on older surfaces (Pelletier et al. 2007).

Amit and Gerson (1986) studied the evolution of pavement soils by examining 15 terraces on a vast sublacustrine delta formed by the Holocene drop in level of Lake Lisan, the precursor of the Dead Sea, Israel. The degree of pavement evolution, determined by percentage cover of the terrace surface, sorting, shape of fragments, length of fragments, and degree of pitting, was shown to be greatest on the oldest terrace and least on the youngest terrace. Al-Farraj and Harvey (2000) showed a similar progression on fans and wadi terraces of different ages in Oman. In Israel, however, a terminal soil thickness was reached early on, with soils of several hundred to about 1000 years old being the same thickness as those 14,000 years in age. By contrast, sites in Oman and on older fan surfaces in the Providence Mountains, California (Sena et al. 1994), showed an increase in the soil thickness over time. These latter sites are of much greater age, however, and have probably experienced several cycles of dust deposition. In the Providence Mountains, the number of tilted clasts also increased with age, as manifested by the degree of concordance of varnish lines and carbonate collars on clasts with respect to the pavement ground surface.

Soils beneath pavements change their hydrologic properties through time. In the primary stages of soil evolution, porosity (30–40%) and permeability (equivalent to 60–80 mm h^{-1} of precipitation) are high, allowing no appreciable runoff (Amit & Gerson 1986). These properties diminish with time. Marked changes in the soil profile occur as fines and salts are introduced into the soil profile, retarding the infiltration of water (Sena et al. 1994; Young et al. 2004). Vegetation growth is retarded by the lack of infiltration as well as salt accumulation in the soils. Runoff is relatively high from well-developed pavement soils (Amit & Gerson 1986), leading to widespread gullying in some locations.

7.8.2.6 Discussion

In summary, it appears that pavements develop by a variety of geomorphic processes that result in stone sorting, surface creep and stone displacement, clast disintegration, and syndepositional lifting of surface clasts by the accumulation of salt- and carbonate-rich aeolian fines (Dohrenwend 1987). These processes operate at different levels of relative importance as the pavements develop through time.

Aeolian processes have significantly influenced soil development on alluvial deposits in deserts. Much of the silt, clay, carbonates, and soluble salts that have accumulated in desert soils are attributable to incorporation of aeolian materials rather than to chemical weathering of soil parent materials. Although aeolian activity during the Quaternary often has been highly episodic, the rate of incorporation of aeolian material has been slow enough to permit development of cumulate soils and to cause uplift of preexisting soil pavements. As pavement development is largely an accretionary process, it is probable that pavement-formation times are related to distance from dust sources (particularly playas) and to the prevailing wind direction (Pelletier et al. 2007).

Varnish data from alluvial fans in the southwestern USA imply that varnish development has occurred on stones exposed continuously since abandonment of the fan surface (no stones reflect emergence of clasts during the Holocene into late Pleistocene pavements, as would develop if there was upward migration of gravel). This suggests that desert pavements: (1) are born at the land surface and (2) remain at the land surface via aeolian deposition and simultaneous development of cumulate soils beneath the pavements. Because pavement development is closely associated with Avk horizon development, and because Avk horizon development is at least partly controlled by aeolian activity, which is, in turn, closely linked to climatic change, it would appear that stone-pavement development is also linked to the effects of climatic change.

8

WATER AS A GEOMORPHIC AGENT

8.1 INTRODUCTION

The fluvial landscape of drylands is by and large a product of highly episodic and discontinuous flow events. Within this environment, the relative roles of surficial runoff and subsurface erosion are related to the infiltration capacity of the terrain (Section 4.3.4). In the subsurface, soil water and groundwater can erode either by flowing through and emerging from a matrix (seepage erosion associated with sapping) or by flowing along conduits (tunnel scour associated with piping). A continuum of processes and forms is found, from surface runoff and high drainage densities on impermeable surfaces (Fig. 8.1), to groundwater flow and low drainage densities on highly permeable substrates (Schumm et al. 1995). Not all processes are mutually exclusive. The overall landform evolution may include coupled fluvial and sapping processes (for example, the amphitheater-headed canyons of the Colorado Plateau) or coupled fluvial and piping processes (for example, the gullies of the American southwest).

Other geomorphic processes also contribute to the fluvial landscape. Between episodic flow events, the bed may be scoured by the wind or blocked by migrating sand dunes. The products of mass movement, both at the channel head and channel sidewalls (from, for example, bank collapse) may accumulate in the channel until high flow events attain sufficient energy to remove them (Fig. 8.2).

8.2 GROUNDWATER SAPPING IN SLOPE AND VALLEY DEVELOPMENT

Groundwater plays a role in mass movement and channel formation by the process known as sapping. Concentrated seepage caused by groundwater convergence slowly erodes material at valley heads or cliff bases, undermining overlying structures, and causing failure and headward retreat (Fig. 8.3). The term spring sapping is often used when a point-source spring is involved, whereas seepage erosion may be employed where the groundwater discharge is less concentrated. Computer modeling suggests that scalloped escarpments develop where groundwater flow is diffuse, whereas elongation into channels or canyons results from higher and more concentrated seepage discharges, often associated with growth updip along fracture systems characterized by higher hydraulic conductivities (Howard 1995). In rocks that are susceptible, chemical weathering renders the rocks even more permeable.

Common morphologic characteristics of valleys in which sapping plays a dominant role include amphitheater-shaped headwalls, relatively constant valley width from source to outlet, high and often steep valley sidewalls, a degree of structural control, short and stubby tributaries, and a longitudinal profile that is relatively straight (Laity & Malin 1985) (Fig. 8.4). These field characteristics are reproduced in simulation models, wherein sapping produces valley systems that are weakly branched, nearly constant in width, and terminate in rounded headwalls (Howard 1995).

The study of groundwater outflow in valley network development was stimulated by images of Martian valleys (Laity & Malin 1985; Gulick 2001). For Earth, an ever-increasing body of research illustrates that groundwater sapping is a global process, which can occur in diverse lithologic and hydrologic settings, including those more humid than discussed here (Baker et al. 1990). In arid settings, sapping has contributed to valley formation in Libya, Egypt, Botswana, the USA (the Colorado Plateau), and northern Chile.

Valleys and escarpments that maintain characteristic sapping forms, but which lack modern seepage, may be relict from previously wetter climates. In Egypt, both fluvial and sapping erosion rates were greater 1–2 Ma, and scarp retreat has decreased progressively as the climate has changed from wet to hyperarid (Luo et al. 1997). Some systems include both active and relict components. On the Colorado

Fig. 8.1 High drainage densities are typical of less permeable surfaces in desert areas. Location: southeastern Jordan. Source: NASA ASTER image, Path 173, Row 39, center 30.93°N, 37.03°E.

Fig. 8.2 Between flow events, bank collapse and other forms of mass movement supply sediments to channels. Location: between Yumen and Anxi, Gansu Province, China.

Fig. 8.3 Groundwater sapping on the Colorado Plateau produces valley headwalls that are high and often very steep. A dark zone of seepage forms at the contact between the Navajo Sandstone and the underlying Kayenta Formation. Groundwater outflow sustains ponds below the headwall that overflow to form small streams. The valleys grow headward by successive slab failures as the headwall is slowly undermined. Location: Explorer Canyon, Utah, USA.

Fig. 8.4 Located on the Colorado Plateau, Long Canyon and Cow Canyon are developed in the Navajo Sandstone. The morphology of these valleys, with theater-shaped heads and relatively constant valley width from source to outlet, is consistent with their formation by groundwater sapping. Surface channels also feed into the canyon heads.

Plateau, sapping is presently active (Laity 1983), but rates of valley development were probably enhanced during wetter intervals of the Pleistocene when recharge rates were two to three times greater than today, and water tables as much as 60 m higher (Zhu et al. 1998). Large valleys formed by sapping processes may be very old. In hyperarid northern Chile, canyons developed by groundwater sapping were already in existence by the late Miocene or early Pliocene (Hoke et al. 2004).

Amphitheater-headed valleys formed by a combination of sapping, fluvial, and mass-wasting processes are prominent geomorphic features of the Colorado Plateau. Many are developed in the Jurassic Navajo Sandstone, a highly transmissive aquifer underlain by essentially impermeable rocks (Fig. 8.3). Groundwater, moving laterally down the hydraulic gradient, emerges above the permeability boundary in concentrated zones of seepage at valley headwalls and sidewalls. The outflow sustains small ponds beneath the headwall that overflow to provide year-round flow to small streams. Small-scale erosional processes slowly reduce the basal support of steep cliffs and the valleys grow headward by successive slab failures (Laity & Malin 1985; Laity 1988). As the systems mature, the valleys increase in width.

Canyon networks developed by sapping lack the randomly branching, space-filling dendritic pattern common to many stream systems developed by surface flow (Fig. 8.4). The patterns reflect the flow of groundwater, and are commonly strongly controlled by structural features, such as bed dip, faulting, and jointing. Not only does fracturing enhance the overall permeability of the surface, but also it acts at depth to increase the transmissivity of the bedrock.

In addition to forming valleys by headward erosion, sapping also contributes to the backwasting of scarps. Slopes are undermined and collapse owing to the weakening of basal support at zones of concentrated seepage or diffuse discharge. These processes are important in scarp retreat in sandstone–shale sequences of the American southwest.

The term 'dry sapping' has been applied to slopes of sandstone, granite, tuff or other massive rocks modified by alveolar weathering or tafoni: although the rock surfaces may be encrusted by salts, they do not appear to be damp. By contrast, larger alcoves formed by 'wet sapping' show wet surfaces on at least a seasonal basis (Laity 1983; Howard & Selby 1994).

Unlike fluvial and piping erosion, valley formation by groundwater sapping is not a weather-dependent phenomenon. It is a sustained process that responds with a time lag to long-term climatic shifts. During dry periods, erosion continues, albeit at reduced rates.

8.3 PIPING PROCESSES IN CHANNEL AND SLOPE EVOLUTION

Soil pipes are an important element in landform development, affecting both slope and channel processes. They are particularly prominent in badland formation, as discussed in Section 6.6.3 (Howard 1994), and in the development and extension of gullies (Higgins 1990). They also contribute sediment to arroyo systems, being common along arroyo walls.

Pipes are subsurface channels, varying in diameter from several centimeters to several meters, which carry both water and sediment. Piping is a more common subsurface process of drainage development than sapping. In general, it is a smaller-scale phenomenon that is discontinuous and weather-dependent. Like sapping, it is usually associated with an underlying permeability barrier that promotes the lateral flow of subsurface drainage.

Pipe holes are also referred to as "pseudokarst" (Parker 1963). They are a widespread phenomenon in silty soils or weakly consolidated siltstone, claystone, or similar bedrock materials in arid or semiarid lands, but are also present in more humid regions. The pipes may enlarge the conduit until the roof collapses (Fig. 8.5).

The initiation of piping structures commonly results from an increase in hydraulic gradient. The active downcutting of base level is an important causative factor for headward erosion by piping (Jones 1971).

Although piping occurs in sands and gravels, piped soils and rock are more prevalent in the presence of swelling clays (Parker 1963; Jones 1971), owing to their susceptibility to cracking during dry periods and their tendency to deflocculate. Periodic high-intensity rainfall and lack of vegetation, or devegetation, are secondary factors (Jones 1971). The piping mechanism results from the dispersion of clay colloids at the walls of the voids of unconsolidated clay and the subsequent movement of the dispersed colloids along the voids. Clay mineralogy is important in that montmorillonites are more easily dispersed than illites, which are in turn more easily dispersed than kaolinites.

Fig. 8.5 Pipe holes form a "pseudokarst" surface in silty soils or weakly consolidated siltstones, claystones, or similar materials. A subsurface pipe carries both water and sediment. As the pipe enlarges, it is common for the roof to collapse, as shown in this photograph. The pipes shown have formed in the Tecopa Lake Beds, California, USA. A deep lake was sustained in this basin between 3 Ma and 160 ka.

Owing to their subsurface location, piping networks are difficult to find and define, and little is known about their form and spatial density (Holden et al. 2002). As such, it is difficult to understand the interlinkages between pipeflow and other flow processes.

8.4 FLUVIAL PROCESSES

The continuous flow in perennial channels fosters feedback throughout the system, with constant mutual adjustments. By contrast, channels in arid and semiarid environments have a more discontinuous operation and lack integration between trunk and tributary streams (Graf 1988). Rainstorms that are infrequent, of short duration, and commonly torrential in nature give rise to ephemeral flow. The streams are typically influent; that is, discharge seeps into the channel bed, contributing to the water table, which is located at depth beneath the surface.

The study of dryland channels has contributed much to our understanding of fluvial geomorphology. Early landmark studies of fluvial systems in the semiarid American southwest examined processes of erosion and deposition (Gilbert 1877; Schumm 1961), the hydraulic geometry of channels (Leopold & Miller 1956), and the development of arroyo and gully systems (Leopold & Madock 1953; Leopold & Miller 1956; Schumm & Hadley 1957). Much of this work was summarized in a benchmark volume by Leopold et al. (1964), which established quantitative relationships between channel processes and forms. Subsequent American research yielded such important concepts in fluvial geomorphology as thresholds (Patton & Schumm 1975; Bull 1980), stream power (Bull 1979; Graf 1983a), and magnitude–frequency relationships (Baker 1977). However, the badlands, arroyos, and alluvial fans of the American southwest are not characteristic of all arid realms.

Distinctive and diverse environments characterize each desert. Australian and South African research has documented fluvial landforms and processes not recorded elsewhere, including the extensive anastomosing channels in the Channel Country of southwest Queensland (Nanson et al. 1988; Gibling et al. 1998; Tooth & Nanson 2000, 2004), and the inland deltas and wetlands of the African continent (McCarthy 1993). Palaeoflood studies in India have shed light on the frequency of large floods in arid systems and the duration of such impacts on channel form (Kale et al. 2000). Research in Israel has provided invaluable insights into sediment transport processes and other aspects of channel form and change (Yair & Lavee 1985; Schick 1988; Yair & Kossovsky 2002), utilizing a stream-monitoring program with a database extending over 30 years (Alexandrov et al. 2003).

8.4.1 CHANNEL MORPHOLOGY AND CHANNEL FLOW

The classification of river channels is based on morphologic characteristics that include sinuosity, degree of braiding or anabranching, and sedimentology (such as coarse or fine load). Traditionally, rivers are classified as straight, meandering, braided, or anastomosing. However, these classes are not mutually exclusive and patterns are difficult to define. Channel configuration may change from place to place along a stream and from one time to another (Graf 1988).

The main factors influencing channel classification are the degree of channel sinuosity and the nature of channel splitting around bars of islands. There are two river forms that are characteristic of desert landscapes: the braided channel, perhaps the most common alluvial channel (Fig. 4.6), and the anabranching channel. Other recognized forms include distributary channels and gullies. Both braided and anastomosing channels are characterized by high width/depth ratios and low sinuosities, which allow greater efficiency of bedload transport (Graf 1988), by either directing a greater proportion of shear stress at the bed or by maintaining relatively high values of particle exposure during flood periods (Reid & Frostick 1997). Low sinuosity maximizes energy slope and reduces frictional resistance of channel bends. Meandering channels are rare in arid lands, although this pattern may develop in allogenic rivers with reliable or perennial flow.

Distributary patterns are common on alluvial fans (Bull 1977) and the large inland deltas in Africa (McCarthy 1993). They form in response to a downstream decrease in discharge, slope, and sediment-carrying capacity. An arroyo is a form not usually considered part of a traditional river pattern classification. It is a trench, which in cross-section has steep wall and flat floors, with a channel in the bottom. There are relatively few descriptions of channels in bedrock.

Braided and anabranching (anastomosing) streams both include multiple channels of varying sinuosity. Braided channels are perhaps the most common form in arid lands, representing a response to highly variable flows, abundant coarse bedload, and easily erodible banks (Graf 1988). The steep channel slope permits large amounts of bedload transport. An anabranching system has a basic planform resembling a braided channel, but islands that divide flow at bankfull separate the individual channels. By contrast, in a braided channel, the islands are bars that are overtopped below bankfull. In anabranching systems, the islands usually are more than three times the width of the river at average discharge. Anastomosing channels are a subset of the anabranching form, and are a group of fine-grained, low-energy rivers with a distinctive alluvial architecture consisting of arenaceous channels and argillaceous overbank deposits.

The best-studied anastomosing channels are those of the Channel Country, central Australia, which drain into Lake Eyre. The alluvial areas lie below 150 m elevation and river gradients are very low, generally less than 0.0002 (Gibling et al. 1998). The primary channels are up to 7 m deep, inset into the muds of the floodplain. Additionally, there are smaller secondary and tertiary channels that operate at different flood levels. The coefficient of variation of annual flows for anastomosing channels is among the highest in the world. During floods, the rivers expand to very great widths. Sheets of water up to 70 km wide on the Cooper River and 500 km on the Diamantina River have been recorded. The flood waves move slowly along the low gradient, with volumes decreasing systematically owing to transmission losses that may exceed 75% (Knighton & Nanson 1994). Although the channels are dry during drought, some water may be retained in waterholes, which are two to three times wider than the associated channels. Channel deposits consist principally of clay to fine sand, with localized medium- to very coarse-grained quartz sand and gravel deposits. Mud aggregates, composed of densely packed clay flakes, are abundant in the channel (Gibling et al. 1998).

8.4.2 Alluvium

The term alluvium refers to material deposited by running water; that is, to fluvial deposits. The coarsest sediments are often deposited on the channel bed or in bars. Finer materials are deposited from suspension, either in slackwater areas of the channel, or in overbank deposits following floods. The nature of deposition, involving cutting and filling, may be very complex.

The sorting, stratification, and structure of deposits varies according to the nature of the stream (braided, meandering, or anastomosing). In general, coarser alluvium is found close to supply points. For example, on alluvial fans, sediment is usually coarsest at the apex of a fan (Fig. 8.6). Other factors that influence sediment size include climatic conditions (for example, coarse debris formed by glacial and periglacial processes in adjoining highlands, either today or in the past) and the degree of input by aeolian processes.

Fig. 8.6 The upper reaches of an alluvial fan are typified by coarse angular debris that have not been carried far from their mountain source. Location: White Mountains, Owens Valley, California, USA.

Although alluvium is normally considered to contain only material deposited by streams themselves, in arid environments there is often a considerable aeolian component, especially where channels lie close to dune fields (Fig. 8.7). Figure 8.8 shows a channel in western China, where well-rounded gravels and cobbles are interbedded with aeolian sands and silts.

The bed material in arid channels is often highly mobile, with banks lacking cohesive sediments. Lateral shifts in channel location are common, most notably on alluvial fans and in alluvial channels inscribed in valley fill. During large floods or low floods of long duration, marked erosion of channel banks occurs. These characteristics should be considered in development. During the 1983 flood in Tucson, Arizona, lateral erosion undermined both homes and businesses (Kresan 1988).

8.4.3 Sediment transport

Streams carry weathered rock either as dissolved load (compounds in solution or colloidal mixtures) or as solid load. The solid load is subdivided into the suspended load, consisting of fine-grained particles that are carried in suspension, and the bedload or traction load, consisting of coarse-grained particles that slide, roll, or bounce along the streambed. The proportion of each type of load carried by streams varies according to the basin's climate, vegetation cover, lithology, and structure. In general, rivers that are subject to large fluctuations in discharge, have sparse vegetation cover, or transport large sediment loads will tend to have a higher percentage of solid load (Laronne & Reid 1993) and will be more likely to have braided channels.

The floodwater chemistry of ephemeral desert rivers remains little studied. Total dissolved concentrations during flash-flood events may decline due to dilution as discharge rises, and then return to former levels as discharge declines (Fisher & Minckley 1978). In some regions, solutes precipitate within channel sediments, resulting in materials cemented by calcite and silica. These chemical sedimentation processes indurate the sediments, reducing bank erosion, and may play an important role in channel development.

Research on sediment transport in deserts is largely based on indirect observations, with direct measurements being rare; and the small number of floods monitored to date limits the inferences that may be made. Furthermore, there is a bias towards measurements of sediment in subhumid and semiarid settings, with few observations from arid or hyperarid areas (Cohen & Laronne 2005). The dissolved and suspended loads of perennial rivers are routinely measured at gauging stations, but this practice is much less common for arid streams, where the runoff is infrequent. An automatic monitoring station

FIG. 8.7 Aeolian processes often contribute to the fluvial landscape in deserts. Between episodic flow events, the bed may be scoured by the wind or blocked by migrating sand dunes. Dune sand is often an important component of the alluvium (see Fig. 8.8). The photograph shows low-flow deposition of silt and clays and the formation of mudcracks, Mojave River, California, USA. In the background are aeolian dunes which reform in the channel during the windy months of late spring.

FIG. 8.8 There is often considerable interaction between aeolian and fluvial processes in deserts. The sidewalls of this channel show well-rounded cobbles and gravels interbedded with fine windblown sands and silt. Location: near the Mingsha Dunes, China.

is the best solution to the problem of rare and inaccessible flood events, but these are rare. As a result, data from flood events in deserts are relatively few and information on suspended sediment concentration has been obtained principally by manual sampling (Alexandrov et al. 2003).

It is estimated that as much as 90% of the sediment carried by an ephemeral stream is suspended load (Alexandrov et al. 2003). Sediment is derived from material entrained by overland flow, eroded from banks, and from bed scour (Fig. 8.9). The long interval between floods allows time for fine sediment to accumulate on slopes and in the channel (for example, by weathering or aeolian deposition), enhancing the suspended sediment component of the total load (Cohen & Laronne 2005). Flows are turbid and the sediment concentration commonly

Fig. 8.9 Summer thunderstorms in a semiarid environment often yield flows with a very high suspended sediment load. Location: Tucumari, New Mexico, USA.

ranges between 30 and 50 g L^{-1}, with values occasionally exceeding 100 g L^{-1} (Tooth 2000). In humid regions, concentrations are often lower than 1%, whereas in drylands they may reach 10–40% (Cohen & Laronne 2005). Suspended sediment concentrations of 600,000 ppm (60%) have been recorded at the US Geological Survey gauging station at the lower end of the Rio Puerco, New Mexico. This is the fourth highest average annual suspended sediment concentration in the world, and is exceeded only by measurements at locations along the Yellow River, China (Gellis et al. 2004).

There are several distinctive characteristics of suspended sediment transportation in drylands. It is usually influenced more by changes in the hydraulic environment than by sediment supply, which usually is not limited. Sediment concentrations are high even at low discharges (Cohen & Laronne 2005). Even at the beginning of a flood event, concentrations are elevated, suggesting flushing of material that has accumulated during low-flow conditions or from sidewall collapse. The suspended sediment concentration is not constant in stream flow, but varies over time and space, even within a single event. In arid zones, it tends to increase downstream because of transmission losses, whereas in humid regions it decreases because of dilution.

Strong inter- and intra-event variability in concentrations are tied to a variety of different synoptic meteorological conditions and variations in source areas. Sediment samples from the northern Negev Desert show that the concentration increases with tributary inflows from loess-rich terrain, and is diluted by flows from sub-basins with more rocky soils. As a result, only 50% of the variance in suspended sediment concentration is explained by variations in discharge (Alexandrov et al. 2003).

The changing chemistry of floodwaters also influences suspended sediment loads. Changes in electrolyte concentrations result in alternations between conditions favoring dispersion or flocculation (Imeson & Verstraten 1981). Suspended sediment concentrations as high as 200 g L^{-1} were recorded in an ephemeral wash in Morocco, where high percentage sodium (%Na) and sodium-adsorption ratio (SAR) values enhanced clay dispersion.

The determination of bedload is even more difficult than that of dissolved or suspended load, and is seldom measured. The competence of a river is a measure of the maximum size of material that a stream can transport. This depends on a number of factors, including channel shape, the shape and sorting of the sediment particles, and the suspended load. The stream's capacity is the maximum amount or mass of sediment that it can transport. Although quantitative data are limited, studies suggest that rivers in arid environments transport large quantities of sediment during flood events, both as suspended load and as bedload (Laronne & Reid 1993; Laronne et al. 1994; Reid & Laronne 1995). Bedload discharge in ephemeral desert streams may exceed that of humid region streams of similar power by several orders of magnitude (Fig. 8.10) (Reid & Laronne 1995).

Bedload is derived principally from unconsolidated bed material that is mobilized at the onset of flow and deposited during flood recession. This 'scour and fill' process operates both in sand- and gravel-bed channels (Tooth 2000). Scour is facilitated by the poor development of protective armor layers in dryland rivers. Owing to their very different hydrologic regimes, perennial and ephemeral gravel-bed streams tend to transport their load in different modes. In perennial streams, the water is constantly moving, and thus can mobilize finer grains located on the bed surface. During low-flow conditions, coarse material is less easily mobilized than fine material, leading to a stratification of sediments in the channel bed, in which coarser materials prevail at the surface, and fine grains, which may be selectively eroded or fall through the crevices of larger particles, are enriched in the subsurface. This stratification may be reversed in arid streams. For example, the bed structure in some Mediterranean ephemeral gravel streams shows surface materials that are finer than those in the subsurface. Almedeij & Diplas (2005) attribute this opposite stratification to different mechanisms of particle size segregation that occur when the water discharge and attendant flow depth vary rapidly during the passage of a flood wave. As discussed in Chapter 4, the hydrograph of an arid stream shows a very rapid rise, indicating a rapid increase in water depth. The shear stresses on the bed become very high, resulting in significant sediment transport as the channel bed is scoured. Initially, there is dilation of the bed material down to the depth of scour. As the water velocity is insufficient to carry all of the grains, the fine particles move to the top of the fluidized grains and the larger grains sink to the channel bed, leading to a segregated deposit. The lack of armoring in dryland streams allows the bed to become mobile even at modest values of flow strength, so that practically all floods are associated with high transport rates (Laronne et al. 1994). As high bedload discharge is possible even in shallow flows, considerable geomorphic work may be done even in small events (Cohen & Laronne 2005). While some authors have observed

FIG. 8.10 Convectional rains over the Bogda Shan, China (4300 m ASL), during the month of August trigger floodwaters that flow towards the Turfan Depression (155 m BSL). Such floods move considerable bedload and contribute to the rounded cobbles that form extensive gobi surfaces. Figure 2.10 shows this area in a satellite view.

that coarse bedload makes a greater contribution to total load in arid areas than humid ones (Cohen & Laronne 2005), it is important to note that in some large dryland catchments, such as Cooper Creek in southwest Queensland, fine-grained sediments dominate the sediment yield (Nanson et al. 1988).

Overall, sediment transport variability in drylands is very high and sediment loads are difficult to predict (Rodier 1985). The sediment tends to move through dryland rivers as a series of waves, owing to temporal variations in sediment supply (for example, bank collapse) or the inherent nature of sediment transport in desert rivers. Thus, both bedload and suspended load may travel downstream in a series of short and irregular pulses. In one reach of the channel scour may be occurring while another section is filling. Thus, there are nonlinear relationships between flood volumes and sediment loads, which leads to great variability in sediment yields (Tooth 2000).

8.4.4 SEDIMENT YIELDS

Sediment yield is a measure of geomorphic activity in deserts. It includes all processes of material removal from a basin, but, in reality, the total yield is difficult to determine. Wind erosion, bedload, and dissolved load are particularly complicated to measure. Thus, sediment yield is usually based on the amount of suspended sediment removed from a basin, commonly estimated by measuring sediment accumulation in reservoirs.

Basin sediment yields are a function of the amount of sediment generated by weathering and mass-wasting processes on slopes and the ability of water to transport them, the contribution of windblown sediments, and the dissolved load. On hillslopes, sediment movement is caused principally by raindrop impact and splash, and by overland flow processes including unconcentrated and concentrated surface wash. Sediment production is affected by rock strength and resistance, the nature and rate of weathering, and the nature of slope processes. Human activity may also play an important role. In many arid regions, burrowing animals, including gerbils, porcupines, and gophers, produce quantities of soil that are readily removed even by shallow overland flow. Windblown sediments may also contribute to the sediment load of a channel (Figures 8.7 and 8.8). The Yellow River, China, carried 99 million t of sediment between 1956 and 1995, much of it contributed as part of a two-phase process in

which the wind carried silts and sands to gullies, channels, and floodplains; and tributaries conveyed the material to the main river (Xu 2006). In the area of the Ulan Buh Desert, sand dunes enter the river directly, and as a result of flow regulation by reservoirs which limits scour, there was as much as 73 cm of channel-bed aggradation between 1986 and 2005 (Ta 2007).

In 1958, Langbein and Schumm published an influential paper wherein they concluded that maximum sediment yields are derived from semiarid areas, with an annual effective precipitation of approximately 300 mm. In more humid areas, vegetation cover increases surface protection and reduces erosion, whereas in drier regions vegetation is sparser, but there is less runoff. The validity of this relationship has been subject to some debate (Walling & Webb 1983). Nonetheless, it is generally agreed that semiarid areas have the potential to generate high sediment yields (Fig. 8.9).

Recent studies have examined how sediment yields differ between disturbed and undisturbed areas, the differences between modern (short-term) and geologic (long-term; >10,000 years) rates of erosion, and the role of lithology and elevation. An examination of the geologic record provides a long-term perspective on sediment yields that integrates both dry and wet periods. It is difficult to extrapolate geologic rates of sediment production from modern yields, as land use, vegetation cover, or climate may have been very different from today. Modern studies monitor the differences in yield between various land surface types or land uses using sediment traps, erosion pins, streamflow monitoring stations, and isotopic examination.

Several studies have examined modern sediment production rates to determine whether they are sufficient to sustain multiple cycles of arroyo cutting and filling. For example, in the Rio Puerco basin, New Mexico, arroyo incision and subsequent backfilling have been repeated in a cycle several times over the past 3000 years. Cosmogenic ^{10}Be and ^{26}Al studies suggest an erosion rate at the Arroyo Chavez sub-basin of around 0.1 mm year^{-1} (100 m per million years) for easily eroded materials in a semiarid climate. The total volume of sediment generated is sufficient to fill modern arroyos in less than 1000 years (Clapp et al. 2001). In areas with less soil and more resistant bedrock, erosion rates are less. Rates in south-central Australia are approximately 0.0006–0.001 mm year^{-1} (Bierman & Turner 1995). In the Rio Puerco basin, sediment yields are lowest

PLATE 1.2 The deserts of continental interiors are arid owing to their distance from the sea and sources of moisture. Such deserts have a much greater range of temperatures than those close to the coast. Winters in the Asian deserts can be very cold, with minima ranging from −30 to −50°C. In this image, the shrinking Aral Sea appears in 2002, filling with seasonal ice, and the deserts of the Kyzylkum and Karakum (also known as Kara Kum) to the southeast and south of the lake, respectively, are blanketed in snow. The diversion of freshwater inflows to the saline Aral Sea for agriculture has led to a considerable loss of lake volume and quality. Source: NASA MODIS Rapid Response Team, NASA/GSFC.

PLATE 1.3 Cool coastal deserts form adjacent to cold ocean currents on the west coasts of continents. The climate is moderated by the proximity to cold waters, which tend to impede convection. Moisture is largely provided by fog, shown here along the southern coast of the Namib Desert. True desert conditions with intense aridity occur in the Namib Desert, a strip 80–150 km wide along the Atlantic coast. To the east, on the right side of the satellite image, is the inland Kalahari Desert. Source: NASA Aqua/MODIS sensor.

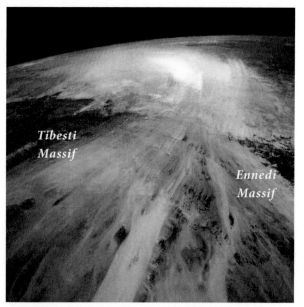

PLATE 2.2 The northeast trade winds prevail across the Sahara and are referred to as the *harmattan*. These dust-generating winds extend to the intertropical convergence zone. This NASA astronaut photograph shows the winds moving between the Tibesti and Ennedi Mountains towards the Bodélé Depression in the Borkou region. In the foreground are vast sand streaks and in the background are light clouds of dust. Yardang systems flank the Tibesti Mountains (dark massif to the center left).

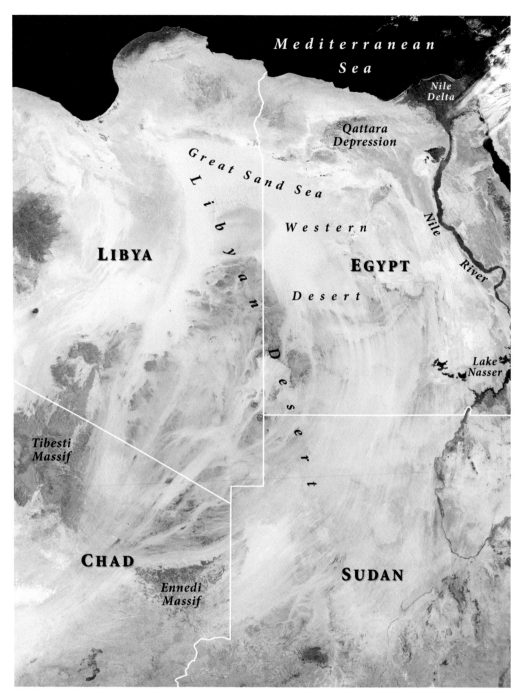

PLATE 2.3 Aeolian processes dominate the geomorphology of the Sahara Desert. The landscape can be divided into sectors of transportation and deposition. Sand seas or ergs represent the depositional sector. This photograph shows the Libyan Desert, where sand can be seen to extend from the north to the south in a broad clockwise arc that represents the flow of the northeast trades. The Nile River carries water through Sudan and Egypt from distant equatorial sources. Large areas of relict fluvial landscape are presently covered by sand. Evidence of previous volcanic activity in the Sahara can be found in the dark-toned Tibesti Massif in northeastern Chad. Source: NASA image.

PLATE 2.7 View of the Middle East including the countries of Egypt, Israel, Jordan, Syria, Lebanon, and Saudi Arabia. The deserts of the Negev and Sinai link arid zones of North Africa, the eastern Mediterranean, and Arabia. The general decrease in rainfall from the north to the south in the Negev Desert is visible in the vegetation cover. The Arabian Peninsula is extremely arid, averaging less than 100 mm annually. The Dead Sea and Sea of Galilee occupy pull-apart basins formed along the Levant Fault. Source: MODIS Land Rapid Response Team, NASA/GSFC.

PLATE 2.10 Arid and semiarid zones of northwest China occur in areas of rugged topography. Meltwater and stormflow from the high-altitude Bogda Shan flow south towards the Turfan Depression (see also Fig. 8.10). Ongoing tectonism is expressed in high erosion and sediment-production rates. The enormous fans transport course, rounded debris, forming the gobi or gravel plains of the region. Wind activity is expressed in the dune field and in the presence of yardangs immediately to the west of the dunes. Source: Landsat-7 image, NASA/U.S. Geological Survey.

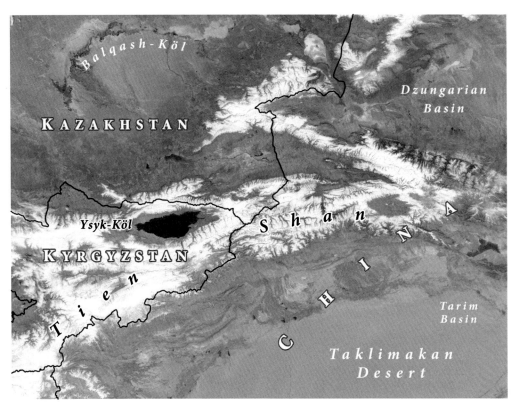

PLATE 3.9 Mountain snow plays an important hydrologic role in desert areas, providing a reliable source of runoff for agriculture and recharging the groundwater system. In this image, snow covers the high mountains of central Asia, most notably the Tien Shan Mountains on the border between northwestern China and Kyrgyzstan. To the south, lies the vast Taklimakan Desert, covered by dunes. Meltwater from the mountains penetrates into the margins of the desert areas. Source: NASA MODIS Rapid Response Team, NASA/GSFC.

PLATE 5.9 Lake systems of the Altiplano of Peru and Bolivia, South America. Lake Titicaca (dark lake) is the largest high-altitude lake in the world. Part of a cascading system, it drains southward into shallow Lake Poopó (Lago de Poopó; greenish-colored salt-water lake) and the salars (salt-crusted lakes) of Coipaso (also known as Coipasa) and Uyuni. The Salar de Uyuni (white-toned surface) is the largest salt flat in the world. The Atacama Desert is located along the coastal plain to the west of the Altiplano. Source: MODIS Land Rapid Response Team at NASA/GSFC.

PLATE 5.10 Lake Eyre is a 9600 km² playa at the lowest point (12–16 m BSL) in Australia. The low-gradient basin surface includes stoney (gibber) plains and continental dune fields (see upper right of image). False-color composite image acquired by Landsat 7 on 29 July 1999. Image from the Goddard Space Flight Center's Landsat Team and the Australian ground receiving station teams.

PLATE 6.9 Cuesta on the Colorado Plateau. The profile consists of a backslope, rim, face, and rampart (or debris-covered slope). The face is nearly vertical and is principally formed of caprock, but may include some of the less-resistant underlying units. The rampart extends from the base of the face to the surrounding plain, and is commonly debris-mantled.

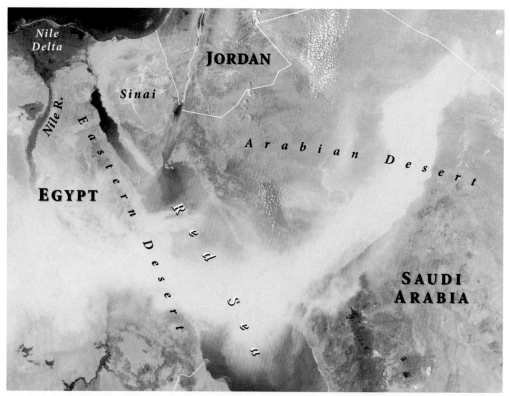

PLATE 10.16 The Arabian Peninsula is one of the world's major sources of dust. In this image, a thick plume of dust crosses Saudi Arabia, the Red Sea, and Egypt, forming an opaque, south-moving front. Source: NASA Aqua/MODI.

from areas that drain extensive Cenozoic volcanic deposits and highest in areas underlain by Mesozoic sandstone and shale (Gellis et al. 2004). In the Arroyo Chavez sub-basin, the geologic rate of sediment formation was found to be similar to the modern sediment yield for most geomorphic units (colluvial slopes and mesa tops) (Gellis et al. 2004).

For areas with high relief and large climatic change, sediment yields may have been greater in the past. For example, in semiarid basins adjacent to glaciated mountain ranges, the bulk of the material was probably delivered during colder periods, when high-discharge streams transported material prepared by glacial and periglacial processes (Zehfuss et al. 2001).

Denudation rates correlate with relief in many arid regions (Jayko 2005). Sedimentation rates in basin epocenters have been determined from core data. In tectonic basins in southeastern California (Searles, Owens, Death Valley, and Panamint), they generally ranged from 0.2 to 0.6 mm year^{-1} between 600,000 and 30,000 years BP, and between 0.6 and 1.6 mm year^{-1} during the past 30,000 years. The volume of material in alluvial fans has also been used to estimate denudation rates, with estimated rates of 0.04–0.23 mm year^{-1} for the Panamint and Death Valley areas (Jayko 2005).

8.5 FLUVIAL LANDFORMS

8.5.1 ALLUVIAL FANS

8.5.1.1 Introduction

Alluvial fans are probably the best studied of all desert landforms. They are not unique to deserts, however, occurring in humid and periglacial regions as well. Additionally, not all deserts have alluvial fans. They are most common in environments with active tectonism, which creates a marked contrast between the mountain front and depositional area. Uplift enhances stream competence and weathering processes provide debris. The debris is emplaced on the fan by water flow, debris flows, or mass movement. Subsequent to its deposition, the debris may be modified by weathering processes that, over time, give a new character to the surface. Most research on fans has been conducted in the Basin and Range Province of the American southwest, although they have also been described in the Sahara, the Middle East, Iran, Mongolia, Pakistan, southern Africa,

Spain, and western Argentina (Harvey 1988; Owen et al. 1997; Milana & Ruzycki 1999). They are less common in areas of low relief, such as the deserts of Australia.

As the name suggests, a fan is a segment of a low cone, with its apex located at the mouth of a canyon or gorge. The plan view of the form is broadly fan-shaped (Fig. 8.11), with the contours bowing downslope. Profiles across the fan are convex. The longitudinal form of a fan may be straight, composed of a series of facets or, more commonly, have a concave upwards profile. In the latter case, the slope is greatest at the fan apex, and decreases down-slope. Thus, the average slope of a fan lessens as the overall water and sediment supply and grain size decrease. The fan slope is generally steeper in arid regions than humid, owing the to coarser sediment load and differences in the transport efficiency (Milana & Ruzycki 1999). In humid areas, "wet" fans develop where water discharge is of greater volume and is more continuous, and water-laid sediments are more common. In arid regions, "dry" fans occur and intermittent flash floods and debris flows add to the flood volume. Where sediment gravity flows (debris flows and mudflows) dominate fan construction, the fan slopes will commonly range from 5 to 15°. Where water flow (either sheetflood or channelized flow) dominates, slopes have lesser radial gradients, on the order of 2–8°, which decrease downstream. Very steep alluvial fans are referred to as alluvial cones. Where fans coalesce to form a depositional piedmont, a bajada is formed.

Fans are constructed by a number of processes including rock falls, rock slides, rock avalanches, sheetfloods, debris flows, and mud flows (Blair & McPherson 1994), which result in different depositional units, including colluvium and bed-load alluvium. In general, fans develop where high-gradient streams leave a confining canyon and discharge onto a floodplain or the floor of an intermontane basin. Deposition of alluvium results, owing to an abrupt loss of competence at the apex of the fan through widening of the channel cross-section, infiltration of water, loss of water depth, and evaporation. The unconfinement of flows induces rapid deposition that results in the angular and poorly sorted textures typical of fans (Blair & McPherson 1994). Large fans may be thicker than 300 m and fan length may exceed 20 km. The debris ranges greatly in size, from clay particles to boulders. On alluvial fans adjacent to the Sierra Nevada, California, boulders attain a diameter as great as 4 m (Zehfuss et al. 2001).

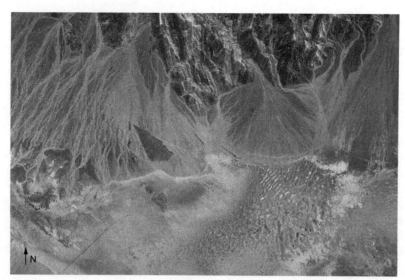

FIG. 8.11 Alluvial fans are common in arid environments with marked tectonism. On the surface of the fan are branching distributary channels. Location: Death Valley fans and dunes near Stovepipe Wells.

The surface morphology of the fan is determined by episodes of deposition, stability, and erosion. On the surface are branching distributary channels, commonly braided in form. Not all of the distributive channels are active contemporaneously. Over time, flood flows are diverted into new channels, in a process of diversion termed avulsion. Avulsion occurs during times of flood, after a period of aggradation near the fan apex, which changes the slope and facilitates the diversion of flow to a new course. The new channel begins to cut into the fan, with maximum incision near the fan apex, where the slope is steepest. Sediment eroded during fan-head entrenchment is deposited near the toe and the channel backfills as continued deposition moves up the fan. When the channel is filled, the deposition at the fan head locally increases the slope, and once again flows are diverted in a new direction, causing another channel avulsion (Schumm et al. 1987). Processes of entrenchment, aggradation, and avulsion may be coupled with pulses of sediment supply from the contributing drainage basin, associated with tectonic activity or climate change.

On many fans, the stream at the apex is incised below the fan in a fan-head trench and emerges onto the surface at a mid-fan intersection point (Hooke 1967). A segmented fan may develop if a stream entrenches the upper fan and transports sediment to the fan margin, reducing the stream gradient. Younger lobes emerge from between areas of older alluvium.

8.5.1.2 Sediment production, transportation, and deposition

Understanding alluvial fan development requires a consideration of sediment production and hillslope stripping, transport, the lithology of debris, depositional processes, and soil and pavement development during depositional hiatuses. The principal controls on sediment production are tectonism and climate. The latter is one of the most active areas of geomorphological investigation, but has been hampered by the paucity of age control in a stratigraphic context (Dorn 1996).

Fan development requires a high degree of sediment production. Climate change influences fan morphogenesis because of its effects on sediment production, water supply, and vegetation cover. The Pleistocene glaciation of the Sierra Nevada, California, helped to prepare immense amounts of debris that were transported by meltwater to create the immense bajada along the eastern base of the mountain range (Zehfuss et al. 2001). Likewise, in Mongolia, the majority of alluvial-fan sedimentation took place during the Pleistocene, whereas fan-head entrenchment and terrace formation occurred during the Holocene (Owen et al. 1997).

Whereas Pleistocene fan aggradation adjacent to high-altitude mountains was enhanced by upland sediment production resulting from frost weathering and glaciation, at lower elevations there were different causative factors and time frames. For much of

the southwestern USA, it is argued that fan aggradation was greatest during the transition to a warmer
and drier Holocene climate, when vegetation was
reduced, enhancing hillslope erosion (Bull 1991;
Reheis et al. 1996; and others). The Holocene transition was also associated with a change to a dominance of summer convective precipitation, and
possibly increased debris-flow deposition, for parts of
the American southwest and northern Mexico (Wells
et al. 1987; Bull 1991; Reheis et al. 1996; Ortega-
Ramírez et al. 2004). Climate change also affected
the lakes that were present in many basins, and thus
the base level of the streams that entered into
them.

Fans are complex systems because they develop
in environments where tectonism is active, providing renewed uplift, stream energy, and sediment
supply. Tectonism not only regulates mass availability in the source area and mass transfer to the alluvial
fan, but also affects the basin geometry and depth of
the sediment infill. High-gradient slopes adjacent to
fault zones give rise to narrow, linear basins with an
abundance of short, first-order channels, whereas
low-gradient slopes are associated with broad palmate
basins and longer drainage courses (Ortega-Ramírez
et al. 2004). Although it is useful to consider both
climate change and tectonism separately, this is difficult to do in areas where fan source areas have
undergone varying degrees of both tectonic and
climate change.

Sedimentary facies in a fan can be characterized
by examining stream-bank exposures and shallow
excavations. In cross-section, the strata may be
arranged in a sheet-like fashion, parallel to the fan
surface, or may contain the cut-and-fill structures of
buried channels. Alluvial fans are major stores of
sediments, trapping the coarse fraction of incoming
sediment: older fan materials are commonly referred
to as fanglomerates, combining the terms fan and
conglomerates.

During periods of stability, soils and desert pavement develop at the surface (Figures 7.7a and 8.12).
In cross-section, more recent deposits may be seen
to overlie older ones with well-developed soils that
developed during stable exposures (Zehfuss et al.
2001). In areas that experienced below-freezing conditions during the Pleistocene, the presence of invo-
lutions, associated with cryoturbation, and ice-wedge
casts, indicate that the fanglomerates were subjected
to permafrost processes (Owen et al. 1997). At
their distal margins, fans may interfinger with
fluvial, aeolian, or lacustrine sediments, forming a

Fig. 8.12 During periods of surface stability, soils and
desert pavement may develop on alluvial fan surfaces.
Weathering processes and aeolian inputs modify such
surfaces, above the reach of the active channel.
Location: Sheep Creek fan, California, USA.

continuous suite of sediments from the mountain
source to the basin center.

Alluvial fan surfaces are constantly modified by
active tectonism, climate change, aeolian inputs, and
weathering processes. Mapped units are distinguished in age based on the development of soil
profiles, the degree of varnish on surface clasts, and
the maturity of pavement formation (Fig. 7.7a).
Aeolian erosion commonly mobilizes fine playa sediments and deposits them on alluvial fans (McFadden
et al. 1987).

The form of the fan is a function of sediment type
and production, climate and vegetation, and tectonism. The type of sediment and hydrologic regime
give rise to two basic fan types: (1) debris-flow dominated and (2) stream-flow dominated. Many fans are

composite in nature (Harvey 1997). Channelized debris flows are associated with distinct levees and terminate in lobate deposits. They develop when masses of poorly sorted sediment, agitated and saturated with water, surge down slopes under the influence of gravity. The interaction of solid and fluid forces influences the motion and gives debris flows a unique destructive power. Debris flows are fluid enough to travel long distances in channels with modest slopes, with peak flow speeds of 10 m s^{-1} or greater, and total sediment concentrations that typically exceed 50% by volume (Iverson 1997).

Debris flows in the White Mountains, Owens Valley, California, are generated during intense rainstorms in the spring and summer. Landslides in high elevation, steep, and water-saturated hillslopes move debris down slope. This material is sheared, dilated, and has water added to become transformed into debris flows (Hubert & Filipov 1989) that move with a surging laminar motion along canyon floors to the fans. The deposits have a sandy mud matrix-supported fabric. Deposits average 50% gravel, 25% sand, 11% silt, and 4% clay. Inversely graded beds are common, with an upper layer that contains cobbles and boulders. Sticks and logs oriented parallel with flow directions indicate the laminar viscous motions of semirigid, high-strength plugs that overrode a high shear-stress layer. The levees that are characteristic of the margins of debris-flow lobes contain the largest clasts in the flows.

Deposition by fluvial processes on fans occurs either by unconfined sheetflow, or in braided channels. The dominant controls on transport mechanisms are the availability of fine sediment (silts and clays) and the water-to-sediment ratio. When there is a high water-to-sediment ratio, and a low concentration of fines, fluvial transport of coarse clasts will dominate, and deposition will occur in fluvial bar and sheet forms. Of course, this water-to-sediment ratio may vary throughout a storm and be affected by transmission losses, therefore resulting in a complex assemblage of depositional facies. Blair (1999) examined two adjoining fans in Death Valley: the Anvil Spring fan built solely by water-flow processes, and the Warm Spring fan, constructed principally by debris flows. Whereas the climate, watershed and fan area, relief, aspect, vegetation, and tectonic setting are essentially the same for each fan, the watersheds are underlain by different rock types, which weather to yield unique sediment suites. The sediment contributing to the Anvil Spring fan ranges from medium sands to boulders, with

little silt and clay. By contrast, the Warm Spring catchment yields an abundance of mud and unsorted sediment, which readily transforms into debris flows.

8.5.2 ARROYOS

Arroyo development in semiarid areas has been studied in southern Africa, India, Spain, and Australia. However, the majority of the work has focused on the western USA, where the episodic instability of dryland rivers has been an important focus of arid lands research (Graf 1983b).

An arroyo has a distinctive form that sets it apart from other stream channels. It appears as a trench developed in alluvium, with a roughly rectangular cross-sectional profile, near-vertical walls, and a flat floor (Fig. 8.13). Arroyos typically range from 5 to 200 km in length (Bull 1997) and are up to 20 m deep and 50 m wide. They are usually dry for most of the year, with flow occurring only in response to heavy rainfall, although arroyos with intermittent and perennial reaches, related to groundwater inflows, are recorded. The terms gully and arroyo are often used synonymously, the distinction being that gullies are smaller in scale (Waters 1991) and have persisted for a shorter period of time (Bull 1997). In a more restrictive use, applied in this text, a gully may be distinguished as having a V-shaped profile and being developed in colluvium. The term arroyo was first used in a geomorphic context in the American southwest by Dodge (1902), who considered them to be a product of stream channel entrenchment associated with changes in land use and climate.

At the upstream end, an arroyo is marked by a headcut or series of headcuts that merge downstream into a single entrenched channel (Leopold et al. 1964). The slope of the channel bed is generally less than that of the original valley floor. As a result, the walls of the entrenched channel gradually decrease in height downslope from the headcut, until the arroyo floor and valley merge at a location known as the intersection point, downslope of which there is alluvial deposition, often in the shape of a fan. Shallow braided channels may radiate across this fan surface. During dry periods, the channel bed may be subject to wind erosion, sometimes creating dunes on the alluvial floodplain (Waters 1991). Locally, marshes develop where the channel intersects the water table (Hendrickson & Minckley 1984).

FIG. 8.13 Arroyos develop in the cohesive, but easily eroded, alluvium of valley floors. In form, they have a roughly rectangular cross-sectional profile, near-vertical walls, and a flat floor. They are usually dry for most of the year. Location: Carrizo Plain, California, USA.

Arroyos develop in the cohesive, but easily eroded, alluvium of valley floors. The sand, silt, and clay are derived from hillslope erosion (Bull 1997; Clapp et al. 2001; Hereford 2002). The clay increases the cohesiveness of the sediment and reduces the infiltration capacity. Owing to numerous episodes of cutting and filling, the alluvial sequence may be quite complex (Patton & Boison 1986). The number of sedimentary units, palaeosols, and erosional surfaces is a function of the areal extent, magnitude, and number of erosional or depositional occurrences, and the degree of stability between such events. Stability allows for the development of soils and the accumulation of archaeological debris on the surface (Waters 1991).

The causes of arroyo entrenchment and valley-fill aggradation have received a great deal of study. Arroyos are inherently unstable and undergo numerous episodes of channel cutting and filling to create a complex alluvial sequence. Periods of stability may be followed by channel entrenchment, backfilling, and a return to stability The Santa Cruz River, Arizona, for example, is thought to have entrenched its floodplain five times since 5500 years BP (Waters 1988). In the Paria River valley, Utah, arroyo cutting occurred at approximately AD 1200–1400 and 1860–1910, with an intervening period of alluviation (Hereford 2002). These cyclic changes have been attributed to fluctuations in climate, tectonic activity, and land use, and to internal geomorphic adjustments.

Documentary records indicate that many alluvial valleys in the American southwest that were unchanneled in the early nineteenth century had well-developed arroyos after the beginning of white settlement. Arroyo development during the historic period (between about 1850 and 1910) coincided with agricultural and pastoral development (Bull 1997). Settlement modified the land use, as grazing animals were introduced, roads, pathways, and buildings were constructed, and artificial irrigation systems developed. A host of hydrologic changes ensued, related to new base levels, reduced infiltration capacities, and to vegetation loss from slopes, the floodplain, and the channel itself (Cooke & Reeves 1976). The changes in surface roughness influenced runoff to channels, sediment yield, and in-channel processes. Additionally, the trampling of cattle affected stream banks and compacted the ground, reducing the infiltration capacity and leading to more runoff from a given rainstorm. The arroyos developed rapidly and produced large amounts of sediment (Gregory 1917; and others). In some areas, arroyos which incised in the late 1800s began to aggrade after approximately 1940. At the same time, suspended sediment loads in the Colorado River basin decreased following reduced sediment production as arroyo cutting declined and sediment storage in channels increased (Gellis et al. 1991). Changes in channel width and depth accompanied sediment storage and floodplain development. For example, the Little Colorado River is now 50% of its width in 1914.

Superimposed on land-use changes were secular trends in climate that are not well understood. Owing to the difficulties in constraining the timing of alluviation and entrenchment, and our uncertainties as to the corresponding climatic conditions, numerous hypotheses have been posed relating climate change to arroyo processes. Arguments have suggested that channel trenching began during drier periods (associated with a decline in stabilizing vegetation, which increases the vulnerability of the land to highly infrequent high-intensity storms) (e.g. Bryan 1925; Antevs 1952; Gonzalez 2001); wetter periods, with higher mean stream discharges, and greater erosive capacity (Huntington 1914); and during periods with a low frequency of small rains and a high frequency of heavy summer downpours (Leopold 1951b). Based on a study of the late Pleistocene to Holocene stratigraphy of valley floors in southern Arizona, Waters and Haynes (2001) ascribed arroyo formation to changing dry–wet cycles. During dry periods, the water table dropped and vegetation cover was reduced. When this cycle was followed by a wet period, arroyo cutting ensued. Historic channel cutting appeared to be coincident with increased El Niño activity, and the associated generation of large, low-frequency flood events (Waters & Haynes 2001; Hereford 2002). In some cases, the introduction of cattle into a region may have exacerbated or accelerated an upstream migration of knickpoints that may have begun in an earlier period (Gonzalez 2001). Alluviation in the Paria River basin of the Colorado Plateau was coincident with the Little Ice Age (about AD 1400–1880) and may have been associated with relatively low-magnitude flood events (Hereford 2002). As the nature of the climate change and land use varied from region to region, it is difficult to determine an overarching set of causes for arroyo development. It may be that different sets of environmental changes lead to similar geomorphic results. As in many areas of geomorphology, it is difficult to resolve this problem without adequate datable material and a well-founded chronology.

Periodic channel entrenchment and aggradation have also been attributed to the inherently episodic nature of channel processes along ephemeral streams, where gradual aggradation by low flows that are not sufficiently competent to move all of the available sediment leads to over-steepening, headcut formation, and channel incision (Schumm & Hadley 1957; Schumm 1973; Patton & Schumm 1975; Bull 1997). Gonzalez (2001) found that meander cutoff processes also trigger incision, locally increasing channel slope and stream power.

In summary, the relative importance of land use, climatic variability, and intrinsic channel processes in arroyo systems remains a contentious subject, with causative factors that vary over time and between channels. Many early papers focused on historic arroyo cutting and the role of human alterations to the landscape. In recent years, an increasing number of studies contend that the principal trigger for cyclic deposition, stability, and erosion is climate change, which influences vegetation type and density and the magnitude and frequency of runoff events (Waters & Haynes 2001; Hereford 2002).

Arroyo cutting has had a number of important impacts in the American southwest. The verdant riverbed marshes, important habitat for flora and fauna, which flourished in the earlier periods, have been drained by arroyos. The removal of as much as 25% of the valley floor material (Cooke & Reeves 1976) has reduced agricultural land and severed transportation links. Furthermore, eroded sediment covers downstream agricultural land with undesirable material that contains large quantities of sand and gravel. Arroyo cutting has caused the displacement of modern towns (Bryan 1925), as well as the abandonment of areas once inhabited by the Anasazi people. Despite intense study, our fundamental lack of understanding of the causes of arroyo cutting and aggradation prevent cost-effective river rehabilitation or restoration programs.

8.5.3 GULLIES

A gully is steep-sided trench or channel that is cut into colluvial slopes by ephemeral flow processes. A gully is deeper than a rill and, being often a meter or more in depth, cannot be eradicated by a plow. The development of gullies is often related to changes in land use that favor greater runoff. These include processes that reduce the infiltration of the soil (for example, trampling by grazing animals or compaction by off-road vehicle use) or reduce the vegetation cover on slopes. Gullies expand upslope by erosion of the gully head. They are a major source of sediment in some small catchments. Collapse of the headcut may lead to temporary storage of sediment in the gully, which may be eroded away in subsequent events (Bradford et al. 1978). Following the removal of this material, headcutting resumes. Gully headcuts vary considerably in morphology, from those that start as rills and then deepen progressively without abrupt changes, to steep arcuate headcuts with plunge pools, connected upslope by a rill (Oostwoud Wijdenes et al. 1999). The form of the headcut may be related to material or environmental properties. In Spain, gullies are developing on marginal lands that are no longer being used for cultivation (Oostwoud Wijdenes et al. 1999).

The development of gullies and their evolution over time has been difficult to evaluate, as field observations of actively eroding gullies are infrequent. For channels with steep headcuts, both mass movement and fluvial processes operate to cause upstream migration and widening. A gully appears to develop as a rill that widens and deepens progressively in a downstream direction: when the channel at some point cuts through a more resistant layer owing to higher shear stresses, a knickpoint develops, which consequently migrates upstream by plunge-pool erosion to create a steep headcut (Oostwoud Wijdenes et al. 1999).

8.5.4 LANDFORM ASSEMBLAGES

Although landforms produced by water and sediment are often discussed separately, they commonly interlink to form landform assemblages. In the largely stable tectonic environments of the Colorado Plateau, there is a general flow of sediment from source areas (often bedrock outcrops) across alluvial fans or hillslope colluvium, to basin floors, where arroyos export material from the basin (Clapp et al. 2001).

In other regions, active tectonism and climate change operate together. In northern Mexico and the southwestern USA, distinctive suites of landforms develop in the many north–northwest-trending extensional basins. Extensional basins are caused by lithospheric stretching and thermal subsidence (Leeder 1999). The result is brittle fracture and normal faulting in relatively shallow parts of the lithosphere. Tectonism throughout the Pliocene and Quaternary produced fault-block mountains with active range-bounding normal fault systems. Sediment is eroded from areas of uplift and deposited in the extensional basins. The sediment loading causes downward flexure of the lithosphere and the onlap of sediment at the basin margins. Recent models relate alluvial fan growth, channel migration, and lake emplacement to stages in extensional basin evolution (Leeder & Jackson 1993; Gawthorpe & Leeder 2000; Ortega-Ramirez et al. 2004). The centers of the developing rift basins are occupied by axial rivers with floodplains, playas or perennial lakes, or aeolian sands; whereas the margins are bounded by alluvial fans whose scale and slope are strongly related to fault activity, type, and tectonic history.

Long-term, large-scale alluvial deposition is punctuated by periods of erosion, resulting in the landform assemblages of incised valleys and river terraces. Erosion may result from basin uplift, base level fall, or climatically induced decreases in upstream sediment supply. Such processes are by no means unique to arid environments. River terraces are parts of floodplains that have become elevated above the bankfull level of the active channel as a result of downcutting. Over time, terraces are themselves subject to degradation by mass wasting, overland flow, and gullying. Different episodes of incision and aggradation can cause the development of a series of terraces of different height and valley fills with a complicated internal structure (Leopold et al. 1964; Bull 1991). The age of terrace surfaces has been estimated using relative dating techniques, including the degree of development of palaeosols or desert pavements, as well as by an analysis of the degradation of terrace scarps.

9

AEOLIAN PROCESSES

9.1 Introduction

Aeolian processes pertain to the action of the wind. They include the transportation of silt- and sand-sized material, the development of landforms by erosion or deposition, and the formation of sedimentary structures (such as ripple marks). This chapter will principally examine the deposition of sand as bedforms and sand sheets, and Chapter 10 will focus on two aspects of wind erosion: the processes and associated landforms of aeolian erosion, and dust entrainment and transport.

Any terrestrial planet with a dynamic atmosphere is subject to aeolian activity, and Earth, Mars, and Venus have received considerable comparative study. On Mars, aeolian activity dominates under present climatic conditions owing to the absence of vegetation and liquid surface water. Aeolian features include wind streaks, ventifacts, mantling wind-blown sediments, duneforms, surface lag deposits, ripples, and yardangs (Greeley et al. 2002). Dust storms are often global in nature and dust devil tracks are clearly evident on high-resolution imagery. On Venus, surface features attributed to aeolian processes include various wind streaks, yardangs, and dune fields (Greeley 1994).

Terrestrial aeolian processes occur where there are abundant small particles, vegetation is sparse or absent, and winds are strong and frequent. Thus, most aeolian activity is limited to hot and cold deserts, to coastal areas, to regions fringing glacial activity, to exposed mountain regions, and to agricultural areas and construction sites where bare soil is exposed to high winds during dry periods. Aeolian landforms are estimated to cover at least 20–25% of the Earth's land surface (Livingstone & Warren 1996).

The nature and intensity of aeolian activity is dependent on many factors including the strength, timing, and direction of the wind, biological conditions, the nature and availability of sediments, atmospheric and surface moisture, vegetation, and topography. Changes in global circulation during the Pleistocene meant that aeolian activity played an even more significant role on Earth then than it does today, as evidenced by dune fields that were more widespread than currently observed (Sarnthein 1978). Stabilized dune fields have recently been examined for their susceptibility to future environmental changes, particularly the effects of global warming (Muhs & Maat 1993), which may cause reactivation of sand.

Aeolian processes are complex, occur on both small and large scales, are difficult to monitor, and commonly are best observed in regions remote from researchers. The ability to test hypotheses both in the field and in laboratory settings contributes to our greater understanding of aeolian processes. Such work was exemplified by the studies of R.A. Bagnold (1941), a British Army officer whose pioneering work in the deserts of Egypt combined fieldwork and wind-tunnel experimentation. His book, *The Physics of Blown Sand and Desert Dunes*, remains a classic and has inspired much modern work. Modern studies of aeolian processes have flourished with improvements in technology: aerial and satellite imagery provide a global perspective (Breed et al. 1979) and allow us to see changes over time, dataloggers and microprocessors allow detailed process studies at the local scale, and scanning electron microscopes allow aspects of the environmental history of a grain or surface to be deduced from microstructures. Today's advances in our understanding of the fundamental meteorological and geological conditions involved in desert dune processes and dynamics, particle transportation and deposition, and mechanisms of erosion have resulted from the scientific collaboration of geomorphologists, engineers, mathematicians, and physicists; and the incorporation of ideas gained from other fields of study, such as the meteorological investigation of airflow over hills (Wiggs 2001). Field studies have incorporated near-surface velocity measurements, using hot-wire or cup anemometers, and synchronous measurements of sand flux using various sand traps. This work has helped constrain the inter-relationships between dune morphology, wind velocity and shear stress, sand transport, and processes of erosion and deposition on dunes. However, field experimental and monitoring work is commonly restricted by limited resources, remote field sites, and the logistical difficulties encountered when operating sophisticated equipment in remote areas. Laboratory experimental work on dune dynamics is handicapped by scaling problems and by difficulties in reproducing the process variability inherent in natural systems. Although aeolian systems have probably received more study than any other branch of desert geomorphology – and despite significant advances in our understanding of aeolian processes over the past few decades – many landforms and processes remain poorly understood.

The study of the aeolian environment is of critical practical importance. Every year, productive agricultural soil is lost to deflation, serious transportation accidents occur owing to loss of visibility, health risks develop owing to the inhalation of fine particulates, land is lost due to dune encroachment, and abrasion by windblown grains has engineering consequences. It is within the context of these important environmental issues that much aeolian research is conducted.

9.2 NEAR-SURFACE FLOW

The movement of aeolian particles is a function of the strength of the wind, its erosivity, and the resistance of the surface to erosion, its erodibility (Wiggs 1997). In order for the wind to transport sedimentary particles and form aeolian bedforms, it must not only be above a certain threshold velocity, but it must also have a sufficient drift potential to prevent the growth of plants that would otherwise stabilize the sand (Tsoar 2001). In addition to wind magnitude, the direction and frequency of flow also affect dune formation.

The atmospheric boundary layer is approximately 1 km in thickness. Its characteristics govern the transport of sand and dust in the near-surface environment, as well as within the atmosphere. Wind in the upper atmosphere is relatively independent of the surface, except for the effects of very large features such as mountain ranges. By contrast, winds close to the surface (the atmospheric boundary layer) are retarded by friction and affected by such features as vegetation, boulders, or topography. Thus, within the boundary layer there is a velocity gradation from

zero at the surface, to a value no longer affected by friction (free stream velocity).

The velocity gradient within the boundary layer produces shear. As air motion is gusty and the surface is irregular, the flow is turbulent. When turbulence is vigorous, momentum from fast-moving layers aloft is transferred to the surface, thereby speeding up its average velocity. This mixing produces a steep velocity gradient and shear stress, the tangential stress exerted by wind blowing over a surface, which carries away grains of sand and dust.

9.2.1 VARIATION IN WIND VELOCITY WITH HEIGHT

Owing to frictional effects, the near surface wind velocity is retarded. There is a thin layer of air immediately above the surface in which the flow is stationary or very slow. Above this thin lowest layer, airflow is turbulent and characterized by fluctuating pressures and velocities and the random formation and decay of many small eddies. Turbulent wind flow is gusty, characterized by bursts of short-duration, choppy, violent wind. Because of the random movement imparted to the air by the turbulent eddies (the air may move upward, sideways, or even in the opposite direction to the net flow), the average velocity, U, is used to describe turbulent flow. Each "layer" of the air has a different average velocity and direction.

The degree of turbulence may be expressed by the *Reynolds number*, *Re*, a dimensionless number defined as:

$$Re = \frac{\rho h U}{\mu}$$

where ρ is the fluid density, h is the thickness of the boundary layer, U is a mean flow velocity, and μ the coefficient of absolute viscosity (a function principally of fluid composition and temperature, and a measure of the molecular interference between adjacent fluid layers). In the atmosphere, the flow depth is always relatively large and the viscosity is low, such that *Re* values are high even in gentle winds and the airflow is turbulent (Livingstone & Warren 1996).

The shearing force, instead of varying directly with the velocity of flow, varies as the square of the velocity. The *shear velocity* or *friction velocity* (u_*) is related to shear stress as follows:

$$u_* = (\tau/\rho)^{1/2}$$

where τ is the shear stress, ρ is the density of air (constant near the surface), and u_* has the dimensions of velocity (m s^{-1}).

Near the surface the wind velocity increases sharply with height. The vertical mass exchange, which characterizes turbulent flow, flattens the velocity profile of the turbulent boundary layer close to the surface. The velocity at any point varies not as the height, but as the logarithm of the height. The assumption of a logarithmic velocity profile does not apply in all natural conditions, but provides a general model for the understanding of wind velocity increase with elevation. The *von Kármán–Prandtl logarithmic velocity profile law* is valid for a neutrally stratified atmosphere (a condition in which the lapse rate equals the dry adiabatic lapse rate) in the absence of sand movement, under strong winds, where the velocity profile is independent of viscosity:

$$u/u_* = (1/k)\ln(z/z_0)$$

where u is the velocity at height z, k is the von Kármán universal constant for turbulent flow (≈ 0.40), and z_0 is the roughness length of the surface. The friction velocity is an expression of the velocity gradient. At the surface the velocity is zero, so the gradient starts at the height of the roughness length, z_0. When the height z is plotted on a logarithmic scale, the velocity gradient becomes linear, with the y intercept representing the aerodynamic roughness length (z_0; Fig. 9.1). Regardless of wind strength, all velocity profiles tend to converge to the same intercept or roughness length and the shear velocity (u_*) increases with increasing wind speed at a given height (z) above the surface.

For smooth sand surfaces, the depth of this layer, the *roughness length*, z_0 (also known as roughness height), of the surface, is approximately $d/30$, d being the grain diameter. Not all surfaces are smooth, however. Roughness elements include ripples, vegetation and crops, rocks, or clods of earth. On such surfaces, the point at which the wind speed is zero is displaced upward to a new reference plane, the height of which is a function of such factors as object height, density, porosity, and flexibility (Oke 1978). Closely spaced roughness elements, such as pebbles in a desert pavement, reduce the erosion of fine sediment by absorbing a portion of the wind shear. On vegetated surfaces, z_0 is displaced considerably upward from its position over bare surfaces.

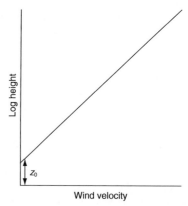

FIG. 9.1 The velocity profile of wind over a smooth, flat surface is linear when the height z is plotted on a logarithmic scale. The y intercept represents the aerodynamic roughness length z_0. From Livingstone, I. and Warren, A. (1996) *Aeolian Geomorphology: An Introduction*. Longman, Harlow. By kind permission of the authors.

In nature, surface winds are often unsteady and multidirectional, atmospheric conditions are unstable owing to surface heating, surface roughness and sediment characteristics vary, and dune morphology induces modifications to the flow. As a result, the assumption of a logarithmic profile is too simplistic for most field conditions (Livingstone & Warren 1996). Topographic obstacles such as dunes or hillslopes, for example, enhance streamline convergence and flow acceleration on the windward face (Fig. 9.2a), causing deviations from the predicted velocity profile (Lancaster 1995). In addition, airflow is typically quite complex, causing bursts of particles or sand, which are often organized into long parallel streamers. Despite the significance of such flow, it remains very poorly understood.

9.2.2 AIRFLOW AND SEDIMENT TRANSPORT OVER HILLS AND DUNES

Airflow over dunes has often been related to flow over low hills, using analytical techniques such as those of Jackson and Hunt (1975) and Taylor (1977). This has led to significant advancements in our understanding of dune dynamics and sediment transport. However, the interactions between airflow, hills or bedforms, and sediment transport are complex. In dune systems, these interactions affect dune form, spacing, and alignment. Where sand traverses bedrock hills, sand blasting processes form ventifacts,

(a)

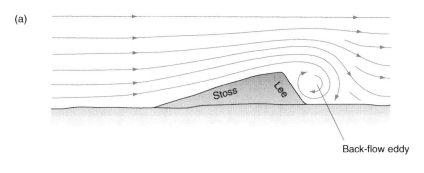

Back-flow eddy

(b)

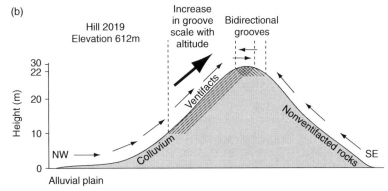

FIG. 9.2 As airflow approaches a hill or dune, the airstream compresses, resulting in an acceleration of flow, and an increase in velocity and sediment transport. (a) Illustrates flow over a dune, with a back-flow eddy developed on the lee slope. After Wiggs (2001). (b) Shows how wind velocity speed-up and increased sand transport cause ventifacts to preferentially develop on hill slopes: in this case, the regional northwest winds are strong, and form ventifacts over two-thirds of the slope; whereas acceleration of weak southeast winds allows ventifacts to form only near the crest. The intensity of ventifaction (groove depth and width) increases with wind velocity up the slope. After Laity (1987).

with the intensity of the process increasing with altitude (Fig. 9.2b) (Laity 1987; see Chapter 10).

On a dune, sediment moves up the stoss (windward) slope, across the dune crest, and then into the lee of the dune. Although most studies have examined the stoss and lee sides of the dune separately, they form part of a continuum of inter-related processes. Changes in the friction velocity (u_*) affect sediment transport rates in both time and space. The resultant patterns of erosion or deposition affect the size and morphology of the dune.

As airflow approaches a hill or dune, the pressure rises slightly because of the increased build up of pressure, or stagnation effect, resulting in a reduction of wind speed. However, as it moves upward over the obstacle, there is a compression of streamlines in the boundary layer that causes winds to accelerate towards the crest of the slope (Fig. 9.2). The increase in wind speed over the stoss slope of the hill amplifies sand transport and produces additional surface shear stress (Lancaster 1985). The magnitude of the

wind-velocity increase towards the crest of the hill is referred to as the speed-up factor (Jackson & Hunt 1975) or the amplification factor (Bowen & Lindley 1977).

The change in wind velocity, or speed-up ratio, can be expressed as:

$$A_z = \frac{\bar{U}_z}{\bar{U}_1}$$

where $\bar{U}_z$ is the mean wind velocity at height z above the hill or dune, and $\bar{U}_1$ is the mean wind velocity at the same height over a flat surface. The magnitude of the effect is on the order of 1.1–2.0 for small- to moderate-sized dunes (Lancaster 1985; Tsoar 1985; Mulligan 1988), with the value reflecting stoss slope morphology, the incident angle of the wind, and dune height and spacing. The largest shear-stress values, pressure changes, form drags, and wind velocities are recorded when the near-surface flow approaches approximately normal to the topography. The rate of change of the speed-up ratio (A_z) depends primarily on the shape and height of the dune.

As the wind speed affects the sand-carrying capacity, determining whether erosion or deposition occurs, there is a self-regulatory process wherein the dune changes its size and morphology to approach a steady state or quasi-steady state (Tsoar 2001). Measured sand flux and wind velocity reach a maximum at the crest. The wind is not only transporting sand eroded at that point, but also sand derived from the length of the stoss slope (Fig. 9.3) (Wiggs 2001). McKenna Neuman et al. (1997) recorded speed-up values ranging from 1.50 to 3.19, with a corresponding increase in sediment flux of one to two orders of magnitude. The dune can grow vertically only as long as the rate of sand supply to the crest is not exceeded by the rate of sand removal. The morphology of the dune will reflect a balance between sand supply, grain size, and wind energy factors, and any changes will modify the shape of the dune and bring it into a new state of balance between sediment input and output.

Sand that is eroded from the stoss slope is deposited on the upper lee slope. Beyond the crest of the dune, the airflow expands and decelerates within a turbulent wake, lowering the transport. Studies of grainfall (suspended fallout) onto the lee slopes of transverse dunes suggest that about 94% of the total deposition is within approximately 1 m of the brink of small dunes and 96% within 2 m (Nickling et al. 2002). Usually the upper part of the lee slope advances faster than the lower part, leading to an increase in the leeward slope angle. Oversteepening of the lee face to an angle of approximately 35° (the angle of internal friction) causes grainflow and the development of "avalanche tongues," which reduce the angle to 32–3°, the angle of repose (Tsoar 2001). Reverse flow plays an important role in maintaining dune form and lee-side dune steepness by returning sediment to the dune (Fig. 9.4). Under certain conditions of incident angle and dune slope and shape (typically a well-developed brink and an incidence angle of 90 ± 15°), a back-flow eddy develops, and flow recycles back up the lee slope (Fig. 9.2a) (Frank & Kocurek 1996; Walker & Nickling 2002). The convergence of air and sand from the stoss slope and the back eddy of the lee slope causes vertical ridge growth, and has been used to explain the development of distinct peaks on star-dune crestlines (T. Wang et al. 2005). The converging air and sediment may extend laterally in three dimensions to form a transverse roller vortex. When the incident air is oblique to the dune, there may be a lateral advection component, and a helical vortex develops that can move sediment along the lee face of the dune. Evidence of this along-dune transport process is provided by the orientation of lee-side ripples, which

Fig. 9.3 Measured sand flux reaches a maximum at the dune crest where the wind velocity is approximately twice that at the toe, driving sediment transport. In this photograph, the dark surface of the dune sand has been wetted by recent rainfall. Strong winds dry the upper layers of the sand, allowing transport. Location: Dumont Dunes, California, USA.

Fig. 9.4 A transverse dune. The most recent wind flow has been from right to left as shown by the orientation of small shadow dunes behind vegetation (parallel to flow), normal aeolian ripples and megaripples (perpendicular to flow), and the form of the dune (less steep windward face and steeper lee face). Reverse flow helps to maintain the form of the dune, as shown by the upslope orientation of linear features to the far left of the scene. Location: Cadiz Dunes, California, USA.

form essentially perpendicular to flow (Walker & Nickling 2002).

Erosion on the windward face and deposition on the lee face, as described above, allows for dune migration, and is characteristic of transverse and barchan dunes. When a steady state is achieved, sand eroded from the stoss slope is deposited in equal volume on the lee side, and the dune advances without substantially changing its shape. Since natural fluctuations in wind strength and direction, underlying topography, and sand supply are typical of all environments, migrating dunes are regarded as being in a quasi-steady state. Detailed studies on dunes also suggest that dune formation is not simply a response to regional wind patterns, but a complex interaction between the dune and airflow, with secondary flow regimes playing an important role (Wiggs 2001).

9.3 WIND PROCESSES

9.3.1 AEOLIAN PARTICLES

9.3.1.1 Particle sizes

Sand is a loose, granular material in the size range of 0.0625–2 mm (62.5–2000 µm), regardless of its

composition or origin (Table 9.1). It is transported primarily by saltation, traction, and impact creep. Dune sands are usually fine-grained and moderately well sorted (Ahlbrandt 1979). The typical dune

TABLE 9.1 Grain-size scale for clastic sediments.

	Grain size		
	mm	**µm**	**Phi (ϕ)**
Pebble	64		−6
Granule	4		−2
Very coarse sand	2	2000	−1
Coarse sand	1	1000	0
Medium sand	0.5	500	1
Fine sand	0.25	250	2
Very fine sand	0.125	125	3
Coarse silt	0.062	62	4
Medium silt	0.031	31	5
Fine silt	0.016	16	6
Very fine silt	0.008	8	7
Clay	0.004	4	8

Values shown are the upper bounds for each class. Silt and clay are commonly grouped together as "dust."
The phi scale for grain size is $\phi = -\log_2 D$, where D is the diameter of the particle (in millimeters), and the base of the logarithms is 2.

quartz-grain size is about 0.1–0.3 mm (Wilson 1973; Ahlbrandt 1979). Finer-grained sediment (silt and clay) is winnowed from active dunes, and dune formation is inhibited when grains are coarser than 0.3 mm. However, sands on the flanks of dunes may be somewhat coarser and more poorly sorted (McKee 1983). Although the general grain-size distribution of dune sands is well known, the effect of grain size on dune size and morphology, and on dune spacing, is not well understood (Kocurek & Lancaster 1999).

Particles larger than sand are moved by traction, by impact creep, or by saltation in rare high-velocity wind events. Granule ripples occur with a mode of 2–4 mm. Coarse grains are rarely moved by wind and 10 mm is usually the size limit for aeolian transportation (Kocurek & Nielson 1986). Larger particles are not moved by the wind.

Dust is composed of silts and clays, less than 0.0625 mm (62.5 μm) in diameter, which are transported in suspension. Coarse dust is approximately 5–60 μm in diameter and is usually transported by dust devils or local dust storms for short distances, seldom greater than 100 km (Middleton 1997). Fine dust is approximately 2–10 μm in diameter and is frequently transported as aerosols, remaining in suspension for long distances for periods of days or weeks until removed from the atmosphere by rainfall. The classifications of coarse dust and fine dust do not correspond to the breakdown sizes for "silt" or "clay." Most silt is of quartz composition and derived from the chipping of the margins of sand-sized particles (Greeley & Iversen 1985). Once grains are reduced to about 1 μm, further reduction in size is retarded. Silt and clay particles usually require stronger threshold velocities than sand as fine particles are held together by cohesion and hard crusts often bind surfaces. As a result, dust is commonly set in motion by the saltation impact of sand, in a "sandblasting" process (Chepil & Woodruff 1963; Gomes et al. 1990; Shao et al. 1993; Gill 1996), or when a surface crust is broken by, for example, off-road vehicles or agricultural implements.

9.3.1.2 Processes of particle formation

The dominant mineral found in dune sand is quartz. However, in some dunes, volcanic fragments or calcium carbonate, gypsum, or feldspar grains dominate or form a large fraction of the grains (McKee 1983). Regional variations in sediment geochemistry, mineralogy, rounding, and texture develop in differ-

ent environmental settings. Aeolian sediments of the Taklimakan Desert, China, have higher ratios of feldspar/quartz and calcite/quartz, finer grain size, fewer rounded grains, and a smaller proportion of frosted quartz grains than those in North African and Saudi Arabian deserts, for example, reflecting the impact of nearby glacial activity (Honda & Shimizu 1998).

Sand production is a result of five major processes: weathering, cataclastic processes, volcanism, precipitation/biological activity, and aggregation (Greeley & Iversen 1985). Weathering is the chemical and physical breakdown of rocks through exposure to the atmosphere, hydrosphere, and biosphere. Mechanical breakdown by salt weathering, for example, produces grains that are rapidly removed by aeolian abrasion and deflation, lowering some basins below present sea level (Corbett 1993) and causing a flattening of relief (Viles & Goudie 2007) (see Section 6.2.2). The end products of weathering are quartz grains of sand and silt size, small muscovite mica flakes, clay minerals, and various soluble carbonates and iron oxides. Most quartz sand is derived from the chemical weathering of granitic rocks. In some deserts, the crumbled mass of feldspar and quartz grains and partly decomposed granite forms a granular surface termed *grus*. Another important source of sand grains is the breakdown of sandstones and sandy conglomerates. Cataclastic processes break, fragment, or crush rocks and particles. Attrition, the reduction of particles in mass, is 100–1000 times more effective in air than in water (Kuenen 1960).

Sands of volcanic origin include both pyroclastic materials and weathered volcanic rocks. The Great Sand Dunes of Colorado are composed mostly of volcanic rock fragments (52%) and quartz (28%) (Wiegand 1977). These grains are initially transported from the San Juan Mountains by fluvial processes and later reworked by the wind.

Chemical and biochemical precipitation also results in sand-size particles. Biogenic carbonate sand, a principal component of coastal dunes, consists of skeletal carbonate debris (foraminiferal tests, echinoderm spines, coral) broken into smaller particles by physical (waves and currents) and biological (boring organisms) processes. Carbonate sands are found principally in tropical settings (McKee 1983). Typically, the biogenic component decays with distance from the coast. In the Wahiba Sands, for example, carbonate grains are too weak to survive transport over 200 km to the northern section of the dune field (Juyal et al. 1997). In India, the dunes near the coast are highly calcareous, but farther inland they are predominantly formed of quartz, although some foraminiferal tests are present (Goudie & Sperling 1988). Cemented carbonate-rich dunes are called aeolianites. They may underlie unconsolidated calcareous dunes as, for example, along the coast of southeastern Oman and the Emirates (Juyal et al. 1997). The calcium carbonate content of the unconsolidated sands ranges from 20 to 60%, whereas the underlying aeolianites have a carbonate content of 25–90%. The carbonate-rich sands were principally derived from coastal beaches during late glacial periods of lower sea level.

Dunes whose sediments are formed principally by evaporative processes are relatively rare. Salt pans and playas act as a source of aeolian sediment, with varying amounts of siliclastic material incorporated (McKee 1966; Svendsen 2003). Gypsum dunes, formed largely of abraded gypsum crystals, may occur along the margins of playas subject to wind scouring. Many dunes in the Chihuahuan Desert of southwestern North America are formed of gypsum or include a significant component of gypsum (Fig. 9.5). The 400 km^2 dune field of White Sands National Monument, New Mexico, adjacent to Lake Lucero, a pluvial lake with a former area of about 466 km^2, is the world's largest gypsum dune field (McKee 1966; Langford 2003). Dunes in the Cuatro Cienegas Basin in Coahuila, Mexico, are also gypsum. Gypsum-oolite dunes also occur, such as in the dunes close to the former shores of Lake Bonneville, Utah. The

gypsum crystals form by chemical precipitation in bottom muds below the water level. When the water level falls, crystals are exposed, removed by wind, and deposited downwind as dunes. The grains are originally tabular or lozenge-shaped and are rounded by abrasion during transport. Ooids are spherical or ovoid in shape and are well-rounded carbonate grains consisting of a detrital nucleus and concentrically laminated cortex. Halite dunes are rare. On the Salar de Uyuni, Bolivia, small horseshoe-shaped parabolic dunes ($\approx$1 m wide) are formed of halite, with minor amounts of windblown dust (Svendsen 2003). The principle factor controlling the formation of these dunes is sediment supply, which is related to the brine table level. When evaporation exceeds precipitation, the halite crystals are formed as 1–3 cm hollow pyramidal crystals, which are reworked by the winds and fragmented into sand-sized grains.

Aggregation is the fifth means of particle formation. Sand-size aggregates are built from smaller particles, forming "composite" grains. During the dry season, cracking of playa deposits of silt and mud, combined with salt crystallization and hydration, cause fragments to be loosened. These can be peeled off as flakes and chunks by the wind, and be abraded to form sand-size pellets. Such aggregates are capable of forming ripples and dunes. Deposits of aeolian clay, referred to as *parna* in Australia, form lunettes, low transverse dune ridges that rarely exceed 30 m in height. Aggregation of fine grains may also result

Fig. 9.5 Dunes of the Chihuahuan Desert of North America are commonly formed of gypsum. The dunes of the 400 km^2 White Sands National Monument, New Mexico, are derived from Lake Lucero, a pluvial lake.

from electrostatic charges and aggregation in the atmosphere has been proposed as a mechanism for settling dust.

9.3.2 PARTICLE ENTRAINMENT (SAND)

The nature of particle entrainment differs according to whether the particles being considered are sand, discussed here, or dust, considered in Chapter 10.

As wind speed increases over a bed of dry sand, the particles begin to oscillate: as the speed increases, the particles may take off suddenly in a vertical, or near-vertical direction. The *fluid or static threshold* is the wind speed at which continuous motion starts without impact from upwind (Bagnold 1941; Greeley & Iversen 1985). The specific shear velocity or wind velocity required to initiate grain motion varies with the average size and sorting of the surface sand population, the density and shape of the grains, grain cohesion, and the presence of surface roughness elements, such as vegetation. Typically, the fluid threshold for roughly spherical quartz grains with a diameter of about 0.25 mm is achieved when the wind velocity at 0.01 m above the surface reaches nearly 5.0 m s^{-1}. However, as natural sands are usually composed of a range of grain sizes and shapes, the fluid threshold velocity also falls within a range of wind speeds.

If particles are introduced from upstream, continuous movement of particles from the initially quiescent surface begins at a lower wind speed, termed the *impact threshold*, which is about 80% of the static value (Bagnold 1941; Greeley & Iversen 1985). At first, small exposed grains are mobilized but their motion disturbs other grains and there is a cascade effect as more and more grains are set in motion. The momentum imparted by saltation impacts can maintain sand movement at shear velocities (or the wind velocity at a specific height) below those at the fluid threshold.

Factors other than wind speed may also play a significant role in the initiation of particle movement. These include the amount of soil moisture (as moisture increases, the threshold speed required for movement also increases), the amount of organic matter, fine particles, or algae and bacteria, the presence of cryptobiotic crusts (which bind the surface) (Chen et al. 2007), and the humidity of the atmosphere (wind erosion may be reduced at higher humidities) (Lancaster & Nickling 1994; Lancaster 1995).

9.3.3 PARTICLE TRANSPORT

9.3.3.1 Modes of transportation

Particles are transported by the wind in one of three modes: suspension, saltation, or traction. Movement by *suspension* occurs under conditions of highly turbulent airflow when particles smaller than about 0.06–0.08 mm (<60–80 μm) in diameter are mobilized. Grains that are lifted or ejected from the surface are subjected to sufficient lift and drag to overcome their tendency to fall back to the surface under their own weight. Suspension represents the dominant mechanism for dust transportation.

Particles that are moved by *saltation* are mostly sand-sized particles, and the particles most easily moved by the wind lie in the range of fine sand (about 0.1 mm or 100 μm) (Greeley & Iversen 1985). Grains move in a bounding motion, receiving their momentum directly from the pressure of the wind on them after they have risen into it. When a surface sand grain is dislodged from the surface owing to the drag and lift forces exerted by the airflow or the momentum imparted by the impact of another moving grain, it travels on a characteristic path (the saltation trajectory) which may reach 1–2 m above the surface and extend several meters downwind. At first the forward motion of the grain is minimal, but the large drag force causes its forward velocity to increase as it rises. At the top of its trajectory, a particle is at it maximum velocity, which is between a half and two-thirds of the wind speed (Anderson & Hallet 1986). As the initial kinetic energy of lift is dissipated, the grain's weight carries it back to the surface, at an angle of between 6 and 16°. Forward velocity is maintained, and upon impact, the grain may bounce if it meets an immobile body and continue to saltate. A saltating grain that falls into a bed of other grains may set others in motion and cause a cascading effect. Once saltation has commenced, wind speed plays little part in determining the height a particle attains, which is more dependent on particle shape and size, being inversely proportional to grain diameter. Shape affects the ability of particles to leave the ground surface. Thus, for the same grain size, an increase in height of 4–15% is attained by the most spherical particles. Maximum particle height is greatest when the grain impacts a hard surface, such as bedrock.

Impacts between the saltating grains and the bed transfer momentum to other grains, which move a short distance. Grain movement that is characterized by a short trajectory and low-impact energy is termed *reptation* (Anderson & Haff 1988).

In a natural population of sand grains there is a range of grain diameters. Larger particles are pushed, rolled, or slid along the surface in a process termed *traction* or *creep*. The term impact creep refers to the movement of larger particles by saltation impact. However, individual grains may move by creep or saltation at different times as wind velocity varies.

Transportation on a dune is more complex than on a level surface and includes saltation or ripple migration (including surface creep), grainfall or the dropping of grains from sand in suspension, and avalanching (either by slumping or grain flowage) (McKee 1983). The interior structure of dunes shows cross stratification. The most common types of structure are tabular-planar or wedge-planar cross strata with relatively high-angle dips. They are the product of lee-side avalanching on dunes.

9.3.3.2 Transport rates

The transport of sand may be estimated by using transport-rate formulae, by employing mechanical traps, or by measuring the impact of saltating grains on a pressure sensor. The computed transport rate is expressed as the volume of sand (in cubic meters) passing over a meter-long line perpendicular to the direction of net drift over a given period, generally 1 year. Most transport-rate formulae incorporate a cubic relationship ($q \propto u_*^3$) between sand transport and the prevailing profile-derived shear velocity or wind velocity at some specific height above the surface (Bagnold 1941). Fig. 9.6 compares empirical transport models.

Transport-rate formulae assume ideal surfaces (horizontal, covered with well-sorted and loose sand, and traversed by steady winds) and a saturated sand flux; whereas, in nature, slope, vegetation, atmospheric and sediment moisture, air temperature (which affects air density and turbulence), and soil crusting also affect transport, and at the foot of some dunes (for example, barchans) the sand flux is often not saturated (Lancaster et al. 1996; McKenna Neuman 2004).

The transport of sand may also be estimated by using both horizontally and vertically aligned mechanical sand traps. Vertical sand traps allow a study of mass transport at varying heights above the surface, but are subject to several limitations: they are undermined by scour, deflect the wind, must respond to changes in wind direction by rotating,

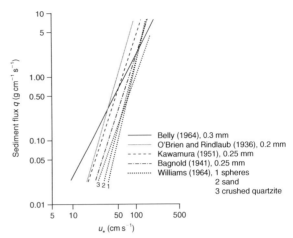

FIG. 9.6 A comparison of sand-sized sediment flux predicted by different transport equations. From Lancaster, N. and Nickling, W.G. (1994) Aeolian sediment transport. In: Abrahams, A.D. and Parsons, A.J. (eds), *Geomorphology of Desert Environments*. Chapman & Hall, London. Fig. 17.4. With kind permission of the authors and Springer Science and Business Media.

need to segregate sand transport by height, and may fill very quickly. The trap itself induces flow distortions, which cause catch efficiency to range from about 50 to 70% (Rasmussen & Mikkelsen 1998). The results from different types of traps are not directly comparable. Electronic devices that measure the impact of saltating grains as a measure of sand transport (such as the Sensit) are presently being employed in some test regions (for example, Lancaster & Baas 1998).

9.4 LANDFORMS OF ACCUMULATION: SAND SHEETS, ZIBAR, AND SAND STRINGERS

Sand deposits occur in bedforms at three scales: ripples, dunes, and large sand bedforms termed draas, a North African term for a large sand hill or megadune (Wilson 1973). A bedform is a systematic morphological feature formed by shearing between flow and cohesionless sediment. In addition, sand sheets may be formed: these are irregular accumulations surfaced by grains that may be too large for saltation. How the sediment accumulates is largely a function of the wind regime, the amount of vegetation, and grain size and grain-size distribution. Other factors, such as cementing agents, may affect the ultimate form of aeolian deposits.

Fig. 9.7 Stabilized sand sheets, up to 2 m in thickness, cover large areas of the Colorado Plateau. In the distance are the Henry Mountains.

Sand sheets are extensive accumulations of low-relief sand accumulation that are common in many sand seas, developing in conditions unfavorable for dune formation. They are low, flat, gently undulating sandy strips with no slipfaces: that is, they lack significant morphological expression (Fig. 9.7). They comprise approximately 40% of aeolian depositional surfaces (Fryberger & Goudie 1981). One of the largest sand sheets is the 100,000 km² Selima Sand Sheet of southern Egypt and northern Sudan, across which move active dunes. Beneath this sand sheet is a buried topography of fluvial origin (McCauley et al. 1982).

Sand-sheet surfaces can be rippled or unrippled, and range from flat, to regularly undulatory, to irregular. Grain size ranges from very fine- to very coarse-grained sand, and a mud or gravel component may be present. The Selima Sand Sheet has a bimodal grain-size distribution, consisting of a fine- to very fine-sized class and a very coarse sand/granule class. The surface may exhibit a one-grain-deep granule armoring (Embabi 1998). The deposits are "low-angle," and consist largely of wind-ripple laminae. Some sand sheets are transitional into sabkhas (sebkhas): low coastal flats in deserts that lie just above the level of high tide.

Although some sand sheets are only thin veneers of sand in transit, deposits up to several meters thick appear in the geologic record. According to Kocurek and Nielson (1986), the accumulation of a significant body of sand without dune formation results from conditions outside the range in which dunes form, or factors that inhibit dune formation. Sand sheets are promoted by, and dunes inhibited by: (1) a high water table, (2) surface cementation or binding, (3) periodic flooding, (4) a significant coarse-grain-size component, and (5) vegetation (Kocurek & Nielson 1986). A high water table, near the coastline or in an interior basin, effectively limits the amount of dry sand available for dune building. Surface cements such as gypsum, anhydrite, and halite, or thin crusts with fines and cyanophyte and fungal binding, limit the quantity of sand available for transport by significantly raising the threshold velocity. Cements may enter the sand sheet via groundwater, seawater flooding during storms, surface runoff, wind, and *in situ* growth of bacteria or fungi. Vertical accretion results from introduction of sand into a sand sheet followed by periodic cementing events. Cementing and binding are most common in sabkhas, coastal ergs, ergs in interiorly draining basins, and ergs adjacent to wadis and playas. Periodic flooding also promotes sand sheets by washing away sand accumulations so that larger dunes never form. Coarse grains, often only a few grains thick, may form armored surfaces, protecting the underlying finer sediment from transport. Vegetation promotes sand-sheet accumulation by causing sediment deposition, lowering the wind velocity, and causing secondary flow around plants that interferes with characteristic flows around dunes. Climatic change probably also plays a role in sand-sheet development (Kocurek & Nielson 1986).

The definition of sand sheet includes the zibar, a coarse-grained, low-amplitude, long-wavelength, migrating bedform covered entirely by wind ripples or megaripples. It is included with sand sheets because it lacks a slipface (Kocurek & Nielson 1986). The low, flat form typical of zibar is governed by the presence of a coarse sand mode in the sediments. The term zibar is derived from the Arabic *zibara*, which means a hard sandy surface that permits the passage of vehicles. On aerial photographs, zibar appears as chevron-shaped, transverse, or linear features. On partially vegetated sand sheets, zibar may also have a parabolic form.

Sand streaks, also known as sand stringers and sand streamers, also lack a well-defined dune form. They consist of linear patches of sand that lack slipfaces and retain a relatively constant width in a downstream direction. The edges are sharply defined and the streaks parallel the dominantly unimodal winds.

9.5 LANDFORMS OF ACCUMULATION: DUNES

9.5.1 INTRODUCTION

The scientific study of dunes began in the late nineteenth century. These early studies were mostly descriptive and concerned with the movement of sand bodies and the external form of dunes. Modern research on dunes concentrates on the nature, timing, and processes of dune formation, dune patterns and classification, dune composition (grain size, type, sedimentary structures) and provenance, the palaeoclimatic implications of dunes, and dune ecology. As in other fields, there is a continuously changing emphasis.

Although sand dunes are found in many climates, from cold periglacial to humid coastal settings, more than 99% of all dunes are found in deserts (Wilson 1973; Tsoar 2001). Modern active inland dunes occur in arid deserts associated with the subtropical high-pressure belt, in some mid-latitude and high-latitude continental interior deserts, and in rainshadow deserts and polar deserts. The formation of dunes requires an abundant supply of sand-sized sediment, strong winds, and conditions that favor deposition. Today, about 10% of the land area between 30°N and 30°S is mantled by active sand dunes (Sarnthein 1978). The percentage of area covered varies considerably by region, however, as shown in Table 9.2 (Breed et al. 1979). Dunes cover almost 30% of the

TABLE 9.2 Percentage of major desert areas covered by dunes.

Location	Dune cover (%)
Australia	31.0
Sahara	28.0
Arabia	26.0
Southern Africa	16.0
Central Asia	4.5
China	3.6
Southwestern USA	0.6

Source: after Breed et al. (1979).

great Saharan and Arabian deserts, but less than 1% of the surface of the North American desert (Fig. 9.8). In the past, sand dunes and associated deserts were much more extensive than they are today, mantling as much as 50% of the land between 30°N and 30°S (Sarnthein 1978). Many of the world's large, active dunes had their origin in the Pleistocene and today are only being modified (Glennie 1970). Additionally, dunes have been identified from even earlier periods in the geologic record, including the early Cambrian, late Devonian, mid Permian, late Triassic, and early Cretaceous (Glennie 1987).

The source of sand for dunes and dune fields varies according to the environment, but is most commonly rivers, lakes, or marine deposits. In some cases, the dunes are located relatively close to the sand source (such as many of the smaller dune fields in California), but in regions such as the Sahara, the sand has probably traveled great distances. Sand will be transported as long as the wind has sufficient energy. Deposition occurs in zones of lower energy such as in topographic basins (for example, the sand seas of the Simpson/Strzelecki desert, Australia, the Western Desert sand sea of Egypt, or the Taklimakan Desert, China) or against topographic obstacles. Dunes that enter shallow basins decelerate their movement (Embabi 1998).

9.5.2 THE DEVELOPMENT OF DUNE FIELDS: PALAEO-AEOLIAN PROCESSES AND EVIDENCE FOR MULTIPLE PHASES OF ACTIVITY

Although the ages of many of the great sand seas of the world are not known, studies of smaller fields suggest that they develop during complex, multi-cyclic events rather than as one single period of input

FIG. 9.8 Dunes cover less than 1% of the North American deserts. The Ibex Dunes, shown here, are a small cluster of star dunes in southernmost Death Valley National Park, California, USA.

(Yang et al. 2003). The formation of vast dune systems requires an enormous amount of time: perhaps up to a million years to form the largest dunes. However, most dune fields lack dateable deposits older than the LGM (Juyal et al. 1997), due to either extensive reworking of earlier deposits or limited available outcrops.

Throughout the Quaternary, sand seas expanded and contracted. Construction occurred during dry periods, when aeolian sands were mobilized and deposited; and stabilized or eroded during wetter periods, characterized by dune consolidation, soil formation, lake expansion, and fluvial erosion. In near–coastal settings, aeolian deposition was also influenced by changes in sea level, which affected the availability of shelf sand and water-table levels. The Wahiba Sand Sea, for example, shows a complex history of erg construction, stabilization, and erosion resulting from late Quaternary climatic and sea-level changes (Radies et al. 2004).

The centers of sand seas usually hold the greatest volumes of sand and may have the most complex dune forms, often consisting of accumulating forms such as linear ridges and star dunes: dunes at the margins are usually simpler in form and smaller in size, including migrating forms such as barchans and transverse dunes (Lancaster 1983; Porter 1986). Through the orientation of dune axes, sand seas often retain the signature of the principal winds that form them (Thomas & Shaw 1991; Lancaster et al. 2002; Kocurek & Ewing 2005). In Arabia, the Rub' al Khali was formed under the influence of the Northeast Trade Winds; the dunes of the emirates and western border of Oman by the northern Shamal winds; and the Wahiba sands by the southwest monsoon (Juyal et al. 1997). Many dune systems include interdune deposits, dominated by fine sand and silt.

The older parts of sand seas are commonly redder in color, as the grains are coated with iron oxides. The red coloration often increases with distance from the sand source (Breed et al. 1979). For example, the Great Sand Sea of Egypt becomes redder to the south, near the end of the sand transport cycle, which is generally thought to be from north to south (Embabi 1998).

The growth of dunes is determined by (1) sediment supply, (2) sediment availability, and (3) transport capacity (Kocurek & Lancaster 1999). Sediment supply refers to sediment of a suitable grain size that provides the source for dune formation. The sediment may be derived from coastal, lacustral or fluvial/alluvial systems, from interdune areas within the dune field, or from the cannibalization of previously stabilized dunes. Sediment availability refers to the susceptibility of the material to entrainment by the wind. It is affected by such factors as soil moisture, surface binding by clays, cementation by carbonate or silica, or surface roughness (cover by vegetation or rocks). The transport capacity of the wind is its sediment-carrying capacity, which varies as a cubic function of friction velocity (u_*). Changes to any one of these factors may cause dune fields to stabilize or grow. For the Mojave Desert, California, for example, it has been suggested that the development of dunes may be largely supply-limited, with sediment transport by fluvial activity to deflation areas (either river beds or shorelines) a critical component of the system (Rendell & Sheffer 1996; Muhs et al. 2003).

Our understanding of the distribution and chronology of past aeolian systems comes from the study of fossil depositional and erosional surfaces and the analysis of dust deposits in ocean sedimentary cores. Palaeodunes have been identified using orbital imagery and the extent of ancient dune fields is now fairly well established. Sand dunes tend to retain sedimentological evidence of both wet and dry climatic changes (Lancaster 1990; Yang et al. 2003). However, the establishment of dune-field chronologies is difficult owing to the poor preservation of dateable material in arid environments. Radiocarbon dates are derived primarily from shells, bone, charcoal, wood, palaeosols, and lake deposits with pollen profiles. Within recent years, thermoluminescence dating has been the principal technique used to establish dune chronologies (Rendell & Sheffer 1996; Kocurek & Lancaster 1999).

Stabilized layers or palaeosols in sand dunes may indicate multiple phases of formation. Sand ramps often contain accumulations of talus derived from adjacent mountain slopes, calcareous deposits (Fig. 6.5), palaeosols, and fluvial and colluvial sediments, suggesting multiple periods of accumulation (Lancaster & Tchakerian 1996). In dunes of the Badain Jaran Desert, up to two dozen calcareous cementation layers are found, with thicknesses of 8–10 cm, and with calcium carbonate as much as 23% by weight. These are considered evidence of periods that were more humid than the present. The dune surfaces are covered by cemented plant root tubes, which have been exposed by deflation (Yang et al. 2003).

Dune studies are often considered in concert with other proxy records of palaeoclimate. Ventifact studies shed additional light on palaeowind processes, providing evidence of transport pathways, but as these features cannot be mapped from satellite or aerial imagery, their study is less conducive to broader-scale interpretations (R.S.U. Smith 1984; Laity 1992, 1995). Marine records have also been used to decipher the continental aeolian record owing to the greater input of dust into the oceans during dry and windy periods (Hesse & McTainsh 2003). Pollen records provide a sense of the direction and intensity of winds, stalagmites record moister conditions, and lake records provide evidence of pluvial periods (Orme 2002).

Inactive (relict, stabilized, fossil, fixed) dunes are identified by several criteria including the amount and nature of vegetative cover, the degree of material induration, the percentage of fines, sediment color, and changes to grain surface textures. As dunes become more stable, the amount of vegetation increases. However, the percentage of cover used to distinguish between active and inactive dunes varies according to dune type, and there is no distinct threshold between aeolian activity and inactivity, but rather a gradient (Tsoar & Møller 1986). A decline in dune activity may result from either an increase in effective precipitation, a decline in windiness, or both.

Older aeolian deposits also show distinctive sedimentological differences from modern active dunes. Active sands are characterized by 95% of all particles falling in the sand-sized range, a bimodal or unimodal size distribution, the presence of bedding structures, and low carbonate content (Thomas 1997). In older sands, there is a higher component of finer material owing to infiltration of dust or *in situ* weathering, the size distribution is altered, the sand grains tend to redden in color, the structure of the sand deposits are disturbed by animal burrowing or plant roots, and carbonate inputs are increased. Tchakerian (1991) used quartz-grain surface characteristics to suggest six late-Quaternary dune-building episodes in the Mojave Desert. Chemical weathering processes were particularly important in post-depositional sediment modification. In the oldest materials, chemical features such as pits and upturned plates predominate, and are associated with the highest percentages of silt and clay (5%) and grain reddening.

Stabilized dunes play an important biological role. Their more indurated nature allows for the development of extensive burrows, free from the potential of collapse. In turn, burrowing redistributes organic matter and may enhance infiltration rates. Stabilized dunes also retain moisture better than active dune systems, promoting vegetation growth.

9.5.3 DUNE REACTIVATION

Dunes that are stable or fossil in nature may become active again as result of changes in climate or human activity. Aeolian systems are very sensitive to small changes and many dunes are near the climatic limits of stability. As a result, only small changes in moisture balance, wind strength, or the nature of land use may be sufficient for reactivation. In the Southern High Plains area, USA, only modest (10–15%) decreases in precipitation and increases in evapotranspiration would mobilize dunes that are presently stable (Muhs & Maat 1993; Muhs & Holliday 2001). Drought conditions in the Great Sand Dunes National Park, Colorado, were shown to increase parabolic dune migration up to sixfold relative to intervening wet years (Marín et al. 2005). Often, more than one factor contributes to reactivation. In Australia, extended droughts, compounded by the overstocking of livestock and the presence of feral rabbits, have caused sand drift that is a threat to commercial pastoralism (Hesse & Simpson 2006).

9.5.4 INTERDUNE DEPOSITS AND LAKES

Interdunal regions are an integral part of the dune system, appearing as essentially flat surfaces between dune forms (Fig. 9.9). In some dune fields they are of significant extent. In the Namib Sand Sea, for example, they average 50% by area (Lancaster & Teller 1988). There are two basic interdune types: non-depositional (deflationary) and depositional. Deflationary processes may expose underlying alluvial sediments and serir deposits are common (Reineck & Singh 1975). Depositional interdune deposits result from lacustrine, fluvial, or evaporitic processes. They include shallow seasonal or ephemeral lakes resulting from local or regional rainfall events, wet areas or ponds associated with high local groundwater levels and seepage from shallow aquifers, or lakes that form at the end points of ephemeral rivers (Lancaster 1990).

Interdunal deposits formed by high-discharge flood events in deserts are relatively common, and illustrate one aspect of the closely knit fluvial/aeolian system. In the northwestern Namib Desert, the Hoanib River breaks into the Skeleton Coast dune field during high-discharge events, which average once every 9 years (Stanistreet & Stollhofen 2002). The interdunal areas adjacent to the main river channel accommodate excess floodwater and reduce stream discharge. Most of the flood deposits are immediately downwind of the active fluvial channel, but some are as much as 1 km away. The interdunal deposits are mudstone layers that act as impermeable

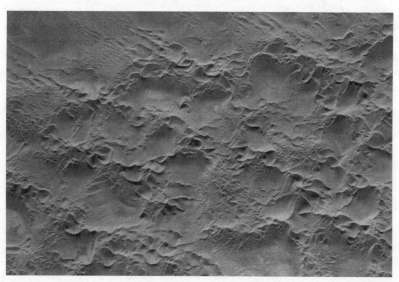

FIG. 9.9 The Issaoune Erg, Algeria. This photograph shows complex dunes on the active southwestern border of the sand sea. The most common forms observed in the middle of the photograph are star dunes. Linear dunes appear in the upper part of the image. As is common in many dune fields, smaller active dunes are located on the margins of the dune field. Interdunal depressions are filled with precipitation during rainy periods, leaving behind light-colored saline deposits when the water evaporates. Source: NASA astronaut photograph ISS010-E-13539.

seals, allowing floodwaters to be perched above the regional water table. Owing to this impermeable base, subsequent floods into the area are more likely to be retained as shallow, ephemeral lakes.

The study of interdune sites aids in the interpretation of sand sea accumulation and mobility. Radiocarbon dates on lacustrine carbonate and mollusk material (Yang et al. 2003), thermoluminescence dating of sand in interdune deposits, and studies of archaeological material in interdune sites help to provide age control. In the Nizzana dune field, Israel, the long-term stability of the interdune areas suggests little lateral movement of the linear dunes over the past several thousand years (Harrison & Yair 1998).

Interdune deposits may be eroded by deflation, or preserved by carbonate cements or burial by migrating dunes (Lancaster 1990). Wind erosion may result in topographic inversion, with former interdune deposits standing as positive-relief features (Harrison & Yair 1998).

In some desert areas, permanent lakes are found within dune fields (Fig. 9.10). More than 100 such lakes, recharged primarily by groundwater, are found in the southern part of the Badain Jaran Desert dune field (Yang et al. 2003). The largest of these, Lake Nuoertu, has an area of 1.6 km^2 and a maximum depth of 16 m. The hypersaline nature of many of the lakes, the presence of higher shorelines and more extensive lacustrine deposits, and the existence of fossil mollusks and ostracods in exposed surface sediments around the lake (with radiocarbon dates of around 5000 years BP) all suggest that lake levels in the region were once higher.

9.5.5 DUNE PATTERNS AND CLASSIFICATION

A sand dune is a hill or ridge of loose sand piled up by the wind. *Simple dunes* consist of individual dune forms, spatially separate from their neighbors (Fig. 9.8). The linear dimension of individual dunes ranges from less than 1 m to several tens of kilometers, whereas the height ranges from a few tens of centimeters to more than 150 m. *Compound dunes* are two or more dunes of the same type which have coalesced or are superimposed. *Complex dunes* are two or more different types of simple dune that have coalesced or are superimposed (Fig. 9.9). Complex dunes and compound dunes are often referred to collectively as draas or megadunes. Dunes may also be linked together to form dune chains or dune networks.

A dune field can be defined arbitrarily as more than 10 dunes occupying an area less than 30,000 km^2, with the area between dunes no greater than the local dune wavelength (Cooke et al. 1993). A sand sea, or erg, exceeds 30,000 km^2 in area. The Great Sand Sea of Egypt and Libya, for example, extends over an area of more than 100,000 km^2, with a sand coverage of more than 75% (Embabi 1998).

FIG. 9.10 Permanent lakes may persist within dune fields, fed by groundwater. In some cases, the lakes supply water for agriculture, as is shown at this location in western China.

Most of the world's active dune sand (85%) is contained in about 58 sand seas (Wilson 1973), the largest of which are found in North Africa, Arabia, and Asia. The borders of dune fields and sand seas are usually sharply defined. The edges tend to have low sediment thickness and to be associated with migrating dune forms (barchans and transverse ridges), whereas the interior regions contain much greater sediment thickness and are associated with dune forms of accumulation (for example, star dunes) (Fig. 9.9) (Lancaster 1983; Porter 1986). Major sand seas usually contain a great variety in dune form: the different assemblages are principally the result of changes in the wind regime over large areas, but also relate to changes in the character of sediments, the availability of sand, and the presence or absence of vegetation, cements, and binding agents. The analysis of orbital imagery indicates regional patterns of dune formation and has stimulated research into explaining the relationships between dune type and wind regime (Fryberger 1979).

Computer models suggest how dunes are self-organizing systems. In the dune-field-evolution computer models of Werner (1995), sand is preferentially deposited in areas of other sand (owing to the loss of momentum) or within the low-energy lee of bedforms. The migration rates are inversely proportional to the size of the patch of sand, so that small features migrate faster than larger ones, causing merging of bedforms. Over time, this results in increased dune size and spacing (Kocurek & Ewing 2005). Adjacent dunes also link laterally and coalesce, causing the development of crescentic ridges (Werner 1995). Variations in the transport direction and duration cause the emergence of basic dune shapes (barchan, transverse, linear, and star dunes; Fig. 9.11). The evolution of the dunes in the model is solely a function of wind direction, with the exception that transverse dunes require more sand than barchans. Figure 9.12 relates dune type to wind-direction variability and sand supply (Livingstone & Warren 1996). In nature, other factors such as water

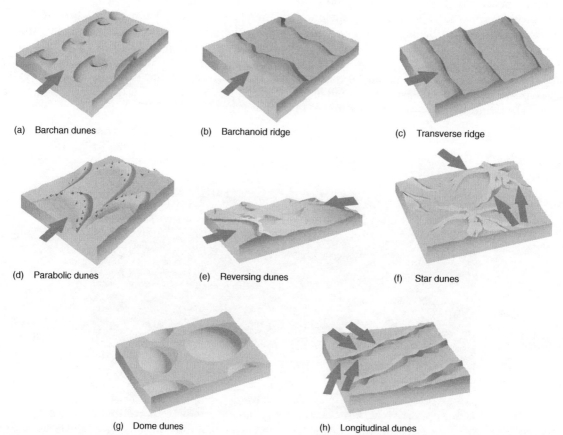

(a) Barchan dunes

(b) Barchanoid ridge

(c) Transverse ridge

(d) Parabolic dunes

(e) Reversing dunes

(f) Star dunes

(g) Dome dunes

(h) Longitudinal dunes

Fig. 9.11 The principal dune types and the direction of the winds that form them. After McKee (1979).

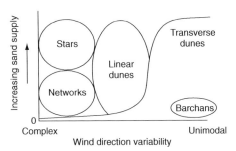

FIG. 9.12 Speculative model of the relationship between the availability of sand and wind direction variability in determining dune type. From Livingstone, I. and Warren, A. (1996) *Aeolian Geomorphology: An Introduction.* Longman, Harlow. By kind permission of the authors.

(particularly as it influences vegetation) or grain size (for example, zibars, rounded bed forms with coarse sands) will influence dune type.

Dune-field patterns can be recognized as *simple*, in which there is a single pattern of dunes (for example, a crescentic dune field), or *complex*, in which there is the spatial superimposition of numerous patterns. In a simple pattern, there is a striking similarity in dune size, orientation, and length, with dune type largely a function of the wind regime, and crest orientation as perpendicular as possible to the constructive winds. Research by Kocurek and Ewing (2005) suggests that simple patterns are developed by a single generation of dune-field construction, whereas complex patterns represent multiple generations of construction. As an example, they point to the work of Lancaster et al. (2002), who recognized the superposition of three sets of linear dunes of different sizes, trends, and ages in western Mauritania. Complex patterns in dunes represent the work of multiple generations of dune-forming events, each with a different wind regime, usually associated with a new influx of sand into the region. In the Gran Desierto, Mexico, for example, "chain of star" patterns develop, in which star dunes are superimposed atop linear dunes. According to OSL dates, the star dunes are much younger in age (after 7 ka) than the linear forms (26–12 ka), representing a younger generation of formation (Kocurek & Ewing 2005). The rapid climatic fluctuations of the late Quaternary have favored the development of complex dune-field patterns.

Many attempts have been made to classify dunes based on a combination of factors including their shape, the number and orientation of slipfaces relative to the prevailing wind or resultant sand drift direction, the degree of form mobility (migrating or

relatively fixed in position), or the processes that form them. Active sand dunes can be classified according to their mode of movement as: (1) migrating, in which the whole dune advances with very little change in shape or dimension as in, for example, barchans and transverse dunes; (2) elongating dunes, such as linear dunes, which become more extended in length with time; and (3) accumulating dunes, wherein the dunes have little net elongation or forward movement, but tend to grow upward with time, as exemplified by star dunes (Tsoar 2001). Wind variability seems to have the strongest influence, with migrating dunes found in a nearly unimodal regime, elongate dunes occurring where winds are bimodal, with the two modes 70–90° apart, and accumulating dunes formed under bimodal or multimodal wind directions, with winds forming an angle of about 180°. Dune nomenclature varies widely, in part because different names have been assigned to similar features from place to place, but also because some dune types have been poorly defined.

Pye and Tsoar (1990) and Tsoar (2001) classified dunes into three different groups based on dune genesis and wind-directional variability. These are as follows.

1 Unvegetated dunes whose development is related to topographic obstacles, which interfere with wind flow: climbing dunes, echo dunes, lee dunes, falling dunes, and cliff-top dunes.
2 Unvegetated dunes that are regarded as self-accumulated (autogenic dunes): barchans, transverse (barchanoid) ridges, unvegetated linear dunes (seif dunes), dome dunes, and star dunes. These dunes accumulate in open terrain owing to changes in bed roughness or aerodynamic fluctuations.
3 Dunes whose development is strongly influenced by vegetation (phytogenic dunes): parabolic dunes, vegetated linear dunes, and coppice or hummock dunes.

Livingstone and Warren (1996) used the term "free" dunes for the "self-accumulated" dunes of Pye and Tsoar (1990). They grouped both vegetated dunes and those related to topographic obstacles as "anchored" dunes, because both of these forms act to immobilize dunes. Regardless of the classification system used, most geomorphologists agree on the fundamental factors that influence dune formation and on the principle dune types.

Linear dunes (vegetated and unvegetated) are the most common dune type found in deserts, followed by transverse dunes. However, there is considerable variation between different regions in the percentage cover of each dune type.

9.5.6 DUNE ACCUMULATION INFLUENCED BY TOPOGRAPHIC OBSTACLES

Topographic obstacles such as boulders, shrubs, escarpments, and hills induce zones of airflow acceleration, deceleration, and enhanced turbulence, and thereby strongly influence the erosion or accumulation of sand. Many dune types, including lee dunes, climbing and falling dunes and sand ramps, are found in deserts of high relief. Often their presence is related to regional or local sand-transport corridors that extend from source areas to depositional sinks (Zimbelman et al. 1995; Lancaster & Tchakerian 1996). The mountains are barriers to sand transport, and accumulate extensive deposits of sand on their piedmont slopes.

9.5.6.1 Lee dunes

Lee dunes form to the lee of topographic obstacles, such as boulders, vegetation, mountains, cliffs, or in the lee of a gap between two obstacles. They are best developed under a nearly unidirectional wind regime, or seasonally in a bidirectional regime. The resulting dunes do not advance or elongate once they have attained a steady-state form. The size and shape of the lee dune are related to those of the obstacle, and its long axis parallels that of the wind flow.

The term shadow dune is commonly used to describe a tapering accumulation of sand formed in the lee of an obstacle (usually a boulder or clump of vegetation) where the wind velocity is locally reduced (Fig. 9.4). Small shadow dunes respond rapidly to recent localized wind fluctuations and are excellent indicators of the direction of the most recent high-wind event. Small shadow dunes are prominent in some Martian rover imagery.

When vegetation forms the anchor for shadow dunes, the height of sand accumulation is determined primarily by the plant width and the angle of repose of the sand. The wider the base of the plant, the higher is the ridge that can be built before the angle of repose is attained (Hesp 1981). This relationship is expressed as:

$$h = (w/2)\tan\phi$$

where h is height, w is width, and ϕ is the angle of repose. The height of the shadow dune is greatest in the immediate lee of the plant, where the reverse flows are greatest, and decreases progressively with distance downwind. Hesp (1981) found that dune length decreases with increasing wind velocity, but for any given velocity the dune length increases (albeit irregularly) with plant width. No relationship was found between dune length and plant height.

Large lee dunes form in association with mountains and escarpments (Breed et al. 1979) and have been described from the Libyan Desert, southern California, and Chad, although they have been little studied. Such dunes are best developed where winds are nearly unidirectional, but may also form in a bidirectional regime if the wind strength is stronger from one direction. Large lee dunes tend to break up downwind into individual barchan dunes at a distance where the topographic obstacle is no longer effective.

The accumulation of sand can also occur in the lee of a gap between two obstacles. The sand flow is accelerated and funneled through the gap, but fans out and decelerates on the lee side, leading to sand deposition.

Falling dunes are also found in zones of calm air to the lee of cliffs or mountains. Irregularities in the cliff line initiate convergence of the streamlines as they pass over the brink, and a general reduction in flow velocity causes deposition of falling dunes along the entire cliff foot.

9.5.6.2 Climbing dunes, sand ramps, echo dunes, and cliff-top dunes

When sand and wind impinge upon a topographic obstacle, the slope of the obstacle affects the nature of deposition. For angles less than about 30°, the sand is generally transported across the obstacle. At angles above 30°, it is trapped on the upwind side of the obstacle base as a sand ramp or climbing dune. In time, these sloping forms provide an avenue for transportation of sand across the obstacle. When the upwind slope exceeds 50–55°, echo dunes form.

Echo dunes develop in front of cliffs and are separated from the scarp by a sand-free zone formed by a reverse wind flow. Their long-term maintenance depends on sand moving laterally along the cliff line by vortices located between the cliff and the echo dune. The sand may ultimately escape through drainage channels or other breaks in the cliff. Wind-tunnel experiments show that the horizontal near-surface wind velocity in front of a vertical obstacle starts to drop at a distance of $d/h = 3.3$ (where d is the horizontal distance upwind from the obstacle and h is the height of the obstacle) and reaches a minimum at $d/h = 0.75$, where the two opposite flow directions meet (Tsoar 1983). Sand accumulates between $d/h = 2$ and 0.3, forming an echo dune with a crestline at $d/h = 0.5$. The echo dune grows upward by sand deposition until a condition of steady state is achieved when its height is about one-third of that of the obstacle.

At lower slope angles, a strong reverse flow does not develop and sand is able to climb the escarpment as a *climbing dune*. A zone of reduced wind velocity is frequently observed just downwind of the crest of escarpments. Consequently, sand often accumulates at such sites, forming *cliff-top dunes*.

Sand ramps are distinguished from climbing dunes by their larger size, lower slope angles, and composite nature. They are low-angle (3–11°) deposits formed principally (>50%) of aeolian sands but, as they lie against hillslopes, they may also incorporate fluvial and talus deposits (Lancaster & Tchakerian 1996). Palaeosol units within the ramps indicate periods of stability and multiple episodes of dune formation. Most Mojave Desert sand ramps, for example, are relict features, mantled by talus and incised by stream channels (Tchakerian 1991). Partial cementation by carbonate stabilizes the ramps, and they often form a preferred site for burrowing animals: such surfaces readily collapse underfoot. With increasing carbonate cementation, the permeability of the dune is reduced, and runoff from the mountains incises the dunes and, in some cases, detaches the ramp from the mountain front.

Sand ramps develop on the windward slopes of mountains and may facilitate the movement of sand through the range. In some areas, such as the Mojave Desert, the distribution of climbing and falling dunes and sand ramps suggests regional sand transport through wind corridors (Zimbelman et al. 1995). Although such movement probably does occur, geochemical analysis of sediments suggests that the sand source for ramps is often derived from discrete, local sources such as alluvial fans, local ephemeral streams, and palaeolake shorelines (Muhs et al. 2003; Pease & Tchakerian 2003).

9.5.7 FORMATION OF SELF-ACCUMULATED DUNES

9.5.7.1 Dune initiation

Field observations show that many small dunes are initiated by chance topographic irregularities or changes in surface roughness, which give rise to spatial variations in the sand-transport rate. Once initiated, the growth of sand mounds is encouraged by positive feedback until a condition of dynamic equilibrium is attained with long-term average wind and sediment transport conditions. Bagnold (1941) suggested that saltating grains bounce off a hard desert surface (pebbles or bedrock) more effectively than over a bed of loose sand. The sand-transport rate over a random sand patch is, therefore, lower than over its surroundings, leading to accretion of the sand patch and extension of its upwind margin (Werner 1995). This process is effective given a constant supply of sand from upwind and conditions of strong winds.

Dunes may also form by being "calved" off other larger dunes, by accumulating around small obstacles like plants (Livingstone & Warren 1996), by deposition in zones of opposing flow (such as reversing dunes on hill crests), or in other ways related to the environmental conditions of a specific locale. There have been relatively few field studies of dune initiation, and a full understanding of this process awaits additional study.

9.5.7.2 Crescentic dunes: barchans and transverse barchanoid ridges

Barchans and transverse dunes are essentially of the same type, forming and migrating under a unidirectional wind regime. The difference between the two is related to the amount of sand: barchans are isolated mounds, whereas transverse dunes are composed of many barchans coalesced into a single, longer dune form (Tsoar 2001). Although barchans are relatively rare in the great sand seas of the world, they are one of the best-known and studied dune types.

Barchans are crescentic dunes whose horns point downwind (Fig. 9.13). The profile typically shows a convex windward slope with a maximum angle of 12° and a steeper leeward slope with a slipface of 33–4°. Some barchan dunes have a separate crest and brink, whereas in others the two coincide.

Barchans show a considerable range in scale. Small, often ephemeral, barchans form where sand is blown over a relatively hard substrate. They are easily destroyed by a change in wind or sand-supply conditions. Barchans that develop in dry riverbeds may also be destroyed by flood flow. By contrast, compound mega-barchans may persist for hundreds of years.

FIG. 9.13 The barchans shown here are examples of simple dunes. In this area of the Moenkopi Plateau, Arizona, they form within a drainage channel during dry periods and are destroyed during flood flows. The sands are recycled, moving downstream by water action and upstream by the wind. Wind flow is from left to right.

Barchans tend to form in vegetation-free areas where sand supply is limited and the winds are almost unidirectional. Winds from a secondary direction may cause asymmetry by elongating one of the horns. Barchans can migrate long distances downwind without experiencing major changes in size or shape. In some cases they act almost as a closed system, with sand prevented from escaping by reverse flows associated with vortices on the leeward side of the dune. However, in most cases, migrating barchans act as open systems in dynamic equilibrium, with the input of sand from upwind equal to the downwind losses from the horn tips.

The rate of barchan advance is inversely related to the brink height (Jimenez et al. 1999). Migration rates in excess of up to 30 m year^{-1} have been recorded for small barchans, but 5–10 m year^{-1} is more typical for larger dunes. Elongate barchan trains consist of many individual dunes, traveling in a relatively straight path. They are often found in corridors of aeolian transport and erosion, and may represent sand in transition from source areas to depositional zones (Mainguet et al. 1980; Corbett 1993). Barchans form a means of sand transport into or out of sand seas and are thus found only on their margins (Mainguet 1984). They often occur as groupings or "swarms."

Where sand supply increases, individual barchan dunes may link up to form a sinuous-crested ridge oriented perpendicular to the strongest wind or resultant sand-drift direction. Many transverse ridges display alternating *barchanoid* (downwind facing) and *linguoid* (upwind-facing) elements viewed in plan, and alternating peaks and saddles viewed in section. The barchanoid element of one ridge is frequently followed in the next ridge downwind by a linguoid element. Such transverse ridges often form the dominant component of complex dune networks that have been referred to as *gridiron* or *fishscale* patterns, *aklé*, or *reticulate networks*. Very large complex and compound barchan and transverse megadunes occur in some areas, including the Algodones dune field of California and the Thar Desert of northwest India. Not all transverse ridges display well-developed barchanoid and linguioid elements: some are almost straight.

Transverse and barchan dunes in a reversing wind regime change their profile dramatically as the slip-face alternates from one side of the dune to the other. Dunes that change their slipface orientation are known as *reversing dunes*.

9.5.7.3 Linear dunes (seif dunes)

Linear dune fields form some of the most extensive landforms in arid regions, forming an estimated 30–60% of sand seas (Fryberger & Goudie 1981;

Lancaster 1982). Linear dunes are characterized by their considerable length, relative straightness, parallelism, regular spacing, and low ratio of dune to interdune areas. Two major types are recognized: unvegetated and vegetated. Lee dunes are also sometimes considered a form of linear dune (Tsoar 1989). Overall, the general term linear dune can be thought of as embracing a continuum of elongate dune types.

The unvegetated linear dune has a triangular profile and a sharp crest that gives it the name *seif* (meaning sword in Arabic) or *sayf* (Tsoar 2001). Seif dunes occur in many African sand seas, where they have often been referred to as *silk* dunes (pl. *slouk*). They are also found extensively in Sinai and other parts of the Middle East. Vegetated linear dunes or sand ridges have a more rounded crest, are active only at the summit and downwind tip, and are more heavily vegetated than seifs (Cooke et al. 1993). They occur widely in Central Australia, the Kalahari, the Thar Desert of northwest India and Pakistan, southern Israel, and Arizona and California in the USA. The vegetated linear dune is discussed further in Section 9.5.8.4.

As the name suggests, linear dunes have high length-to-width ratios (>10:1) (Mainguet 1984). The longest linear dune of the Erg Fachi Bilma, Niger, for example, is 40 km in length and varies in width from 50 to 100 m, with an average length-to-width ratio of 80:1 (Mainguet 1984). Additionally, they are sometimes found in staggered rows, with one linear dune succeeding another, with only a limited interruption. Linear dunes may join each other at low-angle Y junctions.

As with other dune types, simple, complex, and compound varieties can be recognized. Simple linear dunes consist of a single narrow dune ridge. A compound linear dune consists of two or more closely spaced or overlapping dune ridges on the crest of a much larger plinth. The constituent ridges generally rise no more than 40 m above the plinth, which is 0.5–1.0 km wide. Complex linear dunes consist of very large ridges along which are distributed en-echelon peaks that reach a height of 150–200 m in Arabia and 100–200 m in Namibia. Secondary dunes, usually oblique or transverse to the main trend, are often developed on their surfaces. In the Namib Desert, many complex linear dunes have star dunes on their crests (McKee 1983).

Seif dunes have a single, long, continuous, and often sinuous ridge. The crest line rises and falls at regular intervals to form a series of peaks and saddles. This gives the ridge the appearance of a chain of "teardrops" when viewed from the side. Many seif dunes are highly sinuous in plan, but others are relatively straight with less well-developed peaks and saddles. The sinuosity may also vary along a single dune (Bristow et al. 2000). All seif dunes have a sharp crest that shows seasonal or short-term reversal of the slipface. The slipface height is typically one-third to one-half of the height of the dune as a whole. Active seifs are devoid of vegetation except along the parts of the basal plinth.

Linear dunes commonly occur in extensive dune fields, but as yet there have been relatively few attempts to examine the spatial relationships between the dunes (Bullard et al. 1995). The orientation of dunes tends to be very consistent, and there tends to be an allometric increase in dune spacing with height and width, with considerable regularity when viewed on a small scale (Cooke et al. 1993). There is considerable variation in planimetric patterns. Bullard et al. (1995) recognized five pattern classifications for the southwestern Kalahari Desert alone, based on Y junctions, dune length, wavelength, and other factors. The interdune areas commonly show low relief. In between linear dunes of the Namib Desert, for example, the surfaces are broad and gravel-covered (Bristow & Lancaster 2004).

The origin of linear dunes has long been a matter of controversy (McKee 1983). In general, they tend to occur in areas of limited sand availability and reasonably high wind-direction variability (Wasson & Hyde 1983). Net sand transport in linear dunes is parallel to the crest. The wind direction is usually bimodal (Livingstone 1988; Tsoar 1989): in the Namib, for example, winds are from the southwest and south-southwest during the summer, and from the east in the winter (Bristow et al. 2000). The changing angle of the wind helps to explain why the crest is sinuous, as erosion by the wind from one side is offset by deposition on the other. The meandering of the crest influences the specific angle of wind attack, which may vary from around 30–40° to 90° (Tsoar 2001). While the crestline of the dune may "flip" seasonally in response to changing wind directions, the plinth remains relatively stable (Livingstone 1988).

Linear dunes, with their apparently simple morphology and regular spacing, have lent themselves to considerations of dune forms and wind interactions. Bagnold (1953) proposed that horizontal roll vortices would develop under conditions of strong geostrophic wind and strong thermal heating. He suggested that paired helical vortices, moving parallel to the dominant wind direction, might sweep sand out of the troughs and onto the adjacent linear dune ridges. More recent field studies indicate that longitudinal helical vortices are not a dominant factor in the development of large linear dunes (Wiggs 2001), although they may contribute to some linear aeolian features such as sand strips, sand streets, and erosional lineations. Bagnold (1953) and Tsoar (1989, 2001) proposed models by which seifs evolve from barchans by extension of one horn in a bidirectional wind regime with strong winds from a primary direction and gentle winds from a secondary direction. Once a barchan begins to evolve into a seif dune, the longitudinal element elongates faster than the advance rate of the transverse element of the barchan.

Less commonly, seif dunes originate from linear zibar or due to overgrazing on vegetated linear dunes. Anchored dunes (lee dunes) are almost always linear, and thus the presence of obstacles has also been invoked to explain the development of linear dunes. However, obstacles are not always present in areas of well-developed linear dunes, and thus do not appear to be a requirement.

9.5.7.4 Star dunes

Star dunes are characterized by their large size, pyramidal morphology, and radiating, sinuous arms (Figs 9.8, 9.9, and 9.14). Active arms sit upon large basal plinths, which may be partially vegetated. The dunes occur as simple forms with three or more radial arms joined at a single summit, as compound forms with multiple peaks connected by cols, or as complex forms superimposed on linear megadunes. In the Sahara they are known as *demkhas, ghourds, rhourds,* or *oghourds*. Other terms include sand massifs, pyramidal dunes, stellate dunes, sand mountains, and horn or cone-shaped dunes.

Approximately 11% of all dunes are of the star type, forming about 5% of aeolian depositional surfaces. They are an important constituent of the Grand Erg Oriental in Algeria (where 40% of the dunes are of this type), the Erg Fachi-Bilma in Niger, the southeastern Rub' al Khali in Saudi Arabia, the Gran Desierto in Mexico, the Ala Shan Desert in China, and the Namib Desert. Small groups of star dunes are also common in the Basin and Range deserts of North America (Fig. 9.8).

Star dunes are large aeolian bedforms, with a mean width of 500–1000 m and a mean height of 50–150 m, although there are significant variations between deserts. Exceptional star dunes, more than 300 m high, have been reported from Namibia and the Ala Shan Desert. In general, the spacing of star dunes increases with increasing dune height. In some areas star dunes occur in complex star-dune chains. In other areas, there are transitional forms from complex linear dunes or complex transverse dunes to star dunes

Fig. 9.14 Star dunes are pyramidal in form, with radiating, sinuous arms. They often reach to considerable heights. Location: Mingsha Dunes, Dunhuang, China.

Star dunes typically develop in areas of obtuse bimodal or complex (multimodal) wind regimes, whether they are of low, intermediate, or high energy. The net sand-transport rates are low and there is little or no apparent movement (migration) of these dunes. The areas of wind erosion and deposition may alternate with the seasons (T. Wang et al. 2005). The stellate form has generally been explained as a response to sand-transporting winds which blow from different directions at different times of the year, but there are also secondary circulations induced by the star-dune form itself. Star dunes often develop close to the poleward margins of deserts, where the effects of seasonal changes in wind directions are more marked than at the trade-wind-dominated equatorial margins (Tsoar 2001).

Studies of surface airflow and sand transport over individual star dunes in the Namib and Rub' al Khali Deserts, the Gran Desierto, and Dumont dune field, California, have shown that the major arms of the dunes tend to be aligned transverse or slightly oblique to the two major directions of sand transport (Lancaster 1989). The minor arms of the dunes are aligned parallel to these directions and transverse to the secondary wind direction. Nonetheless, much work remains to be done to document the sand-transport dynamics on these dunes.

A close association between the occurrence of star dunes and topographic barriers has been observed (Mainguet 1984). The effect of major topographic features is probably both to increase the complexity of the regional windflow and to generate secondary flows through differential surface heating. Topographic features also create barriers to the sandflow, leading to the accumulation of thick sand deposits that are associated with the occurrence of star dunes.

The processes responsible for the initiation and early development of star dunes are currently uncertain. Nielson and Kocurek (1987) observed that small star dunes in the Dumont dune field formed during winter periods of variable wind regime, but were modified into barchan forms during the summer period of unidirectional winds. This suggests there may be a minimum size for the survival of the star-dune form.

9.5.7.5 Dome dunes

Dome dunes are relatively low (usually 1–2 m high), flat-crested forms, often without slipfaces, which are circular or elliptical in plan (Fig. 9.11). They are estimated to comprise only 1.3% of the dunes in the major sand seas. They are absent in many deserts and are common only in some of the Chinese Deserts (especially the Taklimakan) and parts of Saudi Arabia. McKee (1966) described simple dome dunes in the White Sands dune field of New Mexico which are typically 150–200 m in diameter, 6–10 m high, round or oval in plan, and lack slipfaces. Compound and complex dome dunes of the Taklimakan Desert reach heights of 20–60 m and lengths of 200–900 m (X. Wang et al. 2002).

The wind environment and mechanisms of dome dune formation are not well understood. In the White Sands dune field and the Killpecker dune field of Wyoming, USA, dome dunes occur close to the upwind margin of the dune fields, suggesting that dome dunes develop where winds are sufficiently strong and unidirectional to effectively retard normal upward growth of dune crests. However, dome dunes of the northeastern Taklimakan are developed in areas of low wind energy (X. Wang et al. 2002).

The migration of a 1 m-high dome dune in Namibia was examined by Bristow and Lancaster (2004), who concluded that it moved at an average rate of 4 m year^{-1} for a period of 23 years in a bimodal wind regime. Their study suggested that the dome-dune form is relatively inefficient at trapping sand owing to the lack of a slipface and a lee-side eddy.

The dome dunes at White Sands are composed of coarser, more poorly sorted sand than the other dune types further downwind (McKee 1966). By contrast, dome dunes of the Taklimakan are composed of very fine sand (3.25–3.45ϕ) (X. Wang et al. 2002). The significance of grain size in determining the morphology of dome dunes remains a matter of conjecture.

9.5.8 VEGETATED DUNES

In deserts, vegetation plays a role in forming dunes and in modifying or stabilizing their form. Vegetation-anchored dunes are widespread in the Sahel, southern Africa, Australia, some areas of the Near East, and parts of North America (Thomas & Leason 2005). The variation in cover ranges from grasses to almost complete woodland. Vegetation is only absent from hyperarid and overgrazed regions. However, even in areas of low precipitation, dune-trapping vegetation may be present because of groundwater brought near to the surface by faulting or other subsurface geologic conditions.

In some instances, the presence of vegetation indicates that dunes are partly or wholly relict. Vegetation-anchored dunes are highly responsive to changes in plant cover, and reactivation may result from droughts (Marín et al. 2005; Thomas & Leason 2005), fluctuations in the groundwater table (Laity 2003), and anthropogenic changes to surface cover (Tsoar & Møller 1986; Tsoar et al. 1995).

In general, the mobility of dunes is related directly to wind velocity, but inversely proportional to the vegetation cover. For conditions where vegetation cover is a function of precipitation, Lancaster (1988) has developed a widely employed mobility index (M):

$$M = W/(P/PE)$$

where P is annual rainfall, PE is potential evapotranspiration, and W is the percentage of time the wind is blowing above the threshold velocity for sand transport (4.5 m s^{-1}). Dunes are fully active when the dune mobility index exceeds 200 and completely stabilized by vegetation when M is less than 50. The value may vary seasonally. For example, near Barstow, California, the wind exceeds the threshold for sand transport 41% of the time on an annual basis ($M = 740$); but 60% of the time during the windier months of April and May, when the value is approximately 1071 (Laity 2003).

The role of vegetation in anchoring sand and forming dunes is also considered in Section 11.6.1.

9.5.8.1 Hummock dunes, coppice dunes, or nebkhas

There are a large number of terms applied to irregularly shaped mounds of sand whose surface is wholly or partially vegetated. These include *hummock dunes, coppice dunes, nebkhas,* and *nabkhas.* Shadow dunes, discussed above, are more elongate in form, are anchored only at the upwind end by either vegetation or boulders, and respond rapidly in form to changes in wind direction.

Plants act as an anchor because of the permeability of the vegetation to the wind. They locally lower the wind velocity and promote the deposition of sand. In many cases, the plant can withstand abrasion and continue to grow as the dune grows in size. The dunes are anchored, in that they do not move bodily, but they may show some mobility on their surfaces. They are subject to interdune erosion as wind accelerates between plants and to destruction following the death of the vegetation. The size of the anchored dune relates to the size of the vegetation and to the supply of sediment (Laity 2003) (Fig. 9.15).

In some areas of the world, nebkhas are relatively recent features, associated with environmental degradation. Their development has resulted from

Fig. 9.15 Vegetation-anchored dunes (nebkhas, coppice dunes) grow in relation to the size of the vegetation and the supply of sediment. In addition to sand, they accumulate fine windblown material and some organic matter. Location: Salton Sea region, California, USA.

human-induced changes in land cover, such as grazing and cultivation, which releases aeolian sediment or changes the nature of the vegetation (Nickling & Wolfe 1994; Sefe et al. 1996). For example, the spread of honey mesquite (*Prosopis glandulosa*) into overgrazed rangelands of New Mexico initiated the development of coppice dunes and the erosion of interdunal soil (Hennessy et al. 1985).

Nebkhas may also be destroyed by human activity. In the Mojave Desert, California, they often form where vegetation flourishes owing to high water tables, such as along the Mojave River, where the dunes grow up to 7–8 m in height. Groundwater pumping has caused vegetation death and nebkha destabilization, resulting in sediment release and the development of new aeolian forms, such as sand stringers and barchans (Laity 2003).

9.5.8.2 Parabolic and elongate parabolic dunes

Parabolic and elongate parabolic dunes are common in many coastal dune fields, but they also occur in arid or semiarid regions where some vegetation is present (for example, Saudi Arabia, northwest India (Fig. 9.16), Arizona, and Israel), as well as in colder climates (the Lake Athabasca region of northern Canada and the midwestern USA).

Vegetation is essential to the formation of these dunes, anchoring the less mobile arms and allowing the central section to advance downwind (Wasson et al. 1983). A loss of vegetation may cause parabolic dunes to change form, becoming, for example, active transverse dunes. Conversely, the revegetation of barchan and transverse dunes has been shown to transform these dunes into active parabolics (Tsoar & Blumberg 2002).

Simple parabolic dunes are U- or V-shaped in plan with two trailing arms that point upwind. Many have a large sand mound with a steep lee slipface at the downwind end, whereas others terminate in a low sand ridge or sand lobe without a marked slipface. The outside slopes of the trailing arms are partly or wholly vegetated.

A distinction may be made between a parabolic dune, which is an open, bow-shaped structure that has not migrated; and an elongate parabolic dune, a larger, clearly developed U-shaped dune developed from a migrating spot blowout. Pye (1993) differentiated parabolic dunes according to their length-to-width ratio. Dunes were subdivided into lunate (ratio <0.4), hemicyclic (ratio 0.4–1.0), lobate (ratio 1.0–3.0), and elongate (ratio >3.0) forms. Parabolic dunes tend to form relatively large bodies of sand ranging from hundreds to thousands of meters in length and

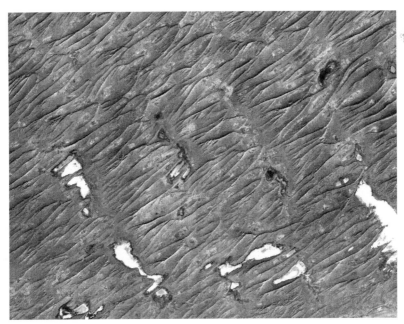

Fig. 9.16 Parabolic dunes of the Thar Desert occur in an area with an annual rainfall of 120–240 mm. Source: NASA/GSFC/METI/ERSDAC/JAROS, and U.S./Japan ASTER Science Team.

width, and tens of meters in height. In coastal dune fields of Queensland, Australia, some elongate parabolic megadunes exceed 15 km in length and 1.5 km in width (Pye 1993). Parabolic dunes are often compound in nature, occurring either as nested, en-echelon, digitate, or superimposed forms. For example, a series of dunes may migrate down the same path, forming a "nested" set of dunes at various stages of development.

Parabolic dunes most commonly develop from blowouts in a vegetated sand surface. Blowouts may be initiated by a change to the vegetation (fire, overgrazing, trampling, or disease), by soil changes (for example, intrusions of saline groundwater into the root zone), or by climatic changes (rainfall or wind). The dune grows downwind as sand is supplied from erosion of the underlying surface. The wind regime is probably nearly unidirectional, paralleling the axis of the dune.

The migration rate of parabolic dunes varies according to climatic factors that influence vegetation cover. The apparent rate of migration at Great Sand Dunes National Park, Colorado, varies widely between dunes and according to the degree of climatic drought (Marín et al. 2005). For a single dune, migration rates varied between less than 1 m year^{-1} during a wet period to a maximum of 70 m year^{-1} during dry conditions. Large dunes appear to migrate faster than smaller ones (Marín et al. 2005).

9.5.8.3 Lunette dunes

A lunette dune is a bow-shaped form composed of sand, silt, and clay that occurs on the downwind margins of ephemeral lakes (Fig. 9.17). The sediment is derived chiefly from deflation of the pan floor, often under saline conditions that encourages flocculation of the mud to silt- or sand-sized particles. Lunette dunes have been studied principally in Australia (Bowler 1973; Dutkiewicz & Prescott 1997; Stone 2006), southern Africa (Marker & Holmes 1995; Lawson & Thomas 2002), and the Southern High Plains of northwest Texas and eastern New Mexico (Holliday 1997). In the Southern High Plains, more than 1100 playa basins have lunettes (Holliday 1997). Stabilized lunettes may be dissected by erosion.

Like parabolic dunes, the arcuate plan form of lunettes points downwind, and sedimentation of the surface of the dune is usually enhanced by the presence of vegetation. However, unlike parabolic dunes, lunettes are rarely transgressive. Sedimentological differences between lunettes are common, and may be related to changing geomorphic conditions. Some lunettes are composed almost entirely of sand, but many contain a high proportion of silt and clay, which are transported in pellet form from the adjacent pans during periods of low water. Clay dunes are less common in southern Africa than Australia (Marker & Holmes 1995). Other lunettes include a low percentage of clay and a high percentage of carbonate or gypsum (Dutkiewicz & Prescott 1997). Over time, the sediments are pedogenically altered, and the original pelletal textures and bedding may be completely lost (Holliday 1997). Factors affecting lunette formation include the strength and unidirectionality of the prevailing winds and the seasonal oscillation of shallow groundwaters (Bowler 1973; Lawson & Thomas 2002).

Luminescence dating of lunettes has advanced our understanding of the late-Quaternary history of arid regions. Lunettes with abundant quartz grains are excellent candidates for thermoluminescence dating because the sediment can be linked directly to the depositional event: unlike typical dune sands, lunettes are not remobilized. In Australia, the formation of lunettes is associated with cold phases of climate from the last interglacial period (≈130,000 years ago) to the present, most notably during the arid conditions of the LGM (Dutkiewicz & Prescott 1997; Stone 2006). In the southwest Kalahari Desert, most lunettes appear to be remnant of past conditions of lower groundwater tables and drier conditions when deflation was more common (Lawson & Thomas 2002).

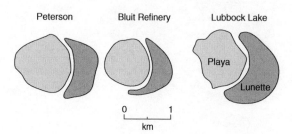

FIG. 9.17 Outlines of playas and their associated lunettes on the Southern High Plains, USA. The lunettes can extend from approximately a quarter to a half of the circumference of the playa and be 10 m or more in height. After Holliday (1997).

9.5.8.4 Vegetated linear dunes

As discussed above, linear dunes form a broad continuum, from sharp-crested seifs to more rounded and vegetated forms. Some authors have referred to both vegetated linear dunes and seifs as longitudinal dunes. In Australia, vegetated linear dunes, widely referred to as *sand ridges* or *parallel ridges*, cover 40% of the continental surface and form an anticlockwise whorl (Fig. 2.15; Pell et al. 2000). Similar forms are found in southern Mauritania, the Kalahari Desert, and in small areas of the southwestern USA (Tsoar & Møller 1986).

Vegetated linear dunes range in height from several meters to a few tens of meters and typically have a rounded cross-sectional profile. The vegetation cover is thickest on the plinth and lower slopes, and is usually sparse or absent on the crest. Some linear dunes appear to be wholly stable at present, indicating that they developed under windier or more arid conditions in the past. Many Australian dunes contain horizons with carbonated rootlets, suggesting formation in zones of dense plant root growth (Lomax et al. 2003).

Vegetated linear dune ridges may run almost parallel without a break for scores of kilometers. In the Simpson Desert, some individual linear dunes exceed 320 km in length. In most cases they are asymmetric in cross-section, though symmetrical profiles are also found. The symmetry can vary from time to time in accordance with the wind conditions. A common feature of vegetated linear dunes is the tendency to form branching networks in which adjacent ridges converge, forming a Y junction, before continuing as a single ridge.

Most of the Australian vegetated dunes display a small deviation from the mean wind direction. Whereas parallelism to the resultant wind direction is a frequent characteristic of seif dunes, a majority of vegetated linear dunes distinguish themselves by extension parallel to the dominant wind direction (Tsoar & Møller 1986). Since the presence of a partial vegetation cover raises the threshold for sand entrainment, only strong winds are able to transport sand and modify the form.

Destruction of the vegetation cover can turn vegetated linear dunes into seifs or *braided linear dunes*, the latter being linear dunes on which small secondary transverse dunes are superimposed (Tsoar & Møller 1986).

9.6 Ripples

Ripples cover the surface of dunes and other dry, sandy, vegetation-free surfaces and are considered the smallest of aeolian bedforms. Unlike dunes, they form and reform quickly as they rapidly reach equilibrium with changing wind conditions. In general they are aligned perpendicular to the wind that forms them. Deflection of sand transport over a sloping surface can cause a deviation in the alignment of the ripple crests with respect to the regional wind direction. On dune lee slopes, for example, ripples often move across the slope under the influence of vortices created by flow separation at the dune crest.

Two main ripple types have been recognized, classified according to their size. *Normal* aeolian ripples have wavelengths of less than 1 cm up to 25 cm, and amplitudes of 0.005–0.010 m (Bagnold 1941; Sharp 1963). They form in sediments of fine sand (Fig. 9.18). Larger ripples, termed *ridges* by Bagnold (1941, p. 149), but more widely known as *megaripples*, have wavelengths of up to 20 m and heights of up to 1 m. Megaripples are often composed of coarse sand and,

Fig. 9.18 Normal aeolian ripples typically have wavelengths ranging from less than 1 to 25 cm. They reform quickly and align perpendicular to the wind. Location: Olancha Dunes, California, USA.

occasionally, granules or small pebbles. In the latter case they are referred to as *granule megaripples* (1–4 mm median grain size) to distinguish them from sand megaripples (Figs 9.19 and 9.20).

The ripple profile is developed into four elements: stoss (windward) slope, crest, lee slope, and trough. The maximum inclination of the stoss slope usually ranges from 8 to 10°, whereas that of the lee slope

FIG. 9.19 Megaripples have longer wavelengths than normal aeolian ripples and are composed of materials ranging from coarse sand, through granules, to small pebbles. In this photograph, the ripple wavelength is about 1 meter. In the background is a barchan.

FIG. 9.20 The coarsest sediment is usually found on the ripple crest. These pebble-sized sediments are from a megaripple at Rogers Dry Lake, California, USA.

ranges from 20 to 30° (Sharp 1963). A dimensionless index, the *ripple index* (*RI*), defined as the ratio of the ripple wavelength (*L*) to ripple height (*h*), is used to distinguish between various types of ripples (formed by wind or water): wind ripples typically have *RI* >10–15, whereas water ripples have lower values. The ripple index varies inversely with grain size and directly with wind velocity. A measure of the lateral continuity of ripples is provided by the ratio of the mean crest length to the mean wavelength: the *horizontal form index* or the *continuity index*. Normal aeolian ripples, whose sinuous crests run in a direction transverse to the local wind direction, typically have horizontal form indices of 10–100.

Aeolian ripples show characteristic grain-size sorting. The mean size of sand comprising ripples is usually coarser than that of the underlying sand body. Within individual ripples, the mean grain size is coarsest at the ripple crest. Within-ripple grain size differences are more pronounced when the parent sand is poorly sorted (Anderson & Bunas 1993).

Normal aeolian ripples develop very rapidly, typically within minutes. The growth of megaripples takes longer and is less well understood. In the California deserts, the location and extent of megaripples appears to vary greatly from season to season, suggesting a response time of months or less.

A fully satisfactory model of ripple development is still lacking, but the topic has received a great deal of recent attention, as reviewed in Livingstone and Warren (1996). An early model (the ballistic theory of ripple formation) was that of Bagnold (1941), who observed a correspondence between calculated saltation path lengths and observed ripple wavelengths in wind-tunnel experiments. Sharp (1963) observed that ripple spacing increases with time, even though the wind velocity remains constant, and therefore is unlikely to be controlled by a characteristic saltation path length. Additionally, most ripple grains move by reptation. Anderson and Bunas (1993) showed that high-energy impacts by saltating grains on the windward side of a ripple preferentially eject more fine than coarse grains, accounting for the coarsening of the heavily bombarded stoss slope.

10

LANDFORMS OF AEOLIAN EROSION AND DESERT DUST

10.1 INTRODUCTION

Aeolian erosion occurs through two principal processes: deflation (removal of loosened material and its transport as fine grains in atmospheric suspension) (Fig. 10.1) and abrasion (mechanical wear of coherent material). The relative significance of each of these processes depends on the nature of the surface materials and the availability and nature of abrasive particles. The landforms that result from aeolian erosion include ventifacts, ridge and swale systems, yardangs, desert depressions (pans) and deflation basins, and inverted relief. Dust is an important product of erosional activity.

The geomorphological significance of wind erosion has been debated over more than a century (Goudie 1989). Remote-sensing images of terrestrial deserts and the Martian surface have been important stimuli for modern research and have provided important information on the extent of large-scale erosion systems as well as on the timing, frequency, and size of dust storms. On Mars, the discovery of ventifacts at the Viking and Pathfinder sites, and more recently in abundance across the regions traversed by the Spirit rover, has spurred investigation of such wind-eroded rocks on Earth (Greeley et al. 2002; Bridges et al. 2004). Their widespread presence suggest that aeolian abrasion is one of the most significant erosional processes on the planet.

Despite this recent interest, landforms of wind erosion have received less attention than those of deposition. Most research involves only general observations on form, materials, and environment and, with few exceptions, short- or long-term measurements of process or detailed environmental analyses are lacking. As with other areas of aridlands investigation, the remoteness of many important field areas has hindered research. Fundamental questions as to the relative role of abrasion and deflation in the formation of yardangs, the mechanisms of microfeature formation on yardangs and ventifacts, the interaction of yardang and ridge systems and wind flow, and the age, evolutionary history, and rates of formation of erosional landforms remain largely unanswered and await further process-oriented research. Such studies will provide an improved basis for understanding the climatic and geomorphic history of arid regions.

10.2 DEFLATION FEATURES: DESERT DEPRESSIONS AND PANS

The role of the wind in excavating both small- and large-scale desert depressions varies in both time and space, as closed depressions may form by a number of different processes, often acting in combination. The famous desert depressions are on a very large scale, the most noteworthy being those of the Western Desert of Egypt (Siwa, Qattara, Baharia, Farafra, Dakhla, and Khargha). Several of these basins have floors that lie below sea level (e.g. Khargha −18 m and Qattara −143 m). Depressions occur along the boundaries of northward-dipping strata and are bounded to the north by escarpments. These depressions are often cited as textbook examples of deflation, with the depth of erosion limited by the groundwater table, which forms a base level (Ball 1927). The role of wind action is suggested by the conformity of depression locations with areas of thinner and more easily breached limestone capping. However, the volume of the depressions, most notably the Qattara Depression (20,000 km³), poses problems for the idea of wind action acting alone. It is probable that many forces acted to excavate the material. The Western Desert experienced a wet climate through much of the Tertiary and Quaternary, punctuated by only brief episodes of aridity (Said 1981, 1983). Albritton et al. (1990) proposed that the Qattara Depression was originally excavated as a stream valley, subsequently modified by karstic activity, and further deepened and extended by

FIG. 10.1 Wind plays an important role in excavating desert depressions. The northeastern Harmattan winds are funneled and intensified as they pass between the Tibesti and Ennedi Mountains in northern Chad, causing episodes of deflation in the Bodélé Depression, one of the world's most active dust-source regions. This image shows a dust storm approaching Lake Chad. Source: NASA Moderate Resolution Imaging Spectroradiometer (MODIS) image.

mass-wasting, deflation, and fluviatile processes. Salt also plays a role in basin evolution, weathering, and preparing materials for deflation. This is a self-enhancing process, for the creation of low areas allows water and solutes to accumulate, forming new salts upon drying. Their crystallization weakens sedimentary cements, preparing grains for deflation (Haynes 1982). Thus, the depressions probably had a polygenetic origin, with wind erosion playing a major role only during the arid phases of the Quaternary.

In the southern Namib Desert, an elongate deflation basin, about 20 km wide and 125 km long, merges to the north with the Namib Sand Sea to form an integrated aeolian system of erosion, transport, and deposition (Corbett 1993). Within the system are barchan trains, yardangs, ventifacts, and sub-basins lowered to below sea level by salt weathering and deflation. Sandy beaches along the southern coast are deflated to provide sediment to 125 km-long barchan trains, with dunes up to 30 m high migrating at rates of 30–50 m year^{-1}. Both short-term (for example, seasonal beach progradation) and long-term (the rise and fall of sea level, which governs coastal morphology and sediment availability) changes in the coastline are linked to

basin-deflation dynamics and the aeolian system as a whole. Thus, the basin represents an aeolian transport corridor, conveying sediment from the coast to the sand sea, and creating erosional features (ventifacts, depressions, and yardangs) en route.

In North America, dunes of White Sands National Monument occupy a large deflation basin (Fig. 9.5) (Fryberger, in Langford 2003). Erosional shorelines, marking wetter episodes, suggest that the basin has deepened, rather than widened, by deflation. Playa-deflation episodes are linked to hydrologic changes associated with increased aridity. During these conditions, the wind erodes down to the climatically lowered water table.

At a much smaller scale are pans, features which are widely distributed in arid lands and especially well developed in southern Africa, on the High Plains of the USA (Fig. 9.17), and in western and southern Australia (Goudie & Wells 1995; Marker & Holmes 1995; Dutkiewicz & Prescott 1997; Holliday 1997) (see also Section 5.2.2). Pans may occur in interdunal basins, as palaeodrainage depressions aligned along former river courses, and by excavation on the floors of former lakes. Pans may also result from the deflation of interdunal swales and the noses of parabolic dunes (Goudie 1999). Lunette dunes are common to the lee of pans, but are not ubiquitous features.

Pans may be stratigraphically complex, as they develop by both depositional and erosional processes, a result of changing climatic and hydrologic conditions through time. In the High Plains, Holliday (1997) proposed that deflation occurred during dry periods. A rise in the water table during wet intervals resulted in the deposition of lacustrine carbonate in the deepest basins. Accretion also resulted from the deposition of calcareous sandy loams during dry periods.

There are several predisposing factors for pan development. As in the case of larger depressions, a single mode of genesis is unlikely, and several processes have probably contributed to pan development and maintenance. Pan initiation and growth appear to depend on materials that are susceptible to deflation, developing preferentially in poorly consolidated sediments, shales, and fine-grained sandstones. Aridity is an important factor as a paucity of vegetation enhances wind flow and permits sediment deflation. Space Shuttle photography indicates that pans are source regions for dust storms (Fig. 10.1) (Middleton et al. 1986). Water that accumulates in the depressions evaporates to leave salts, which further retard the development of vegetation and make sediments more susceptible to deflation as a result of fine-grained debris produced by salt weathering (Haynes 1982; Goudie 1989). As the depressions are enlarged, they become more attractive to grazing animals which, drawn by the lack of cover and the availability of water and salt licks, further disrupt the surface by trampling, rendering it more prone to erosion (Goudie 1989).

10.3 VENTIFACTS

The term ventifact is used to describe wind-eroded rocks of varying scale, form, and material composition. In size, ventifacts range from small pebbles to large boulders. Bedrock outcrops may also be extensively abraded. On Earth, ventifacts are relatively common, but often overlooked, features of arid, beach, and periglacial environments. The Martian surface, with its lack of vegetation, active wind systems, supply of sediment, and the presence of stones is the site of innumerable ventifacts, as witnessed by ground-based digital images taken by the twin Mars Exploration Rovers (MERs) Spirit and Opportunity during the 2003–4 series of missions to Mars. These rovers demonstrated that aeolian activity, previously evident on space imagery in the form of dunes and yardangs, has also affected the surface at a smaller scale. Such discoveries have provided the impetus for additional studies on Earth (Greeley et al. 2002; Bridges et al. 2004).

Ventifact formation is influenced by factors similar to those affecting dunes: wind frequency, magnitude, and persistence, as well as sediment supply, basin geomorphology, and vegetation cover. Ventifacts occur where strong winds are combined with abundant moving sand, such as near lake shorelines, downwind of alluvial rivers (Fig. 10.2), adjacent to dune sands, and in corridors of present or former sand transit (Laity 1994, 1995). They are best developed where the wind is accelerated, as in topographic saddles or near ridge crests.

In North America, where ventifacts have been studied extensively, most ventifacts are fossil in nature, characterized by weathered, dulled, fretted, or partly exfoliated faces (H.T.U. Smith 1967; R.S.U. Smith 1984; Laity 1992, 1995) and by rock varnish and rock coatings such as silica glaze and oxalate coatings (Dorn 1995). Published works date back to the mid-nineteenth century (Blake 1855; Gilbert 1875). Blackwelder (1929) examined fossil ventifacts along the now semiarid eastern base of the Sierra Nevada, where sediment derived from a glacial environment and blown by strong westerly winds abraded deposits of till. Early research provided valuable observations on the role of sand as an abrading agent, the nature of surface microfeatures, and the role of topography in enhancing wind action. In California, ventifacts commonly occur in proximity to dune systems, including the Dumont Dunes, Kelso Dunes, and dunes of the Devils Playground (H.T.U. Smith 1967); in areas that are presently sand-free, but which acted as corridors of sand transit from source areas to depositional zones during earlier climatic periods (Laity 1995); and adjacent to former lakes, where sand was blown from the shoreline as water level dropped, as seen to the west of Silver Lake (Fig. 10.3) and south of Owens Lake, California.

Aspects of ventifact form enable a determination of palaeowind direction, while their presence provides clues to the nature of earlier environments (Laity 1992). Erosional forms include smoothed and polished surfaces, facets, pits, flutes, grooves, helical forms, and etching. Facets are relatively plane surfaces cut at right angles to the wind, regardless of the original shape of the stone. They join along a sharp ridge or keel and the number of keels (kante) is used

FIG. 10.2 Ventifacts are wind-eroded rocks formed by abrasion. In this photograph, windflow is from left to right, abrading the windward side of the rock. The sand is segregated into "ribbons" during this event, with exposed rocks being sand blasted and covered rocks protected. As seen in the lower left, sand is often projected off the ventifact face into higher elevations of the airstream, where it may travel until it impacts downwind ventifacts. The wind speed is about 22 m s^{-1}. Location: Mojave River sink, California, USA.

FIG. 10.3 Heavily fluted ventifacts near the crest of a hill above Silver Lake, California, USA. Wind speed and sand transport are highest in elevated regions, increasing the intensity of the abrasion.

to describe the stones as einkante, zweikanter, or dreikanter (one-, two-, or three-ridged). Martian rocks show very strong single-facet development, perhaps because the rocks are less subject to move- ment than terrestrial rocks, which sometimes develop multiple facets owing to displacement by animal activity or mass movement. Rocks with two facets develop in areas with bidirectional wind flow.

Pitted surfaces are indented by closed depressions, often of irregular shape. They occur on surfaces that are inclined at high angles to the wind (55–90°). As the angle between the face and the wind decreases, a transition develops between pits and deep flutes. Flutes are scoop-shaped elongate forms, open at one end and closed at the other, forming a shape that looks like a modified arrowhead (Fig. 10.3). The flutes become shorter and steeper as the angle of inclination of the surface steepens. Grooves are longer than flutes and open at both ends. They are best developed on gently inclined surfaces. Grooves and flutes show remarkable parallelism with the strongest winds in a region. Helical forms are uncommon, commencing as shallow grooves, then deepening and spiralling in a downwind direction, to terminate at a sharp point. They appear to form in zones of high wind velocity and have been observed in marble and basalt. Etching refers to the selective erosion of less resistant strata or foliations in a rock (Laity 1994).

The ability to determine wind direction based on aspects of ventifact form allows the inference of palaeowind direction (Sharp 1949; R.S.U. Smith 1984; Laity 1992). Ventifacts are eroded on their windward face, with the scale of features (such as flutes or pits) increasing up the ventifact surface owing to the greater kinetic energy delivered as both wind speed and particle flux increase. Lee-side abrasion is not observed. Ventifacts that have more than one abraded face occur in areas of bidirectional or multidirectional winds: these can be either regional winds, or winds that change direction within single storms. For example, on 5 March 2000 a large Pacific front moved into the Mojave Desert. On a ridge-top ventifact site near the Mojave River sink, prefrontal winds were from the southeast, changing to westerly as the front passed. Consequently, both faces of the ventifacts were eroded in a single storm event, but at different times of the day.

Abrasion rates are determined by the time-dependent particle flux (Sharp 1980), the velocity of the wind, the strength of the rock, and the size, shape, and density of the impacting grains (Greeley & Iversen 1985). Although dust particles have been invoked as abrasive agents, they lack sufficient mass and are easily deflected past the obstacle. Sand is a much more effective agent of abrasion, and although sand-blasted rocks appear macroscopically smooth (for example, to the eye and to touch), they are nonetheless very rough at the microscale. Scanning electron micrographs of the smooth and polished surfaces of basalt and marble ventifacts show that cleavage fracture of crystalline material is widespread across rock surfaces (Fig. 10.4). Microgouging and rubbing of the surface also develop as grains skid up the face of the rock (Laity 1995).

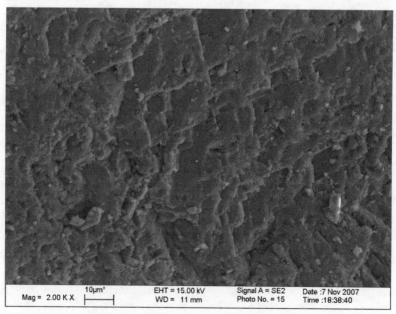

Fig. 10.4 A scanning electron micrograph of a marble boulder (ventifact) abraded by sand shows a surface developed by impact-generated cleavage fracture. Original magnification: ×2000.

The effective range of sand abrasion on level ground rarely exceeds 1 m. However, abrasion heights of 2 m or greater are observed on rocks on hill crests. During strong windstorms, the acceleration of airflow discussed in Section 9.2.2 moves sand up the slopes, raising the overall level of the sand. Additionally, sand is propelled from ventifact facets high into the airstream and impacts downstream rocks at heights greater than a meter (see Fig. 10.2).

10.4 YARDANGS AND RIDGE AND SWALE SYSTEMS

Yardangs and ridge and swale systems are closely related features of aeolian erosion. Whereas yardangs occur in all deserts except those of Australia, ridge systems appear to be limited to the Saharan Desert. Although related, the two systems differ in their aerodynamic form and scale, in the probable degree of material hardness, and relative roles of abrasion and deflation (Laity 1994).

Ridges and swales are best exemplified by vast systems to the east, south, and west of the Tibesti Mountains, Chad (Fig. 10.5). In orbital views, the systems appear to be continuous for hundreds of kilometers, sweeping in a broad arc around the mountains. On aerial photographs, the discontinuous nature of the systems is evident, with the largest ridges not longer than 4 km (Mainguet 1972). In some areas the ridges reach up to 1 km in width.

On satellite images the ridges appear dark, owing to a well-developed patina of rock varnish on the upper surfaces, and the corridors show as lighter-colored lineations because of constant abrasion by sand (Fig. 10.5). The development of these ridge and swale systems is a function of: (1) extensive exposure of sandstone, (2) a dense network of joints that allows the wind to be channelized, and (3) a unidirectional wind, charged with sand. The circum-Tibesti region of the Sahara is deformed and fractured along two major axes, northeast–southwest and northwest–southeast, and the wind has exploited the fracture system, gradually enlarging corridors, the size of which is a function of the deviation between the wind direction and joint trends. The trade-wind circulation is deflected around the Tibesti Mountains, and in northern Chad, to the east of the Tibesti, sand-laden winds blow 8 months out of 12 from the northeast, reaching maximum daytime velocities of 6–8 m s^{-1}. Sand abrasion occurs in the corridors, but is ineffective on the middle and upper slopes of the ridges, which are marked by a basal fringe area of eroded rock. Barchan dunes, whose

ISS014E06304

FIG. 10.5 Wind-eroded ridge systems that follow the deflection of the trade winds around the Tibesti Mountains. The system crosses the Aorounga impact crater, shown near the center of the photograph. The dark lines are ridges and the lighter lineations are sand. Source: Image Science and Analysis Laboratory, NASA Johnson Space Center. Astronaut photo record ISSO14E06304.

axes parallel those of the ridge systems and the resultant winds, often occupy wider swales. There is no evidence that deflation plays a significant role in ridge formation.

The channelized topography has a wavelength that, in cross-section, appears to be relatively constant for any given group of ridges and swales (Mainguet 1970). The periodicity is determined to some degree by the fracture density, but topography and wind strength also play a role, with the largest ridges found on the more elevated parts of the terrain.

Yardangs are more streamlined than ridges, often described as resembling an inverted ship's hull (Fig. 10.6), although in many cases the yardangs are flat-topped. The windward face of the yardang is typically blunt-ended, steep and high, whereas the leeward end declines in elevation and tapers to a

Fig. 10.7 An extensive field of yardangs in the central portion of the Dasht-e Lut, Iran. Source: NASA Advanced Spaceborne Thermal Emission and Reflection Radiometer (ASTER) image, 13 May 2003.

Fig. 10.6 This yardang at Edwards Lake, California, USA, exhibits a streamlined form, high length-to-width ratio, and sharp ridge.

point. However, the yardang takes on many forms (Whitney 1983; Halimov & Fezer 1989). Measurements of yardang length-to-width ratios commonly average 3:1 or greater, an elongate form which minimizes the resistance to the wind. Yardangs form parallel to one another, typically occurring as extensive fields (Fig. 10.7). They may occur as tight arrays, separated from one another by either U-shaped or flat-bottomed troughs, or as widely spaced, highly streamlined features on wind-beveled plains.

Yardangs occupy only a small part of the Earth's surface, as they require conditions of great aridity, nearly unidirectional or seasonally reversing winds, and in some cases, a favorable material and some assistance from weathering to form. In Africa, yardangs occur in the Arabian Peninsula, Egypt, Libya, Chad, Niger, the Namib Desert, and possibly along the coast of Mauritania. Asia has several yardang fields, including those of the Taklimakan Desert in the Tarim Basin, the Qaidam depression of Central Asia, and the Lut Desert, Iran. Yardangs are found along the coastal desert of Peru in South America and as minor groups in North America. Relict yardangs are present in the semiarid Ebro Depression of Spain (Gutiérrez-Elorza 2002). Yardangs are also widespread on Mars.

Yardangs have received attention as curiosities and geologic oddities since the late 1800s in the Namib Desert (Stapff 1887), Egypt (Walther 1891), and China (Hedin 1903). The term yardang was introduced by Hedin (1903, p. 350), who attributed their formation to initial erosion by running water, and subsequent resculpturing by the wind. More recent interest was fostered by the discovery of large yardang fields on Mars (Ward 1979) and by the availability of aerial photography and satellite imagery (Mainguet 1972; Ward 1979; Halimov & Fezer 1989).

There has been little detailed study of yardangs and much remains to be determined about their formation. Meteorological data are not available and thus the velocity of winds within the corridors and the frequency of sand-transporting winds are not known. Our understanding of wind flow around individual yardangs is based primarily on laboratory and theoretical determinations (Ward & Greeley 1984; Whitney 1985), rather than instrumentation. The long-term evolutionary development of yardang fields, the role of fluvial erosion and jointing in initial channelization of winds, and the rates of formation are poorly understood. The role of abrasion and deflation in determining the ultimate form of yardangs also remains uncertain, and probably varies according to the rock type.

Yardangs develop where wind processes dominate over fluvial processes, and are thereby limited to extremely arid deserts. Plant cover and soil development are generally minimal. Many yardang fields develop where strong unidirectional winds occur throughout much of the year (Hobbs 1917; McCauley et al. 1977; Corbett 1993), but others, such those in the Lut Desert, develop in regions of seasonally opposing winds. In areas of reversing winds, it appears that one wind is usually dominant, and the opposing wind is lighter and less frequent.

Yardangs form in a broad range of geologic materials, including sandstones, limestones, dolomites, claystones, granites, gneisses, schists, volcanic ignimbrites and basalts, and lacustrine sediments. Variations in yardang form result from differences in material composition and structure. Where the wind erodes horizontally layered rocks of different resistances, tabular forms with protruding shelves or stair-step profiles are formed. Inclined or vertically layered sediments will be eroded into ridges characterized by grooves and fins (Blackwelder 1934).

Wind-tunnel experiments and measurement of mature, streamlined yardangs suggest an ideal length-to-width ratio of 4:1 (Ward & Greeley 1984), independent of scale. This value varies according to variations in lithology, wind strength, and direction, and the length of time the yardangs have been in existence. The ideal proportions are probably approached only after a long period of erosion. In Peru, well-developed streamlined yardangs have ratios ranging from 3:1 to 10:1 (McCauley et al. 1977). Dimensions vary from a few meters, to yardangs that have attained heights of as much as 200 m and are several kilometers long (Mainguet 1968). Long megayardangs, up to 10 km in length, are developed in basaltic lava flows of Holocene age in the Payun Matru Volcanic Field in the southern Andes, Argentina (Inbar & Risso 2001).

The inter-yardang spaces have been referred to as troughs, couloirs, corridors, swales, and boulevards. Where the inter-yardang space is narrow, troughs appear to be U-shaped; as they widen the corridors become flattened (Blackwelder 1934). Although attention is often focused on the yardang, most of the geomorphic activity is concentrated in the troughs, which may be totally engulfed in sand, or be only partially sand-covered (Mainguet 1972), show low transverse ridges of fine gravel (Blackwelder 1934), or ripple trains that diverge at the head and converge in the downwind direction around the yardang flanks (McCauley et al. 1977). Corridors may be occupied by migrating barchan dunes (Gabriel 1938; Hagedorn 1971; Mainguet et al. 1980). The rocks in the corridors often carry numerous marks of aeolian erosion, including longitudinal striations, and shallow erosional basins, meters or tens of meters in length, that occur either as isolated forms or in groups (Mainguet 1972).

Yardangs may form by a combination of abrasion and deflation and, in softer materials, be further modified by fluvial erosion and mass movement. The role of each of these processes in the ultimate form varies according to yardang lithology and structure and regional climatic history. The role of abrasion in the formation of yardangs is evident in the overall shape and color of the lower slopes. The windward end and lateral slopes are commonly lighter in color than the upper slopes, and polished and fluted by sand blasting to a height of 1 or 2 m (Hobbs 1917; Hagedorn 1971; Grolier et al. 1980). In the sandstone yardangs of the Sahara there is little active removal of material from the ridge summits, as evidenced by the dark patina of varnish. Limestone yardangs in Egypt commonly have flat tops that are irregular in form and retain some of the weathered surface that pre-dates erosion (El-Baz et al. 1979). In the southern Namib Desert, however, crystalline dolomite yardangs 10–15 m in height are streamlined from the base to the top (Corbett 1993). In this region, both active and relict yardangs occur. Where abrasion is ongoing, the yardang surfaces are smoothly polished and fluting is evident: relict forms have rough surfaces colonized by lichen, with fluting strongly altered by solution weathering. In many yardang fields, abrasion exerts a strong control on the overall form, causing undercutting of the steep windward face and lateral slopes. Where abrasion is the most important formative process, ventifacts may be found in association with yardangs.

Researchers differ in their opinion as to the relative importance of abrasion and deflation in forming yardangs. The small yardang field at Rogers Lake, California, for example, has been attributed both to deflation (McCauley et al. 1977) and to abrasion (Blackwelder 1934; Ward & Greeley 1984). Deflation appears to be more important in less-indurated lacustrine material, such as siltstones, as demonstrated by yardangs 30–50 m high and up to 1.5 km long on the Pampa de la Averia, Peru, that possess smooth, streamlined shapes from base to crest (McCauley et al. 1977). The aerodynamic form is the primary evidence for the role of deflation, as there have been no field studies of this process.

Fluvial erosion acts in several ways to aid in the development of yardang fields. At the outset, yardang fields may be initiated along stream courses that are enlarged and modified by the wind. Referring to the yardangs of Lop Nor, China, Li Daoyuan (AD 466–527) defined yardangs as "rills cut by water is blown by wind subsequently" (sic) (Xia 1987). Subsequently, the inter-yardang troughs may be fluvially eroded and the yardang slopes gullied. Intense gullying, earthflows, and solution on the flanks and summits characterize the yardangs of the Lut Desert of Iran (Krinsley 1970). These 60 m-high yardangs, formed in playa sediments, have upper slopes well above the range of effective abrasion, and deflation is too slow a process to remove evidence of fluvial activity. The gullies may be relics of earlier wetter climates. It is thought that yardangs cannot persist in an increasingly humid climate, as more frequent fluvial erosion destroys the yardang form, and vegetation limits wind erosion. An exception appears to be a cluster of about 50 relict yardangs of the Ebro Basin, Spain, which retain their form despite an annual rainfall of about 400 mm and flanks that are partially covered by vegetation (Gutiérrez-Elorza 2002).

Mass movement has received little attention as a modifying process, although the presence of slump blocks alongside yardangs appears to be common, particularly near the undercut nose of the yardang.

Conceptual models of yardang development are not well advanced. Yardang fields may be initiated in gullies and fractures that are aligned parallel to the prevailing wind and become enlarged (Hedin 1903; Blackwelder 1934; Ward & Greeley 1984). As air enters the corridor, the streamlines are compressed and the wind accelerates, with sand abrading the bottom and sides of the trough. During the early stages of formation, wind and occasional fluvial erosion erode the passages more rapidly than the yardangs, causing the passages to deepen and the yardangs to grow (Halimov & Fezer 1989). As abrasion is more intense at the yardangs prows, the ridges become shorter over time, and may eventually develop into conical hills, mesas, and pyramids (Halimov & Fezer 1989). When the yardangs are lowered to a height of 1–3 m, sand blasting affects the entire form, and the crests become more rounded and streamlined and, in some cases, the hills vanish (Halimov & Fezer 1989). Brookes (2001) noted a spatial transgression of yardang formation over time, and observed that in rare cases older yardangs may "piggyback" on new yardangs being created in a renewed erosional phase.

Little is known about the rate of yardang formation. However, research in the eastern Sahara and China provides some time constraints and suggests that the development of mature yardangs can be relatively rapid (several thousand years) in soft materials. Playa muds in the eastern Sahara were deposited in a moist period between 9800 and 4500 years BP, and scoured in the hyperarid period that followed, leaving yardangs several meters in height standing as erosional remnants (Haynes 2001). This erosion was facilitated by a drop in the water table, which otherwise would have formed a base level to wind erosion (Haynes 1980). Four-meter-high yardangs in the Bodélé Depression, Chad probably formed within 1200–2400 years (Washington et al. 2006).

There has been very little research on yardang formation in harder materials. Basaltic yardangs, 2–3 m in height, have formed in the Payun Matru

Volcanic Field, Argentina, in less than 10,000 years (Inbar & Risso 2001). Bedrock megayardangs along the coast of Namibia may have taken millions of years to form (Goudie 2007).

10.5 DESERT DUST

Dust is a highly significant component of the desert system, affecting not only local geomorphological and biologic processes and atmospheric conditions, but influencing the atmosphere, oceans, and land surfaces far from its original source area. Atmospheric dust is derived mainly from aeolian erosion in tropical and subtropical arid and semiarid areas, with the Sahara-Sahel region in northern Africa and the Gobi-Taklimakan in central Asia being the most important source areas. Every year, 1–2 billion t of dust enter the atmosphere, representing about half of the total tropospheric aerosols. In some regions, such as Australia, the annual sediment load carried

TABLE 10.1 Maximum mean Aerosol Index (AI) values for North African/Arabian dust sources, derived from TOMS on the Nimbus 7 satellite.

Location in North Africa	AI value
Bodélé Depression (Sahara)	>30
Mali and Mauritania (West Sahara)	>24
Arabia	>21
Eastern Sahara (Libya)	>15

Source: Goudie and Middleton (2001).

by winds is larger than that carried by rivers (Knight et al. 1995). Spread by the global winds, dust has been identified far from its source areas, in soils, deep-sea cores, and ice cores. Far from the desert, dust impacts urban areas, including the cities of eastern Asia, coastal Australia, and western Europe. Dust can be hazardous to human and environmental health (Griffin et al. 2001), and to civilian and military transportation. Micro-organisms and chemicals hitchhike on small particles and are associated with environmental problems, including coral reef decline (Shinn et al. 2000), red tide events, and fungal outbreaks on commercial crops (Griffin et al. 2001).

To assess the impact of dust on the global environment, to characterize global dust sources, and to measure dust transportation, researchers model airmass trajectories, use records of dust-storm occurrences from weather stations, make lidar measurements of dust, measure wet and dry dust deposition at the surface, and use satellite imagery, in particular Total Ozone Mapping Spectrometer (TOMS) data (Fig. 10.8). The TOMS on the Nimbus 7 satellite allows the observation and tracking of large dust plumes (Prospero et al. 2002; Washington & Todd 2005; Koren et al. 2006). It detects ultraviolet-absorbing aerosols in the atmosphere and an Aerosol Index (AI) has been developed which is linearly proportional to the aerosol optical thickness (Fig. 10.9 and Table 10.1). Although TOMS provides a powerful tool, it is affected by persistent clouds, biomass burning, and air pollution, thus limiting its applications in certain areas of the world, particularly India and China. In India, aerosols include those from fossil fuels associated with rapid industrialization, marine aerosols from neighboring seas, aerosols from seasonal forest fires, and desert dust (Badarinath et al. 2007). Dust can be distinguished in TOMS data by its particle size (larger than smoke) and

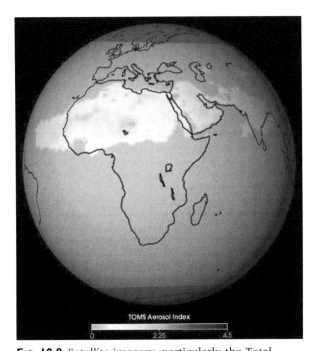

FIG. 10.8 Satellite imagery, particularly the Total Ozone Mapping Spectrometer (TOMS) data, have allowed researchers to observe and track large dust storms. This image shows dust blowing from the Sahara Desert into the Atlantic Ocean and dust above the Rub' al Khali and Nafud Deserts of the Arabian Peninsula. Source: April 2000 image, TOMS science team, NASA Goddard Space Flight Center.

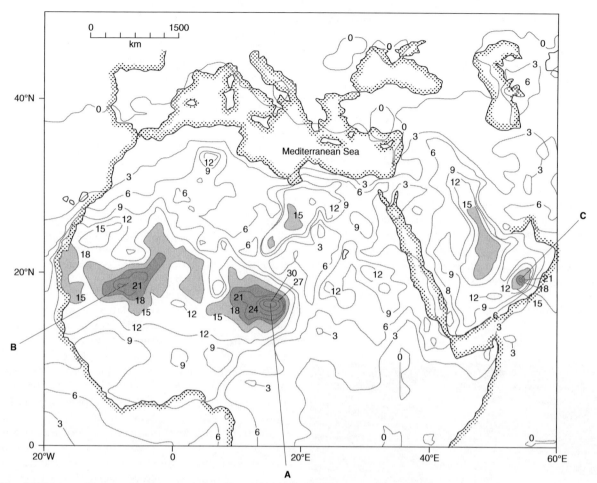

Fig. 10.9 Map of Aerosol Index (AI) values for the Sahara Desert. Area A is the Bodélé Depression, the world's most intense dust source (AI values >30); B is the western Sahara (AI values >24), including Mauritania, Mali, and southern Algeria; and C is Arabia (>21) in the region of the Oman/Saudi Arabia border. From Goudie, A.S. and Middleton, N.J. (2001) Saharan dust storms: nature and consequences. *Earth-Science Reviews* **56**, 179–204. By permission of Elsevier.

absorbing efficiency in the ultraviolet spectral band (Badarinath et al. 2007). Dust plumes are shown to cover a larger area, be more persistent, and occur more frequently than particulates attributed to air pollution (Husar et al. 1997).

10.5.1 Definitions

Dust emission into the atmosphere occurs as sporadic events that vary in magnitude, timing, and particle concentration and size. Dust episodes are commonly classified according to their effect on visibility. *Blowing dust* is material being entrained within sight of the observer, but not obscuring visibility to less than 1000 m; *dust storms* are blowing dust events associ-

ated with visibility reductions to less than 1000 m; and *dust events* are defined by a reduction of visibility to less than 11 km (Middleton 1997). Dust that results from mechanical means, such as traffic on unpaved roads or from mining operations, is generally referred to as *fugitive dust*. A *dust devil* is a short-lived and localized column of dust that does not travel for any great distance.

10.5.2 Environmental role and impacts of dust

Dust plays many environmental roles, both positive and negative, depending on the magnitude of the event, the distance that the dust travels, and the chemical constituents that are present. Effects can be subdivided into those that are principally local and those that are at a regional or global scale. At a local scale, dust influences soil, desert varnish, and stone-pavement formation (see Chapter 7); affects plant photosynthesis if leaves are coated; causes automobile and airplane accidents and contributes to machinery problems; reduces property values; and causes human health problems and animal suffocation. At a regional or global scale, continental and marine ecosystems, soil development, the Earth's climate, and air quality and visibility are affected.

10.5.2.1 Effects on marine and terrestrial ecosystems

Mineral dust that falls to the Earth's surface along its transport path contributes iron and other micro-nutrients to terrestrial and marine ecosystems (Baker et al. 2006; Reynolds et al. 2006). In the ocean, mineral aerosol particles enhance biological productivity. Widespread atmospheric transport of mineral dust provides important nutrients to phytoplankton and zooplankton in areas of the oceans that are far removed from supplies provided by continental river inflows (Buseck & Pósfai 1999; Baker et al. 2006). Saharan dust, for example, is an important nutrient source for the North Atlantic Ocean (Goudie & Middleton 2001).

The terrestrial ecosystem also benefits from long-range dust movement. Each year, 40 million t of dust are transported from the Sahara to fertilize the Amazon basin, enhancing the health and productivity of the rainforest (Swap et al. 1992; Koren et al. 2006). About half of this dust is derived from a single small source region, the Bodélé Depression, located northeast of Lake Chad (Figs 2.1 and 10.1). Rainforest soils are typically shallow, nutrient-poor, and lacking in soluble minerals owing to leaching by heavy rains. About 50 Tg of dust reach and fertilize the Amazon basin, acting to balance its nutrient budget (Koren et al. 2006).

Desert dust events also inject large pulses of microorganisms and pollen into the atmosphere. It is postulated that such events may play a role in transporting pathogens or expanding the biogeographical range of some organisms (Kellogg & Griffin 2006). The transport of pathogens is of particular concern, as intercontinental dust may help spread plant and animal diseases. Saharan dust has been implicated as an efficient carrier for disease-spreading spores, which can occasionally decrease the vitality of coral reefs in the Caribbean (Shinn et al. 2000). To date, there

have been no confirmed reports of human infectious diseases related to the long-distance transport of desert dust (Kellogg & Griffin 2006). However, Caribbean air samples indicate that during African dust events the number of cultivatable airborne microorganisms is two to three times greater than at other times (Griffin et al. 2001).

10.5.2.2 Relationship to soil development and earth surface processes

Dust plays an important role in earth surface processes, stabilizing dunes, contributing to soils (Sections 7.2 and 7.4.1), enhancing rock weathering (Section 6.2.5), and pavement-formation processes (Section 7.8.2) and, when deposition is extensive and long-lived, forming loess. Loess is a non-stratified, weakly indurated, aeolian sedimentary deposit. In a typical loess, more than 80% of the particles are of silt size. Loess covers up to 10% of the Earth's land surface, with the thickest and most extensive deposits in the mid-latitude zones, including China, central Asia, central Europe, the Ukraine, Argentina, and the Great Plains of North America. The Chinese Loess Plateau is composed of more than 90% aeolian mineral dust (Zhang et al. 1994). Thick loess deposits provide an important environmental record of past climatic change. The loess and intercalated palaeosols in the Chinese Loess Plateau were derived from arid and semiarid regions and transported by northwest and westerly winds during the last glacial cycle (Lu & Sun 2000).

Dust also provides important parent material for other soils, including those in the Mediterranean region (Yaalon 1997; Fiol et al. 2005; Buis & Veldkamp 2008) and on remote Caribbean islands (Muhs et al. 1990). The biogeochemical significance of dust is a function of the initial nutrient status of the local bedrock. Bedrock typically weathers slowly in drylands, and if the bedrock is low in nutrients, the input of imported dust becomes increasingly important (Reynolds et al. 2006). The aeolian dust component of soils on the Colorado Plateau is commonly 20–30% and enriches sandy sediments in numerous elements, including phosphorus, magnesium, sodium, potassium, molybdenum, and carbon (Reynolds et al. 2001).

However, severe wind erosion can disrupt the soil nutrient cycle. Plant nutrients are concentrated on suspension-sized particles, and thus wind-eroded soil is rapidly depleted. Experiments in the Jornada Basin, south-central New Mexico, USA, show that this process takes about 10 years (Okin & Gillette 2004). Depleted soils may allow new species to gain a competitive edge, thereby changing plant communities.

10.5.2.3 Impact of dust on climate, weather, and air quality

Atmospheric aerosols are an important component of the Earth's weather and climate system. In the atmosphere, the role of dust depends on such factors as dust concentration, mineralogical composition (through the refractive index), size distribution, and vertical distribution of the particles. On a global scale, dust can affect the radiative budget of the Earth by absorbing and scattering solar and terrestrial radiation, and by perturbing atmospheric circulation patterns (Miller & Tegen 1998). On a regional scale, satellite imagery shows that dust exerts a strong radiative effect in the tropical and subtropical belt of marine regions located downwind of deserts. When a dust storm originating in the Thar Desert reached Hyderabad, India, radiation in the 0.3–3 μm band was reduced by up to 62% (Fig. 10.10) (Badarinath et al. 2007). During Chinese dust events, the incoming solar flux at the surface in Beijing has decreased by 40% (Zhou et al. 1994), and at Kwangju City, South Korea, the mean light extinction has been esti-

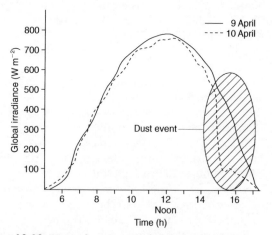

FIG. 10.10 Diurnal variation of global solar flux in Hyderabad, India, derived from LICOR pyranometer. Ground-level solar flux in the mid- to late afternoon was reduced by about 62% on 10 April relative to 9 April 2006, as a dust plume from the Thar Desert reached the city. From Badarinath, K.V.S. et al. (2007) Case study of a dust storm over Hyderabad area, India: its impact on solar radiation using satellite data and ground measurements. *Science of the Total Environment* **384**, 316–32. By permission of Elsevier.

mated at more than 44% (Kim et al. 2001). In April 1998, a Chinese dust plume crossed the Pacific Ocean and moved across the west coast of the USA, causing the sky to appear milky white, and decreasing direct normal solar radiation by 25–35% (Husar et al. 2001). A 1999 dust storm that originated in northwest Africa reduced visibility in the Canary Islands to 200 m (Criado & Dorta 2003). Although these long-range effects are significant, closer to source regions near-zero visibility may develop, impairing vehicular traffic and frequently causing accidents.

Atmospheric dust acts as condensation nuclei that facilitate cloud-formation processes. As a result, dust affects convectional activity and cloud formation, modifying raindrop size and rainfall amount (Maley 1982). An unusual consequence of intense dust storms is the "blood rains" that develop when rain, or occasionally snowfall, combine with fine dust particles. The characteristics of the event depend on the amount of dust, the size of the aerosols, and their chemical composition. This phenomenon is well known in northern and central Europe, but is most common in the Mediterranean countries such as Spain, where blood rains occur on average 3 days a year (Criado & Dorta 2003). Dust also plays a role in neutralizing rainwater acidity. Owing to the calcium carbonate content of soils in arid and semiarid regions, rainwater pH commonly shifts from 4–5 to 6–8 (Losno et al. 1991).

Incursions of atmospheric dust have important effects on air quality. To date, there has been relatively little research on the effects of dust events on human respiratory and cardiovascular health. However, Meng and Lu (2007) show that persons living in nonindustrial zones of China, but close to dust-source regions, experience an increase in daily hospitalizations for upper respiratory tract infections, pneumonia, and hypertension during the 3–6 days following dust events. Additional health effects related to dust emanating from the USA, China, and the Aral Sea region are discussed in Chapter 13.

Enormous numbers of livestock deaths can also result from dust inhalation. The dust storm of 5 May 1993 killed 120,000 animals in China, and that of 14–16 April 1998 resulted in 110,000 animal deaths (Shao & Zhao 2001). An intense 3-day dust storm in November 1910 in northwestern Turkmenistan killed 29,000 of the 30,000 cattle that grazed the area. In a secondary effect, the destruction of pasture grasses caused thousands of other sheep, goats, cattle, and horses to die in the following winter (Orlovsky et al. 2005).

10.5.2.4 Dust storms and vehicular accidents

Dust storms pose a very potent danger to motor vehicles. The ratio of individuals killed to those injured is 68% higher for dust accidents than other accidents (Burritt & Hyers 1981). In a typical dust accident, a blinding cloud of dust is emitted from a nearby source and passes over the highway (Fig. 10.11). Visibility is near zero within the cloud and accidents occur because drivers react differently to the conditions: accelerating, veering, or decelerating. As a result, there is a mix of vehicles taking different paths at different speeds, favoring chain-reaction accidents or several closely spaced accidents. Day (1993) reports that drivers also indicate that dust on the highway reduces the traction between the tires and the road surface, creating ice-like conditions. Cars are more likely to be involved in accidents than trucks, perhaps because truck drivers are sitting high enough to be above the blinding, locally derived dust.

The relationship between dust storms and vehicular accidents has been most closely examined for the state of Arizona, USA, an area where frequent dust storms pose a hazard to large numbers of travelers on the major highways between Phoenix and Tucson. Most accidents (84%) occur late during the summer season, when thunderstorm development and accompanying gusts are most prevalent. The remainder occur in winter owing to high winds associated with the passage of cold frontal systems (Brazel & Hsu 1981). Dust-related accidents commonly occur late in the day when traffic volumes tend to be high and thunderstorm activity greatest. In Arizona, large parcels of silty, abandoned or tilled farm fields often lie adjacent to highways.

Evidence indicates that dust accidents occur when the site of deflation is very near the highway and the dense short-lived dust moves across the driving lanes while being dispersed within a general dust cloud. Although dust warning systems are sometimes employed, they have been relatively unsuccessful owing to inappropriate driver response. Drivers are often caught unawares, because the general visibility may not appear to be dangerous.

10.5.3 DUST ENTRAINMENT, TRANSPORT, AND DEPOSITION

The entrainment of dust from the ground surface is controlled by wind speed and gustiness, the nature of the soil, and the presence of any surface obstacles to windflow. Entrainment is a process that refers to how surface sediment is incorporated into a fluid flow (air or water), as part of the process of erosion. Sediment is entrained when forces acting to move a particle are greater than those resisting movement. Once entrained, dust is transported by suspension, wherein the turbulent flow structures of the air carry the particles without contact with the surface. Turbulent eddies in the atmosphere are able to keep dust suspended for many days and over great distances.

FIG. 10.11 Dust storms pose a very potent danger to motor vehicles, reducing visibility to near zero within the cloud. In this photo, dust plumes cross a road during a high-wind event east of Turfan, China.

Intense dust storms reduce visibility to near zero in and near the source regions. In general, there is an exponential decline in dust concentration with height above the surface. Dust that travels great distances is initially raised by high-wind events to altitudes of several kilometers, and is chiefly transported above the atmospheric boundary layer by large-scale atmospheric circulations. Dust from central Asian deserts has traveled as far as 5000 km, crossing the Pacific Ocean to North America (Arimoto 2001; Zdanowicz et al. 2006), and dust from the Sahara is regularly observed on the western side of the tropical Atlantic.

Dust is continuously deposited on the Earth's surface through wet and dry deposition processes and has been studied using dust traps (Reheis et al. 1995). Several studies suggest that the mineral aerosol assemblages vary according to the source. Soils from the northern Sahara Desert, for example, are low in quartz and have more calcite and palygorskite than soils from other parts of the Sahara. Using a mineralogical approach and isotopic methods may help establish the provenance of atmospheric dust (Arimoto 2001).

The dust load of the atmosphere varies over both short and long time periods. Terrestrial and marine records indicate that LGM atmospheric dust loads were up to 10 times more than are presently recorded downwind of major source areas, although this increase was neither spatially uniform nor ubiquitous (Kohfield & Harrison 2001).

10.5.3.1 Climatic factors in dust entrainment

The threshold wind velocity required to entrain dust varies considerably according to the grain-size distribution and sorting characteristics of the particles, cohesion of the surface, and the presence of roughness elements or vegetation. Portable wind-tunnel experiments on many different geomorphological surfaces by Gillette et al. (1980, 1982) show a wide range of threshold velocities, from about 0.4 to 3.39 m s^{-1}. Those for disturbed surfaces are about two to six times less than for undisturbed surfaces. Under field conditions, the mean wind velocity associated with dust-raising events in the western Sahara varied at individual stations from 6.5 to 13.0 m s^{-1} (Helgren & Prospero 1987). Blowing dust events with visibilities of 10 km or less in Arizona required average wind speeds of greater than 11 m s^{-1} and wind gusts of greater than 16 m s^{-1} (Hall 1981).

Rainfall is critical to dust-storm development insofar as it influences surface moisture content, vegetation cover, weathering processes, and human habitability of a region. Semiarid regions (with a mean precipitation of about 100–200 mm) were once thought to produce more dust than arid or hyperarid regions, because weathering and fluvial activity produce and concentrate fine sediment, and because human cultivation and activity act to disrupt sediment surfaces (Goudie 1983b; Pye 1987). However, more recent research shows that the most vigorous source for dust on the planet is the Bodélé Depression, with an estimated mean annual rainfall of less than 10 mm (Fig. 10.9) (Washington & Todd 2005).

The occurrence of drought plays a major role in enhancing dust production. Correlations between dust storm activity and drought have been observed for Africa (Prospero et al. 2002) and for Australia (Yu et al. 1992). In the USA, dust production increases by about an order of magnitude during drought (Gillette & Hanson 1989). Even small amounts of soil water strongly limit wind erosion. Spring dust emissions in Arizona were shown to decline following a wet El Niño-related fall and winter season (MacKinnon et al. 1990). In industrial plants and mines, spraying with water is commonly used to suppress dust emissions.

The seasonal distribution of rainfall is as important as amount in predicting dust generation. In the Mojave Desert, California, winter annuals reach their maximum size and coverage in April–May, when wind speeds are highest. Dust storms arise when precipitation totals are severely below normal and the annuals do not germinate. In the Negev Desert, wet seasons with low rainfall are followed by dry seasons with high dust-storm activity because protective crusts are not as well developed following low rainfall (Offer & Goossens 2001).

10.5.3.2 Surface factors: vegetation, crusts, and the availability of sand

In addition to the meteorological conditions discussed previously, surface factors, with their inherent spatial and temporal variability, are critical in determining dust emission rates. Vegetation and its seasonal patterns are important controls on dust-storm development. Vegetation augments surface stability, thereby reducing wind erosion and dust emissions. Plants increase surface roughness (and thereby reduce wind speed); their root systems hold soil particles together; they increase soil moisture

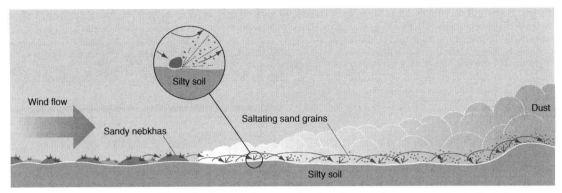

FIG. 10.12 Dust may be entrained by saltating sand grains that break up soil aggregates by ballistic impact.

temporarily by shading the surface; and they promote the development of crusts and soils as they stabilize surface movement. At a broad scale, studies of dust-storm frequency suggest that the highest occurrences are in very arid regions with bare ground (60–80 days year^{-1}), followed by zones with shrubland (20–30 days year^{-1}), and grassland (2–4 days year^{-1}) (Engelstaedter et al. 2003). Dust-storm activity is inversely correlated with net primary productivity and the leaf area index and therefore loss of vegetation increases dust emissions. In central and eastern parts of northern China, declines in springtime vegetation cover are associated with more frequent dust storms. In the more arid areas of northwestern China, vegetation is already at a minimum and has little seasonal change, therefore showing much less correlation with dust-storm activity (Zou & Zhai 2004).

In areas that lack vegetation, the soil can be resistant to the wind in the presence of crusts or armored surfaces, or where there are particle cements such as salts or organic breakdown products. Wind-tunnel experiments show that even low concentrations of salts cement soil particles and reduce erosion (Rice et al. 1997), although the effect is not as great as that of moisture. Roughness elements, such as gravel covers or crop stubble, also act to reduce erosion. However, low densities of roughness elements may actually reduce the threshold velocity and activate erosion by inducing scour.

Where naturally crusted soils are present, dust may be entrained by saltating sand grains, which break up soil aggregates by ballistic impact and overcome the otherwise strong cohesive forces associated with small particles (Figs 10.12 and 10.13) (Shao et al. 1993). Studies at Mono and Owens Lake, California, reveal that abrasion of the surfaces by salta-

FIG. 10.13 Lineations developed on a small playa surface by sand abrasion.

tion, not direct deflation of puffy efflorescences, is the key mechanism for dust generation (Cahill et al. 1996; Gill 1996). Indeed, the wind-velocity threshold for dust suspension at these sites (7–10 m s^{-1}) is comparable with that for sand-sized grains. As a result of this sand-blasting effect, a strong correlation often exists between dust-particle concentration and horizontal wind speed. The production of dust by sand blasting has been noted in other areas as well, including the northern Sahara (Gomes et al. 1990). The particle size that results from sand blasting may be considerably finer than that derived from the entrainment of loose soil aggregates (Gomes et al. 1990).

Destruction of crusts exposes unconsolidated fines to wind erosion (Fig. 10.14). This can occur by the trampling of livestock or by off-road vehicle traffic. In the Sahel, the monsoon season (May to September) favors crust formation. During the dry season, these crusts are destroyed in cultivated fields by cattle trampling (Rajot et al. 2003).

10.5.3.3 Anthropogenic activity

Human impacts on dust production are well documented (Gill 1996). The threshold wind speeds for disturbed surfaces are usually considerably lower than those for undisturbed surfaces. Susceptible areas have bare, loose, and mobile sediments containing substantial amounts of sand and silt, but little clay. Humans influence dust production chiefly by breaking surface crusts and removing vegetation. Dust emissions increase in association with overgrazing, fuel wood reduction, off-road vehicle use, the development of transportation, communication, and power-line corridors, construction-site activity, and the cultivation of soils. The highest dust emissions in the USA occur where cultivation and other intensive land uses disrupt the soil. Abandoned or fallow fields are an important cause of dust-related traffic accidents in Arizona and California. In Australia, accelerated erosion caused by land-use changes following European colonization has degraded soils (McTainsh & Leys 1993). Similarly, cultivation of the loess terrain of China is blamed for much of the intense dust-storm activity in this region (Xu 2006), although Xuan and Sokolik (2002) suggest that dust storms on the Loess Plateau are of relatively low frequency. It is unclear whether the increasing trend in dust export from North Africa is related to human-made climatic change or to growing Sahelian desertification due to human pressure on the land (Prospero 1996). However, it is important to note that the largest and most persistent dust-source areas in Africa are located in regions that are largely uninhabited (Middleton & Goudie 2001; Prospero et al. 2002).

Dune destabilization by human activities also increases the dust load in the atmosphere. Stabilized dunes may contain up to 15–30% clay and silt, resulting from *in situ* weathering and atmospheric inputs of dust (Pye & Tsoar 1987). Many areas of the world have experienced reactivation of formerly stable dunes by human and climatic influences, thus increasing dust output (Middleton 1997). During

FIG. 10.14 The same location shown in Fig. 10.13. Beneath the thin crust is a fine, loose dust that is easily disturbed, in this case, by kicking the surface. Such crusts are easily disturbed by livestock or off-road traffic.

high-wind events, long plumes of dust blow along some stretches of the Mojave River, California, as a result of dune destabilization caused by devegetation and the use of the riverbed as a playground for off-road vehicles.

Lakebeds, exposed when water levels decline owing to the diversion of tributary streams for agricultural or urban use, are also important sources of dust. Well-documented examples include the fine and alkaline dusts derived from Owens and Mono Lakes, California, and the desiccated regions of the Aral Sea. Issues related to anthropogenic sources of dust are discussed further in Chapter 13.

10.5.4 CLIMATIC EVENTS ASSOCIATED WITH BLOWING DUST: SCALES OF ACTIVITY

Dust-generating winds occur at several scales. Synoptic-scale weather systems associated with the pressure gradient may cause high winds and dust storms that last for hours on end. These occur principally in higher-latitude semiarid lands. The passage of cold fronts associated with mid-latitude depressions is probably the most important cause of dust storms (Pye 1987). At the convective scale, thunderstorm downdrafts spread horizontally outward as shallow fast-moving density currents with strong winds. The turbulent conditions raise impressive dust storms termed *haboobs*, whose duration is typically an hour or less (Idso 1967; Brazel & Nickling 1986). Haboobs and frontal systems can lift dust into higher-level jet streams.

At a local scale are dust devils, rotating updrafts of buoyant air that form over strongly heated surfaces during fair-weather daytime conditions. Dust devils tend to be most common during the hottest time of the day when the lapse rates in the lower troposphere are most unstable (super-adiabatic conditions) (Ferri et al. 2003). They are usually tens of meters in diameter (up to 150 m), rise to about 100 m, and have maximum horizontal velocities of approximately 25 m s^{-1} (Greeley et al. 2003). Unlike tornadoes, there is no preferred direction of dust-devil rotation. Both the lifetimes and path lengths of dust devils are short. They typically last only minutes, with smaller dust devils having shorter durations than large ones. Although dust devils raise only a small proportion of the total dust on Earth, they appear to play a more significant role on Mars (Greeley & Iversen 1985; Balme et al. 2003; Ferri et al. 2003; Greeley et al. 2003). Martian dust devils

appear on average to be larger than their terrestrial counterparts, ranging from 15 m across and 350 m high, to several kilometers across and up to 8 km high. Their tracks are superimposed on the Martian terrain as linear, meandering, or looping features that probably result as the surface is swept free of fine particles (Balme et al. 2003).

10.5.5 FREQUENCY OF BLOWING DUST: INTERANNUAL, SEASONAL, DIURNAL

The frequency of dust events varies interannually, seasonally, and diurnally in response to climatic, surface, and anthropogenic conditions. The climatic causes of interannual variations are not fully understood, but may be related to changes in the activity of latitudinal versus meridional circulation. For northern China, a plot of the frequency of dust storms since AD 300 suggests an increase in frequency after AD 1100, and five periods of major dust storms, most fewer than 100 years in duration, up until the late nineteenth century. The frequency of strong dust storms has generally increased since the 1950s (Shi et al. 2004), owing to a warmer and drier climate and to increased human pressures on the land. By contrast, in central Asia, there appears to have been a decline in overall dust activity in the period since 1980–5 (Orlovsky et al. 2005). The large interannual changes in the quantities of dust emitted from Africa appear to be tied to rainfall, with larger quantities being emitted during droughts than during periods of more normal precipitation (Fig. 10.15) (Nicholson et al. 1998; Prospero & Lamb 2003). During the peak of the North African drought (1983–4), dust transport was the greatest observed since measurements began in 1965 (Prospero et al. 2002). The number of dusty days in Gao (Mali) increased from about 20 to more than 250 during the Sahelian droughts of 1972–3 and 1982–4 (N'Tchayi et al. 1994).

Most regions report seasonality in dust-storm activity. In the Mojave Desert, late-winter and spring months (February–May) are the time of greatest dust activity (Bach et al. 1996). These events characteristically occur during passing cold fronts, associated with westerly to northwesterly winds. Twenty percent of dust storms occur in April, a period when high winds coincide with grounds that have dried after the winter rains. In China, 53% of dust events occur in the spring, 26% in summer, 15% in winter, and 6% in fall (Wang et al. 2004b). There are distinct regional variations in seasonality: spring and winter dust

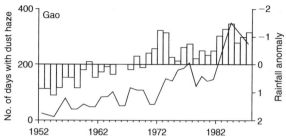

Fig. 10.15 Frequency of dust occurrence (the number of days with dust haze; left-hand axis, line) at Gao, Mali (1952–87) related to Sahelian rainfall anomalies (right-hand axis, bars). The rainfall on the right axis is shown as an anomaly, expressed as the regionally averaged, standardized departure from the long-term mean divided by the standard departure. Note that dust loading increases during periods of drought (a negative departure). From Nicholson, S.E. et al. (1998) Desertification, drought and surface vegetation: an example from the West African Sahel. *Bulletin of the American Meteorological Society* **79**. By permission of the American Meteorological Society.

storms are most common in northeast China; in the Taklimakan they occur mainly in spring and summer; whereas for the Tibetan Plateau, Hexi corridor, and west Inner Mongolian Plateau regions they occur in spring, summer, and winter. Most, but not all, Chinese dust storms are associated with cyclonic activity and cold air outbreaks. Dust outbreaks in Turkmenistan also peak in the spring, related to high winds associated with energetic spring cyclone activity, surface desiccation as temperatures rise, and low precipitation values (Orlovsky et al. 2005).

Other factors influencing dust-storm frequency include elevation and surface materials. In Turkmenistan, dust frequency increases from once in 10 years at an elevation of 2000 m, to 100 events in 10 years in piedmont valleys (elevations less than 600 m). The surfaces generating the highest number of events are sandy deserts (for example, 62 days a year in the central Karakum Desert). High wind speeds, often related to topographic enhancement, and overgrazing also contribute to dust frequency (Orlovsky et al. 2005).

Dust is also marked by diurnal variability. In many, but not all areas, the most intense dust-raising activity occurs during the afternoon hours when solar heating is greatest and the turbulent boundary layer is present (Gomes et al. 1990; Middleton 1997). Figure 3.10 shows a typical diurnal wind profile for a period of strong winds in the central Mojave Desert

of California. Winds are strong in the early afternoon through early evening, and diminish to near-calm conditions by early morning, with dust emissions mirroring this pattern. Orlovsky et al. (2005) report dust storms reaching a daylight peak between 1100 and 1500 hours in Turkmenistan. For other regions, the pattern is quite different. In the Bodélé Depression of the Sahara, the diurnal pattern is pronounced, reaching a critical velocity in the early morning and weakening towards evening (Koren et al. 2006).

10.5.6 DUST-SOURCE AREAS

Global-scale studies using remote sensing techniques indicate that the most active dust-source regions are the lower zones of inland drainage basins, particularly where the floodplains of inland-draining river systems and dune fields merge (McTainsh 1989; Prospero et al. 2002). The best geomorphic environments for dust appear to be those that: (1) have a relatively recent history of aridity, (2) are located in zones where fluvial processes remain active, (3) have deep deposits of Quaternary alluvium, largely deposited during pluvials (for example, the Bodélé Depression in the Lake Chad Basin; Figs 10.1 and 10.9), and (4) have strong topographic relief (Prospero et al. 2002). Old arid terrains, characteristic of much of Australia, are not good sources: the Lake Eyre basin remains the most active region owing to ongoing fluvial inputs (Knight et al. 1995). In mountainous regions of Iran, Afghanistan, Pakistan, and China, closed intermountain basins are persistent dust sources. These topographic lows are regions where fine sediments are transported and ephemeral lakes form, thus providing an easily eroded source of particles. However, some basins are more active sources than others. Gross changes in lake hydrology, related in part to the frequency and extent of inundation, can lead to large variations in regional dust emissions (Mahowald et al. 2003).

Within a given topographic basin, small regions ("hot spots") are likely to contribute the majority of airborne dust, despite their relatively small size. Owing to complex chemical and physical changes to the surface that are not yet fully understood, dust storms in vegetated terrains commonly display distinct plumes that emanate from relatively small areas (Gillette 1999).

At a global scale, the largest and most persistent sources of dust are located in the Northern Hemisphere, in a broad belt extending from the west coast of North Africa, through the Middle East, through Central Asia to China. Southern-Hemisphere sources are less significant, but desert emissions and deposits are well documented for Australia (Hesse & McTainsh 2003), and loess is recorded in the drylands of South America (Zárate 2003), although this area is not an important modern source of dust.

North Africa is considered to be the largest source of mineral dust in the world, contributing at least 1 billion t of dust per year to the global atmosphere (d'Almeida 1986). This enormous quantity of dust indicates that aeolian deflation and abrasion play a key role in shaping the landscape of this region (Goudie & Middleton 2001). Two important source areas are the Bodélé Depression, the most intense dust-source region in the world (Goudie & Middleton 2001; Koren et al. 2006), and a region in the western Sahara that includes eastern Mauritania, western Mali, and southern Algeria (Middleton & Goudie 2001) (see Table 10.1). For these regions, natural factors are more significant than anthropogenic activities in generating dust.

Maximum North African dust transport occurs during the summer months although there is also considerable movement during the winter. In summer, large dust outbreaks occur in association with strong convective disturbances in West Africa at about 15–20°N, with the source regions being the Sahara and sub-Sahara (Sahel). There is a seasonal shift in the locus of dust-storm activity (Prospero et al. 2002).

The movement of Saharan dust throughout the global atmosphere has received considerable attention. During the summer (approximately May–November) trade winds carry Saharan dust towards the Caribbean and North America. Satellite imagery shows that it takes about 1 week for the dust to cross the north Atlantic Ocean and reach the Caribbean. The transportation of dust from the Sahara over the North Atlantic is extremely strong, with dust concentrations ranging as high as 5–80 $\mu g\ m^{-3}$ (Savoie & Prospero 1977). The transportation mostly takes place above the trade-wind inversion layer. In winter (around December–April), the dust flow moves towards South America. Dust leaving the Sahara also travels northwards into Europe, impacting air quality; and northeastward into the Middle East. The cycle of Saharan dust lifting, transport, and deposition is episodic in nature, affected by changes in storm tracks, changes in atmospheric vertical stability, and in the precipitation regime. Falls of rain containing dust in the Mediterranean area are often associated with southerly winds (Sirocco, Khamsin) that originate in desert depressions, inside which convective mesoscale systems collect the dust and inject them into the synoptic scale systems. Falls of rain and dust are also recorded in regions to the north of the Italian Alps, although with less frequency (Conte et al. 1996).

In southern Africa, recent research has drawn attention to the importance of the Etosha and Makadikai Pans to regional dust loadings (Bryant 2003). The Etosha Pan has a surface area of 5000–6000 km^2, lies at an elevation of approximately 1080 m ASL, and is underlain by up to 450 m of sediments. Dust emissions are comparable to those of other large Southern-Hemisphere sources, but significantly lower than those of North African sources (Bryant 2003).

The Arabian Peninsula is a globally important region of intense dust-storm activity (Fig. 10.16). Dust storms are most common in the spring and summer months, and a dust haze is commonly observed over the Arabian Sea during the summer months (Prospero 1981; Middleton 1986). The area of greatest entrainment is the Lower Mesopotamian Plains. Material may be transported for great distances, particularly into the Indian Ocean. In addition to local dust sources, fine materials are transported into the region from the Sahara and from northwest India and Pakistan (Middleton 1986).

Dust storms over India are common, with a maximum frequency during the premonsoon season (March through May). The primary source regions appear to be eastern Pakistan and the Thar Desert of western India (Badarinath et al. 2007).

Northern China is the second largest source of atmospheric dust in the world (Xuan & Sokolik 2002). Wind erosion occurs in arid, semiarid, and subhumid regions, where it is estimated that close to 17% of the land surface generates dust (Shi et al. 2004). Dust storms originate from the Hexi Corridor and western Inner Mongolia Plateau, the Taklimakan Desert, and the central Inner Mongolia Plateau (Wang et al. 2004b). There is a substantial transport of mineral aerosol from Asia to the Central Pacific, with a pronounced spring peak. Asian dust storms show a great deal of interannual variation, which may be related to large-scale climate phenomena. Hara et al. (2006) suggest that the Asian dust-transport path may correlate with the ENSO, the center of mass moving southward in La Niña years and northward in El Niño years.

In the USA, maximum dust production occurs in the spring, and minimum production in the summer.

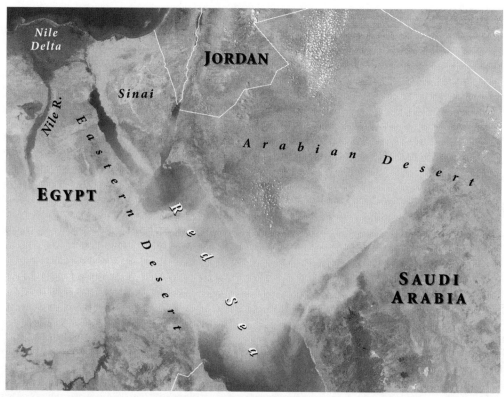

FIG. 10.16 The Arabian Peninsula is one of the world's major sources of dust. In this image, a thick plume of dust crosses Saudi Arabia, the Red Sea, and Egypt, forming an opaque, south-moving front. Source: NASA Aqua/MODI. See Plate 10.16 for a color version of this image.

The most productive source area is in and around the panhandles of Texas and Oklahoma, where land-management practices have a strong influence on the amount of dust emitted (Gillette & Hanson 1989; Lee et al. 1993). During the 1930s, the Southern High Plains formed part of the infamous Dust Bowl.

In Australia, large quantities of dust were entrained in Quaternary dust storms during the formation of the Australian dune fields. This material formed soils (parna) in much of southeastern Australia (Butler 1956), and contributed to marine sediments in the Tasman Sea, off the southeast Australian coast, and the Indian Ocean, adjacent to northwestern Australia. These deposits result from two general Australian dust paths, associated with the westerly winds in the south and the easterly trade winds in the north (McTainsh 1989). The most active modern dust-source areas lie in the Lake Eyre Basin and the western sector of the Murray-Darling Basin (Hesse & McTainsh 2003). Large events can reach New Zealand, tingeing the snow red (Knight et al. 1995), and half way to South America.

11

PLANT COMMUNITIES AND THEIR GEOMORPHIC IMPACTS

11.1 INTRODUCTION: CHARACTERISTICS OF DESERT ECOSYSTEMS

Arid-land ecosystems are characterized by extremes of temperature and wind velocity, spatial and temporal variations in rainfall, and low plant and animal productivity. Whereas temperature, solar radiation, and soil nutrient inputs are relatively predictable from year to year, precipitation varies greatly in time and space, yet it is following rainfall that the life of both plants and animals is activated (Gupta 1979). In addition to the quantity of rainfall, the seasonal timing is often critical for growth and reproduction (Mourelle & Ezcurra 1996). Episodic precipitation often occurs in high-intensity rain pulses during which there is little opportunity for infiltration. Rain days that measure more than 0.1 mm are rare, with intervals of days to years between rain pulses. The effective rainfall (that which is available for plant growth) is a function not only of precipitation amount and intensity, but also ambient temperature, relative humidity, wind speed, and soil infiltration rate. Thus for plants, the only reliable sources of water may be dew, fog, and groundwater at shallow depths.

Floral and faunal communities vary not only with substrate, but with altitude, and thus a diverse landscape is apparent, particularly in regions of strong relief. The hydrologic properties of the surface often differ greatly over short distances, and this becomes apparent after the first flush of ephemeral vegetation appears following widespread storms. In 2004–5, heavy rains brought a carpet of blooms to Death Valley, but the flowers were not evenly distributed (Figs 1.5 and 11.1). Higher-altitude areas bloomed first and more vigorously; bedrock slopes with a cover of blown sand retained moisture better than mature desert pavements or saline clay slopes, and blooms were intense in the former and absent in the latter; and washes that had flooded during the August 2004 summer rains had lost their seed banks and were barren, but stream beds that had not recently flowed bloomed vigorously.

Although it is often suggested that deserts are relatively simple ecosystems marked by low diversity and the absence of strong biological controls and feedback, recent research suggests that deserts are relatively complex and biologically rich (Polis 1991). Deserts support a surprisingly diverse fauna and flora including almost every taxon of terrestrial plant and animal. Even some mesic taxa (algae, mosses, ferns, isopods, and a few species of salamanders and amphibians) occur in deserts (Fig. 4.4). However, foliage height diversity is considerably less in deserts than in forests, providing fewer spatial niches for birds and other species. There is a general decrease in plant and animal species as the environment becomes more arid, although this has not been documented quantitatively. On a local scale, plant-species richness is highest in depressions or areas that receive runon from adjacent areas (Noy-Meir 1985; Dunkerley 2002; Ludwig et al. 2005).

The paucity of shared taxa and the disparity in dominant life forms between deserts strongly suggest that the floras have evolved largely in isolation. Nonetheless, there are certain characteristics common to all desert vegetation. The first is scarcity, with a closed cover seldom maintained, and the percentage of cover becoming more open and discontinuous with increasing aridity. Deserts have low biomass, the total amount of living plant material above and below ground. The wormwood and saltwood deserts of Central Asia have a biomass 100 times less than an equivalent area of temperate forest. The second characteristic is the strong seasonality of the vegetation. Plant productivity is extremely variable in both time and space, ranging from zero to several hundred grams per square meter (Noy-Meir 1973). It is strongly correlated with rainfall, which is the main input for plant growth. Water is thus a species-limiting and productivity-limiting factor.

Fig. 11.1 Heavy rains brought a carpet of blooms to Death Valley in the winter of 2004–5. Plant productivity in deserts is extremely variable in both time and space and is strongly correlated with rainfall.

Deserts exhibit spatial heterogeneity on several scales. Local edaphic factors, such as differences in mineral composition, nutrient reserve, organic content, and capacity to hold water, affect the distribution and abundance of plants. On a larger scale, rain resulting from thunderstorms results in a patchy distribution of moisture, and runoff contributes to a mosaic pattern of soil moisture and subsequent productivity. Topographical differences translate into highly heterogeneous desert habitats, including well-watered riparian communities of high species abundance and diversity. Noy-Meir (1985) suggests that plant cover and production may be greater in areas where hard substrates and steep topography cause increased runoff (Fig. 11.2), concentrating flow in certain habitats, than in landscapes where water infiltrates more uniformly, such as sandy plains.

Water falling on limestone and dolomite outcrops in the Middle East runs off to infiltrate into pockets of soil and rock crevices where plants germinate and grow. These microhabitats show little seasonal change in their moisture content, allowing them to harbor plant species which otherwise would require more humid conditions (Orshan 1986). By contrast, depressions that accumulate saline or alkaline runoff support almost no life. Slope aspect influences the amount of solar radiation received and soil moisture, causing differences in plant communities and percentage cover (Parker 1991; Gutiérrez et al. 2000).

Precipitation and runoff may vary along geographic gradients or clines. In the Namib Desert, for example, rainfall increases, but water from fog condensation decreases, eastward and inland from the Atlantic coastline. In the southwestern USA, rain changes from predominantly frontal winter systems in the Mojave Desert, California, to dominantly summer convective rains to the east in the Sonoran and Chihuahuan Deserts. Predictability of rainfall may also vary along these clines.

11.2 ADAPTATIONS TO DESERT CONDITIONS

Deserts are biologically stressful environments. Plants have acquired two principal strategies, avoidance and tolerance, that allow them to cope with heat and drought to ensure that neither internal temperatures nor tissue dehydration attain deadly levels. The majority of the flora in deserts are evaders, surviving stressful periods by living permanently or temporarily in cooler and/or moister microhabitats.

The sparseness of vegetation in deserts contrasts with more humid regions, evident in satellite images of the gross primary productivity (grams of carbon per meter squared). Furthermore, the structure of desert plant communities is different. In the vertically structured tropical rainforest biome, for example, there is intense competition for light. By contrast, desert plants compete for water, and are distributed in ways that maximize this limited resource, existing either as widely spaced individuals or being distributed in vegetation patches or bands that garner runoff from intervening bare patches.

11.2.1 ADAPTATIONS TO TEMPERATURE

Desert plants are able to function at higher temperatures than those in more mesic environments.

Fig. 11.2 In areas of hard substrates and steep topography, runoff is concentrated. Plants often take advantage of moister microhabitats such as pockets of soil in rock joints. The MacDonnell Ranges cycad (*Macrozamia macdonnellii*) grows only in the central Australian ranges. It grows mainly on rocky slopes where its roots can tap deep into moisture-trapping rock crevices. The plant is slow growing and some individuals may be up to 300 years old. Photograph courtesy of Lloyd Laity.

Herbaceous plants can tolerate temperatures of 50–55°C, some cacti 65°C, and many crustose lichens 70°C or more (Tivy 1993). The cactus *Melocactus peruvianis*, which grows in the rock deserts of Peru, is found on granite slabs that attain a surface temperature of between 60 and 70°C. The temperature of the central water tissue of the cactus was 45°C (Rauh 1985).

Plants have several adaptations to reduce their temperature. Some species change their leaf orientation to minimize the surface area or the time to which they are exposed to maximum heat input. The light coloration of many desert plants also acts to maximize the reflection of light (Fig. 11.3) (Wendler & Eaton 1983), as does the surface growth of hairs and spines that absorb or reflect radiation and create a thin boundary layer of air to insulate the underlying surface. Cooling by transpiration is probably most effective in cacti and small-leaved shrubs owing to their relatively large surface area-to-volume ratio. An increase in stem diameter helps to ameliorate daytime temperature maxima for some columnar cacti, such as *Lophocereus schotii* (Nobel 1980).

For most species, the availability and seasonal distribution of water is the most important factor governing plant growth and distribution in deserts, with temperature causing only a secondary effect (Noy-

Meir 1973; Latorre et al. 2002). Production is rarely affected by temperatures that are too high, as long as moisture is available. By contrast, temperatures that are too cool limit growth. In high-altitude or high-latitude deserts, such as parts of central Asia or the Great Basin Desert, there may be no root or shoot growth during the colder months, even with abundant moisture from rain or snow (Noy-Meir 1973).

Freezing temperatures limit the range of species, and this factor is very important for the columnar cacti. The latitudinal and altitudinal distribution, species richness, and morphology of columnar cacti are strongly related to the incidence of freezing temperatures. In Argentina, the strongest predictor for columnar growth form and species richness is the number of frost-free days (Fig. 11.4) (Mourelle & Ezcurra 1996), suggesting the high sensitivity to freezing of this growth form. For the Sonoran Desert, Shreve (1911) suggests that the saguaro, *Carnegiea gigantea*, is limited to those latitudes and altitudes where freezing temperatures of the tissues do not persist for 24 hours. Furthermore, this cactus is observed to increase its diameter from the southern to the northern part of its range, in apparent response to the decreasing minimum temperatures. This strategy raises the minimum nocturnal temperature at

FIG. 11.3 Plants have several adaptations to reduce their temperature in desert environments. The light color of many plants acts to maximize the reflection of light. Desert holly (*Artiplex hymenelytra*) is a small shrub whose leaves are pale green to nearly white. It is found in dry alkaline washes in the southwestern USA and parts of Mexico.

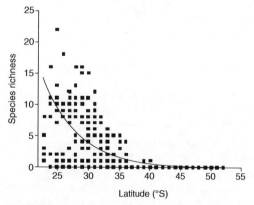

FIG. 11.4 Relationship between species richness of columnar cacti and latitude for the Southern Hemisphere (Argentina). From Mourelle, C. and Ezcurra, E. (1996) Species richness of Argentine cacti: a test of biogeographic hypotheses. *Journal of Vegetation Science* **7**. By permission of Opulus Press.

the stem apex, helping the plant to avoid freezing injury (Nobel 1980). However, it may also be a response to the increase in annual rainfall, allowing more water storage per unit surface area, and allow-

ing a greater period of stomatal activity. Not all species of columnar cacti have the same thermal adaptations, and other factors, such as long spines, affect the temperature regulation of plants, appearing to provide some thermal insulation (Nobel 1980; Mourelle & Ezcurra 1996).

11.2.2 WATER USE BY PLANTS

Over the course of a year, temperature, radiation, and nutrient inputs vary in a continuous manner, whereas precipitation comes in discontinuous packages. Water is vital for plant growth as: (1) protoplasm (living material) functions only in the presence of water and most tissues die if their water content falls too low, (2) plant nutrients are dissolved in soil water, and water is the medium by which they enter the plant and move from cell to cell, (3) water is a raw material in photosynthesis, and (4) water regulates the temperature of the plant by its ability to absorb heat and because transpiration helps to lower plant temperatures.

Water availability governs productivity and limits the distribution and abundance of many species, thus restricting membership in desert communities. Low water availability results in rates of primary productivity that are one to three orders of magnitude below those of other habitats (Louw & Seely 1982; Ludwig 1987). Under dry conditions, plants tend to grow little, if at all, whereas with adequate to heavy rains, relatively luxuriant growth takes place (Fig. 11.1). Thus, productivity tends to mirror precipitation. When heavy rains followed a 13 year dry period in the Namib, standing biomass increased by 600% (Seely & Louw 1980).

Owing to precipitation variability and to differences in soil and rock type, there are profound seasonal and local differences in the availability of moisture for plants. Wet periods are relatively short, and during this time evapotranspiration approaches its potential and is not limited by soil water availability. To cope with dry periods, desert plants have evolved several strategies to reduce water loss. Diurnal stomatal activity can be regulated. Many plants have xeromorphic adaptations such as thick cuticles, sunken stomata, and surface hairs. Succulents (cacti and euphorbia) have well-developed storage tissues, low cuticular transpiration, and rapid stomatal closure. The roots of some desert plants can exert a greater suction pressure (over 100 bar; 10,000 kPa) than others, removing water from the soil more effectively than mesic plants. Some plants can rapidly produce new roots after rainfall. Others, such as the creosote bush and cacti have shallow, lateral roots that can exploit rainstorms that only wet the upper surfaces of the soil (Fig. 11.5).

Plants may be classified as xerophytes, mesophytes, or phreatophytes according to their water requirements. A *xerophyte* ("dry plant") can survive and reproduce in an environment in which water is limited. Xerophytes dominate desert environments. They have a number of metabolic and morphological responses to water scarcity, including both drought avoidance and tolerance. Adaptations include the presence of deep roots (for example, the taproots of *Welwitschia mirabilis* of the Namib Desert), the ability to store water (such as in cacti and succulents), the ability to use fog and dew as water sources (as in the *lomas* vegetation of South America; Rauh 1985), and the ability to complete the life cycle in a short period of time when rains are available (ephemeral plants). *Fluctuating* xerophytes reduce photosynthetically

Fig. 11.5 Succulents adapt to dry periods by having well-developed storage tissues, low cuticular transpiration, and rapid stomatal closure. The teddy-bear cholla (*Opuntia bigelovii*) is found on well-drained slopes and alluvial fans at elevations below 1065 m. The shallow surface roots of cacti are well adapted to exploit rain events that only wet the upper surface of the soil. The teddy-bear or jumping cholla has long, thin spines that are easily attached to flesh or clothing and the branch segments readily detach from the plant. As the plant does not reproduce well by seed, it relies on these segments being dispersed and eventually rooting. Photograph location: Joshua Tree National Monument, USA.

active biomass and transpiring surfaces by having
deciduous leaves and stems or producing smaller
leaves and stems during drought. As a result, residual
activity is maintained, but water and energy losses
are low, and the plant survives on minimal soil moisture. Ocotillos (*Fouquieria splendens*) shed their leaves
to reduce transpiration, but continue stem photosynthesis, enabling the plant to leaf out quickly after
precipitation (Fig. 11.6). Depending on the rains, a
plant may gain and lose its leaves six or more times
a year (Bowers 1993). *Stationary* xerophytes maintain a nearly constant biomass. The creosote bush
(*Larrea tridentata*) can survive severe dehydration
and live up to a year without rain. Succulents and
cacti typically have shallow root systems, and thus
have little access to external moisture once the upper
layers of the soil desiccate. They rely on their internal
water reserves.

A *mesophyte* is a plant that flourishes under conditions that are neither very wet nor very dry. Such
plants have limited distribution in the deserts, being
largely confined to areas near streambeds. The *phreatophyte* ("well plant") lifts water from the zone of
saturation and sends roots down to the capillary
fringe immediately overlying the water table or to
the water table (Meinzer 1927). The plant is thus
able to obtain a secure supply of water (Fig. 11.7).

11.2.3 Reproduction

There are numerous strategies exhibited by desert
plants to ensure successful reproduction. Reproduction tends to be synchronized with favorable times,
and suppressed during periods of drought. Seed
dissemination may be delayed until after several
days of rains. To ensure that there are always
some seeds in reserve, germination may be spread
over several years. Some seeds will not germinate
until a critical amount of water is present in the
soil. Others require severe mechanical abrasion, as
occurs during transport by flash floods, in order to
stimulate germination. Some employ hygrochasy,
the impedance of long-distance seed dispersal, ensur-
ing that seeds remain within the same habitat as
the parent.

Whereas seeds are the only mechanism for the
survival of annuals, the succession of perennials
depends mainly on the vegetative state and seeds are
only a subsidiary method of reproduction. Although
the seed bank is rapidly replenished with ephemeral
seeds following heavy rainfall, studies in western
South America suggest that perennial seeds take up
to 1 year longer (Jacsik 2001). The mass germination
and emergence of perennial plants usually requires
a significant rainy period. *Agave deserti* germinated in

Fig. 11.6 Fluctuating xerophytes reduce photosynthetically active biomass and transpiring surfaces during
drought, maintaining residual activity, but minimizing water losses. The ocotillo (*Fouquieria splendens*) grows up
to 6 m in height and has cane-like, spiny, mostly unbranched stems arising from the ground. It sheds its leaves
and flowers during drought periods, but chlorophyll in the bark allows the plant to continue to photosynthesize.

Fig. 11.7 Phreatophytes secure water from the zone of saturation in the soil. This vegetation is limited in distribution in deserts, being largely confined to areas near stream channels, along faults, or on the margins of lakes. This 1917 photograph of the Mojave River shows a river dominated by broadleaf mesophytic trees such as *Populus* (cottonwoods) and *Salix* (willows). In the American southwest, much of this riparian ecosystem, including that shown in this photograph, has diminished in extent owing to groundwater pumping and stream diversion. Photograph: Old Camp Cady area, Mojave River, David G. Thompson, Photograph 15 from the USGS Photographic Library, Denver, Colorado.

only 1 of 17 years in the northwestern Sonoran Desert, requiring above-normal winter/spring rains (Jordan & Nobel 1979). *F. splendens* germinated in 3 of 8 years in the northern Sonoran Desert, following heavy and unusually early summer rains (Bowers 1994). Shrub and cactus recruitment in the Sonoran Desert often appears to be related to El Niño years (Barbour & Diaz 1973; Bowers 1997).

Recently emerged seedlings often suffer high mortality as a result of seasonal drought and predation (Bowers 1996). Many young plants develop under nurse shrubs, which provide shade and a moister environment. Adult trees and shrubs are generally less vulnerable to desiccation and herbivory than seedlings and, once established, can persist for a long time and provide a microclimate for subsequent seedlings (Holmgren et al. 2001). Grazing of palatable shrubs that serve as nurse plants for cacti in the Sonoran Desert has been shown to cause disequilibrium in the age structure of cacti populations. On ungrazed sites, young plants are able to replace old ones as they die, and the age subpopulations are more or less in equilibrium (Bowers 1997). For barrel and columnar cacti, it is essential that water is available during the seedling stage, until the plant can reach a sufficient size to attain a lower surface area-to-volume relationship (Mourelle & Ezcurra 1996).

11.2.4 Nutrient cycling

The decomposition of dead organic matter occurs as a result of physical and biological processes. In deserts, microbial activity by bacteria and fungi is less important in decomposition than in temperate forests, as such activity is limited to the short periods when the soil is moist. Instead, physical fragmentation and detritivore consumption remove most plant materials (Noy-Meir 1985). Wind and runoff act to break down and transport dead organic matter, with the finely comminuted material accumulating around the base of shrubs and in washes and surface depressions. Additionally, invertebrate detrivores (macrodecomposers) such as termites, ants, mites, and isopods are relatively common in deserts.

Nutrients tend to be concentrated around shrubs and in the uppermost soil layer. Elsewhere, soil fertility is low (Noy-Meir 1985). The terms "fertility island" or "island of fertility" have been used to describe the high biological activity surrounding the shrub microenvironment (Schlesinger & Pilmanis 1998). Schlesinger et al. (1996) examined soil fertility under *Larrea tridentata* in the Mojave Desert, California, where individual plants of great age have existed under dry conditions for thousands of years. Nitrogen, phosphate, chloride, and potassium were found to be more concentrated under shrubs than in shrub interspaces. Soil erosion can cause a significant loss of these nutrients.

By concentrating the biogeochemical cycling, shrubs concentrate the biodiversity of animals, including those at higher trophic levels, such as birds and lizards. The intershrub spaces are relatively devoid of biotic activity (Schlesinger & Pilmanis 1998).

11.2.5 SALT ADAPTATION

Desert soils, particularly those that are fine- or medium-textured, may be saline in nature, characterized by accumulations of soluble salts such as sodium chloride, sulphates, and carbonates. Where leaching is incomplete, salts that are washed out of the upper soil may be deposited at about 10–100 cm below the surface (Noy-Meir 1973). If the groundwater table is near the surface, the capillary fringe can translocate salts near to or onto the ground surface. The salt tolerance of plants is often quite variable and species restricted to saline soils are rare (Caldwell 1974).

The osmotic potential regulates water movement through the walls of living organisms, such as plant roots. Where the soil has a high concentration of salts, the uptake of water by plants and germinating seeds is limited. Roots exposed to saline water may not be able to absorb water. Even though the soil may be saturated, if the relative concentration of water molecules in the root cells exceeds that in the salt water outside, selective diffusion tends to cause water to move from the roots into the soil (water moves from areas of low solute concentration to areas of high solute concentration: the purer water "tries to dilute" the impure water by moving into the stronger solution). Thus, many soils that are high in salts are devoid of plant life.

The halophytes are a group of salt-tolerant plants, growing in highly saline soils and on the edges of saline lakes, which have devised a number of mechanisms to cope with salinity. Cellular ion concentrations in plants like *Salicornia* may be reduced by increased water uptake by cells, resulting in succulence, which is associated with an increased thickness of plant organs (Fig. 11.8). The greater cell water content per cell surface area dilutes electrolyte concentrations in the cell protoplasm (Caldwell 1974). Other plants have developed the ability to avoid salt, either by the process of exclusion, whereby they block from entry ions that are toxic in high

FIG. 11.8 Halophytes are salt-tolerant plants that grow in highly saline soil or along the margins of saline lakes. This photograph illustrates the iodine bush (*Allenrolfea occidentalis*), a shrub with round, succulent, grey-green jointed stems that is extremely salt tolerant. Note the saline soil adjacent to the plant. Location: Death Valley, California, USA.

concentrations, or else by excretion of toxic salts through glandular cells when concentrations become high. Salt-excreting glands have been described in several halophytic genera, including *Tamarix* and *Atriplex* (Caldwell 1974). Another adaptation is to evade salt by regulating the plant's life cycle to ensure that germination and vegetative growth take place during the rainy season, when the salt hazard is less extreme.

11.3 PLANT COMMUNITIES AND ECOTONES

A community refers to different populations of organisms that live and interact with each other within a prescribed area. In North America, for example, there are four principal desert regions that can be separated on a floristic basis. The warm deserts, the Chihuahuan and Sonoran, have a 52% similarity in genera, but share only 25% of genera with the cold Great Basin Desert. The boundary between cold and warm deserts lies across southern Nevada and Utah, and is a broad zone of transition in which the dominant *Artemesia* shrub communities of the cold deserts are replaced by the *Larrea*-dominated shrub communities of the warm deserts. However, other important shrubs, such as shadscale (*Atriplex confertifolia*) extend across this boundary and integrade into both desert types (Smith et al. 1997). An ecotone is a region of ecological tension in which one vegetation type, for example the steppe climate of Eastern Europe, is replaced by another type, the desert climate of Middle Asia. The ecotone is the semidesert, where both vegetation types occur side-by-side under the same macro-climatic regime. Micro-climatic conditions, or dissimilar edaphic conditions at different locations, give the advantage to one vegetation type or the other.

There are several classifications of desert vegetation. Evenari (1985b) focused on functional traits, subdividing plants into poikilohydrous (which survive air drying) and homoiohydrous (which require relatively constant levels of tissue hydration) types. Homoiohydrous plants were subdivided into arido-passive (which survive drought with metabolically inactive plant parts) or arido-active (which remain metabolically active during drought). Shantz (1927) categorized plants as drought-escaping (annuals), drought-evading (succulent tissues or water derived from groundwater), or drought-resisting (metabolic adaptations that allow cellular function to be maintained under extreme stress). Not all plants fit neatly

into these categories. For example, evergreens in deserts remain metabolically active during seasonal drought (drought-resistors), but may become deciduous during extended droughts (drought-endurers). Other classifications emphasize physiological ecology, a process-based approach in which water relations, photosynthetic gas exchange, and stress physiology are examined (Smith et al. 1997). The primary growth forms of desert plants, including evergreen shrubs, deciduous shrubs, crassulacean acid metabolism (CAM) succulents, perennial grasses, phreatophytes, annuals, and poikilohydric plants (Smith et al. 1997) are considered in the following sections, which, owing to space constraints, provide examples of principally North American species.

11.3.1 EVERGREEN SHRUBS

Evergreen shrubs obtain all of their water from the unsaturated zone and maintain at least a partial leaf canopy during climatic drought. They are classified into three broad types: (1) sclerophyllous shrubs (with small tough leathery leaves and a thick waxy cuticle that reduces water loss), (2) leafless shrubs with green stems, and (3) xerohalophytes with green stems (Gupta 1979). The aboveground strategies of such plants include sclerophyllous leaf structure, control of cuticular and stomatal transpiration, high tolerance of heat and water stress, and high water-use efficiency. Below ground, many have a well-developed root system in which surface lateral roots extend well beyond the plant canopy to utilize near-surface soil moisture derived from typical rain events, while a central taproot extends to deeper depths to exploit soil moisture derived from major rainfall events (Drew 1979). In North America, the most common examples are big sagebrush (*Artemisia tridentata*) and creosote bush (*Larrea tridentata*), species that are often dominant through much of their range. Other notable examples include the Joshua tree (*Yucca brevifolia*), and leafless shrubs with green stems, such as *Ephedra* spp. (Mormon tea).

The genus *Larrea* consists of five species of evergreen shrubs, four in South America and one in North America. Creosote bush is the most abundant perennial in North American deserts and can form almost pure stands (Fig. 11.9). It reaches its greatest development on alluvial fan surfaces, and prefers substrates that are high in gravel content and calcareous throughout the soil profile (Hallmark & Allen 1975). On younger or periodically disturbed surfaces

Fig. 11.9 The creosote bush (*Larrea tridentata*) is a common North American species that is dominant throughout its range. A sclerophyllous shrub, it grows 1–3 m tall and is most common on the well-drained soils of bajadas and plains, where it may form almost pure stands. Within any given region, the plant spacing is quite regular and most plants within a stand appear to be of similar age.

such as washes, other species such as *Acacia*, *Ambrosia*, or *Encelia* are dominant (Smith et al. 1997).

11.3.2 DROUGHT-DECIDUOUS SHRUBS

Deciduous shrubs are usually drought-deciduous or winter-deciduous. The drought-deciduous forms are the most abundant. They are considered drought-evaders because they lose their primary photosynthetic surface in response to drought. They include both xerophytic and mesophytic vegetation. Some forms exhibit relative mesophytic behavior in the wet season, putting on leaves; and become dormant and exhibit xerophytic responses in the dry season. In the Chihuahuan and Sonoran Deserts, the shrub *Fouquieria splendens* is able to develop several crops of leaves each year in response to rain (Smith et al. 1997). Some of the desert perennials that exhibit deciduous leaf canopies in the dry season also possess photosynthetic stems and twigs. All of the aboveground parts of the palo verde trees, *Cercidium microphyllum* and *Cercidium floridum*, are photosynthetic. A leaf canopy is maintained for only a short period prior to stress-induced leaf fall. The green bark contains stomata that permit normal carbon dioxide and water exchange with the atmosphere. Chlorophyll concentrations are actually higher in the stems than in the leaves. Broadleaf mesophytes are most

common in moist microhabitats such as washes and typically lose their leaves during the summer months.

11.3.3 CAM SUCCULENTS

There are several unique aspects of the functional ecology of desert succulents. First, they use the crassulacean acid metabolism (CAM) photosynthesis pathway to conserve water by storing it within their structure and transpiring at night. Succulent plants open their stomata at nighttime and assimilate carbon dioxide, storing it as malic acid, which is converted into carbohydrates. The plants separate in time the light-requiring process of photosynthesis from carbon dioxide uptake. During daylight, the stomata are closed to retard transpiration and desiccation. Photosynthesis is conducted by the C_3 pathway, in which the carbon dioxide from the atmosphere is converted into a three-carbon molecule called 3-phosphoglyceric acid. Second, cacti, euphorbia and yuccas have a succulent morphology, characterized by extensive water-storage tissues with very large cells.

CAM succulents are not well represented in all arid regions, but reach their greatest development in semideserts and in temperate coastal deserts. In North America, cacti and agaves reach their greatest

Fig. 11.10 CAM succulents reach their greatest development in semideserts and temperate coastal deserts. The Sonoran Desert favors agaves and large columnar cacti (shown here) owing to a bimodal rainfall regime and lack of freezing temperatures. The photograph illustrates coastal vegetation near San Telmo, Baja California, Mexico.

diversity and cover in the Sonoran Desert and parts of the Chihuahuan Desert. The Sonoran Desert favors agaves and large columnar cacti with its semiarid climate, predictable summer rainfall in a bimodal rainfall regime, and lack of freezing temperatures (Fig. 11.10). Both cacti and agaves are excluded from a wide range of desert habitats by their susceptibility to freezing temperatures (Shreve 1911; Nobel 1980). The Mojave Desert is more arid and lacks summer rainfall, and the Great Basin Desert is too cold. Although cacti are not true xerophytes, they respond rapidly to rainfall events and are able to endure drought and high temperature stress. Cacti are also common in the South American deserts. In a study of species richness patterns for 223 cactus species in Argentina, Mourelle and Ezcurra (1996) found that environmental factors operating at the regional scale account for a large proportion of the geographic variation. Species richness was greatest in the northwestern areas of the country, influenced by latitude and altitude (which affects the incidence of freezing temperatures), the proportion of annual rain falling in summer, fire, and past environmental conditions.

11.3.4 Perennial grasses

Although the climates of grasslands and savannahs vary greatly, they are usually characterized by seasonal water stress, high solar radiation, and semipermanent areas of high pressure. In North America, grasslands once occupied significant portions of the warm Sonoran and Chihuahuan Deserts, but their area has decreased substantially as a result of livestock grazing. Grasses are second only to shrubs in their cover of the cold Great Basin Desert, where plant productivity is constrained by cold winters and warm, dry summers. The big sagebrush/bunchgrass communities have a long evolutionary history with recurring fire (Wright & Klemmedson 1965), burning during the late summer and early fall when the aboveground parts of the grasses have dried. Characteristics of grasses that make them tolerant of grazing include deciduous shoots and belowground nutrient reserves, small stature, and rapid photosynthesis and growth (Smith et al. 1997).

11.3.5 Phreatophytes

Phreatophytes are a group of plants that have adapted to the desert environment by the development of long taproots that penetrate into, or near to, the water table (the "phreatic surface"). They thus survive in deserts by being largely independent of soil water derived from incident precipitation. Arborescent phreatophytes have taproots that may be tens of meters in length. The mesquite (*Prosopis* spp.) of the American southwest may extend its roots to depths of 10–30 m. *Acacia* trees in Botswana have root depths of more than 50 m, and support green leaves at the end of the dry season in areas where borehole observations mark the water table at a depth of 14–30 m below the surface (De Vries et al. 2000). Maximum root development of phreatophytes is in the capillary fringe, as the soil beneath the water table is oxygen-poor.

Phreatophytes grow near stream channels or springs, along fault zones, or on the margins of lakes. They dominate perennial river systems of the southwestern American deserts, where they include broadleaf mesophytic trees such as *Populus* (cottonwoods) and *Salix* (willows) (Fig. 11.7). These are obligate phreatophyes: plants that are always in contact with the groundwater. Facultative phreatophytes are in contact with the water only seasonally, and thus will have deciduous leaves and/or more xerophytic leaves. Additionally, there are halophytic shrub taxa that are indicative of shallow, saline groundwater, such as *Artiplex* (Smith et al. 1997). Finally, there are a number of perennial herbaceous monocotyledons, including *Distichlis* and *Phragmites*, which form marshes in the bottomlands of warm deserts.

The genus *Prosopis* is particularly widespread. In the New World, mesquite has been studied extensively owing to its significance as a rangeland invader (Buffington & Herbel 1965; Schlesinger et al. 1990; Bahre & Shelton 1993; Fredrickson et al. 1998) (see Chapter 13) and its pivotal role in many desert ecosystems. *Prosopis* attains both a shrub-like form and an arborescent form, although it is unclear whether these morphological differences are the result of genetic or environmental factors. In some areas, such as the lowland Sonoran Desert, it is usually restricted to desert wash habitats; whereas in the Chihuahuan Desert it has invaded former desert grasslands and forms extensive dune fields. In addition to having a deep taproot, allowing it to take advantage of streamside shallow groundwater, the plant has extensive lateral roots, which enable it to function in the dunelands where the water table is at much greater depths.

Phreatophytic and riparian communities are highly vulnerable to hydrologic changes. Increased groundwater pumping in the USA and Mexico, as well as impoundment and stream diversion, have had significant effects on this community, reducing the availability of water and causing widespread loss (Lines 1999; Laity 2003). As early as 1928, Bryan

noted that significant changes in riparian plant associations in the arid southwest had resulted from decreases in the water table (Bryan 1928a).

Seasonal variations in streamflow are considered critical for the health of riparian communities, but are often reduced by impoundments. Periodic flooding scours the streambed of senescent vegetation, provides soil moisture for seedling development, and leaches out streambed salts. In many floodplains of the American southwest, saltcedar (*Tamarix ramosissima*) has invaded, assuming greater dominance owing to its superior salt and drought tolerance relative to native phreatophytes (Robinson 1958; Sala et al. 1996; Cleverly et al. 1997).

11.3.6 DESERT ANNUALS

Annual plants tend to evade dry conditions and maintain no photosynthetic parts during rainless periods. They become established, utilizing water in the upper 2–10 cm of the soil to germinate, and use the 10–30 cm-deep transient store of moisture to complete their growth and reproduction before evaporation dries the soil (Noy-Meir 1973). The growth of annuals is more sensitive to phenological controls than other plant types, as they must establish all of their above- and belowground biomass in response to relatively specific temperature and moisture conditions. For example, winter-active species in the Mojave Desert exhibit maximum germination at a daytime temperature of 10°C, whereas summer-active annuals require a temperature of 30°C (Smith et al. 1997). As a group, desert annuals are not tolerant of the water- and heat-stress extremes exhibited by other vegetation forms, and have adapted traits to develop an entire new plant and produce seeds within a short period of time. Rapid growth is supported by low root/shoot biomass ratios, high stomatal conductances, and high photosynthesis rates, traits that are not compatible with the long-term conditions of deserts.

Annual plants are a majority of the flora in most deserts. During a year of favorable precipitation, they germinate, develop vigorously, produce a large number of flowers and fruits, set seed, and die (Dillon & Rundel 1990; Polis et al. 1997; Gutiérrez et al. 2000; Jacsik 2001). This replenishes the seed content of the desert soil. The seeds do not germinate until the next wet year. With the proper combination of temperature, quantity of rainfall, timing of rainfall, and light, germination is widespread, growth is rapid,

Fig. 11.11 Annual plants are a majority of the flora in most deserts, developing vigorously and achieving high densities during years of favorable precipitation. This photograph shows wildflowers on an alluvial fan in Death Valley during March 2005. The normally dry playa in the distance has filled with water.

and the annuals are the most abundant and showiest plants in the desert, blooming as living carpets of flowers (Fig. 11.11). The roots of annuals are often shallow and poorly branched, to take advantage of water in the upper parts the soil (Evenari et al. 1971). This water may be lost to evaporation within several weeks, but is sufficient for roots of fast-growing annuals to complete their life cycle.

Germination of seeds may be restricted by different types of controls, such as dormancy, hardseededness, or the need for scarification, heat and cold treatments, or leaching. Seeds from desert annuals are present throughout the year, and they form the base of a diverse food chain, eaten by ants, rodents, and birds, which in turn are eaten by secondary consumers such as snakes, lizards, birds, and mammalian predators. The seeds are able to remain dormant for many years in the soil, and not all seeds will germinate at once, even in a wet year. Thus, if seeds that have germinated do not survive to flower and set fruit, the remaining seeds in the soil preserve the species. Seed densities for deserts of the southwestern USA typically exceed 3500 m^{-2}, with densities as high as 229,000 m^{-2} reported (Inouye 1991). The seed density in the soil increases after rainy periods. On xeric slopes in semiarid Chile, the peak density of seeds was 4500 m^{-2} in 1996, and 24,000 m^{-2} during 1997, an El Niño year. On a mesic slope the equivalent values were 3000 and 17,000 seeds m^{-2}

(Gutiérrez et al. 2000). The strong response of the ephemeral growth and seed banking to the ENSO event demonstrates the significance of this phenomenon in sustaining the ephemeral vegetation of the region. Slope also appears to play a role in seed banking, with steeper slopes having fewer seeds than more level slopes. In some cases, land slips caused by heavy rains also decrease seed densities on steep slopes. Because seeds are better than plants at surviving extended droughts, unpredictability of rainfall can favor annual plant growth. The more arid a climate is, the larger is the proportion of annuals in the vegetation (Went 1979).

There is considerable variability in the life cycle of annual plants. Year-to-year differences exist in the time between germination and death, plant size at maturity, the number of flowers produced, plant density, diversity, and productivity (Inouye 1991). The range in productivity between relatively wet and dry years can exceed an order of magnitude. During an El Niño year (1997) in semiarid Peru, the cover of ephemerals increased from 8 to 43% on a xeric slope and from 26 to 75% on a mesic slope. Species that were absent in 1996 emerged in 1997, with some of them becoming dominant (Gutiérrez et al. 2000). The minimum amount of rainfall required to stimulate germination in desert annuals is about 25 mm, although this may differ somewhat for summer and winter annuals (Went 1979).

In years of abundant rain, plant densities of nearly 1000 m^{-2} have been recorded, with as many as 187 species of annuals found in 1 ha (Shmida & Ellner 1984). This diversity is maintained because in normal years deserts do not maintain a closed canopy of perennial vegetation, and thus the open spaces between plants can be colonized by ephemerals. Where deserts have distinct summer and winter rainy seasons, the majority of annual plants grow during only one of these growing seasons, rather than both seasons. Availability of water in time and space also contributes to plant diversity, with small changes in soil type and topography amplifying spatial differences in rainfall received. Because of differences in germination requirements, annual precipitation variability can cause species to appear in different proportions from year to year.

Predation also can contribute to diversity, reducing the extent to which competitively dominant annual species can displace poorer competitors. Small rodents reduce shrub cover by browsing and increase soil disturbance, which facilitates the colonization of the disturbed patches by annuals (Gutiérrez et al. 2000). Seed-eating rodents tend to consume more of the relatively large seeds (>1 mg), whereas ants collect more of the relatively small seeds (<1 mg). Grazing by cattle may also lead to lower species diversity.

11.3.7 POIKILOHYDRIC PLANTS

Poikilohydric plants, such as lichens and algae, equilibrate with the relatively dry conditions of desert air, maintain all structures under extreme changes in hydration, and exhibit complete physiological recovery upon rehydration. They can rapidly absorb water from rain or dew, and even the atmosphere when the relative humidity is above 70% (Evenari 1985b). They survive prolonged desiccation without damage; enter an anabiotic state (metabolically inactive) and tolerate extremes of heat, cold, and desiccation; rapidly switch their metabolic state when water is available; and can go through repeated cycles of anabiosis and metabolic activity without injury (Evenari 1985b). Their distribution is constrained by the availability of microhabitats that provide protection from desiccation and result in a favorable number of hydration events. Nonetheless, they are very common on soil and stones in deserts (Noy-Meir 1973).

Many poikilohydric plants occur in protected microsites such as rock crevices or under rock overhangs. Most lichens and mosses are restricted to polar-facing slopes that are well-shaded, allowing plants to remain hydrated after moisture-providing events. In the Negev Highlands of Israel, for example, lichen cover is densest on northern and western exposures owing to longer wetness duration following dew and rain (Kidron 2005). Poikilohydric plants only reach a moderate level of importance where a reliable source of moisture is available through dew or fog, such as along the coastal deserts of Baja California (Rundel 1978; Nash et al. 1979; MacMahon & Wagner 1985), or the Peru–Chile Desert. In the latter area, lichens form dark-colored, cushion-like stands in sandy lomas near the coast, where not disturbed by cattle. Crustose lichens cover the stems of columnar cacti, as well as rocks and stones near the ocean (Rauh 1985). In interior warm deserts, hydration occurs only after episodic rainfall events, and poikilohydric plants are only a minor vegetational component.

11.3.8 EXOTIC PLANTS

In most parts of the world, the modern flora and fauna contains large numbers of introduced species. Exotic plants can be very disruptive to the functioning of natural ecosystems and can affect human land use. Although well adapted to the new environment, many introduced species lack predators, allowing them to spread and displace native flora. In Australia, the prickly pear cactus (*Opuntia stricata*) became a major pest that destroyed grazing land until it was largely eradicated by the larvae of an introduced moth.

In North American deserts, saltcedar (*T. ramosissima*), an Asian desert shrub, has invaded floodplains, forming monotypic stands that displace native vegetation (Cleverly et al. 1997). *Tamarix* is a deciduous salt-pumping shrub or tree that salinizes the floodplain ecosystem that it invades. Salinity levels increase on flow-regulated rivers that are no longer subject to overbank flooding and leaching processes, thereby precluding the growth of native seedlings, and promoting the invasion of more salt-tolerant *Tamarix* (Sala et al. 1996; Glenn et al. 1998). It also tends to initiate a fire cycle in ecosystems with little prior history of fire, accelerating the invasion cycle. *Tamarix* invasion has been so successful that two of the historical riparian gallery forest dominants (*Populus fremontii* and *Salix gooddingii*) are locally becoming extinct along river courses (Sala et al. 1996). To date, *Tamarix* has been difficult to eradicate.

Another introduced species, cheat grass (*Bromus tectorum*), is a ubiquitous annual grass that expanded its range rapidly in the western USA and which now threatens the integrity of native shrub steppe communities because of its opportunistic response to fire (Klemmedson & Smith 1964; Young & Evans 1978; Brown & Minnich 1986). Not only is cheat grass spread by fire, but it promotes burning, causing an irreversible decline in native species such as sagebrush (*Artemesia* spp.), and the replacement of a species-rich shrubland by a cheat-grass-dominated grassland. In turn, this has led to a decrease in faunal diversity.

11.4 SUCCESSION IN DESERT PLANT COMMUNITIES

Succession refers to a directional sequence of changes in species composition and other plant community characteristics, such as productivity, biomass, and diversity, that occur as vegetation colonizes an open site and then changes through time. The pathway is not necessarily deterministic, as there may be multiple endpoints or unpredictable, contingent change (Vale 1988).

Succession in deserts has been little studied and remains poorly understood. Research results have been complicated by the interference of livestock grazing in areas of secondary succession, or, in Australia, by introduced lagomorphs (rabbits). Long-term studies of vegetation dynamics and climate change have been made possible in the North American deserts by the study of packrat middens that record immigration of plant species from the end of the Pleistocene to a few millennia ago (Betancourt et al. 1990). Even after their establishment, species continue to respond to minor climate change by expanding and retreating across local habitat boundaries. Some studies of vegetation succession compare vegetation on disturbed surfaces (abandoned fields, ghost towns, energy corridors, pipeline construction sites, freeways, areas of off-road vehicle impacts, and military training camps) with nearby "undisturbed" sites. The stability of the control site itself is often undetermined.

A study of vegetation dynamics at MacDougal Crater in the Sierra del Pinacate Reserve, Sonora,

Mexico, benefited by a setting which combined protection from domestic livestock and other human impacts, with the availability of multiple observations through time using a series of exactly matched photographs (beginning in 1907), and a series of detailed permanent plot maps (Turner 1990). The Pinacate region is formed of hundreds of basaltic lava flows, cinder cones, craters, and two volcanic peaks. Annual precipitation is around 100 mm (Turner & Brown 1982), with slightly more precipitation falling in summer than winter. In the middle of the crater is a playa that is occasionally flooded. June average daily temperatures exceed 30°C, and maximum temperatures reach 56°C in summer (Ezcurra & Rodrigues 1986). Vegetation is dominated by woody perennials *C. microphyllum*, *Encelia farinosa*, *Prosopis* spp., and *L. tridentata*, and the columnar cactus *Carnegiea gigantea*. During the first half of the twentieth century, populations of *L. tridentata* declined 50–90%, and *Cercidium* declined 60%, with little or no recruitment since that time. In the early 1970s there was a 200-fold increase in *Prosopis* in the crater center, and *Encelia* increased in population markedly from insignificant levels in the early 1960s. These surges in establishment probably resulted from intense rainfall resulting from the entry of Tropical Storm Norma (1970) and Hurricane Joanne (1972) into the region. Analysis of the age distribution of the *Carnegiea* population reveals three peaks of establishment during the period 1790–1960. For the crater as a whole, periods of high mortality were correlated with prolonged drought, and establishment surges for some species related to periods of unusually heavy precipitation. However, Turner (1990) suggests that the rapid shifts in plant populations are more closely linked to establishment surges than to mortality events. This study provides evidence that plant communities in this desert region are not stable, buffered by long life and slow growth, but are highly responsive to changes in climate regime. Comparable studies in other desert areas of the world are necessary to establish whether similar rapid turnover of perennial plants occurs in the absence of interference by humans.

Studies of secondary succession in the southwestern USA suggest that the rates and patterns of succession vary considerably, and are affected by drought, grazing, and the degree of disturbance. Carpenter et al. (1986) examined secondary succession in Mojave Desert scrub vegetation by studying 20 abandoned homestead fields in the Lanfair Valley, about 50 km northwest of Needles. The fields were located at elevations that included three vegetation types: creosote bush scrub (1100 m elevation), Joshua tree grassland (1280 and 1430 m), and sagebrush/juniper scrub (1615 m). The study showed that the ability of plants to recolonize the fields varied according to species and elevation. At 1100 m, significant invasion by *Larrea* had occurred, and in most cases density, cover, and canopy size were nearly equivalent to the control sites. At the 1280 and 1430 m elevations, disturbed sites showed almost no recolonization by arboreal species, there was significantly less total cover, and the site was far from reaching predisturbance composition. At the 1615 elevation, the abandoned fields once again had virtually no arboreal species, 37% of the species showed significant cover differences between the disturbed field and control sites, and the sagebrush/juniper scrub was even further from reaching the predisturbance composition than the Joshua tree grassland. The work suggests that short-lived perennials initially invade the sites and are replaced slowly by increasing populations of long-lived climax shrubs. The estimated time for the reestablishment of predisturbance vegetation is 65–130 years (Carpenter et al. 1986).

When changes in the initial conditions of the area are marked, successional convergence is not always possible, and the invasion of a new species may be a permanent change. This was demonstrated in a study of secondary plant succession over a 70 year period in two abandoned mining camps in the western Great Basin Desert by Knapp (1992). An introduced annual, *Bromus tectorum*, now dominates the disturbed areas. This species has changed the fire, grazing and competition dynamics of the vegetation, and these factors, in addition to changes in the soil substrate, prevent successful return to predisturbance conditions.

11.5 DUNE COMMUNITIES

A psammophile ("sand loving") is a plant restricted to a sandy substrate. There are many sandy habitats in deserts, ranging from mobile dunes, in which the plinth may be stable and conducive to plant growth, but the upper sections too mobile for vegetation; to stabilized dunes, where sand is anchored by denser grass and shrub associations; to sand flats or plains, where there is little sediment movement and vegetation is present; to very shallow sand layers covering

underlying rocks. Trees are uncommon, but occasionally occur in favorable situations, such as interdune depressions or on some stabilized dunes (Bowers 1984).

There are several limitations to plant growth imposed by sand: (1) the moisture regime, (2) the high mobility of sand, and (3) the lack of nutrients in nearly pure quartz deposits. In general, water tends to infiltrate readily into sand, producing little if any runoff, and percolating more deeply than into other soils. The upper sand layers dry rapidly following rainfall, limiting seed germination and plant establishment. However, the relatively coarse grain size and texture of sands reduce capillary action and the loss of deep moisture to direct evaporation. Sand holds water at less negative matrix potentials than fine-textured soils, so that even small amounts of water in sand are readily available to plants. As a result, sandy areas are often considered to be relatively mesic habitats for plant growth, and the plant cover and productivity are often higher than on other surfaces (Bowers 1984; Orshan 1986; Danin 1991). In areas with very low precipitation, sand may be too mobile for vegetation, and no plants are found. Strong sand-carrying winds abrade exposed aerial parts of plants, bury entire plants, or expose roots. Plant coverage and sand stability increase with rainfall. Once a plant is established, it induces changes in the substratum by trapping fine-grained particles and adding organic material, thus improving the soil and facilitating the subsequent germination and growth of new plants (Danin 1991).

The adaptations of psammophytic plants to mobile sand include: (1) the ability to germinate under low or transient soil moisture conditions, (2) rapid initial root elongation to compensate for the drainage of water, (3) the ability to stay emergent above accumulating sand, (4) the ability to withstand sand blasting, which may create surface lesions, (5) the ability to withstand root exposure due to erosion, (6) a root system that can take advantage of water at all depths, (7) the development of vine-like lateral roots that enhance anchorage, (8) tolerance of drought and low nutrient levels, and (9) a symbiotic association with mycorrhizal fungi or with nitrogen-fixing bacteria to provide nutrients such as nitrogen and phosphorus (Bowers 1984; Bendali et al. 1990). The first three factors relate to the initial germination and establishment of plants, which are hindered by changes in sand level associated with erosion or burial. Seeds may become buried so deeply that newly germinated seedlings exhaust their energy reserves before they emerge (Bowers 1996). To counter this problem, many dune plants develop large seeds containing abundant food reserves. Following emergence, the seedlings are very vulnerable to burial or excavation. On the Algodones Dunes, California, only 29% of 94 perennial seedlings were observed to survive a 6-week period, a much lower percentage than for plants on rocky substrates, where about 60–80% survived during a similar time period (Bowers 1996).

Once established, the plant must have additional survival strategies. To accommodate sand accumulation, many plants form adventitious roots and develop new aerial structures at a level corresponding with the new soil surface. Plants in Tunisia were observed to grow until wind-borne sand ceased to accumulate (Bendali et al. 1990). Honey mesquite (*Prosopis glandulosa*) is an important dune-forming plant in the American southwest. As sand accumulates, the stems elongate, and adventitious roots develop, allowing the plants to seek out water and nutrients in successive layers of accumulating sand. Elongation is frequently accompanied by gigantism, an adaptation permitting the plants to grow quickly enough to outstrip sand accumulation (Fig. 9.15) (Danin 1996). Along the Mojave River, California, abundant near-surface groundwater allowed mesquite-anchored nebkhas to grow to 7 or 8 m tall, 9 m long, and 9 m wide, and accumulate up to $450 \, m^3$ of sand (Thompson 1929; Laity 2003).

Sand deflation poses serious risks to plants, undercutting their support and exposing roots to desiccation and abrasion. Up to 1–2 m of roots are commonly left unprotected in deflationary zones of the Negev and Sinai Deserts, and the most common psammophiles have roots with a resistant corky bark (Danin 1991). Some vegetation can withstand only parts of the sand cycle. For example, *Ammophila arenaria*, a pioneer plant on dune crests in Israel, can tolerate sand transport and burial, but not erosion (Tsoar & Blumberg 2002).

Over time, a sandy surface may become stabilized by the addition of vegetation, the incorporation of finer materials (silts and clays), and the development of a cryptobiotic crust (see Chapter 7). According to Danin (1991) and Le Houérou (1986), this process occurs in areas where the rainfall exceeds 100 mm. The value of 150 mm is considered the threshold level for stabilized dunes by Wilson (1973) and Tsoar and Møller (1986). The biogenic stabilization of sand is outlined in Table 11.1.

TABLE 11.1 The biogenic stabilization of dune sand.

1 Mobile sand is present.
2 Biogenic stabilization begins as plants germinate following a prolonged pulse of rainfall.
3 Perennial grasses and plants locally decrease wind speed, leading to deposition of finer particles. Litter accumulates and evaporation is lessened owing to shading and a reduced windspeed.
4 New species with high plasticity establish themselves, replacing the pioneer plants, and leading to more stability.
5 Additional fine particles are trapped and the dust content in the soils increases. This binds the soil and increases available soil moisture and fertility.
6 When the silt and clay content in the upper soils reaches 1.5–2%, cyanobacteria are found, and cryptobiotic soils develop.
7 The cryptobiotic crusts bind the surface layer of soil and further promote stability.

After Tsoar and Møller (1986) and Danin (1991).

11.6 VEGETATION TYPE AND DENSITY AND RELATIONSHIP TO GEOMORPHIC PROCESSES

11.6.1 THE ROLE OF VEGETATION IN THE EROSION AND DEPOSITION OF SAND

Vegetation plays a critical role in aeolian processes: (1) it reduces wind velocity, and as the transport of sand is related to the cube of wind velocity, this means a large decrease in the volume of sand transported, (2) it increases the threshold velocity for sand movement, (3) it traps sand, and therefore less passes through a region, and (4) it increases the organic and fine non-organic component of dunes, enhancing moisture retention and sand cohesion. With the removal of vegetation, the wind impinges directly on the sand grains and mobilizes them so that more sand will move and at lower wind velocities than when the region was vegetated.

Studies of the relations between vegetation cover and sand transport show that sand flux increases exponentially as vegetation cover is lost (Li et al. 2005) (see Table 11.2) and that when vegetation cover is reduced below 15–20% there is clearly a threat for dune reactivation or encroachment (Wasson & Naninga 1986; Wiggs et al. 1995). Dune destabilization often occurs near settlements as, for example, in the Rajasthan Desert, India, where localized grazing by goats and camels creates mobile

TABLE 11.2 Area of moving sand and vegetation cover, Inner Mongolia, China.

Classification	Degree of degradation	Vegetation cover (grassland)	Number of species	Daily wind erosion rate (g m^{-2})	Area of moving sand
Fixed	Least	50%	20	67	10%
Semi-fixed	Light	30–50%	17	357	10–25%
Semi-shifting	Moderate	10–30%	10	906	25–50%
Shifting	Severe	<10%	8	3120	>50%

After Li et al. (2005).

dunes that pose a threat to agricultural land (Wasson & Naninga 1986). There is little movement of sand as shrub cover increases to 35–40% (Wasson & Naninga 1986). Studies by Li et al. (2005) indicate that wind erosion rates decrease linearly with increasing vegetation cover, canopy height, plant density, aboveground biomass, and species richness. Wind velocity (above 20 cm) was reduced by 50–60% for a vegetation cover of 30–40% (Li et al. 2004). The threshold wind shear velocity for transport was increased by a factor of two for vegetated surfaces on the Owens River Delta, California (Lancaster & Baas 1998).

In a natural setting, there are many aspects of vegetative cover that are difficult to quantify, including vegetation structure and porosity (shrubs or grass), the percentage of live and dead cover, plant geometry, seasonal leaf cover, and plant size and distribution (small plants, closely and evenly distributed, or large plants, widely spaced). Further studies will be required to determine additional details of plant and sediment transport relations and to develop predictive models.

Vegetation assumes an important role in dune form (Tsoar & Blumberg 2002), with plants acting as anchoring elements. The permeability of the plant structure allows air and sediment through, but velocity is slowed, and deposition promoted. At a small scale, sand accumulations such as sand shadows develop. These ephemeral features change orientation rapidly in response to fluctuating wind patterns and are indicative of local wind direction (Hesp 1981).

The nebkha (or nabkha; coppice dune, vegetated sand mound, hummock dune, shrub dune, phytogenic hillock, or rhebda) is a large, vegetation-anchored dune. Nebkhas are anchored, in that they do not move bodily, but they have mobile surfaces. Nebkhas range from largely isolated single mounds to extensive groupings, and vary in height from less

than a meter to several meters. Their size and form depends on the size, density and growth habit of the vegetation, the nature and supply of sediment, and the wind direction and intensity. To form a large dune, the tree must be able to withstand the sandy environment, be able to keep up with sand accumulation, and have a relatively long life. Honey mesquite (*P. glandulosa*), for example, is thought to live for 200–500 years or more. Nebkhas harbor a rich fauna of rodents, lizards, snakes, and insects (Le Houérou 1986). Owing to the accumulation of organic material in nebkhas (leaves, bark, and twigs shed under the shrub, often in a stratification pattern), they are sometimes bulldozed to create a terrain richer in nutrients for crop development (Le Houérou 1986).

In some areas of the world, nebkhas result from environmental degradation, as human-induced land cover changes (chiefly grazing and cultivation) release aeolian sediment to accumulate around vegetation. The spread of *P. glandulosa* into the arid overgrazed rangelands of New Mexico, for example, initiated the development of coppice dunes and the erosion of interdunal soil (Buffington & Herbel 1965; Hennessy et al. 1985; Bahre & Shelton 1993). Similarly, nebkhas in Mali, West Africa (Nickling & Wolfe 1994), and in north-central Botswana (Sefe et al. 1996) formed as a consequence of anthropogenic and climatic stresses and are linked to desertification processes. Nebkhas in Botswana are transient landforms, persisting as long as sediment is in adequate supply and the anchoring vegetation is maintained by precipitation. The dunes are thus affected by drought periods. By contrast, the nebkhas of the Mojave River are representative of a "wet aeolian system" (Kocurek 1998), where the water table lies at or close to the surface. In this area, *P. glandulosa* is a phreatophyte and is not tied to local rainfall events. As a result, the dunes are persistent and grow to large sizes. However, their stability is also strongly dependent on human activity insofar as it affects the level of the groundwater table (Laity 2003) (see Chapter 13).

Parabolic dunes also require vegetation for their formation, and are therefore more common in humid and cold areas than arid and semiarid regions. Extensive parabolic dunes, reaching tens of meters in height, are found in the Thar Desert of India (Wasson et al. 1983), the Jafurah Sand Sea in eastern Saudi Arabia (Anton & Vincent 1986), the Kalahari Desert (Eriksson et al. 1989), and Arizona (Hack 1941). The most commonly propounded mechanism for parabolic dune advance is that vegetation protects the less mobile arms from wind erosion and the central part advances downwind to form a dune with a U-shaped form, in which the arms point upwind (Livingstone & Warren 1996). Tsoar and Blumberg (2002) proposed that parabolic dunes may also develop by an increase in the crestal vegetation cover of transverse dunes and barchans in a unidirectional wind regime. According to this model, sand accumulates at the dune crest as it becomes densely vegetated. In turn, erosion changes the windward slope to a convex form and wind streamline convergence erodes and undermines the crestal vegetation, pushing forward a parabolic apex and leaving behind trailing arms. This process was observed to occur in the Mediterranean coastal dunes of Israel. In 1944, dunes were principally of the transverse/barchan type and lacked vegetation owing to agricultural land use and grazing. As a result of relocation of citizens, the vegetation rebounded from 4.3% cover in 1944, to 8.4% in 1974, and 17.0% in 1995. During this time period, the dunes advanced an average of 14.1 m and changed form to become parabolic.

Vegetated linear dune systems form extensive fields in southern Africa (Thomas & Leason 2005), Sahelian Africa, and Australia, and smaller, but still significant, fields in North America (Hack 1941), some Near-East drylands (Tsoar & Møller 1986), and China (Hesp et al. 1989). Both wind energy and vegetation cover need to be considered in order to understand linear dune activity. The relatively stable linear dunes of the Kalahari Desert (mean rainfall from 150 to 350 mm) become active on their upper slopes and crest when vegetation is lost (Wiggs et al. 1995).

Owing to the profound influence of humans on vegetation, many dunes are in the process of becoming either stabilized (Tsoar & Blumberg 2002) or destabilized (Kumar & Bhandari 1992; Laity 2003; Li et al. 2005). Extensive degradation of sandy steppes

in China has resulted from overgrazing by livestock (Li et al. 2005). Table 11.2 illustrates how the wind erosion rate and area of moving sand increase as vegetation declines. As the surfaces become more degraded, the fine sand, silt, and clay content decreases, as does the soil organic matter and total nitrogen content. In the Thar Desert, India, fossilized sand dunes were reactivated as human and livestock populations increased and generated greater demand for food, fodder, and fuelwood (Kumar & Bhandari 1992). Domestic animals removed all palatable species, and then reduced less desirable shrubs to woody branches. In severely grazed regions, unpalatable species expanded in density and cover. In the Negev Desert, Bedouin nomads used linear dunes for cultivation and livestock grazing and removed shrubs for firewood and shelters. Over time, vegetation destruction caused a change in the dune morphology towards linear braided and seif dunes (Tsoar & Møller 1986). Similarly, the concentration of livestock near boreholes in the Kalahari Desert accelerated linear dune crestal disturbance and reactivation (Thomas & Leason 2005). Not all dune reactivation is related to human activity. Other impacts include lightning-induced vegetation fires that cause localized and short-term dune reactivation (Thomas & Leason 2005), with an up to threefold increase in dune surface activity (Wiggs et al. 1995).

11.6.2 THE ROLE OF SLOPE AND ASPECT IN PLANT DISTRIBUTION

Many desert environments have diverse lithologies, and rock type and texture can change markedly over relatively short distances. The uplands may be steep and rocky, the bajadas moderate in slope and composed of coarse-textured soil, and the lower flats gently inclined and formed principally of fine-textured soil. In turn, there is a moisture gradient associated with topography, and this spatial variation in moisture and substrates plays a major role in determining vegetation patterns (Phillips & McMahon 1978; Bowers 1980; Parker 1991).

Steep rocky upland sites in Organ Pipe Cactus National Monument, Arizona, support a diversity of shrubs, trees, and cacti, whereas alluvial flats are dominated by one or two species (Parker 1991). The coarser soils on the upper bajadas, composed of alluvial sediment, allow more rapid infiltration of water and hold it at higher potentials than the fine-textured lowland soils. Lowland soils have stronger capillary action, promoting drying of the soil and movement of salts to the surface. These xeric sites are not well suited to plants with shallow root systems (Shreve 1964).

Variations in slope aspect affect insolation. In semiarid regions of Peru, the equatorial-facing slopes are more xeric and herbs are scarce, but cacti, bromeliads and sclerophyllous shrubs are abundant. The more mesic polar-facing slopes have more herbaceous plants, but lack cacti and bromeliads, and have a different assemblage of shrubs (Gutiérrez et al. 2000). In North America, equatorial-facing (south) slopes generally are characterized by greater heat stress and lower moisture availability than polar-facing (north) slopes and there are marked compositional differences with aspect (Parker 1991). Shreve (1964) observed that these aspect-related differences were greater on granitic than volcanic substrates in the Sonoran Desert. This compositional variation may in part reflect the tolerances of different species to mass movement of surficial slope material (Pérez 1987; Parker 1991).

11.6.3 EFFECTS OF VEGETATION ON STREAM-CHANNEL PROCESSES

Many dryland plant species are found only within or close to river channels. The higher moisture levels support moderate to dense communities of perennial trees, shrubs, and grasses (Tooth & Nanson 2000). Riparian vegetation has the potential to significantly influence hydrology and dryland river form, process, and behavior. To date, studies have been largely confined to the USA and Australia. The impact of invasive *Tamarix* species in the American southwest (Hadley 1961; Robinson 1965; Graf 1978, 1979; Blackburn et al. 1982; Cleverly et al. 1997; Glenn et al. 1998) and the role of flow regulation have received the most attention (Graf 1999; Merritt & Cooper 2000).

There are twin aspects to the study of vegetation and river channels: (1) the means by which floods affect vegetation succession, species composition, and age structure, and (2) the means by which bottomland vegetation affects flood magnitude, duration, and landform change.

11.6.3.1 How floods and fluvial landforms affect vegetation

Floods, and their attendant effects on river-channel process and form, may initially damage or destroy vegetation, but they set the stage for subsequent succession. Plants are broken by the force of floodwaters or the impact of floating wood debris, killed by prolonged saturation of the root zone, and destroyed by erosion of the substrate in which they are rooted (Bendix & Hupp 2000). Riparian species are adapted to a disturbance regime. However, as each species varies in its ability to withstand flooding, the severity of flooding affects the spatial pattern of species composition.

For some riparian species, periodic flooding allows persistence through time with a consistent composition, because repeated disturbance holds successional change in abeyance. Species that are subject to high stream power during flooding have two principal strategies: (1) flexibility, so as to bend with the flow (Bendix & Hupp 2000; Osterkamp & Friedman 2000), or (2) large size, so that mature individuals are resistant to flood damage (such as the Fremont cottonwood (*P. fremontii*) in the American southwest) (Bendix 1999). The type of vegetation varies according to the fluvial landform (for example, depositional bar, channel shelf, floodplain, or terrace). Vegetation on the flood-prone channel shelf tends to have a shrub growth form with small, highly resilient stems and the ability to sprout rapidly from flood-damaged stumps (Bendix & Hupp 2000). Floodplain species are less tolerant of destructive flooding, but endure periods of inundation. Terrace species are generally intolerant of repeated flood damage or inundation. In arid regions, the landform position becomes doubly critical, because it affects access to moisture in addition to determining the nature of the flood regime.

The channel type may also play a role in vegetation dynamics. As gradient increases, there is often a shift to a braided channel pattern (Graf 1988), a relatively unstable form, with shifting islands and bars. Powerful floods (owing to the steep gradient) are combined with an unstable substrate, making a difficult environment for the establishment of vegetation. Braided channels are usually free of mature vegetation.

The colonization of riparian sites is intimately tied to the flood regime. Most riparian plants germinate in alluvium that is deposited during floods, as the strong water flows eliminate pre-existing vegetation and provide newly disturbed sites. Energy conditions determine the texture of the new substrate. Floods also play a role in dispersing propagules to new sites (as either clonal segments or seeds). For plants to gain a foothold in the disturbed site, the timing is critical with respect to air temperature, water availability, and substrate. There must be enough water for the plant to survive, but it must be at an elevation that is high enough to survive any ongoing fluvial processes (Bendix & Hupp 2000).

Plant diversity in arid land streams is poorly understood, although it is thought to be tied to the supply of water and disturbance patterns. In arid channels, there may be a reach-to-reach variation between perennial, intermittent, and ephemeral flow. In the American southwest, channel and riparian vegetation of semiarid areas with perennial or intermittent flow is dominated by phreatophytes, such as tamarisk, willow, and cottonwood; and where streamflow is ephemeral, by xerophytes such as mesquite and paloverde. Many phreatophytes have flexible stems that bend with the flow during floods and thus do not increase flow resistance dramatically; whereas stems of riparian-zone forests of mesquite and paloverde often are not flexible enough to withstand large floods and may contribute to landform damage if toppled during infrequent floods. The relationship between the degree of disturbance and plant diversity at a riparian site remains uncertain. In an analysis of Californian streams, Bendix (1999) found that species richness increased in a downstream direction, hand-in-hand with stream power.

11.6.3.2 How vegetation affects dryland river-channel process and form

Studies show that vegetation can exert a significant influence on channel form and process, but this influence is complex and not yet fully understood.

Bottomland vegetation affects flood magnitude, duration, and the onset of landform change in two principal ways. First, it provides root cohesiveness to otherwise unconsolidated alluvium (Graf 1978), thereby impeding near-surface disturbance (Tooth & Nanson 2004), and second it provides resistance to flow that generally reduces velocity and unit stream power, and increases flood duration (Osterkamp & Friedman 2000). Vegetation affects the progression of floods by increasing the hydraulic roughness. Dense riparian vegetation affects the value of n in the widely used Manning equation, increasing it by as much as an order of magnitude. Flood velocities may be substantially decreased as vegetation grows, or abruptly increased if the vegetation is destroyed. In a series of complex feedbacks, the type of vegetation is influenced by the nature of

the substrate, and the plants in turn affect depositional processes by lowering stream velocity. Overall, the potential for intense runoff and flood damage is greatest where vegetation offers little flow resistance and curve numbers are high, conditions most likely met along channels in semiarid regions (Osterkamp & Friedman 2000).

The spread of tamarisk into channels of the Colorado Plateau, USA, has caused dramatic changes in the fluvial landscape (Graf 1978, 1979; Glenn et al. 1998). Tamarisk colonizes moist sand surfaces and grows rapidly. Shifting sand bars, beaches, and islands that were once sparsely vegetated are now covered by dense thickets. As surfaces stabilize, the islands and bars grow in extent, narrowing channel widths. The decrease in channel width causes more frequent overbank floods and greater sediment deposition. Before the introduction of the hardy tamarisk, native plants such as salt grass and dwarf willow were swept away by floods. Additional anthropogenic variables affecting the fluvial morphology of the Colorado Plateau are changes in sediment input to the Green-Colorado River system that stemmed from arroyo cutting (which commenced in the 1800s) and the period of dam building in the 1930s (Everitt 1979). According to Glenn et al. (1998), the major factor in the replacement of native riparian species by tamarisk has been flow regulation, which restricts the overbank flooding that periodically washes salts out of the soils. Undisturbed river stretches have soil salinity levels two to three times less than disturbed stretches (Busch et al. 1992). As tamarisk is more tolerant of salt than native species, monotypic stands of tamarisk now dominate most of the former cottonwood (*P. fremontii*) and willow (*Salix gooddingii*) habitats of the lower Colorado River.

11.6.3.3 Flow regulation and riparian communities in arid lands

The riparian vegetation of developed or developing nations has been altered substantially by human activities such as damming, irrigation diversion, and groundwater pumping (Dynesius & Nilsson 1994; Graf 1999). These effects have been most intensively studied in the deserts of the southwestern USA. Natural streamflow variability is considered a primary organizing force in riparian habitats (Richter & Richter 2000), but large, main-stem dams have modified the flow patterns of nearly all of the major rivers in the western USA. Some dams store more than 3.8 years of mean runoff, thereby having a significant impact on river flow. They also act to partition the watersheds (Graf 1999).

In the immediate vicinity of a dam, effects include the inundation of streams and floodplains, the trapping of fluvial sediment, and the creation of new lakeshore vegetation types. Downstream of the dams, there is a reduction in peak flows; reduced extent, frequency, and duration of flooding; changes in the seasonal timing of flooding; changes in the growing season flow volume; modified sediment erosion and deposition rates; and changes in patterns of water and nutrient availability on floodplains. Stream-channel geometry, meander rates, channel pattern, and the creation of fluvial landforms are affected.

The impact of dams on riparian vegetation have been documented in a number of studies. Merritt and Cooper (2000) compared the regulated Green River and the free-flowing Yampa River in northwestern Colorado, USA. Vegetation along the Yampa River is characterized by a continuum of species along a gradual environmental gradient. Meanwhile, the Green River has changed from a meandering pattern to a shallow, braided channel during the 37 years of flow regulation, resulting in the progressive loss of a *Populus*-dominated riparian forest, and its replacement by drought-tolerant desert shrublands. Within the channel, fluvial marshes have developed. Research by Richter and Richter (2000) suggests that it is not only the magnitude of flood flow that is responsible for significant geomorphic reorganization of a river, but also the duration of the flood. For the Yampa River, the duration of flooding at or above $209 \text{ m}^3 \text{ s}^{-1}$ (125% of bankful discharge) drives lateral channel migration, which is responsible for initiating ecological succession in the riparian forest.

Recognizing that dam operations have adversely affected biodiversity by eliminating both low and high flows, water-resource managers are modifying water-release practices. The experimental releases of floodwater from Glen Canyon Dam into the Grand Canyon of the Colorado River are examples of flow modification for the benefit of native species or riverine ecosystems.

12

ANIMAL COMMUNITIES

12.1 INTRODUCTION: ENVIRONMENTAL REQUIREMENTS

The basic problem for animals living in desert environments is to maintain body temperature below, and hydration above, critical limits. In order to deal with the heat and aridity of deserts, animals have evolved an array of strategies that can be grouped into three categories: behavioral, morphological, and physiological. Behavioral strategies include being nocturnal, crepuscular (late afternoon), or fossorial (burrowing below ground), and staying in the shade during the heat of the day (Fig. 12.1). Foraging is largely restricted to cooler periods of the day, when the increase in metabolic activity has the least impact on water and heat balances (Noy-Meir 1974). Morphological adaptations to the heat include mammalian body sizes that are small with long appendages, so as to better radiate heat; and light-colored plumage or pelage to reflect away incoming radiation. Physiological adaptations are less common. They include aestivation (dormancy during summer), the concentration of urine, the deposition of fat in tails or humps, salt glands that secrete salt without the loss of fluids, and an absence of sweat glands. Mammals minimize water loss by efficient kidneys that concentrate urine, and birds and reptiles excrete almost dry uric acid.

The adaptations to desert environments vary according to animal groups (Evenari 1985b). Mollusks generally have a high heat tolerance, reflective shells, an activity pattern restricted to wet periods (during which they take up sufficient water and food to survive a long anabiotic phenophase), the synchronization of reproduction with favorable environmental conditions, and relative longevity that counterbalances the high mortality of offspring. Arthropods, such as ants and beetles, stay in burrows during the hot part of the day and restrict their activity to the cooler parts of the day, use dew as water sources (Broza 1979), have relatively low metabolic rates, and can survive for long periods without food.

Reptiles are exceptionally well adapted to the arid environment, which explains their wide diversity in deserts (Fig. 12.2). They produce uric acid instead of urine, have hard-shelled eggs, gain heat directly from the sun, and retreat into the shade or underground during the hottest time of the day. Desert fish show a high degree of endemism and those that survive in the few permanent water sources are able to tolerate wider ranges of temperature and salinity than those in more mesic environments (Figs 12.3 and 12.4).

Most birds are diurnal and do not use burrows for shelter. They have body temperatures that are 3–4°C warmer than mammals, and they allow their body temperatures to increase an additional 2–4°C in response to heat, water, or exercise stress. This creates a higher temperature gradient between the bird and the atmosphere, driving radiative, conductive, and convective heat transfer mechanisms (Wolf 2000). Birds also maintain their heat balance by the insulating effects of plumage, by cooling processes (soaring at high altitudes or assuming special wing postures when resting), and maintaining a pattern of activity that is largely restricted to the early morning and evening hours.

Large herbivorous desert mammals usually require access to drinking water, whereas carnivores can apparently do without, with prey constituting their water source. However, the Arabian oryx can survive indefinitely without drinking water, relying on preformed water in forage (Mésochina et al. 2003). With the exception of some large grazing animals, such as gazelles and ibex, most species are active only at night. The water balance may be restored rapidly by a large drinking capacity, feces are fairly dry, urine flow rates are reduced, and many species may survive for long periods without water. Food reserves may be stored as fat during favorable conditions. Rodents live in burrows or caves, where temperatures are cooler and more humid, and also are active mainly in the morning and evening, except during the cool,

Fig. 12.1 In order to avoid the heat of the day, many animals of the desert engage in behavioral strategies such as being nocturnal or burrowing. Foraging in the cooler parts of the day has the least impact on water and heat balances. The ring-tailed cat (*Bassariscus astutus*) is a nocturnal omnivorous species of the southwestern American deserts most closely related to the raccoon.

Fig. 12.2 Reptiles are exceptionally well adapted to the desert environment, and have a wide diversity, which includes lizards, snakes, and tortoises. They generally retreat into burrows during the warmest time of the day. Horned lizards (*Phrynosoma*) have a mottled coloration that serves as camouflage and lizards remain still to avoid detection. This individual lives on a stabilized, debris-mantled sand dune in southernmost Death Valley.

Fig. 12.3 Although Death Valley National Park is one of the hottest and driest places in North America, it is nonetheless home to 1000 plant species and 440 animal species. The Salt Creek pupfish (*Cyprinodon salinus salinus*) is one of several pupfish endemic to Death Valley National Park. It is adapted to the warm and saline waters of Salt Creek, a perennial shallow stream. Large populations build up during favorable conditions of high water, as the fish reproduce several times a year. Reproductive males are deep blue on the sides (upper fish in the photograph), and the females are a less conspicuous brown.

Fig. 12.4 Salt Creek in Death Valley. Sections of this creek provide habitat for an endemic pupfish (Fig. 12.3). Fish populations decline during low flow periods and periodic flash floods. The creek also provides habitat for mollusks.

rainy season. Plants are the water source for rodents, although some species can drink highly saline water (Evenari 1985b).

12.1.1 ADAPTATIONS TO AIR AND SOIL TEMPERATURE, FIRE, AND THE GASEOUS ENVIRONMENT

The most common adaptations to high air temperatures in deserts are seeking shade, moving subsurface to burrows, or rising to higher altitudes (as in the case of birds). Shade can be sought at the base of steep slopes, in rock crevices, and beneath trees. Large grazing animals, such as antelope, require the shade of trees in summer. Their numbers are threatened by habitat destruction (tree cutting) as well as overhunting, with 65% or more of all desert antelope species endangered (Mésochina et al. 2003).

Burrowing is the most common strategy for dealing with hot and fluctuating air temperatures at the surface (Figs 12.5 and 12.6). As discussed in Chapter 3, subsurface temperatures at the depth of animal burrows are out of phase with those at the surface, with temperatures coolest at depth when they are warmest at the surface. A burrowing trait, plus nocturnal behavior, allows an animal to survive the extremes of the desert environment. Not only is the burrow temperature relatively cool, but also the temperatures are quite stable throughout the day and the humidity is moister than at the surface. Evaporative water loss for animals, such as tortoises, is much less in the burrow environment. The burrowing trait is a worldwide one, being found across taxonomic and community boundaries (Kinlaw 1999).

Mammals excavate the majority of burrows, although reptiles also include many members who utilize tunnels built by other animals or build their own. The architecture of mammal burrows may be complex, with many forks and chambers. Some species are primary excavators ("tunnelers"), strong and accomplished burrowers that create large and extensive underground systems, such as the antbear (*Orycteropus afer*) in southern Africa (Kinlaw 1999). North American tortoises (*Gopherus agassizii*) can also build burrows that are simple but deep (Fig. 12.6). Secondary modifiers inhabit and modify burrows built by the primary excavator. A third category is animals that occupy burrows created or modified by others, but play no role in creating the structure.

FIG. 12.5 Burrowing is the most common adaptation to high air temperatures in deserts. Subsurface temperatures are out of phase with those of the air, being cooler in the day and warmer at night. The relative humidity in burrows is also greater than at the surface. Mammals excavate the majority of burrows, but other species inhabit or modify burrows built by others. In this photograph, a large number of burrows have been excavated in the stabilized sand of a coppice dune. Burrowing affects soil properties, including macroporosity, nutrient heterogeneity, texture, and bulk density.

There are several types of burrows that serve different functions: residence burrows, food-storage burrows near food sources, refuge burrows, and escape burrows. Some animals may utilize more than one type, such as the desert tortoise (*G. agassizii*) that has both a winter den, often 2–3 m deep in channel banks, and a shallower summer den (Ernst & Barbour 1972). Burrows tend to be persistent landscape features as they are often occupied by

Fig. 12.6 The desert tortoise (*Gopherus agassizii*) spends at least 95% of its life in burrows, where it is protected from predators as well as the surface heat of summer and freezing temperatures in winter during its dormant period, November through March. This inhabited burrow is in stabilized sand-sheet deposits of the Cady Mountains, near Lake Manix.

several generations of animals (Whitford & Kay 1999). Burrowing is associated with morphological adaptations, such as webbed feet in lizards or head adaptations in snakes.

In addition to serving a thermoregulatory function, burrows offer animals protection from seasonal fires. Complex burrow systems with escape exits offer protection from predators that enter for the first time, and are unfamiliar with the architecture. Burrows may also serve a food-storage function, allowing animals to survive the cyclic nature of food availability in arid environments. Within the burrows, some seeds absorb additional water from the high humidity, gaining up to 25% additional mass, and serving as a source of water for rodents (Morton & MacMillen 1982).

In addition to having lower temperatures and higher humidities than the surface environment, burrows are of interest because the soil impedes the movement of oxygen and carbon dioxide. Measurements of the gaseous environment show elevated carbon dioxide levels and depressed oxygen levels, with the effect dependent on the gaseous permeability of the soil, the geometry of the burrow, and the metabolic intensities of the animals and soil microorganisms. A unique burrowing species is the sandswimming mole, who moves through loose sand rather than cohesive soil. The Namib Desert golden mole (*Eremitalpa granti namibensis*) is a solitary and blind insectivore that emerges at night to run across the surface of the sand in search of food. During the day, it remains buried in the soft sand, with a low resting metabolism, rebreathing the interstitial gas that surrounds it. Although the partial pressure of oxygen is less than at the surface, the dryness and uniformity of the Namib dune sand results in a continuity of pore spaces and significant porosity, which allows oxygen diffusion toward the buried animal (Seymour & Seely 1996).

12.1.2 MOISTURE PARAMETERS

Moisture in desert environments is uncertain and limited and water is critical to the metabolic functioning of all animals. Animals, including insects, are composed of 65–75% water by weight. For mammals, a loss of 15–20% of body water can prove fatal. Water is lost from the skin as well as excrement and other secretions. There are many physiological and behavioral adaptations to these limitations. Physiologically, water-loss rates are lower for animals that live in arid environments. A desert rattlesnake can survive up to 2–3 months without water, because the daily loss of water is only 0.5% of body weight (McDonald 2003). Kangaroo rats (*Dipodomys spectabilis*) do not sweat and produce very small quantities of dry feces and highly concentrated urine. They eat young green vegetation in the spring and store and hydrate seeds in humid burrows before consuming them (Nagy 2004).

Body water may be replenished in a number of ways: by drinking where free water exits (ponds, seeps, and springs, and dew on surfaces); by the direct absorption through the skin (for insects); by eating succulent leaves and fruits that may contain up to 90% water; and by oxidation of dry foods (for example, the kangaroo rat, *D. spectabilis*) (McDonald 2003).

The mobility of animals allows flexible behavioral adaptations to foraging and seeking water. The distribution and availability of water also influences breeding behavior. For example, birds in North American deserts tend to have more access to year-round water than Australian birds, owing to the presence of mountain ranges that offer humid habitats and permanent streams. They therefore maintain breeding seasons similar to other temperate-zone birds. The more pervasive aridity and lower altitudes of the Australian deserts result in an altered breeding strategy, where birds synchronize their migration and breeding readiness according to erratic and sporadic rainfall cues (Roshier et al. 2002).

Although the presence of quasi-permanent waterholes is important for wildlife, large mammals are more affected and limited by the quality and quantity of precipitation-dependent forage. The survival of pronghorn fawns and bighorn lambs in southwestern Arizona in the period following the winter rains depends on the amount and timing of seasonal vegetation growth ("green up") for food. Vegetation growth following late-summer monsoonal rains is also critical. The amount and spatial distribution of rainfall are therefore determinants of seasonal and annual animal activities and movements (Comrie & Broyles 2002).

For species that do not engage in long-range foraging activity, dew and fog play an important ecological role in some deserts, providing a source of water for small animals and plants. Dew may be vital for desert arthropods, such as isopods, ants, and beetles (Broza 1979; Moffett 1985). Soil faunae, such as nematodes, are also sensitive to dew and fog deposition (Steinberger et al. 1989).

Many animal species rely on small permanent water sources and springs for their survival. These are often intimately tied to fluctuations in the water table. The increasing human population in deserts and use of groundwater resources threatens many species. Groundwater pumping has already dried up many water sources. Additionally, humans affect aquatic species through flow regulation of streams. The construction of dams alters aspects of flood frequency, water discharge, and water temperature that are critical to animal survival.

12.2 Effects on geomorphic processes

Biogeomorphology refers to the influence of plants, animals, and micro-organisms on the development and evolution of landscapes. The impact of animals on dryland geomorphic processes has received less attention than that of vegetation, although the evidence of its importance is rising (Butler 1995). The significance of vegetation in geomorphic processes is discussed in Chapter 11.

The principal geomorphic impacts of animals in desert environments appear to be on slope processes and soil development. The term *bioerosion* has been used to indicate some of these impacts (Naylor et al. 2002). In hyperarid environments, animal impacts are particularly pronounced, owing to the low rates of other slope processes (Boelhouwers & Scheepers 2004). Although the density of animals is lower owing to the sparse vegetation cover, the extreme aridity causes the animals to be very mobile in search of food, and to concentrate their numbers and impacts around waterholes. In semiarid environments and savannah landscapes, the impact of large animals has been compounded by the concentration of wild animals in reserves and by increasing numbers of domestic animals (Thomas 1988). Animals have direct biotic effects, by changing vegetation composition, plant structure, seed dispersal, percent litter, and overall cover. Their abiotic effects are largely related to trampling and compaction, which may affect stream banks or pan margins (Butler 1995). It is likely that the effects of large migratory herds may have had a considerable geomorphic impact in the past, as they followed preferred migration routes.

12.2.1 Slope processes, surface stability, and soil development

12.2.1.1 Surface movement and animal tracks

Animal tracks are particularly pronounced in arid and hyperarid environments, where they are clearly delineated. Boelhouwers and Scheepers (2004) studied the impact of antelope along the Skeleton Coast, Namibia. On level surfaces, tracks may develop, but show no lasting geomorphic impact. However, on scarps, antelope may initiate gully erosion. The animals descend slopes at an angle to the maximum gradient. Initial disturbance of the surface displaces the gravel armor, exposing fine sediment for deflation, and breaks up the gypsum slope at the scarp edge. A scarp recess develops that is further deepened by trampling. This tends to concentrate runoff, which may eventually lead to gully development by fluvial processes.

Scree slopes are particularly susceptible to animal movement, as the sediment is often very near the angle of repose. Grazing animals (sheep, cattle, and goats) are likely to have widespread impacts in semiarid environments. Studies in Turkey suggest that animal movement displaces clasts, with larger fragments moving the longest distance downslope. This leads to both vertical and lateral sorting of sediments (Govers & Poesen 1998).

Often the impacts of animal movement are imperceptible to casual observation and can be monitored only through long-term studies. An experiment in Death Valley in which a 40 cm by 40 cm area of pavement was completely cleared of stones showed stone displacement and partial repaving after 61 months (Haff & Werner 1996). This study showed that animal disruption causes desert pavement surfaces to remain dynamic even in maturity.

12.2.1.2 Biopedturbation and burrowing

Biopedturbation (soil disturbance by animals) and burrowing affect numerous ecosystem processes from the patch to the landscape scale. Biopedturbation includes digging to excavate roots, bulbs, and tubers, to reach insect larvae, or to access food caches (Whitford & Kay 1999). The act of burrowing displaces large quantities of sediment to the surface, as evidenced by the formation of surface mounds that dot the landscape. In some arid and semiarid regions, this ejected material may cover up to 15–20% of the land's surface (Whitford & Kay 1999). In any given ecosystem, a large number of species may produce soil disturbance. A study in the northern Chihuahuan Desert by Jackson et al. (2003) found contributions from badger, skunks, kangaroo rats, ground squirrels, pocket mice, jack rabbits, lizards, subterranean termites, grass cicadas, and ants.

Animals affect a number of soil properties, including texture, bulk density, macroporosity, nutrient heterogeneity, and pedogenesis. Often, displaced material is of lower bulk density than the soil as a whole, and is more prone to wind or water erosion. The loss of this material may be offset by the capture of windblown soil or suspended material in overland flow. The ejecta from deep burrows may differ markedly in chemistry and texture from the surrounding soils. The concentration of nitrogen-rich feces around burrow openings and food storage and defecation chambers at depth may locally improve the nutrient status of the soil (Dean & Milton 1991). Small mammals bring soluble nutrients from deep soil layers (10–200 cm) to the surface, and bring insoluble materials up to the zone of weathering (Whitford & Kay 1999). Bragg et al. (2005) estimate that porcupines (*Hystrix indica*) annually displace 1.6 m^3 of soil per hectare. These activities make considerable contributions to soil development over long time periods.

12.3 HYDROLOGIC IMPACTS

Burrows have a hydrologic impact, increasing infiltration rates and allowing recharge to greater depths in the soil. The effect of burrowing on water infiltration is greatest under shrubs (Whitford & Kay 1999). As a result of infiltration, soil moisture may be positively correlated with burrows. The rate of infiltration in porcupine burrows in a study area in South Africa was twice that of the surrounding undisturbed soil (Bragg et al. 2005).

Burrowing and biopedturbation also impact the sediment yields in drainage basins. Whereas the increased infiltration rates tend to decrease surface runoff and sediment yield, the disturbed soil and mounds also act as sediment sources (Neave & Abrahams 2001). Sediment concentration from runoff plots in the degraded grassland environments of the Jornada Basin, New Mexico, USA, was positively correlated with the average diameter of smallmammal disturbances (Neave & Abrahams 2001). In the Negev Desert, porcupine foraging contributes to soil erosion and may produce from 40 to 100% of the available sediment in small watersheds (Yair & Rutin 1981). Wilby et al. (2001) classified the porcupines as ecosystem engineers, "organisms that directly or indirectly modulate the availability of resources... to other species, by causing physical changes in biotic or abiotic materials."

The hydrologic impacts of animals also include the trampling and modification of stream banks, and the reduction of infiltration owing to the compaction of the soil. When trampling occurs along tracks, it may lead to enhanced soil erosion and preferential drainage development (Butler 1995; Boelhouwers & Scheepers 2004).

12.4 EFFECTS OF THE GEOMORPHIC ACTIVITY OF ANIMALS ON PLANT COMMUNITIES

The changes that animals make to the surface cause variations in soil bulk density, water infiltration and storage, and soil nutrient stores that directly affect plant communities. Some plant species emerge from lost or forgotten seed caches. Foraging may reduce certain species, for example perennial grasses, allowing the invasion of others, such as annual forbs. Foraging has also been shown to be important for the germination and renewal of geophytes (perennial plants that resprout from underground storage organs) (Bragg et al. 2005). Relatively shallow excavations, whether foraging pits or resting hollows for animals, affect ecosystem properties and processes via the materials that accumulate in them, often forming nutrient-rich germination locations for some species (Whitford & Kay 1999). Indian crested porcupine (*H. indica*) diggings, for example, tend to be cooler in temperature and to collect runoff water, creating a more mesic environment that favors plant germination and growth (Alkon 1999). The disturbance by burrowing animals may kill existing vegetation and create colonization opportunities for competitively inferior plant species (Martinsen et al. 1990). Animal mounds may thus harbor different species than the surrounding terrain, with some ejecta piles benefiting pioneer species or "weeds." The net result of animal activity is to maintain spatial and temporal heterogeneity in arid and semiarid landscapes (Jackson et al. 2003).

The heterogeneity in soil texture, structure, and hydrologic properties is manifest in differing burrow densities. Studies in the northern Mojave Desert, for example, show that burrow density varies according to soil age and according to whether the soil is located beneath a plant canopy or in the intercanopy space (Shafer et al. 2007). For younger soils, burrow density is twice as high underneath canopies.

Animals that exist in high densities in an area over a long period of time have profound impacts on the environment and are an essential part of the ecosystem, modulating resource flows (Jones et al. 1994). Although some of these species (for example, the porcupine) are regarded as agricultural pests and are therefore hunted or otherwise exterminated, their ecological role in natural vegetation dynamics is an important one that should be recognized in conservation efforts.

13

DESERTIFICATION AND THE HUMAN DIMENSION

13.1 DESERTIFICATION: INTRODUCTION AND TERMINOLOGY

Desertification refers to changes in dryland productivity, stability, and species composition. The term was first employed by Lavauden (1927) and Aubreville (1949), who used it to describe the transformation of productive land into desert as a result of human activity in Africa (Le Houérou 2002). Renewed interest in the concept of desertification resulted from a series of droughts and famines in the Sahelian region that began in the late 1960s (Herrmann & Hutchinson 2005). In its present usage, desertification is a poorly defined term, with many shades of meaning (Thomas & Middleton 1994). Inconsistency in terminology reflects an incomplete understanding of the causes and nature of change, and a lack of agreement as to the permanence and endpoints of ecosystem disruption. Desertification is difficult to determine in part because there is an absence of measurable criteria.

In general, the term *desertification* has been used to describe conditions of decreasing productivity and long-lasting, possibly irreversible, desert-like conditions. The United Nations Environment Program (1992) defined desertification as "land degradation in arid, semiarid and dry sub-humid areas resulting from various factors including climatic variations and human activities." Mainguet (1991) concluded that desertification "is the ultimate step of land degradation: irreversibly sterile land, meaning irreversible in human terms and within practicable economic limitations."

A distinction is often made between desertification and degradation. The terms desertification and desertization imply an expansion of desert-like conditions to regions that climatically would not be classified as desert. Le Houérou (2002) defines *desertization* as "irreversible arid land degradation resulting in desert-like land forms and landscapes in areas where they did not occur in the recent past." *Degradation*

refers to decreases in productivity, increases in water or wind erosion, or unfavorable changes in species composition, but does not indicate that the changes are permanent or the result desert-like. The changes should be reversible with good weather and a little time. However, the use of these terms varies greatly by author.

Another way to look at desertification is as a gradational process, not an end point. Dregne (1984) suggested that landscape degradation occurs in a number of stages, from slight to very severe, with only the latter resembling a wasteland. In very severe cases, restoration may not be economically feasible. In his survey of North American desertification, he noted that the driest zones are only slightly affected, whereas the semiarid zones are most degraded. The reason for this is that very dry areas are usually too inhospitable to humans to be extensively damaged. This trend towards the greatest damage in semiarid zones is also notable in other areas of the world.

The classification system (drought, desiccation, and degradation) of Warren (1996) incorporates natural climatic variability, the ability of an ecosystem to recover from a climatic perturbation, and the degree of human intervention. *Drought* is considered a short-term and natural decline in productivity. As rainfall variability in arid regions is inherently large, dryland ecosystems and economic systems are generally geared to sustain interannual or decadal droughts with little or no damage. To avoid the difficult conditions, wild and domestic animals, as well as humans, may migrate temporarily. Vegetation may retreat from drier sites, but survive in damper ones where seed banks are retained. Recovery is complete during moister conditions. *Desiccation* is a term applied to drought over a more extended time period, which destroys natural or cultural communities. In the Sahel, rainfall began to decline in the late 1960s and never fully recovered. With respect to vegetation, desiccation may destroy local seedbeds, so that it may take many years to reestablish plant communities. Economic systems may be changed on a massive scale, and there may be significant shifts in population and changes to life style (Warren 1996). Viewed from a human perspective, *degradation* is often defined as a reduction in the land's ability to produce crops. As opposed to drought, direct intervention by humans, rather than climatic change, appears to be largely responsible, although there are probably exceptions. Degradation may also include changes to the vegetation as a result of grazing, firewood collection, or a drawdown in the water table. Water erosion, wind erosion, and salinization may singly, or in concert, play a role in loss of productivity. A change in climate is not necessary.

In many arid environments there is a positive feedback to land degradation: the demand for natural resources remains high even during drought periods, thereby causing increased pressure on the land, and the biophysical systems are unable to return to previous levels of productivity. In the Lake Chad basin, Africa, human activities have amplified water losses originally caused by drought: as rainfall for crops decreased, the amount of water withdrawn for irrigation increased, enhancing a drop in lake level.

Desertification can take place even in the absence of drought if there is increased pressure on the land by humans and livestock. A resource decline has occurred in north-central Botswana, even though there has been no decline in average rainfall over the past few decades. River discharge is insufficient for modern population levels; the water table is declining and the quality of well water is deteriorating as it becomes locally more saline; the density and species types of vegetation are failing to return to earlier levels, with an increase in unpalatable species; and bare soil is exposed, leading to wind erosion and the formation of deflation hollows (Sefe et al. 1996).

Individual countries may choose to define desertification in a way that expresses the dominant regional concerns. This definition may evolve over time. Chinese scientists in 1984 stated that "Man's excessive economic activities destroy the ecological balance of arid and semiarid areas and, as a result, blown sand and dunes have extended into these areas. This environmental change is called desertification" (cited in Ding et al. 1998). This definition focused on aeolian activity and did not directly address issues of long-term loss of biological productivity. By 1996, the meaning was enlarged to include land degradation from salination and water erosion. More recently, defined areas of desertification have included some semi-moist regions, soil erosion has been given more

significance than wind-drift sand landforms, and the loss of ecological balance has been included in some Chinese definitions of desertification (Ding et al. 1998).

Although desertification influences many of the world's marginal environments, its effects appear to be most pronounced and extensive in China and Africa. It is estimated that Chinese deserts are increasing at a rate of 1560 km^2 year^{-1}, mainly from anthropogenic effects (Mitchell & Fullen 1994). In human terms, Africa seems to be most critically affected. During the intense drought of 1984, 40–50% of the livestock perished, 1 million people starved (Graetz 1991), and 10 million became environmental refugees (Warren 1996). Although emergencies such as these are precipitated by drought, they represent the end result of years of declining land productivity. The accumulated damage is particularly hard on the inhabitants, as Africa is already home to the majority of the world's most disadvantaged countries. Although desertification is not a new process, its effects are increasingly damaging to the dry rural economies of the region.

13.2 CLIMATE CHANGE AND DESERTIFICATION

The contribution of climate change to the desertification process is very difficult to separate from human effects, as the processes of drought and desertification often work together. Desert margins are areas of extreme variability, with decades of good years followed by short, cruel, droughts. The lack of good climatic data makes understanding long-term trends difficult. Additionally, vegetation phenology presents a major challenge to efforts to study degradation and desertification in arid lands. Phenology refers to the timing of recurring natural phenomena in the life cycles of animals and plants (such as the timing of flowering and fruiting). The climatic variation that is characteristic of deserts causes chronic vegetation stress. Complex mosaics of plant phenological states at both landscape and regional scales, and the spatial and temporal variation of live and senescent vegetation, complicate assessment of ecosystem conditions. Thus, quantitative spatial information on vegetation and biogeochemical changes associated with human land use remains elusive.

In order to address the problem of desertification, both remote sensing and field study approaches have been utilized. Many large-scale studies of drought, lacking adequate rainfall data, have used surrogate measures such as the Normalized Difference Vegetation Index (NDVI), derived from satellite data.

Advances in monitoring technology have helped modify our perspective on the nature and causes of Sahelian desertification. It is likely that the extent of desertified lands in Africa was exaggerated in early studies (Nicholson et al. 1998). Satellite imagery helped to establish that significant displacements of the desert's margin result from natural climatic variation, and that the Sahara is not marching relentlessly southward into the Sahel (see also Section 3.4.2). Over time, the boundary see-saws over as much as 300 km in response to rainfall variations. Data from satellites that carry an Advanced Very High Resolution Radiometer (AVHRR) are used to calculate a "greenness" index, which shows little long-term change in the biomass of green plants produced per unit of water ("water-use efficiency") (Nicholson et al. 1998). Although there may be some land degradation (such as a change to less palatable shrubs in overgrazed areas near villages and wells, or a change in soil texture or fertility), there appears to be no decline in overall photosynthetic activity. Thus, Nicholson et al. (1998) suggest that for the Sahel desertification is largely significant on a local scale.

In other areas, satellite imagery clearly demonstrates human activity on a regional scale, with devegetation by grazing readily apparent. From space, the international borders between Egypt (the Sinai Desert) and Israel (the Negev Desert), Namibia and Angola, and the USA and Mexico are striking owing to the differences in grazing intensities. In the visible range, the albedo of the overgrazed Sinai is 0.4, whereas that of the Negev is 0.12 (Otterman 1974; Otterman et al. 1985). Similarly, differences in grazing intensities along the border between the USA (Arizona) and Mexico (Sonora) have led to albedo changes and, possibly, to climatic effects (Balling et al. 1998; Couzin 1999). After accounting for altitude and latitude, Balling (1988) found that the higher-albedo, more overgrazed Mexican border region was on average 2.5°C warmer than the adjacent areas of Arizona. Recently, Michalek et al. (2001) examined 25 km of border and found that the differences in albedo and temperature were less compelling. They reported considerable spatial heterogeneity: in some locations, the American grasslands were more heavily grazed, and in others, the Mexican lands showed more impact. However, their research confirmed that areas with more vegetation had a lower albedo and radiant temperature than overgrazed areas.

Whereas climate change clearly affects the land surface, desertification and land-surface degradation may also influence climatic processes. Changes in land use are thought to have increased the amount of dust in the desert atmosphere (Goudie & Middleton 1992). The role of dust in changing precipitation or temperature in arid lands is controversial. Dust modifies both incoming shortwave radiation and outgoing longwave radiation, and therefore has the potential to cause either a heating or a cooling effect, depending on other factors such as cloud cover and surface albedo (Nicholson 2001). In large quantities, dust may reduce precipitation efficiency by coalescence suppression effects (Rosenfeld et al. 2001). It may also inhibit the formation of convective clouds because of radiative cooling and increased subsidence; suggesting that dust causes, rather than results from, decreased rainfall. However, dust loading of the Sahelian atmosphere has tended to follow trends in precipitation (Nicholson et al. 1998).

Although the idea is controversial, it has been argued that devegetation leads to further desiccation of the land through various biophysical feedback effects (Otterman 1974; Charney 1975; Charney et al. 1975). Charney (1975) postulated that devegetation increases the reflectivity of the ground, cooling the air that lies above it and enhancing subsidence, thus reducing rainfall. Furthermore, a reduction in trees lessens water supplied to the atmosphere via evapotranspiration. These processes, although potentially significant, are not yet fully understood (see also Sections 3.4.2 and 3.4.3). They are significant in that they imply that desertification exacerbates or causes drought through a positive feedback mechanism. The impact would depend on the degree of albedo change, which to date appears to be largely localized in extent, and therefore insufficient to cause a significant difference in rainfall (Hulme et al. 2001).

13.3 ANTHROPOGENIC CAUSES OF DESERTIFICATION

There is a closely linked, interacting set of processes that leads to desertification. Pressures on the land are related to population increases (from both high birth rates and immigration), to changing life styles which increase the demand for water and energy, to wars, to poor land-use strategies and inappropriate technology and, in some cases, to a change from a nomadic population to a more sedentary one (Abahussain et al. 2002). The resulting anthropogenic stresses include: (1) physical restructuring (land-use changes, dams, logging, earth-moving operations, watering points for cattle, dredging, artificial-lake creation, etc.), (2) the introduction of exotic species, (3) the discharge of toxic substances, (4) deforestation, overcultivation, and overstocking, and (5) changes to the hydrologic system (drawdown of groundwater, intensive irrigation, etc.) (Rapport & Whitford 1999). Manifestations of desertification include changes in speciation, water erosion (rilling or gullying), wind erosion (deflation and dust production), loss of topsoil, and soil salinization. The result is an ecosystem marked by less biodiversity, reduced primary and secondary production, the dominance of exotic species, and a lowered capacity to return to an original state.

The exact nature of human impact differs according to the nature of the desert environments and the associated political and economic systems. Impacts are broadly subdivided into those that are largely rural in nature (for example, the effects of grazing, woodcutting, or irrigation); those that involve the impact of technology on the landscape (such as transportation and communication corridors, military and off-road vehicles, and oil and mining production facilities); and urban effects (most notably, the depletion of regional water resources). This chapter briefly reviews some of these effects in the remaining sections and concludes with a case study, focusing on the environmental crisis in the Aral Sea region.

13.3.1 THE RURAL ENVIRONMENT: OVERGRAZING AND WOODCUTTING; DEVEGETATION AND BIOLOGICAL FEEDBACKS

Grazing is the most common form of land use in drylands, extending over $3.75 \times 10^7 \text{ km}^2$ or 25% of the total global land surface (Dregne 1983). It affects vegetation cover and biomass, floral and faunal community structure, soil compaction and erosion, water availability, and nutrient status. The impact of livestock is a function of stocking density and management practices. Livestock tended by nomadic herders are spread in a relatively even fashion across the landscape and lead to less degradation of semiarid regions, especially grasslands, than pastured livestock. However, nomadism is increasingly uncommon and rangelands have deteriorated throughout the world. It is estimated, for example, that 94% of the rangelands of the Arab region, including northern Africa and the Middle East, have been subject to some form of desertification (Abahussain et al. 2002).

A number of recent studies suggest that overstocking can result in positive feedbacks that give rise to new ecosystem states (Schlesinger et al. 1990; Asner et al. 2003). Ecosystems tend to be stabilized by a series of feedbacks, that in undisturbed environments tend to be negative, thus maintaining the status quo. Periodic disturbances by natural events, such as fire, drought, or insect infestations are usually only a temporary setback, from which recovery is rapid. In disturbed settings, feedbacks are more likely to be positive, reinforcing the trend, and causing desertification to accelerate (Schlesinger et al. 1990; Nicholson 1998). Drought conditions enhance this latter tendency. Numerous studies suggest that arid ecosystems do not recover fully from debilitating anthropogenic stresses (Schlesinger et al. 1990; Rapport & Whitford 1999).

Overgrazing in the southwestern USA has resulted in the encroachment of woody vegetation, in particular mesquite, into former productive grasslands (Fredrickson et al. 1998). The problems, initiated following the introduction of wells and commercial cattle ranching, were exacerbated by changing fire frequency. The Chihuahuan Desert, for example, changed from historically dominant grasslands to modern shrub/dune systems supporting little herbaceous vegetation (Fig. 13.1) (Buffington & Herbel 1965). The first records of shrub and tree invasions appeared after the drought of 1891–3 (Bahre & Shelton 1993). Today, indigenous desert grasslands

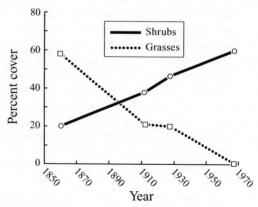

FIG. 13.1 Changes in the percentage of land area occupied by grass and shrubs, Jornada Experimental Range, Chihuahuan Desert, USA. After Buffington and Herbel (1965).

are fragmented, persisting as remnant patches amidst shrublands, shrub/grass mosaics, and coppice dunes. Deterioration of the vegetative cover has led to an increase in wind and water erosion, the formation of nebkhas, and the development of gullies.

Studies of ecosystem processes in the Jornada Experimental Range, southern New Mexico, suggest how positive-feedback processes lead to the desertification of formerly productive land (Schlesinger et al. 1990), even without long-term changes in climate. In 1885, the Jornada Plain was treeless and waterless, but covered with rich, nutritious grass (Fredrickson et al. 1998) that supported approximately 800,000 cattle and 5 million sheep. Over time, grazing increased the spatial heterogeneity of available water, nitrogen, and other soil resources, leading to shrub invasions. Domestic livestock destroyed grass cover during moderate drought and lowered the competitive potential of grasses; trampling compacted the soil and reduced infiltration; livestock dispersed shrub seeds; and greater runoff resulted in heterogeneous erosion and distribution of soil moisture, which was better exploited by shrubs.

The establishment of shrubs on the Jornada Plain tended to concentrate soil resources beneath the canopy in so-called islands of fertility, while wind and water eroded soil from the intershrub area and transported it to new locations. The bare areas between shrubs increased soil temperature, retarding the accumulation of organic nitrogen, and decreasing relative humidity. As soil-water recharge became less predictable, shrubs were better able to exploit water in deep soil layers (Schlesinger et al. 1990). Changes in species composition and nutrient conditions took place over a relatively short time frame (a few decades), even though the mechanisms that caused the changes were in place much earlier. Once the process began, it accelerated, and there has been little success in restoring grazing lands. As a result, the land today supports a small fraction of the commercial cattle herds once present (Rapport & Whitford 1999). An irreversible threshold appears to have been crossed (Schlesinger et al. 1990): the increased heterogeneity of soil resources triggered a positive-feedback mechanism that reinforced the new ecosystem state via changes in physical properties, such as soil temperature and structure, and biochemical cycles, including the availability of water and nutrients.

In Australia, the introduction of sheep and cattle into a semiarid environment also caused very rapid rates of environmental change, perhaps the greatest of any during the Holocene (Pickard 1994). Both domestic and feral herbivores have grazed the land for at least 160 years and have contributed to extensive soil erosion and reservoir sedimentation. By the 1890s, drought and overstocking had removed the feed, and up to 50% of the sheep had died from starvation (Pickard 1994). As in North America, the very high initial stocking rates were not sustainable.

The loss of herbaceous plants has also been documented in South America. In the Monte Desert, Asner et al. (2003) used imaging spectroscopy to compare ungrazed (the UN Nacunan Man-and-Biosphere Reserve) and adjacent grazed areas. Relative to the protected ecosystem, values of soil organic carbon and nitrogen storage were 25–85% lower in areas subjected to long-term grazing. The total carbon and nitrogen capital of dryland ecosystems is a broad indicator of its productivity, material cycling, and biological carrying capacity. Soil organic carbon is an integrator of ecosystem processes: a function of plant-litter inputs, microbial decomposition, nutrient dynamics, faunal activity, and hydrological processes. The loss of nonphotosynthetic vegetation (senescent, woody, or dead tissues) due to grazing has impaired carbon and nitrogen cycles that are key to the biogeochemical functioning of this arid region. The percentage cover of bare soil is also higher in grazed areas, affecting hydrological processes. In ungrazed areas, litter and standing senescent herbaceous plants act as a barrier to soil evaporation and dust generation. As grazing proceeds and the soil desiccates, only deeply rooted woody species survive, enhancing their proliferation, and the loss of herbaceous and other shallow-rooted species (Asner et al. 2003).

Woodcutting for fuel is another form of devegetation. In Africa, this is often a problem around villages, leading to a "halo effect" seen on aerial images of the Sahelian region. In North America, woodcutting has not had a particularly destructive effect on the environment, as these fuels were long ago supplanted by oil, kerosene, and coal (Dregne 1984).

13.3.2 URBANIZATION AND TECHNOLOGICAL EXPLOITATION

Although issues like grazing remain an important public concern, technological developments have created new challenges in the desert environment. Drylands are being rapidly exploited for agriculture, minerals and oil, urban development, tourism, and

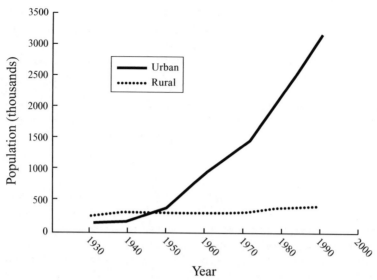

Fig. 13.2 Many of the world's deserts are undergoing rapid urbanization. This graph shows that while the rural population in Arizona, USA, has remained approximately stable, the urban population is skyrocketing. The result is air and water pollution and an increased demand for water. Groundwater pumping lowers water tables and threatens both native riparian habitats and the agricultural potential of the land. From Fredrickson, E. et al. (1998) Perspectives on desertification: south-western United States. *Journal of Arid Environments* **39**, 191–207. By permission of Elsevier.

military purposes. Impacts have included increased levels of wind and water erosion, ground subsidence due to groundwater overdraft, the spread of salinity as a result of irrigation, ecological effects consequent to the diversion and damming of rivers, the desiccation of inland lakes because of inter-basin water transfers, and rising levels of air pollution.

Rapid population growth and urbanization are putting a large stress on some areas (Fig. 13.2). In the Arab regions, 24% of the population lived in cities in 1950, around 56% in 2000, and an estimated 71% is projected to inhabit urban areas by 2030. From 1950 to 2000, the population grew from 77 million to 288 million, at an annual growth rate of 3%. In some countries, almost all of the population is urban, including Kuwait (97%), Bahrain (90%), and Saudi Arabia (83%) (Fig. 13.3) (Abahussain et al. 2002). As cities grow, they encroach on agricultural lands. To meet the greater demand for quality food in urban areas, there is a heavy use of fertilizers and pesticides, causing land and water pollution.

The demand for water in rapidly growing cities in the southwestern USA has resulted in significantly lower water tables, threatening both native riparian habitats and the agricultural potential of the land. Owing to the high cost of extracting deep groundwater, some agricultural land has been abandoned.

Unfortunately, native vegetation cover does not rebound easily, owing to changes in soil texture and chemistry that result from farming. The land is thus subject to water and wind erosion that further deplete its productivity. In the deserts to the east of Los Angeles, California, suburban development is rapidly replacing arable land and agricultural water rights have been purchased to fuel urban growth.

13.3.3 OFF-ROAD VEHICLES AND MILITARY VEHICLES

The impact of off-road vehicular traffic on desert surfaces has received little attention beyond North America. In the USA, the use of off-road vehicles for recreational purposes is a small but very significant source of land degradation. Off-road driving compacts the soil, increases soil erosion, and initiates channelization and gully formation. Regulations have been issued to limit and regulate such traffic, but compliance is sometimes a problem. Some areas have been set aside as "sacrifice zones" for intensive off-road vehicle use. Ultimately these areas become so eroded that drivers seek new areas for enjoyment.

Military vehicles, such as tanks in war or combat training exercises, mechanically disrupt surfaces and

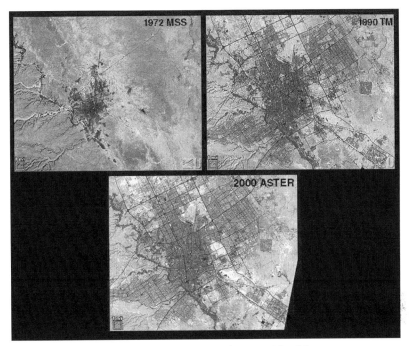

Fig. 13.3 Rapid population growth and urbanization are putting considerable stress on some areas. In Arab regions, the population has been growing at an annual rate of 3% and the population is increasingly urban. In Saudi Arabia, 83% of the population lives in cities. This image shows the growth of the national capital, Riyadh, from 1972 through 2000, during which time the population grew from about a half million to more than 2 million. Source: NASA/GSFC/METI/ERSDAC/JAROS, and U.S./Japan ASTER Science Team.

change the nature, thresholds and rates of aeolian sediment transport. During military maneuvers in the Tularosa Basin, south central New Mexico, USA, tanks eroded the margins of coppice dunes, releasing sand for transport, and enhanced interdune deflation of finer particles by breaking up the surface crust (Marston 1986).

13.3.4 INCREASES IN DUST-STORM ACTIVITY AND THE EFFECT ON HUMANS AND THE ENVIRONMENT

Changes in land use have resulted in marked increases in global dust-storm activity (Goudie & Middleton 1992). The dust commonly has negative impacts on human activities and the environment, modifying cloud formation, affecting air temperatures, degrading air quality and reducing visibility, transporting pathogens, and inducing respiratory problems (see also Section 10.5.2). Human impacts on dust generated have been reported in Australia, Africa, eastern and western Asia, and North America.

Of increasing concern in dust studies is the separation of anthropogenic and non-anthropogenic con-

tributing factors. According to Tegen et al. (1996), 50 ± 20% of the total atmospheric dust originates from disturbed soils, including those affected by cultivation, deforestation, erosion, and changes in vegetation cover due to drought. The nature of the disturbance varies regionally. In North America, off-road vehicle use, lake drainage, and agriculture are the principal contributing factors. In Inner Mongolia, a change from nomadic pastoralism to crop cultivation and livestock rearing encouraged a threefold increase in the human population and a ninefold rise in the animal population. During periods of drought, crops fail and the sandy terrain becomes a major source of dust.

In China, issues of desertification have focused largely on wind erosion (Fig. 13.4). Desertification has enlarged the erodible area in and around the Taklimakan and Gobi Deserts, increasing dust storm activity over the past decade. It is estimated that human factors account for 78% of total Chinese wind erosion. Wind-tunnel studies suggest that erosion is accelerated by a factor of 10 by cultivation, 1.14 by overgrazing, and 22.8 by overcutting (Shi et al. 2004).

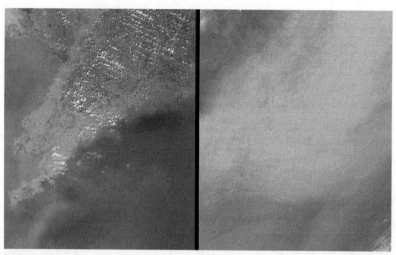

FIG. 13.4 In China, wind erosion and dust-storm activity are critical environmental concerns. It is estimated that 78% of Chinese dust is related to human causes, including cultivation, overgrazing, and overcutting. Large dust storms have effects well beyond the continent and transport not only dust, but also anthropogenic pollutants. The paired image covers the Liaoning region of China and parts of northern and western Korea. The image on the left is a clear day (23 March 2002) and is contrasted with an extremely dusty day (8 April 2002) on the right. Dust from the 2002 Asian storms was detected as far away as western North America. Source: NASA/ GSFC/LaRC/JPL Team.

Large Chinese dust storms have effects well beyond the continent, drifting across the Pacific Ocean in the upper westerlies to eventually arrive in North America. Dust generated during the period 6–9 April 2001 covered vast areas of China and Mongolia, reached the Korean Peninsula on 8 April, Japan on 9 April, and North America on 12 and 13 April, reducing visibility throughout the storm's path (Liu et al. 2003). The total annual dust emissions from China range from approximately 25 to 43 million t (Xuan et al. 2000). Of this, 30% is redeposited in the immediate desert area, 20% is deposited downwind within continental China, and 50% is transported for long distances, being deposited in the oceans and islands downwind, including Japan (Zhang et al. 1997). Mass concentrations of aerosols in Beijing during a spring dust storm in 2001 were 6000 mg m^{-3}, more than 30 times the value for a normal day (Kar & Takeuchi 2004). Dust events are recorded in Seoul, South Korea, on an average of 4 days year^{-1}, with the frequency appearing to rise in recent times (Chun et al. 2001).

Dust that emanates from northern China threatens air quality, human health, transport, and industry in China, Japan, and South Korea. Health issues include respiratory problems and eye infections in China and South Korea, and a suspected link between dust storms and foot-and-mouth disease in South Korea. Heavy dust loads have led to traffic accidents and closed airports, including Beijing international airport. Dust particles on computer microchips threaten the South Korean electronic industry. The trajectory of spring dust events moves the air through heavily industrialized regions in China and South Korea (Chun et al. 2001). The finer dusts (<1 μm) capture anthropogenic pollutants, and nitrogen oxide and sulphur dioxide from industrial effluents enrich dust particles blown into Japan and contribute to soil acidification (Kar & Takeuchi 2004). Closer to the source, dust can have a devastating impact on livestock. The Chinese dust storm of 5 May 1993 killed 120,000 animals and that of 14–16 April 1998 was responsible for 110,000 deaths. Despite the extent and severity of Chinese dust emissions, wind erosion is often controlled on a local basis, whereas an overall national strategy is necessary.

The effect of human activity on dust storms has been well documented throughout arid and semiarid areas of the western USA. One of the best-studied sites is Owens Lake, at the southern end of the Owens Valley, probably the largest source of PM$_{10}$ particles (particles 10 μm or less in diameter that can cause respiratory health problems) in the Western Hemisphere (Gill 1996) (Fig. 13.5). In addition to

Fig. 13.5 Dust-storm activity on Owens Lake, California, USA, has been a major health concern and impacts the visibility in several national parks. The dry lake floor is probably the largest source of PM_{10} particles in the Western Hemisphere. It was desiccated when the Los Angeles Department of Water and Power diverted inflowing tributaries of the Owens River to provide water for Los Angeles. The dust storms are most frequent in the spring and fall. Los Angeles is presently in the process of mitigating the dust emissions by a number of control mechanisms.

fine dust particles, an additional health concern is arsenic, with concentrations as high as 400 ng m^{-3} in air samples (Reid et al. 1994). Dust affects visibility in nearby national parks (Sequoia, Kings Canyon, and Death Valley), national forests, and wilderness areas, and has suspended operations at China Lake Naval Weapons Center, near China Lake. Owens Lake was desiccated in the mid-1920s after the Los Angeles Department of Water and Power (LADWP) began diverting the tributaries of the Owens River to provide water for Los Angeles. Evaporation left a 280 km^2 dry alkaline lakebed, underlain by a water table that is a few centimeters to several meters below the surface. Frequent surface flooding occurs in the winter and spring following storms. Salt crusts result from hydration and dehydration of salts brought to the lakebed surface by groundwater discharge or by precipitation/evaporation recycling of surficial evaporite deposits (Cochran et al. 1988). Efflorescent crusts on the eastern and southern sides of the lake are vulnerable to wind-blown saltating particles (St. Amand et al. 1986; Cochran et al. 1988). The largest and most frequent storms occur in the spring and fall. Wind flow in the Owens Valley is highly turbulent. Steep valley walls and variations in valley width help to create surface eddies approxi-

mately 5–9 km in diameter (Reid et al. 1994). Dust plumes are entrained 2000 m or higher over the lakebed in as little as 5 km (St. Amand et al. 1986; Reid et al. 1994). Owing to the violations of Environmental Protection Agency standards, Los Angeles has been required to mitigate the dust emissions. In a phased implementation process, three dust-control mechanisms are being used: shallow flooding, managed vegetation, and gravel. Shallow flooding involves applying water to an area until it is inundated with a few centimeters' depth or the soil becomes thoroughly saturated to the surface (LADWP 2005).

Although most human activities increase the amount of blowing dust, there are a limited number of conditions in which dust-storm frequency has decreased. In the Tedjen region of southern Turkmenistan, this decline coincided with the development of irrigated agriculture (Orlovsky et al. 2005).

13.4 WATER RESOURCES: A RURAL AND URBAN PROBLEM

Water is often the most important priority in the socio-economic development of arid regions. It limits population growth and distribution, which in deserts are concentrated on a small percentage of the land. The principal water sources are exogenetic rivers, storage reservoirs and lakes, and non-renewable groundwater resources. The reasonable allocation of these resources is constrained by a lack of good hydrologic data and an inadequate understanding of water balances and the long-term consequences of water use. Short-term and short-sighted responses to water needs and use have resulted in numerous environmental problems that are explored in the following sections.

13.4.1 GROUNDWATER WITHDRAWAL

Owing to the paucity of surface water flow in arid regions, communities increasingly rely on groundwater withdrawal to meet the needs of rapidly growing populations, which require water for basic needs, such as drinking and sewage systems or, in more affluent regions of the world, to maintain life styles transferred from more humid areas, such as maintaining suburban lawns, golf courses, and backyard swimming pools. Unfortunately, groundwater is a limited resource.

Groundwater is critical not only for human use, but also for the maintenance of plant and animal habitats. Its exploitation at rates far beyond those subject to natural recharge has caused widespread environmental damage. In the Mojave Desert, for example, groundwater withdrawal along the Mojave River for agriculture, domestic use, industry, and recreational lakes has caused the death of native riparian vegetation, the destabilization of dunes, and enhanced dust-storm activity (Laity 2003). Much of the groundwater is fossil, recharged more than 20,000 years ago, when the climate was cooler and wetter than at present (Izbicki et al. 1995). At present, the main source of groundwater recharge is winter flood flows along the Mojave River. In the 1920s, there was year-round stream flow, ponds, and over 3000 ha of tules, willows, cottonwoods, mesquite, and other plants growing in the Mojave River to the east of Barstow (Fig. 11.7) (Thompson 1929; Blaney & Ewing 1935). The mesquite that flourished in response to the high groundwater table anchored large coppice dunes. Today, groundwater, which was at or near the surface in the 1930s, is 16–30 m below the surface, well beyond the reach of vegetation. In addition to causing the death of mature phreatophytes, groundwater decline also prevents the establishment of riparian seedlings that survive the summer on underground water. Dieback of the vegetation, in combination with the strong regional winds, has caused destabilization of large floodplain nebkhas (Fig. 13.6), and actively migrating sand is encroaching on homes and agricultural

FIG. 13.6 A decline in groundwater owing to overpumping in the vicinity of the Mojave River, California, USA, has resulted in the death of mesquite that once anchored large coppice dunes. As the sand is eroded from these destabilizing dunes, it forms migrating barchans. Downwind, sand encroaches on homes (Fig. 13.7) and agricultural properties.

FIG. 13.7 Sand released from the floodplain of the Mojave River has buried this home.

properties (Fig. 13.7). Additionally, blowing dust in the lower Mojave Valley commonly obscures downwind areas. Although the Mojave River Pipeline is being constructed to bring water from the California Aqueduct to replenish groundwater within the basin, it is unlikely to bring the groundwater table back to its original level, and the short-term recovery of the environment is doubtful (Laity 2003).

Another consequence of groundwater withdrawal is land subsidence. This has been well documented in the Las Vegas Valley area, Nevada, by leveling studies conducted along survey lines, some of which were first established in the early 1930s during the construction of Hoover Dam (Lake Mead). Las Vegas is the most rapidly growing metropolitan area in the USA, with a population of over 1 million people living in an area where the mean annual precipitation is 100 mm and summer temperatures reach a maximum of 47.2°C. Evapotranspiration is between 1520 and 1830 mm per year (Devitt et al. 1992). At present, 20% of the water supply is from groundwater and the remainder from an apportionment of the Colorado River. Withdrawals of groundwater have exceeded annual recharge since 1946 (Maxey & Jameson 1948) and, although reduced after 1968, still exceed recharge by a factor of two to three. As a result of this long-term overdraft, groundwater levels have declined more than 90 m in some areas, and heavily pumped areas show surface deformation and major changes in surface elevation, with as much as 2 m of subsidence in the valley center. Much of the subsidence is preferentially focused on pre-existing faults. Spatially associated fissures have grown up to 3 m in width and 4.5 m in visible depth

(Bell et al. 1992), affecting buildings, highways, buried pipelines, and electrical lines. To mitigate the problems, the Las Vegas Valley Water District initiated a program in 1988 to artificially recharge the aquifer with treated Colorado River water during the winter months. Water-conservation practices have curbed water use and the number of liters of water used per person per day has dropped annually from 1945 to the present. However, owing to the explosive population growth, the total residential use of water continues to grow (Morris et al. 1997).

13.4.2 DEPLETION OF RIVER FLOW AND LOSS OF SEDIMENT

The diversion of river water for irrigation and the employment of dams deplete or regulate river flow and change the sediment load of streams. The Colorado River once carried 18.5 km^3 of annual flow, moving 125–160 million t of suspended sediment to be deposited in the Gulf of California. Owing to intensive water extraction by the USA and Mexico, no water or sediment presently reaches the sea (Schwarz et al. 1991). The Nile River has also suffered sediment starvation owing to dam construction. Today, it only transports 8% of its former load below the Aswan High Dam. In Asia, Africa, and western North America, inland lakes have declined in area and volume owing to the depletion of inflows, as water is removed for irrigation or to fuel urban growth. One of the most severe changes has been to the Aral Sea, a region that is explored in the latter part of this chapter.

13.4.3 EFFECTS OF IRRIGATION: WATERLOGGED SOILS AND SALINIZATION

The development of large modern irrigation schemes has caused widespread soil and land degradation within and beyond desert areas. Waterlogged soils and salinization are the "twin evils" of irrigation. Increased salinity is most pronounced where irrigation raises the water table to near the ground surface, and high evaporation rates cause mineral salts to be precipitated from the water (Fig. 13.8). Attempts to reduce salinity by lowering the water table and draining and leaching the soil often alter the calcium/ sodium balance to the point that soils become very alkaline. The result is a dispersion of clay particles and a breakdown of organic matter that cause a loss of soil structure and permeability, forming black alkali soils that are intractable. Increases in soil salinity and alkalinity levels reduce primary productivity and render the soil virtually sterile for all but a few specialized life forms.

The problems of salinization and waterlogging in dryland soils are global in extent. Large areas of irrigated land are affected in the Arab region, including 54% of the cultivated area in Saudi Arabia and 93% in Egypt. Yield reduction in Egypt averages 25% (Abahussain et al. 2002). Many of the irrigated lands of the southwestern USA and northern Mexico are affected to some degree by waterlogging and salinization. The principal irrigation source is the Colorado River, whose waters are allocated among seven American states and Mexico according to a complicated set of compacts, court decisions, and treaties. All of the available water is extracted and the riverbed runs dry before its terminus. The worst problems of salinization occur in the upper reaches of the Colorado River, in the Mexicali valley, and along the Pecos River in Texas (Dregne 1984). Nonetheless, there has been little abandonment of irrigated land, although there has been some decline in the economic yield. However, water that is returned to the rivers (for example, the Rio Grande) is more saline than before.

In California, much of the prime farmland is undergoing rapid urbanization, creating increased pressure on desert agriculture. Although the desert climate allows a year-round growing season, there are numerous problems, including invasions by pests and fungi (Stephens 1997). An important agricultural area is the Imperial Valley, which extends from southern California into Mexico. Water from the Colorado River is fed to the Imperial Valley via the All American Canal, and thenceforth delivered to fields through an extensive system of canals. The Colorado River has a relatively high salt content (700–850 ppm), and high rates of evapotranspiration lead to a rapid build-up of salts in the root zone. As a result, growers typically apply more water than the

FIG. 13.8 Salinized soils in arid areas reduce agricultural productivity or require the area be abandoned. Location: western China.

crop requires, with the excess leaching salts out of the soils and into a subterranean system of porous tile drains. The excess field drainage flows to the Salton Sea. Although drip-irrigated fields conserve water, they create their own set of problems. Salt accumulates in the soil and must be periodically flushed by surface irrigation. Silt from the unfiltered Colorado River water also clogs drip-delivery systems (Stephens 1997).

Salinization and changing water levels can lead to problems beyond the agricultural zone. Dam construction on the Nile River affected the water level and capillary fringe, causing the migration of salts into major archeological sites, including temples and other structures. The extension of irrigated areas on the Indus Plain has caused the disintegration of bricks at the Harappan site of Mohenjo-Daro in Pakistan by efflorescences of sodium sulphate (Goudie 1977).

An unexpected effect of irrigation may be an increase in the insect population. Irrigation provides a constant food supply for insects and results in continuous reproduction rather than a short discontinuous cycle.

13.4.4 DESERT LAKES AFFECTED BY HUMANS

Desert lakes affected by humans fall into several different categories: (1) naturally occurring lakes, such as Owen's Lake, California, the Aral Sea of central Asia, or Lake Chad in Africa, which have generally suffered from a decline in volume owing to extraction of water to meet increasing economic needs, (2) large reservoirs created by damming or diversions, designed for flow regulation, hydroelectric power generation, or irrigation (Lake Mead, Nevada; Lake Nasser, Egypt), and (3) "accidentally created lakes" such as the Salton Sea, California, sustained by irrigation waters.

The shallowing and degradation of freshwater and salt lakes and inland seas are major environmental problems in arid regions. The growth of human population, combined with climate changes that are at least partially attributable to anthropogenic pollution, exert considerable stress on closed or semi-enclosed seas and lakes. Lakes have suffered changes in their hydrologic balance (evaporation rates and inputs from surface and groundwater sources); reductions in the quality of water resources, including deterioration of geochemical balances (increased salinity, oxygen depletion); disruptions to ecosystems (eutrophication, decrease in biological diver-

sity); and the exposure of lakebeds to the atmosphere, resulting in toxic dust emissions. The degradation of arid lake systems is exemplified by Mono and Owen's Lake, California, where water has been abstracted for long-distance transfer to the Los Angeles metropolitan area; the Dead Sea, whose level has dropped by 14 m since 1977 and whose present salinity is about 340 g L^{-1}; Lake Chad, which had shrunk to less than one-tenth of its size in 1963; and the Salton Sea, which is beset by a number of environmental problems.

The Salton Sea is located in the Imperial Valley, California. One of the most arid regions in the USA, with an average annual rainfall of 50–70 mm, it is also one of the most agriculturally productive owing to the warm climate and irrigation water supplied from the Colorado River (de Vlaming et al. 2004). The present-day Salton Sea occupies a basin that was previously occupied by a series of late-Quaternary lakes, collectively referred to as Lake Coahuilla. It was accidentally created in 1905 when Colorado River floodwaters washed out canal headgates south of Yuma, Arizona, causing the river to change course. The breach was repaired, and by 1907 the river had returned to its normal channel, but the Salton Sea remained. As its name suggests, it is a saline lake, with salinity values 30% greater than the ocean as a result of the dissolution of preexisting basin salts, high evaporation rates, and the discharge of saline tailwater from irrigation in the Imperial Valley. The lake has no outlet and today is sustained principally from agricultural drainage. However, it is plagued by elevated salinity, rising lake elevations, high nutrient loading and elevated levels of selenium from irrigation drainwater (Stephens 1997). Two rivers sustain the lake, the Alamo River and the New River. The headwaters of the New River are in Mexico, where the river receives the discharge of industrial pollutants, urban runoff, agricultural runoff, and poorly treated sewage. The Alamo River consists almost entirely of runoff from irrigation, and discharges toxic concentrations of insecticides into the Salton Sea National Wildlife Refuge (de Vlaming et al. 2004). As a result, there have been massive die-offs of fish (principally tilapia, an introduced African species) and birds, including the deaths of over 140,000 eared grebes (*Podiceps nigricollis*) in 1992. Outbreaks of fish diseases occur because of environmental stresses that include pollution, overcrowding, and high temperatures and salt levels (Fig. 13.9). Algal blooms develop in the warm summer months, stimulated by nutrients in agricultural runoff

Fig. 13.9 The Salton Sea of the Sonoran Desert has experienced massive fish die-offs owing to environmental stresses. Most of these fish, shown littering the shoreline, are introduced African tilapia. However, the lake also provides an important habitat for migratory birds. Increasing levels of salinity, selenium, industrial pollutants, and sewage have resulted in the loss of thousands of birds.

and sewage (Stephens 1997). Their death and decomposition give an unpleasant odor to the shores.

The Salton Sea National Wildlife Refuge at the southern end of the lake remains a key stop on the Pacific flyway for more than 1 million migratory birds annually and provides an important habitat for five endangered species. Measures to conserve tailwater runoff have helped to control the level of the Salton Sea, but only exacerbate the salinity problem, as tailwater had helped to dilute elevated concentrations of salt in the drainage water. Several solutions to the problems have been proposed, including evaporation ponds to concentrate the salts in one area of the sea, a pipeline to the Gulf of California to circulate water from the Sea of Cortez through the Salton Sea, and the use of agricultural waste water to create new wetlands which might act as a biological filter to partially treat the waste water (Stephens 1997).

As an illustration of the complexity of environmental issues resulting from water loss to lakes, the next section will examine the Aral Sea in central Asia. Since 1960, the lake level has dropped by about 23 m and the salinity has increased fourfold. The Aral Sea crisis is not only about water, but also air quality, nutrition, climate, the economy, and the health care system.

13.5 CASE STUDY: THE ARAL SEA

The Aral Sea, located in Uzbekistan and Kazakhstan, is a completely enclosed sea (lake) with a large inland catchment area (Fig. 1.2). Most of the surrounding land is desert with an annual rainfall of less than 90 mm. The Kyzylkum Desert lies to the southeast and the Karakum Desert to the south. The region has a strongly continental climate: temperatures may exceed 45°C in summer and fall below freezing in winter. The sea is a terminal basin with no outflow that is fed by two large rivers: the Amu Dar'ya and Syr Dar'ya. The Syr Dar'ya River, with its source in the Tien Shan Mountains of Kyrgyzstan, flows into the northern end of the lake. The Amu Dar'ya rises in the Pamir Mountains of Tajikistan and Afghanistan and flows into the southern shores. Input is dominated by large spring discharges of water fed by the snowfields and glaciers of the rivers' headwaters.

Flooded since the Pliocene, and reaching its maximum extent at the beginning of the Holocene, the Aral Sea has been subject to repeated major advances and recessions over the past 10,000 years, with variations even within historic time. Evidence of these fluctuations includes marine fossils, relict shore terraces, archaeological sites, and historical records. The level of the water surface has fluctuated by as much as 20–40 m in response to cyclical variations in river discharge caused by climate change and natural diversions of the Amu Dar'ya River (Micklin 1988). In 1911, when continuous instrumental observations were established, the lakebed was stable at about 53 m ASL, and remained so until 1960 (Peneva et al. 2004). During this time, it was the fourth largest lake in area on Earth. The salinity of the sea was about 1%: during periods of low lake levels, salt was precipitated in shallow bays and blown away by the wind; and during high water periods, overflows carried water to the Caspian Sea (Waltham & Sholji 2001).

Since 1960, the lake surface has shrunk from 70,000 to 25,000 km², a 60% decrease in area, coupled with an 80% decrease in volume (Fig. 13.10) (Micklin 1988). Towns such as Muynak, which once sat on the shores of the Aral Sea, are now more than 100 km away from an ever-receding sea. Presently, the Aral Sea is divided into two parts: the Large and Small Seas, fed separately by the Amu Dar'ya and Syr Dar'ya rivers, respectively. The Small Sea appears to be in a quasi-stable state. However, the level of the Large Sea continues to decline. Preservation of the Large Sea depends not only on the water volume delivered by the Amu Dar'ya, but also on the nature of local topographic relief. With areal shrinkage, the Large Sea has, in its turn, divided into two, nearly isolated parts; eastern and western. Water comes to the western part only through the eastern. Further water-level decline will lead to complete isolation of the West Sea from the East Sea. As a result, the West Sea, which now has more water than the East, may eventually disappear owing to evaporation: in 2000, the southern passage connecting the eastern and western basins became dry. Systematic measurements on sea level and water inflow were interrupted by the transition from control of the region by the Soviet Union to the newly independent states. Many stations that once recorded sea level have now fallen dry. Nonetheless, based in part on TOPEX/ Poseidon altimeter data, Peneva et al. (2004) believe that there has been some slowdown in sea-level drop, caused by increasing discharge of groundwater into the lake.

The principal cause of the decline in the Aral Sea has been consumptive use of the two main tributaries to the sea, with a steadily declining discharge of river inflow. The environmental problems began with forced cotton cultivation in the former Soviet

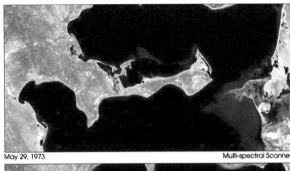
May 29, 1973 — Multi-spectral Scanner

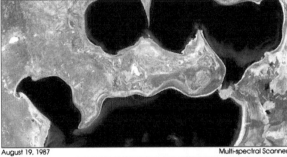
August 19, 1987 — Multi-spectral Scanner

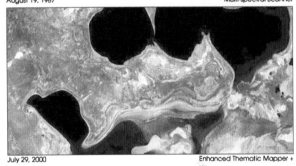
July 29, 2000 — Enhanced Thematic Mapper +

FIG. 13.10 The Aral Sea is an immense inland lake that forms a terminal basin for two large rivers, the Amu Dar'ya and Syr Dar'ya. Consumptive use of these tributaries for irrigation has led to a steady decline in the levels of the sea. Over the past 60 years, more than 60% of the lake has been lost. This image shows a sequence of images in 1973, 1987, and 2000. The exposure of the lakebed to the wind has resulted in a significant decline in air quality and a reduction in crop yields as salt-laden air is deposited on arable land. Image: USGS Eros Data Center based on data provided by the Landsat science team.

Union, during which period inefficient agricultural organization and production, associated with water mismanagement, caused a lowering of the Aral Sea. Today, the Amu Dar'ya and Syr Dar'ya are tapped intensively for irrigation water to produce cotton and rice, with both crops requiring immense quantities of water. The irrigation canals are unlined and uncovered, and this inefficient, poorly maintained system means that little water reaches the sea. The single largest feature is the Karakum Canal, which taps into the Amu Dar'ya River and stretches 1370 km into the Turkmenistan desert (Waltham & Sholji 2001). Much of the diverted water is lost to evapotranspiration and to groundwater recharge. Many fields have become salinized and the water contaminated by salt and agricultural chemicals. Each year up to 400 km² of arable land is abandoned due to soil degradation.

13.5.1 LAKE-BOTTOM EXPOSURE AND SALT AND DUST STORMS

As lake surface area and volume declined, and the water table lowered, vast areas of the former lakebed (more than 1 million ha in 1990) were exposed to the winds. Most of the seabed is covered by sandy or silty sand deposits, which contain between 10 and 12% salt (Stulina & Sektimenko 2004). The exposed bottom of the Aral Sea gives rise to "white" dust storms, which combine both salt and dust (Orlovsky et al. 2005). These arise with increasingly frequency, displacing more than 43 million t of dust annually (Micklin 1988), which are deposited over a region extending from the Black Sea coast to the Arctic (Kotlyakov 1991). Dust-deposition rates south of the Aral Sea are amongst the highest in the world (O'Hara et al. 2000; Wiggs et al. 2003). The dust plumes include sodium and calcium salts as well as pesticide residues. Even though systematic spraying by pesticides has ceased owing to economic constraints, the soil remains contaminated by the organophosphate phosalone.

The dust presents a significant threat to human health for the 5 million people living in the region, and some studies suggest that diseases related to dust have markedly increased, including cancer of the respiratory tract (Micklin 1988). In addition to the Aral Sea bed, additional dust sources lie to the south and west, including the Karakum desert (Wiggs et al. 2003).

13.5.2 ECOSYSTEM DAMAGE

Prior to 1960, the Aral Sea was a vital part of the central Asian ecosystem. The delta was a splendid wetland with reed beds and lakes rich in wildlife. The reduced Aral Sea has become more saline, its biota increasingly impoverished, and its biological produc-

tivity reduced. By the early 1980s, 20 of the 24 species of fish that once inhabited the lake had disappeared, devastating the commercial fishing industry that had sustained the canneries at Aral'sk and Muynak and provided important nutritional elements to the region. The last indigenous fish disappeared in 1985 (Waltham & Sholji 2001).

The reduction in river inflow devastated the deltaic ecosystems that were once of such ecological value in a desert environment. Native plant communities which fringed delta arms, such as the *Tugay* forests, dense stands of phreatophytes mixed with tall grasses and shrubs, have been more than halved in area as floodplain inundation was reduced and the groundwater table dropped. Only 38 of the 173 animal species that once inhabited the delta have survived. In 1960, 650,000 muskrat skins were harvested annually, but by the late 1980s that number had dropped to 2400 (Micklin 1988).

13.5.3 CLIMATIC ALTERATION

Local or regional climate may be altered by the desiccation of inland water bodies. Bordering the Aral Sea, 40,000 km² of former lake bottom is exposed as sparsely vegetated land and evaporite deposits, changing the thermal, moisture, and radiative properties of the surface. Fluctuations in the surface area, volume, and temperature of the sea have affected the pathways of moisture into the atmosphere, decreasing precipitation (Kotlyakov 1991; E. Small et al. 2001).

Lakes similar in size to the Aral Sea are believed to influence climate over a distance of several hundred square kilometers – the well-known "lake effect" – the result of the seasonal contrast in temperatures between land and water bodies. Heat stored in a lake causes its surface temperature to lag behind that of the land, moderating both summer and winter temperatures. The reduction of the lake's size lowered its thermal capacity and increased the continentality of the regional climate, causing greater extremes of temperature (Kotlyakov 1991). Sea surface temperatures have increased by up to 5°C in the spring and summer, and decreased by up to 4°C in the fall and winter. As a result, spring and summer air temperatures are warmer, and fall and winter temperatures are cooler. Additionally, there is a greater diurnal range of air temperature, a shortening of the growing season, a decrease in relative humidity, and more severe "droughtiness" (E. Small et al. 2001). The annual net rate of precipitation minus evaporation ($P–E$) became more negative by about 15%, with the greatest changes occurring in the summer. Although the natural variability of the climate system often hinders attribution of causes of climate change, the seasonal variations in the local temperature trends are consistent with the change to a shallower, less extensive lake. Furthermore, the most statistically significant changes are clustered closest to the lake, suggesting that the observed local and regional climate changes are a result of anthropogenic modification (E. Small et al. 2001).

13.5.4 HEALTH CONCERNS

The principal direct impacts on human health in the region surrounding the Aral Sea are the salinization of the water table, pesticides in the environment and food chain, and dust storms and reduced air quality. Diseases and health conditions associated with these environmental changes include elevated rates of hypertension, respiratory conditions, heart disease, anemia, various cancers, and kidney diseases (I. Small et al. 2001b; Wiggs et al. 2003). In the area immediately south of the sea, the incidence of childhood pneumonia is the highest in the former Soviet Union. Fifty per cent of all childhood diseases in Turkmenistan are respiratory in nature (O'Hara et al. 2000). In addition, there is an epidemic of tuberculosis, probably related to the lowering socioeconomic status of the residents. Other suspected adverse health effects include problems at the maternal–fetal interface (I. Small et al. 2001b).

Water quality for human consumption is poor. Overirrigation and water mismanagement have caused a rise in the water table and an increase in the total dissolved solids (TDS) in well water. Groundwater quality ranges from 0.5 to 6 g L^{-1} TDS, 20 times the average North American values. Values in the lake itself are as high as 140 g L^{-1}. Drinking water is unpalatable and more than 65% of water does not reach standards applicable to chemical contamination. Owing to the deteriorating health and economic conditions, there are now more than 100,000 "environmental migrants" who have been displaced from the region (I. Small et al. 2001a).

In order to resolve the hydrogeological, economical, ecological, and health crises in the Aral Sea region, at least 300 projects have been proposed. It is said that "if every expert brought a bucket of water, the Aral Sea would be filled again" (I. Small et al. 2001a). Proposals include restoring the Aral Sea at the expense of the Caspian Sea or north-flowing Russian rivers. However, responses to the Aral Sea situation explicitly or implicitly assume that it cannot be preserved or restored to its former level. Human responses now are geared toward diminishing the possible consequences of its shrinkage.

13.6 DISCUSSION

The topic of desertification is not a simple one. The term is ill-defined and its application varies according to author and to regional interests. At the most fundamental level, questions exist as to whether desertification exists and, if it does, how it might be defined, measured, and assessed (Herrmann & Hutchinson 2005). The topic is a fluid one, related to our changing understanding of climate variability, vegetation responses, and social processes and responses. As such, it is difficult to specify the degree to which semiarid regions have been impacted. It is likely that the process should be considered a gradational one, with severity ranked according to the chance that a region can fully recover to its original state.

The case studies discussed in this chapter suggest that, by changing the properties and scale of naturally occurring land-surface elements, human influences can significantly affect weather, climate, and the chemistry and aerosol loading of the atmosphere, at least in the short term. Larger-scale escalating effects, such as global warming, may feed back into the climatic processes of deserts. A doubling in carbon dioxide in the global atmosphere has been predicted to cause a 17% increase in the area of desert land (Schlesinger et al. 1990). The ability of some species, such as birds, to physiologically respond to increased temperatures remains doubtful (Wolf 2000). It is likely that the makeup of many desert communities will thus inexorably alter. The loss of meltwater from glacier systems will have profound effects on human, faunal, and floral communities throughout the interior deserts of Asia.

Satellite imagery suggests that, in terms of biomass loss, most desertification occurs at a local or regional scale. Nonetheless, regional environmental management practices are different enough to allow the recognition of international borders from space, including those of Egypt/Israel, Namibia/Angola, and Mexico/USA. Satellite data are of value owing to the poor rainfall records in dryland areas, but they present an incomplete measure of desertification as they provide little information on how species coverage may have changed over time. The preservation of biomass alone is not sufficient if a change occurs that causes less-desirable species to invade and proliferate in a habitat.

Positive feedbacks characterize both human and environmental systems in drylands. In southwestern North America, valuable grassland has been replaced by shrub and dune systems in an apparently irreversible change. Many other changes in deserts are irreparable over the span of a human lifetime, including the loss of valuable topsoil by wind erosion, the salinization of soil, land subsidence, the loss of riparian systems and the animal species that rely upon them, the drawdown of the water table by the mining of fossil water, the loss of vast inland seas, and the contamination of lakes and the soil by herbicides and pesticides. Unfortunately, the effects of desertification are not always limited to the immediate surroundings. Particularly in the case of dust and pollution, the impact may be global in nature,

affecting the economies and public health of surrounding countries and the ecosystems of distant lands.

From a human perspective, desertification is an important issue because it heightens the effects of climatic variability (droughts) and political crises (wars). One impact is to make agricultural yields less predictable, affecting the food security of people living in a region. The result is often migration, causing suffering and death for hundreds of thousands of people.

As with other global environmental concerns, dryland degradation and species and habitat loss are accelerating rapidly within our lifetimes. Unfortunately, there are very few solid quantitative data to understand the process, particularly on a regional scale. Renewed scientific inquiry, the dissemination of such studies to a broader audience, and greater political and economic awareness of the scarcity of water and resources in the desert environment are essential if these unique areas of the world are to be preserved.

REFERENCES

Abahussain, A.A., Abdu, A.S., Al-Zubari, W.K., El-Deen, N.A., and Abdul-Raheem, M. (2002) Desertification in the Arab Region: analysis of current status and trends. *Journal of Arid Environments* **51**, 521–45.

Abd El Rahman, A.A. (1986) The deserts of the Arabian Peninsula. In: Evenari, M., Noy-Meir, I., and Goodall, D.W. (eds.), *Hot Deserts and Arid Shrublands, B*. Elsevier, Amsterdam, pp. 29–54.

Abel, P.I. and Hoelzmann, P. (2000) Holocene palaeoclimates in northwestern Sudan: stable isotope studies on mollusks. *Global and Planetary Change* **26**, 1–12.

Abeliovich, A. and Shilo, M. (1972) Photooxidative death in blue-green algae. *Journal of Bacteriology* **111**, 682.

Abrahams, A.D. and Parsons, A.J. (1991) Relation between infiltration and stone cover on a semiarid hillslope, southern Arizona. *Journal of Hydrology* **122**, 49–59.

Abrahams, A.D., Parsons, A.J., and Luk, S.H. (1986) Resistance to overland flow on desert hillslopes. *Journal of Hydrology* **88**, 343–63.

Abrahams, A.D., Howard, A.D., and Parsons, A.J. (1994) Rock-mantled slopes. In: Abrahams, A.D. and Parsons, A.J. (eds.), *Geomorphology of Desert Environments*. Chapman & Hall, London, pp. 173–212.

Adams, D.K. and Comrie, A.C. (1997) The North American Monsoon. *Bulletin of the American Meteorological Society* **78**, 2197–211.

Adams, J.B., Palmer, F., and Staley, J.T. (1992) Rock weathering in deserts: mobilization and concentration of ferric iron by microorganisms. *Geomicrobiology Journal* **10**, 99–114.

Adams, K.D. (2003) Estimating palaeowind strength from beach deposits. *Sedimentology* **50**, 565–78.

Agassi, M., Shainberg, I., and Morin, J. (1981) Effect of electrolyte concentration and soil sodicity on infiltration rate and crust formation. *Soil Science Society of America Journal* **45**, 848–51.

Ahlbrandt, T.S. (1979) Textural parameters of eolian deposits. In: McKee, E.D. (ed.), *A Study of Global Sand Seas*. US Geological Survey Professional Paper 1052. US Government Printing Office, Washington DC, pp. 21–51.

Ahnert, F. (1960) The influence of Pleistocene climates upon the morphology of cuesta scarps on the Colorado Plateau. *Annals of the Association of American Geographers* **50**, 139–56.

Albritton, C.C., Brooks, J.E., Issawi, B., and Swedann A. (1990) Origin of the Qattara Depression, Egypt. *Bulletin of the Geological Society of America* **102**, 952–60.

Alexandrov, Y., Laronne, J.B., and Reid, I. (2003) Suspended sediment concentration and its variation with water discharge in a dryland ephemeral channel, northern Negev, Israel. *Journal of Arid Environments* **53**, 73–84.

Al-Farraj, A. and Harvey, A.M. (2000) Desert pavement characteristics on wadi terrace and alluvial fan surfaces: Wadi Al-Bih, U.A.E. and Oman. *Geomorphology* **35**, 279–97.

Al-Farraj, A., and Harvey, A.M. (2004) Late Quaternary interactions between aeolian and fluvial processes: a case study in the northern UAE. *Journal of Arid Environments* **56**, 235–48.

Ali, Y.A., and West, I. (1983) Relationships of modern gypsum nodules in sabkhas of loess to composition of brines and sediments in northern Egypt. *Journal of Sedimentary Petrology* **53**, 1151–68.

Al-Juaidi, F., Millington, A.C., and McClaren, S.J. (2003) Merged remotely sensed data for geomorphological investigations in deserts: examples from central Saudi Arabia. *Geographical Journal* **169**, 117–30.

Alkon, P.U. (1999) Microhabitat to landscape impacts: crested porcupine digs in the Negev Desert highlands. *Journal of Arid Environments* **41**, 183–202.

Allan, R.J. (1988) El Niño Southern Oscillation influences in the Australasian region. *Progress in Physical Geography* **12**, 4–40.

Alley, N.F. (1998) Cainozoic stratigraphy, palaeoenvironments and geological evolution of the Lake Eyre Basin. *Palaeogeography, Palaeoclimatology, Palaeoecology* **144**, 239–63.

Allison, R.J. (1988) Sediment types and sources in the Wahiba Sands. *Journal of Oman Studies*, special report no. 3, 161–8.

Allison, R.J. (1997) Middle East and Arabia. In: Thomas, D.S.G. (ed.), *Arid Zone Geomorphology: Process, Form and Change in Drylands*, 2nd edn. John Wiley & Sons, Chichester, pp. 507–21.

Almedeij, J. and Diplas, P. (2005) Bed load sediment transport in ephemeral and perennial gravel bed streams. *EOS* **86**, p. 429, 434.

Altinbilek, D. (2004) Development and management of the Euphrates-Tigris Basin. *Water Resources Development* **20**, 15–33.

Amit, R. and Gerson, R. (1986) The evolution of Holocene reg (gravelly) soils in deserts-an example from the Dead Sea region. *Catena* **13**, 59–79.

Amit, R., Gerson, R., and Yaalon, D.H. (1993) Stages and rate of the gravel shattering process by salts in desert reg soils. *Geoderma* **57**, 295–324.

An, Z. (2000) The history and variability of the East Asian paleomonsoon climate. *Quaternary Science Reviews* **19**, 171–87.

Anderson, K., Wells, S., and Graham, R. (2002) Pedogenesis of vesicular horizons, Cima Volcanic Field, Mojave Desert, California. *Soil Science Society of America Journal* **66**, 878–87.

Anderson, R.S. and Hallet, B. (1986) Sediment transport by wind: toward a general model. *Geological Society of America Bulletin* **97**, 523–35.

Anderson, R.S. and Haff, P.K. (1988) Simulation of eolian saltation. *Science* **241**, 820–3.

Anderson, R.S. and Bunas, K.L. (1993) Grain size segregation and stratigraphy in aeolian ripples modeled with a cellular automaton. *Nature* **365**, 740–3.

Antevs, E. (1935) The occurrence of flints and extinct animals in pluvial deposits near Clovis, New Mexico, Part II – Age of the Clovis lake clays. *Proceedings of the Academy of Natural Sciences of Philadelphia* **87**, 304–12.

Antevs, E. (1938) Postpluvial climatic variations in the southwest. *Bulletin of the American Meteorological Society* **19**, 190–3.

Antevs, E. (1952) Arroyo cutting and filling. *Journal of Geology* **60**, 375–85.

Anton, D. and Vincent, P. (1986) Parabolic dunes of the Jafurah desert, Eastern Province, Saudi Arabia. *Journal of Arid Environments* **11**, 187–98.

Arakawa, H. (1969) Climates of Northern and Eastern Asia. Landsberg, H.E. (ed.), *World Survey of Climatology*, vol. 8. Elsevier, New York.

Arimoto, R. (2001) Eolian dust and climate: relationships to sources, tropospheric chemistry, transport and deposition. *Earth-Science Reviews* **54**, 29–42.

Ash, J.E. and Wasson, R.J. (1983) Vegetation and sand mobility in the Australian desert dunefield. *Zeitschrift für Geomorphologie NF Supplementband* **45**, 7–25.

Asner, G.P., Borghi, C.E., and Ojeda, R.A. (2003) Desertification in central Argentina: changes in ecosystem carbon and nitrogen from imaging spectroscopy. *Ecological Applications* **13**, 629–48.

Assouline, S. (2004) Rainfall-induced soil surface sealing: a critical review of observations, conceptual models, and solutions. *Vadose Zone Journal* **3**, 570–91.

Atwater, M.H. (1977) Urbanization and pollution effects on the thermal structure in four climatic regions. *Journal of Applied Meteorology* **16**, 888–95.

Aubreville, A. (1949) *Climats, forêts et désertification de l'Afrique tropicale*. Société des Editions Geographiques, Maritimes et Coloniales, Paris.

Bach, A.J., Brazel, A.J., and Lancaster, N. (1996) Temporal and spatial aspects of blowing dust in the Mojave and Colorado deserts of Southern California, 1973–1994. *Physical Geography* **17**, 329–53.

Badarinath, K.V.S., Kharol, Shailesh Kumar, Kaskaoutis, D.G., and Kambezidis, H.D. (2007) Case study of a dust storm over Hyderabad area, India: its impact on solar radiation using satellite data and ground measurements. *Science of the Total Environment* **384**, 316–32.

Bagnold, R.A. (1941) *The Physics of Blown Sand and Desert Dunes*. Methuen, London.

Bagnold, R.A. (1953) The surface movement of blown sand in relation to meteorology. *Research Council of Israel Special Publication* **2**, 89–96.

Bahre, C.J. and Shelton, M.L. (1993) Historic vegetation change, mesquite increases, and climate in southeastern Arizona. *Journal of Biogeography* **20**, 489–504.

Bailey, H.P. (1981) Climatic features of deserts. In: Evans, D.D. and Thames, J.L. (eds.), *Water in Desert Ecosystems*. Dowden, Hutchinson, & Ross, Stroudsburg, PA, pp. 13–41.

Baker, A.R., Jickells, T.D., Biswas, K.F., Weston, K., and French, M. (2006) Nutrients in atmospheric aerosol particles along the Atlantic Meridional Transect. *Deep-Sea Research II* **53**, 1706–19.

Baker, V.R. (1977) Stream-channel response to floods, with examples from central Texas. *Geological Society of America Bulletin* **88**, 1057–71.

Baker, V.R., Kochel, R.C., Baker, V.R., Laity, J.E., and Howard, A.D. (1990) Spring sapping and valley network development. In: Higgins, C.G. and Coates, D.R. (eds.), *Groundwater Geomorphology; the Role of Subsurface Water in Earth-Surface Processes and Landforms*. Geological Society of America Special Paper **252**, 235–65.

Baldridge, W.S. (2004) *Geology of the American Southwest*. Cambridge University Press, Cambridge.

Ball, J. (1927) Problems of the Libyan Desert. *Geographical Journal* **70**, 209–24.

Ballantine, J.-A.C., Okin, G.S., Prentiss, D.E., and Roberts, D.A. (2005) Mapping North African landforms using continental scale unmixing of MODIS imagery. *Remote Sensing of Environment* **97**, 470–83.

Balling, Jr, R.C. (1988) The climatic impact of a Sonoran vegetation discontinuity. *Climatic Change* **13**, 99–109.

Balling, Jr, R.C., Klopatek, J.M., Hilderbrandt, M.L., Moritz, C.K., and Watts, C.J. (1998) Impacts of land degradation on historical temperature records from the Sonoran Desert. *Climatic Change* **40**, 669–81.

Balme, M.R., Whelley, P.L., and Greeley, R. (2003) Mars: dust devil track survey in Argyre Planitia and Hellas Basin. *Journal of Geophysical Research* **108**, 5-1–9.

Barber, R.T. and Chavez, F.P. (1983) Biological consequences of El Niño. *Science* **222**, 1203–10.

Barbour, M.G. and Diaz, D.V. (1973) Larrea plant communities on bajada and moisture gradients in the United States and Argentina. *Vegetatio* **28**, 335–52.

Barnett, T.P., Adam, J.C., and Lettenmaier, D.P. (2005) Potential impacts of a warming climate on water availability in snow-dominated regions. *Nature* **438**, 303–9.

Bartov, Y., Stein, M., Enzel, Y., Agnon, A., and Reches, Z. (2002) Lake levels and sequence stratigraphy of Lake Lisan, the Late Pleistocene precursor of the Dead Sea. *Quaternary Research* **57**, 9–21.

Bayly, I.A.E. (1999) Review of how indigenous people managed for water in desert regions of Australia. *Journal of the Royal Society of Western Australia* **82**, 17–25.

Beaumont, P. (1968) Salt weathering on the margin of the Great Kavir, Iran. *Bulletin of the Geological Society of America* **79**, 1683–4.

Beaumont, P. (1998) Restructuring of water usage in the Tigris-Euphrates Basin: the impact of modern water management policies. *Yale Forestry and Earth Science Bulletin* **103**, 168–86.

Beck, W., Donahue, D.J., Jull, A.J.T. et al. (1998) Ambiguities in direct dating of rock surfaces using radiocarbon measurements. *Science* **280**, 2132–5.

Bell, F.C. (1979) Precipitation. In: Goodall, D.W., Perry, R.A., and Howe, K.M.W. (eds.), *Arid-Land Ecosystems: Structure, Functioning and Management*. Cambridge University Press, Cambridge, pp. 373–92.

Bell, J.W., Price, J.G., and Mifflin, M.D. (1992) Subsidence-induced fissuring along preexisting faults in Las Vegas Valley, Nevada. *Proceedings, Association of Engineering Geologists*, 35th Annual Meeting, Los Angeles, CA, 66–75.

Belly, Y. (1964) Sand movement by wind. U.S. Army Corps of Engineers, Coastal Engineering Research Center, Technical Memo **1**, Addendum III.

Belnap, J. (2003) The world at your feet: desert biological soil crusts. *Frontiers in Ecology and the Environment* **1**, 181–9.

Belnap, J. and Gardner, J. (1993) Soil microstructure in soils of the Colorado Plateau: the role of cyanobacterium *Microcoleus vaginatus*. *Great Basin Naturalist* **53**, 40–7.

Belnap, J., Welter, J.R., Grimm, N.B., Barger, N., and Ludwig, J.A. (2005) Linkages between microbial and hydrologic processes in arid and semiarid watersheds. *Ecology* **86**, 298–307.

Bendali, F., Floret, C., Le Floc'h, E., and Pontanier, R. (1990) The dynamics of vegetation and sand mobility in arid regions of Tunisia. *Journal of Arid Environments* **18**, 21–32.

Bendix, J. (1999) Stream power influence on southern Californian riparian vegetation. *Journal of Vegetation Science* **10**, 243–52.

Bendix, J. and Hupp, C.R. (2000) Hydrological and geomorphological impacts on riparian plant communities. *Hydrological Processes* **14**, 2977–90.

Benson, L.V. (2004) The tufas of Pyramid Lake, Nevada. *U.S. Geological Survey Circular* **1267**.

Benson, L.V. and Thompson, R.S. (1987) Lake-level variation in the Lahontan Basin for the past 50,000 years. *Quaternary Research* **28**, 69–85.

Benson, L.V., Currey, D.R., Dorn, R.I. et al. (1990) Chronology of expansion and contraction of four Great Basin lake systems during the past 35,000 years. *Palaeogeography, Palaeoclimatology, and Palaeoecology* **78**, 241–86.

Benson, L.V., Currey, D., Lao, Y., and Hostetler, S. (1992) Lake-size variations in the Lahontan and Bonneville basins between 13,000 and 9,000 ^{14}C yr B.P. *Palaeogeography, Palaeoclimatology, Palaeoecology* **95**, 19–32.

Benson, L., Kashgarian, M., and Rubin, M. (1995) Carbonate deposition, Pyramid Lake subbasin, Nevada: 2. Lake levels and polar jet stream positions reconstructed from radiocarbon ages and elevations of carbonates (tufas) deposited in the Lahontan basin. *Palaeogeography, Palaeoclimatology, Palaeoecology* **117**, 1–30.

Benson, L.V., Burdett, J.W., Kashgarian, M., Lund, S.P., Phillips, F.M., and Rye, R.O. (1996) Climatic and hydrologic oscillations in the Owens Lake basin and adjacent Sierra Nevada, California. *Science* **274**, 746–9.

Ben-Zvi, A. (1996) Quantitative prediction of runoff events. In: Issar, A.S. and Resnick, S.D. (eds.), *Runoff, Infiltration, and Subsurface Flow of Water in Arid and Semi-Arid Regions*, vol. 21. Water Science and Technology Library, Kluwer Academic Publishers, Norwel, MA, pp. 121–30.

Berger, I.A. (1997) South America. In: Thomas, D.S.G. (ed.), *Arid Zone Geomorphology: Process, Form and Change in Drylands*, 2nd edn. John Wiley & Sons, Chichester, pp. 543–62.

Berger, I.A. and Cooke, R.U. (1997) The origin and distribution of salts on alluvial fans in the Atacama Desert, Northern Chile. *Earth Surface Processes and Landforms* **22**, 581–600.

Bernal, J.M. and Solís, A.H. (2000) Conflict and cooperation on international rivers: the case of the Colorado River on the US-Mexico border. *Water Resources Development* **16**, 651–60.

Berndtsson, R., Nodomi, K., Yasuda, H., Persson, T., Chen, H., and Jinno, K. (1996) Soil water and temperature patterns in an arid desert dune sand. *Journal of Hydrology* **185**, 221–40.

Betancourt, J.L., Van Devender, T.R., and Martin, P.S. (eds.) (1990) *Packrat Middens: the Last 40,000 Years of Biotic Change*. University of Arizona Press, Tucson, AZ.

Bierman, P.R. and Gillespie, A.R. (1991) Accuracy of rock varnish chemical analyses: implications for cation-ratio dating. *Geology* **19**, 196–9.

Bierman, P.R. and Turner, J. (1995) ^{10}Be and ^{26}Al evidence for exceptionally low rates of Australian bedrock erosion and the likely existence of pre-Pleistocene landscapes. *Quaternary Research* **44**, 378–82.

Birkeland, P.W. (1984) *Soils and Geomorphology*. Oxford University Press, New York.

Birkett, C.M. (2000) Synergistic remote sensing of Lake Chad: variability of basin inundation. *Remote Sensing of the Environment* **72**, 218–36.

Blackburn, W.H., Knight, R.W., and Schuster, J.L. (1982) Saltcedar influence on sedimentation in the Brazos River. *Journal of Soil and Water Conservation* **37**, 298–301.

Blackwelder, E. (1925) Exfoliation as a phase of rock weathering. *Journal of Geology* **33**, 793–806.

Blackwelder, E. (1929) Sandblast action in relation to the glaciers of the Sierra Nevada. *Journal of Geology* **37**, 256–60.

Blackwelder, E. (1934) Yardangs. *Geological Society of America Bulletin* **45**, 159–65.

Blair, T.C. (1999) Cause of dominance by sheetflood vs. debris-flow processes on two adjoining alluvial fans, Death Valley, California. *Sedimentology* **46**, 1015–28.

Blair, T.C. and McPherson, J.G. (1994) Alluvial fans and their natural distinction from rivers based on morphology, hydraulic processes, sedimentary processes, and facies assemblages. *Journal of Sedimentary Research Section A: Sedimentary Petrology and Processes* **A64**, 450–89.

Blake, D.W., Krishnamurti, T.N., Low-Nam, S.V., and Fein, J.S. (1983) Heat low over the Saudi Arabian desert during May 1979 (Summer MONEX), *Monthly Weather Review* **111**, 1759–75.

Blake, W.P. (1855) On the grooving and polishing of hard rocks and minerals by dry sand. *American Journal of Science* **20**, 178–81.

Blake, W.P. (1905) Superficial blackening and discoloration of rocks especially in desert regions. *Transactions of the American Institute of Mining Engineers* **35**, 371–5.

Blaney, H.F. and Ewing, P.A. (1935) *Utilization of the Waters of Mojave River, California*. US Department of Agriculture, Washington DC.

Boelhouwers, J. and Scheepers, T. (2004) The role of antelope trampling on scarp erosion in a hyper-arid environment, Skeleton Coast, Namibia. *Journal of Arid Environments* **58**, 545–57.

Booth, W.E. (1941) Algae as pioneers in plant succession and their importance in erosion control. *Ecology* **22**, 38–46.

Bornhardt, W. (1900) *Zur Oberflächengestaltung und Geologie Deutsch-Ostafrickas*. Reimer, Berlin.

Bowden, A.R. (1978) Geomorphic perspective on shallow groundwater potential coastal northeastern Tasmania. *Australian Water Resources Council*, Technical Paper **36**. Australian Government Services, Canberra.

Bowen, A.J. and Lindley, D. (1977) A wind-tunnel investigation of the wind speed and turbulence characteristics close to the ground over various escarpment shapes. *Boundary-Layer Meteorology* **12**, 259–71.

Bowers, J.E. (1980) Flora of Organ Pipe Cactus National Monument. *Journal of the Arizona-Nevada Academy of Science* **15**, 1–11, 33–47.

Bowers, J.E. (1984) Plant geography of southwestern sand dunes. *Desert Plants* **6**, 31–54.

Bowers, J.E. (1993) *Shrubs and Trees of the Southwest Deserts*. Southwest Parks and Monuments Association, Tucson, AZ.

Bowers, J.E. (1994) Natural conditions for seedling emergence of three woody species in the northern Sonoran Desert. *Madroño* **41**, 73–84.

Bowers, J.E. (1996) Seedling emergence on Sonoran Desert dunes. *Journal of Arid Environments* **33**, 63–72.

Bowers, J.E. (1997) Demographic patterns of *Ferocactus cylindraceus* in relation to substrate age and grazing history. *Plant Ecology* **133**, 37–48.

Bowler, J.M. (1973) Clay dunes: their occurrence, formation and environmental significance. *Earth-Science Reviews* **9**, 315–38.

Bowler, J.M. (1976) Aridity in Australia: age, origins and expressions in aeolian landforms and sediments. *Earth-Science Reviews* **12**, 279–310.

Bowler, J.M. and Wasson, R.J. (1984) Glacial age environments of inland Australia. In: Vogel, J.E. (ed.), *Late Cainozoic Palaeoclimates of the Southern Hemisphere*. Balkema, Rotterdam, pp. 183–208.

Bradford, J.M., Piest, R.F., and Spomer, R.G. (1978) Failure sequence of gully headwalls in western Iowa. *Soil Science Society of America Journal* **42**, 323–8.

Bradley, W.C. (1963) Large-scale exfoliation in massive sandstones of the Colorado Plateau. *Geological Society of America Bulletin* **74**, 519–28.

Bragg, C.J., Donaldson, J.D., and Ryan, P.G. (2005) Density of Cape porcupines in a semi-arid environment and their impact on soil turnover and related ecosystem processes. *Journal of Arid Environments* **61**, 261–75.

Brakenridge, G.R. (1978) Evidence for a cold, dry full-glacial climate in the American Southwest. *Quaternary Research* **9**, 22–40.

Brazel, A. and Hsu, S. (1981) The climatology of hazardous Arizona dust storms. In: Péwé, T.L. (ed.), *Desert Dust: Origin, Characteristics, and Effect on Man. Geological Society of America Special Paper* **186**, 293–303.

Brazel, A.J. and Nickling, W.G. (1986) The relationship of weather types to dust storm generation in Arizona (1965–1980). *Journal of Climatology* **6**, 255–75.

Breckle, S.W. (1983) Temperate deserts and semi-deserts of Afghanistan and Iran. In: West, N.E. (ed.), *Temperate Deserts and Semi-Deserts*. Ecosystems of the World, vol. 5. Elsevier, Amsterdam, pp. 271–319.

Breed, C.S., Fryberger, S.G., Andrews, S. et al. (1979) Regional studies of sand seas using Landsat (ERTS) imagery. In: McKee, E.D. (ed.), *A Study of Global Sand Seas*. US Geological Survey Professional Paper 1052. US Government Printing Office, Washington DC, pp. 305–98.

Bridges, N.T., Laity, J.E., Greeley, R., Phoreman, J., and Eddlemon, E.E. (2004) Insights on rock abrasion and ventifact formation from laboratory and field analog studies with applications to Mars. *Planetary and Space Science* **52**, 199–213.

Bristow, C.S. and Lancaster, N. (2004) Movement of a small slipfaceless dome dune in the Namib Sand Sea, Namibia. *Geomorphology* **59**, 189–96.

Bristow, C.S., Bailey, S.D., and Lancaster, N. (2000) The sedimentary structure of linear sand dunes. *Nature* **406**, 56–9.

Broecker, W.S. and Liu, T. (2001) Rock varnish: recorder of desert wetness? *GSA Today* **11**, 4–10.

Brookes, I.A. (2001) Aeolian erosional lineations in the Libyan Desert, Dakhla Region, Egypt. *Geomorphology* **39**, 189–209.

Brookfield, M. (1970) Dune trends and wind regime in Central Australia. *Zeitschrift für Geomorphologie Supplementband* **10**, 121–53.

Brotherson, J.D. and Rushforth, S.R. (1983) Influence of cryptogamic crusts on moisture relationships of soils in Navajo National Monument, Arizona. *Great Basin Naturalist* **43**, 73–8.

Brown, D.E. and Minnich, R.A. (1986) Fire and changes in creosote bush scrub of the western Sonoran Desert, California. *American Midland Naturalist* **116**, 411–22.

Brown, W.J., Wells, S.G., Enzel, Y., Anderson, R.Y., and McFadden, L.D. (1990) The late Quaternary history of pluvial Lake Mojave-Silver Lake and Soda Lake Basins, California. In: Reynolds, R.E., Wells, S.G., and Brady, R.H.I. (eds.), *At the End of the Mojave: Quaternary Studies in the Eastern Mojave Desert*. Special Publication of the San Bernardino County Museum Association, San Bernardino, CA, pp. 55–72.

Broza, M. (1979) Dew, fog and hygroscopic food as a source of water for desert arthropods. *Journal of Arid Environments* **2**, 43–9.

Bryan, K. (1922) Erosion and sedimentation in the Papago country. *US Geological Survey Bulletin* **730**, 19–90.

Bryan, K. (1925) Date of channel trenching (arroyo cutting) in the arid southwest. *Science* **62**, 339.

Bryan, K. (1928a) Change in plant associations by change in ground water level. *Ecology* **9**, 474–8.

Bryan, K. (1928b) Niches and other cavities in sandstone at Chaco Canyon, New Mexico. *Zeitschrift für Geomorphologie NF* **3**, 125–40.

Bryan, K. (1954) The geology of Chaco Canyon, New Mexico, in relation of the life and remains of the historic peoples of Pueblo Bonito. *Smithsonian Miscellaneous Collection* **122**.

Bryant, R.G. (2003) Monitoring hydrological controls on dust emissions: preliminary observations from Etosha Pan, Namibia. *Geographical Journal* **169**, 131–41.

Bryant, R.G. and Rainey, M.P. (2002) Investigation of flood inundation on playas within the Zone of Chotts, using a time-series of AVHRR. *Remote Sensing of Environment* **82**, 360–75.

Buffington, L.C. and Herbel, C.H. (1965) Vegetational changes on a semidesert grassland range. *Ecological Monographs* **35**, 139–64.

Buis, E. and Veldkamp, A. (2008) Modelling dynamic water redistribution patterns in arid catchments in the Negev Desert of Israel. *Earth Surface Processes and Landforms* **33**, 107–22.

Bull, W.B. (1977) The alluvial-fan environment. *Progress in Physical Geography* **1**, 222–69.

Bull, W.B. (1979) Threshold of critical power in streams. *Geological Society of America Bulletin* **90**, 453–64.

Bull, W.B. (1980) Geomorphic thresholds as defined by ratios. In: Coates, D.R. and Vitek, J.D. (eds.), *Thresholds in Geomorphology*. Allen & Unwin, London, pp. 259–63.

Bull, W.B. (1991) *Geomorphic Responses to Climate Change*. Oxford University Press, New York.

Bull, W.B. (1997) Discontinuous ephemeral streams. *Geomorphology* **19**, 227–76.

Bullard, J.E., Thomas, D.S.G., Livingstone, I., and Wiggs, G.F.S. (1995) Analysis of linear sand dune morphological variability, southwestern Kalahari Desert. *Geomorphology* **11**, 189–203.

Burke, K. and Wells, G.L. (1989) Trans-African drainage system of the Sahara: was it the Nile? *Geology* **17**, 743–7.

Burney, D.A. (1993) Late Holocene environmental changes in arid Southwestern Madagascar. *Quaternary Research* **40**, 98–106.

Burns, G., Billard, T.C., and Matsui, K.M. (1990) Salinity threat to Upper Egypt. *Nature* **344**, 25.

Burritt, B.E. and Hyers, A. (1981) Evaluation of Arizona's highway dust warning system. In: Péwé, T.L. (ed.), *Desert Dust: Origin, Characteristics, and Effect on Man. Geological Society of America Special Paper* **186**, 281–92.

Busch, D., Ingraham, N., and Smith, S. (1992) Water uptake in woody riparian phreatophytes of the southwestern United States: a stable isotope study. *Ecological Applications* **2**, 450–9.

Busche, D. (2001) Early Quaternary landslides of the Sahara and their significance for geomorphic and climatic history. *Journal of Arid Environments* **49**, 429–48.

Buseck, P.R. and Pósfai, M. (1999) Airborne minerals and related aerosol particles: Effects on climate and the environment. *Proceedings of the National Academy of Sciences USA* **96**, 3372–9.

Butler, B.E. (1956) Parna – an aeolian clay. *Australian Journal of Science* **18**, 145–51.

Butler, D.R. (1995) *Zoogeomorphology: Animals as Geomorphic Agents*. Cambridge University Press, Cambridge.

Butzer, K.W. (1959) Environment and human ecology in Egypt during predynastic and early dynastic times. *Bulletin of the Society of Egyptian Geographers* **32**, 43–88.

Buxbaum, C.A.Z. and Vanderbilt, K. (2007) Soil heterogeneity and the distribution of desert and steppe plant species across a desert-grassland ecotone. *Journal of Arid Environments* **69**, 617–32.

Cáceres, L., Gómez-Silva, B., Garró, X., Rodríguez, V., Monardes, V., and McKay, C.P. (2007) Relative humidity patterns and fog water precipitation in the Atacama Desert and biological implications. *Journal of Geophysical Research* **112**, G04S14, doi:10.1029/2006JG000344.

Cahill, T.A., Gill, T.E., Reid, J.S., Gearhart, E.A., and Gillette, D.A. (1996) Saltating particles, playa crusts and dust aerosols at Owens (dry) Lake, California. *Earth Surface Processes and Landforms* **21**, 621–39.

Caldwell, M.M. (1974) Physiology of desert halophytes. In: Reimold, R.J. and Queen, W.H. (eds.), *Ecology of Halophytes*. Academic Press, New York, pp. 355–78.

Campbell, I.A. (1974) Measurements of erosion on badlands surfaces. *Zeitschrift für Geomorphologie Supplementband* **21**, 122–37.

Campbell, I.A. (1992) Spatial and temporal variations in erosion and sediment yield. In: *Erosion And Sediment Transport Monitoring Programmes In River Basins*. Proceedings of the International Symposium, Oslo. International Association of Hydrological Sciences, Publication 210. International Association of Hydrological Sciences, Wallingford, Oxon, pp. 244–59.

Campbell, I.A. (1997) Badlands and badland gullies. In: Thomas, D.S.G. (ed.), *Arid Zone Geomorphology: Process, Form and Change in Drylands*, 2nd edn. John Wiley & Sons, Chichester, pp. 261–91.

Carpenter, D.E., Barbour, M.G., and Bahre, C.J. (1986) Old field succession in Mojave Desert Scrub. *Madroño* **33**, 111–22.

Cayan, D.R. and Douglas, A.V. (1984) Urban influences on surface temperatures in the southwestern United States during recent decades. *Journal of Climate and Applied Meteorology* **23**, 1520–30.

Cayan, D.R. and Peterson, D.H. (1989) The influence of north Pacific atmospheric circulation on streamflow in the West. *Geophysical Monograph* **55**, 375–95.

Cereceda, P., Osses, P., Larrain, H. et al. (2002) Advective, orographic and radiation fog in the Tarapacá region, Chile. *Atmospheric Research* **64**, 261–71.

Chadwick, O.A., Nettleton, W.D., and Staidl, G.J. (1995) Soil polygenesis as a function of Quaternary climate change, northern Great Basin, USA. *Geoderma* **68**, 1–26.

Chafetz, H.S., Akdim, B., Julia, R., and Reid, A. (1998) Mn- and Fe-rich black travertine shrubs; bacterially (and nanobacterially) induced precipitates. *Journal of Sedimentary Research* **68**, 404–12.

Charney, J. (1975) Dynamics of deserts and drought in the Sahara. *Quarterly Journal Royal Meteorological Society* **101**, 193–202.

Charney, J., Stone, P.H., and Quirk, W.J. (1975) Drought in the Sahara: a biogeophysical feedback mechanism. *Science* **187**, 434–5.

Chauhan, S.S. (2003) Desertification control and management of land degradation in the Thar Desert of India. *The Environmentalist* **23**, 219–27.

Chawla, S., Dhir, R.P., and Singhvi, A.K. (1992) Thermoluminescence chronology of sand profiles in the Thar desert and their implications. *Quaternary Science Reviews* **11**, 25–32.

Chen, F.-H., Shi, Q., and Wang, J.-M. (1999) Environmental changes documented by sedimentation of Lake Yiema in arid China since the Late Glaciation. *Journal of Paleolimnology* **22**, 159–69.

Chen, J. and Xue, L. (2003) Is ground-water in arid inland drainage basins a resource for sustainable development? *Journal of Arid Environments* **55**, 389–90.

Chen, Y. (1983) A preliminary analysis of the processes of sediment production in a small catchment on the Loess Plateau. *Geographical Research* (China) **2**, 35–47.

Chen, Y., Tarchitzky, J., Brouwer, J., Morin, J., and Banin, A. (1980) Scanning electron microscope observations on soil crusts and their formation. *Soil Science* **130**, 49–55.

Chen, Y.N., Wang, Q., Li, W.H., and Ruan, X. (2007) Microbiotic crusts and their interrelations with environmental factors in the Gurbantonggut desert, western Chiina. *Environmental Geology* **52**, 691–700.

Chepil, W.S. and Woodruff, N.P. (1963) The physics of wind erosion and its control. In: Norman, A. (ed.), *Advances in Agronomy*, vol. 15. Academic Press, New York, pp. 211–302.

Cherkauer, D.S. (1972) Longitudinal profiles of ephemeral streams in south-eastern Arizona. *Geological Society of America Bulletin* **82**, 353–66.

Chiew, F.H.S., Piechota, T.C., Dracup, J.A., and McMahon, T.A. (1998) El Nino/Southern Oscillation and Australian rainfall, streamflow and drought: links and potential for forecasting. *Journal of Hydrology* **204**, 138–49.

Christenson, G.E. and Purcell, C. (1985) Correlation and age of Quaternary alluvial-fan sequences, Basin and Range province, southwestern United States. In: Weide, D.E. (ed.), *Soils and Quaternary Geology of the Southwestern United States. Geological Society of America Special Paper* **203**, 115–22.

Chun, Y., Boo, K.-O., Kim, J., Park, S.-U., and Lee, M. (2001) Synopsis, transport, and physical characteristics of Asian dust in Korea. *Journal of Geophysical Research* **106**, 18461–9.

Clapp, E.M., Bierman, P.R., Nichols, K.K., Pavich, M., and Caffee, M. (2001) Rates of sediment supply to arroyos from upland erosion determined using *in situ* produced cosmogenic ^{10}Be and ^{26}Al. *Quaternary Research* **55**, 235–45.

Clauer, N., Chaudhuri, S., Toulkeridis, T., and Blanc, G. (2000) Fluctuations of Caspian Sea level: beyond climatic variations? *Geology* **28**, 1015–18.

Clements, T., Merriam, R.H., Stone, R.O., Reade, E.L., and Eymann, J.L. (1957) A study of desert surface conditions. *Headquarters Quartermaster Research and Development Command*. Environmental Protection Research Division Technical Report EP-53. Environmental Protection Research Division, Natick, MA.

Cleverly, J.R., Smith, S.D., Sala, A., and Devitt, D.A. (1997) Invasive capacity of *Tamarix ramosissima* in a Mojave Desert floodplain: the role of drought. *Oecologia* **111**, 12–18.

Cloudsley-Thompson, J.L. (1984) Introduction. In: Cloudsley-Thompson, J.L. (ed.), *Key Environments: Sahara Desert*. Pergamon Press, Oxford, pp. 1–15.

Cochran, G.F., Hiveve, T.M., Tyler, S.W., and Lopes, T.J. (1988) *Study of Salt Crust Formation Mechanisms on Owens (dry) Lake, California*. Publication 41108. Los Angeles Department of Water and Power, Water Resources Center, Los Angeles, CA.

Coe, M. and Foley, J. (2001) Human and natural impacts on the water resources of the Lake Chad basin. *Journal of Geophysical Research* **106**(D4), 3349–56.

Cohen, H. and Laronne, J.B. (2005) High rates of sediment transport by flashfloods in the Southern Judean Desert, Israel. *Hydrological Processes* **19**, 1687–1702.

Cole, G.A. (1968) Desert limnology. In: Brown, Jr, G.W. (ed.), *Desert Biology*, vol. 1. Academic Press, New York, pp. 423–86.

Cole, K.L. (1986) The lower Colorado River Valley: a Pleistocene desert. *Quaternary Research* **25**, 392–400.

Collon, P., Kutschera, W., Loosli, H.H. et al. (2000) ^{81}Kr in the Great Artesian Basin, Australia: a new method for dating very old groundwater. *Earth and Planetary Science Letters* **182**, 103–13.

Comrie, A.C. (2000) Mapping a wind-modified urban heat island in Tucson, Arizona (with comments on integrating research and undergraduate learning). *Bulletin of the American Meteorological Society* **81**, 2417–31.

Comrie, A.C. and Broyles, B. (2002) Variability and spatial modeling of fine-scale precipitation data for the Sonoran Desert of south-west Arizona. *Journal of Arid Environments* **50**, 573–92.

Condé-Gaussen, G., Rognon, P., and Federoff, N. (1984) Piegeage de poussières éoliennes dans des fissures de granitoides du Sinai oriental. *Compte Rendus de l'Academie des Sciences de Paris II* **289**, 369–74.

Conil, S. and Hall, A. (2006) Local regimes of atmospheric variability: a case study of Southern California. *Journal of Climate* **19**, 4308–25.

Conkling, H. and Gleason, G. (1934) *Mojave River Investigation*. Bulletin 47. California State Department of Public Works, Division of Water Resources, Sacramento, CA.

Conte, M., Colacino, M., and Piervitali, E. (1996) Atlantic disturbances deeply penetrating the African continent: effects over Saharan regions and the Mediterranean basin. In: Guerzoni, S. and Chester, R (eds.), *The Impact of Desert Dust Across the Mediterranean*. Kluwer Academic Publishers, Dordrecht, pp. 93–102.

Cooke, R.U. (1970) Stone pavements in deserts. *Annals of the Association of American Geographers* **60**, 560–77.

Cooke, R.U. and Warren, A. (1973) *Geomorphology in Deserts*. University of California Press, Berkeley.

Cooke, R.U. and Reeves, R.W. (1976) *Arroyos and Environmental Change in the American Southwest*. Oxford University Press, Oxford.

Cooke, R.U., Warren, A., and Goudie, A. (1993) *Desert Geomorphology*. UCL Press, London.

Cooley, M.E., Harshbarger, J.W., Akers, J.P., and Hardt, W.F. (1969) *Regional hydrogeology of the Navajo and Hopi Indian Reservations, Arizona, New Mexico, and Utah*. US Geological Survey Professional Paper 521-A. US Government Printing Office, Washington DC.

Corbett, I. (1993) The modern and ancient pattern of sand-flow through the southern Namib deflation basin. In: Pye, K. and Lancaster, N. (eds.), *Aeolian Sediments: Ancient and Modern*. International Association of Sedimentologists Special Publication 16. Blackwell Scientific Publications, Oxford, pp. 45–60.

Cornet, A.P., Delhoume, J.P., and Montaña, C. (1988) Dynamics of striped vegetation patterns and water balance in the Chihuahuan Desert. In: During, H.J., Werger, M.J.A., and Willems, H.J. (eds.), *Diversity and Pattern in Plant Communities*. SPB Academic Publishing, The Hague, pp. 221–31.

Couzin, J. (1999) Landscape changes make regional climate run hot and cold. *Science* **283**, 317–18.

Criado, C. and Dorta, P. (2003) An unusual 'blood rain' over the Canary Islands (Spain). The storm of January 1999. *Journal of Arid Environments* **55**, 765–83.

Croke, J. (1997) Australia. In: Thomas, D.S.G. (ed.), *Arid Zone Geomorphology: Process, Form and Change in Drylands*, 2nd edn. John Wiley & Sons, Chichester, pp. 563–73.

Crowley, J.K. and Hook, S.J. (1996) Mappping playa evaporite minerals and associated sediments in Death Valley, California, with multispectral thermal infrared images. *Journal of Geophysical Research* **101**, 643–60.

Cullen, H.M. and deMenocal, P.B. (2000) North Atlantic influence on Tigris-Euphrates streamflow. *International Journal of Climatology* **20**, 853–63.

Currey, D.R., Atwood, G., and Mabey, D.R. (1983) Major levels of Great Salt Lake and Lake Bonneville. *Utah Geological and Mineral Survey*, Map 73. Utah Geological and Mineral Survey, Salt Lake City, UT.

Czys, R.R. (1995) Progress in planned weather modification research: 1991–1995. *Reviews of Geophysics* **33**, 823–35.

Dabous, A.A. and Osmond, J.K. (2001) Uranium isotopic study of artesian and pluvial contributions to the Nubian Aquifer, Western Desert, Egypt. *Journal of Hydrology* **243**, 242–53.

Dahan, O., Shani, Y., Enzel, Y., Yechieli, Y., and Yakirevich, A. (2007) Direct measurements of floodwater infiltration into shallow alluvial aquifers. *Journal of Hydrology* **344**, 157–70.

d'Almeida, G.A. (1986) A model for Saharan dust transport. *Journal of Climate and Applied Meteorology* **25**, 903–16.

Dan, J., Yaalon, D.H., Moshe, R., and Nissim, S. (1982) Evolution of reg soils in Southern Israel and Sinai. *Geoderma* **28**, 173–202.

Danin, A. (1983) *Desert Vegetation of Israel and Sinai*. Cana, Jerusalem.

Danin, A. (1991) Plant adaptation in desert dunes. *Journal of Arid Environments* **21**, 193–212.

Danin, A. (1996) *Plants of Desert Dunes*. Springer, New York.

Danin, A., Bar-Or, Y., Dor, I., and Yisraeli, T. (1989) The role of cyanobacteria in stabiization of sand dunes in southern Israel. *Ecologia Mediterranea* **15**, 55–64.

Davies, C.P. and Fall, P.L. (2001) Modern pollen precipitation from an elevational transect in central Jordan and its relationship to vegetation. *Journal of Biogeography* **28**, 1195–1210.

Davis, W.M. (1892) The convex profile of bad-land divides. *Science* **20**, 245.

Davis, W.M. (1938) Sheetfloods and streamfloods. *Geological Society of America Bulletin* **49**, 1337–1416.

Day, R.W. (1993) Accidents on interstate highways caused by blowing dust. *Journal of Performance of Constructed Facilities* **7**, 129–32.

Dean, W.R.J. and Milton, S. (1991) Patch disturbances in arid grassy dunes: antelope, rodents, and annual plants. *Journal of Arid Environments* **20**, 231–7.

Denny, C.S. (1965) *Alluvial Fans in the Death Valley Region, California and Nevada*. United States Geological Society Professional Paper 466. US Government Printing Office, Washington DC.

Denny, C.S. (1967) Fans and pediments. *American Journal of Science* **265**, 81–105.

Derbyshire, E. and Goudie, A.S. (1997) Asia. In: Thomas, D.S.G. (ed.), *Arid Zone Geomorphology: Process, Form and Change in Drylands*, 2nd edn. John Wiley & Sons, Chichester, pp. 487–506.

Desconnets, J.C., Taupin, J.D., Lebel, T., and Leduc, C. (1997) Hydrology of the HAPEX-Sahel Central Super-Site: surface water drainage and aquifer recharge through the pool systems. *Journal of Hydrology* **188–9**, 155–78.

Descroix, L., Viramontes, D., Estrada, J., Gonzalez Barrios, J.-L., and Asseline, J. (2007) Investigating the spatial and temporal boundaries of Hortonian and Hewlettian runoff in Northern Mexico. *Journal of Hydrology* **346**, 144–58.

de Smet, K. (1998) Status of the Nile crocodile in the Sahara desert. *Hyrobiologia* **391**, 81–6.

Devitt, D.A., Morris, R.L., and Bowman, D.C. (1992) Evapotranspiration, crop coefficients and leaching fractions of irrigated desert turf-grass systems. *Agronomy Journal* **84**, 717–23.

de Vlaming, V., DiGiorgio, C., Fong, S. et al. (2004) Irrigation runoff insecticide pollution of rivers in the Imperial Valley, California (USA). *Environmental Pollution* **132**, 213–29.

De Vries, J.J. and Simmers, I. (2002) Groundwater recharge: an overview of processes and challenges. *Hydrogeology Journal* **10**, 5–17.

De Vries, J.J., Selaolo, E.T., and Beekman, H.E. (2000) Groundwater recharge in the Kalahari, with reference to paleo-hydrologic conditions. *Journal of Hydrology* **238**, 110–23.

Diem, J.E. and Comrie, A.C. (2001) Air quality, climate, and policy: A case study of ozone pollution in Tucson, Arizona. *The Professional Geographer* **53**, 469–91.

Dillon, M.O. and Rundel, P.W. (1990) The botanical response of the Atacama and Peruvian Desert floras to the 1982–1983 El Niño event. In: Glynn, P.W. (ed.), *Global Ecological Consequences of the 1982–1983 El Niño-*

Southern Oscillation. Elsevier Oceanography Series 52. Elsevier, New York, pp. 487–504.

Ding, D., Bao, H., and Ma, Y. (1998) Progress in the study of desertification in China. *Progress in Physical Geography* **22**, 521–7.

Dixon, J.C. (1994a) Aridic soils, patterned ground, and desert pavements. In: Abrahams, A.D. and Parsons, A.J. (eds.), *Geomorphology of Desert Environments*. Chapman & Hall, London, pp. 64–81.

Dixon, J.C. (1994b) Duricrusts. In: Abrahams, A.D. and Parsons, A.J. (eds.), *Geomorphology of Desert Environments*. Chapman & Hall, London, pp. 82–105.

Dodge, R.E. (1902) Arroyo formation. *Science* **15**, 746.

Dohrenwend, J.C. (1987) Basin and Range. In: Graf, W.L. (ed.), *Geomorphic Systems of North America*. Centennial Special, vol. 2. Geological Society of America, Boulder, CO, pp. 303–42.

Dohrenwend, J.C. (1994) Pediments in arid environments. In: Abrahams, A.D. and Parsons, A.J. (eds.), *Geomorphology of Desert Environments*. Chapman & Hall, London, pp. 321–53.

Dolman, A.J., Gash, J.H.C., Goutorbe, J.-P. et al. (1997) The role of the land surface in Sahelian climate: HAPEX-Sahel results and future research needs. *Journal of Hydrology* **188–9**, 1067–79.

Dorn, R.I. (1983) Cation-ratio dating: a new rock varnish age determination technique. *Quaternary Research* **20**, 49–73.

Dorn, R.I. (1988) A rock varnish interpretation of alluvial fan development in Death Valley, California. *National Geographic Research* **4**, 56–73.

Dorn, R.I. (1995) Alterations of ventifact surfaces at the glacier/desert interface. In: Tchakerian, V.P. (ed.), *Desert Aeolian Processes*. Chapman & Hall, London, pp. 199–217.

Dorn, R.I. (1996) Climatic hypotheses of alluvial-fan evolution in Death Valley are not testable. In: Rhoads, B.L. and Thorn, C.E. (eds.), *The Scientific Nature of Geomorphology: Proceedings of the 27th Binghamton Symposium in Geomorphology*. John Wiley & Sons, New York, pp. 191–220.

Dorn, R.I. and Oberlander, T.M. (1982) Rock varnish. *Progress in Physical Geography* **6**, 317–67.

Dorn, R.I. and Dragovich, D. (1990) Interpretation of rock varnish in Australia: case studies from the arid zone. *Australian Geographer* **21**, 18–31.

Dorn, R.I., Clarkson, P.B., Nobbs, M.F., Loendorf. L.L., and Whitley, D.S. (1992) New approach to the radiocarbon dating of rock varnish, with examples from drylands. *Annals of the Association of American Geographers* **82**, 136–51.

Dragovich, D. (1993) Fire-accelerated boulder weathering in the Pilbara, Western Australia. *Zeitschrift für Geomorphologie NF* **37**, 295–307.

Drake, N.A., Heydeman, M.T., and White, K.H. (1993) Distribution and formation of rock varnish in southern Tunisia. *Earth Surface Processes and Landforms* **18**, 31–41.

Dregne, H.E. (1983) *Desertification of Arid Lands*. Harwood Academic, New York.

Dregne, H.E. (1984) North American deserts. In: El Baz, F. (ed.), *Deserts and Arid Lands*. Martinus Nijhoff Publishers, The Hague, pp. 145–56.

Drew, M.C. (1979) Root development and activities. In: Goodall, D.W. and Perry, R.A. (eds.), *Arid-Land Ecosystems: Structure, Functioning and Management* 1. Cambridge University Press, Cambridge, pp. 573–606.

DuBois, S.M. and Smith, A.W. (1980) The 1887 Earthquake in San Bernardino Valley, Sonora: historic accounts and intensity patterns in Arizona. *Arizona Bureau of Geology and Mineral Technology, Special Paper* no. **3**.

Duffield-Stoll, A.Q. (1994) *Zzyzx: History of an Oasis*. California State University, Northridge, CA.

Duller, G.A.T. and Augustinus, P.C. (2006) Reassessment of the record of linear dune activity in Tasmania using optical dating. *Quaternary Science Reviews* **25**, 2608–18.

Dunkerley, D. (2000a) Measuring interception loss and canopy storage in dryland vegetation: a brief review and evaluation of available research strategies. *Hydrological Processes* **14**, 669–78.

Dunkerley, D.L. (2000b) Hydrologic effects of dryland shrubs: defining the spatial extent of modified soil water uptake rates at an Australian desert site. *Journal of Arid Environments* **45**, 159–72.

Dunkerley, D.L. (2002) Infiltration rates and soil moisture in a groved mulga community near Alice Springs, arid central Australia: evidence for complex internal rainwater redistribution in a runoff-runon landscape. *Journal of Arid Environments* **51**, 199–219.

Dunkerley, D.L. and Booth, T.L. (1999) Plant canopy interception of rainfall and its significance in a banded landscape, arid western NSW, Australia. *Water Resources Research* **35**, 1581–6.

Dunkerley, D.L. and Brown, K.J. (1997) Desert soils. In: Thomas, D.S.G. (ed.), *Arid Zone Geomorphology: Process, Form and Change in Drylands*, 2nd edn. John Wiley & Sons, Chichester, pp. 55–68.

Dunkerley, D. and Brown, K. (1999) Flow behaviour, suspended sediment transport and transmission losses in a small (sub-bank-full) flow event in an Australian desert stream. *Hydrological Processes* **13**, 1577–88.

Dunkerley, D.L. and Brown, K.J. (2002) Oblique vegetation banding in the Australian arid zone: implications for theories of pattern evolution and maintenance. *Journal of Arid Environments* **51**, 163–81.

Dutkiewicz, A. and Prescott, J.R. (1997) Thermoluminescence ages and palaeoclimate from the Lake Malata-Lake Greenly complex, Eyre Peninsula, South Australia. *Quaternary Science Reviews* **16**, 367–85.

Dutton, R.W. (1988) Introduction, overview and conclusions. The scientific results of the Royal Geographical Society's Oman Wahiba Sands Project 1985–1987. *Journal of Oman Studies*, Special Report no. **3**, 1–17.

Dynesius, M. and Nilsson, C. (1994) Fragmentation and flow regulation of river systems in the northern third of the world. *Science* **266**, 732–62.

Eden Again Project (2003) *Building a Scientific Basis for Restoration of the Mesopotamian Marshlands*. Findings of the International Technical Advisory Panel Restoration Planning Workshop. The Iraq Foundation, Irvine, CA.

Edgell, G. (1989) Evolution of the Rub al Khali desert. *Earth Science* **3**, 109–26.

Edmunds, W.M., Fellman, E., and Goni, I.B. (1999) Lakes, groundwater and palaeohydrology in the Sahel of NE Nigeria: evidence from hydrogeochemistry. *Journal of the Geological Society, London* **156**, 345–55.

Eissa, S.M., El-Ziyadi, S.M., and Ibrahim, M.M. (1975) Autoecology of the jerboa, *Jaculus jaculus*, inhabiting Al-Jalia desert area, Kuwait. *Journal of the University of Kuwait (Science)* **2**, 111–22.

El-Baz, F. (1998) Sand accumulation and groundwater in the eastern Sahara. *Episodes* **21**, 147–51.

El-Baz, F., Breed, C.S., Grolier, M.J., and McCauley, J.F. (1979) Eolian features in the Western Desert of Egypt and some applications to Mars. *Journal of Geophysical Research*, **84**, 8205–21.

Eldridge, D.J. (1993) Cryptogams, vascular plants, and soil hydrological relations: some preliminary results from the semiarid woodlands of Eastern Australia. *Great Basin Naturalist* **53**, 48–58.

Embabi, N.S. (1998) Sand seas of the western desert of Egypt. In: Alsharhan, A.S., Glennie, K.W., Whittle, G.L., and Kendall, C. (eds.), *Quaternary Deserts and Climate Change*. Balkema, Rotterdam, pp. 495–509.

Engel, C.G. and Sharp, R. (1958) Chemical data on desert varnish. *Geological Society of America Bulletin* **69**, 487–518.

Engelstaedter, S., Kohfeld, K.E., Tegen, I., and Harrison, S.P. (2003) Controls on dust emissions by vegetation and toporgraphic depressions: An evaluation using dust storm frequency data. *Geophysical Research Letters* **30**, doi:10.1029/2002GL016471.

Enzel, Y. (1990) *Hydrology of a Large, Closed Arid Watershed as a Basis for Paleohydrological Studies in the Mojave River Drainage System: Soda Lake and Silver Lake Playas, Southern California*. Unpublished PhD dissertation, Department of Earth and Planetary Sciences, University of New Mexico, Albuquerque, NM.

Enzel, Y. and Wells, S.G. (1997) Extracting Holocene paleohydrology and paleoclimatology information from modern extreme flood events: an example from southern California. *Geomorphology* **19**, 203–26.

Enzel, Y., Ely, L.L., and Mishra, S. (1999) High-resolution Holocene environmental change in the Thar Desert, Northwestern India. *Science* **284**, 125–8.

Enzel, Y., Knott, J.R., Anderson, K., Anderson, D.E. and Wells, S.G. (2002) Letter to the Editor. Is there any evidence of Mega-Lake Manly in the eastern Mojave Desert during Oxygen Isotope Stage 5e/6? A comment on Hooke R.L. (1999). Lake Manly (?) shorelines in the eastern Mojave Desert, California. *Quaternary Research* **57**, 173–6.

Epstein, E. and Grant, W.J. (1973) Soil crust formation as affected by raindrop impact. In: Hadas, A., Swartzendruber, D., Ritjema, P.E., Fuchs, M., and Yaron, B. (eds.), *Physical Aspects of Soil Water and Salts in Ecosystems*. Springer-Verlag, Berlin, pp. 195–201.

Ericksen, G.E. (1981) *Geology and Origin of the Chilean Nitrate Deposits*. US Geological Survey Professional Paper 1188. US Government Printing Office, Washington DC.

Ericksen, G.E. (1983) The Chilean nitrate deposits. *American Scientist* **71**, 366–74.

Eriksson, P.G., Nixon, N., Snyman, C.P., and Bothma, JduP. (1989) Ellipsoidal parabolic dune patches in the southern Kalahari desert. *Journal of Arid Environments* **16**, 111–24.

Ernst, C. and Barbour, R. (1972) *Turtles of the United States*. University Press of Kentucky, Lexington, KY.

Evans, J.P., Smith, R.B., and Oglesby, R.J. (2004) Middle East climate simulation and dominant precipitation processes. *International Journal of Climatology* **24**, 1671–94.

Evans, M. and Mohieldeen, Y. (2002) Environmental change and livelihood strategies: the case at Lake Chad. *Geography* **87**, 3–13.

Evenari, M. (1985a) The Desert Environment. In: Evenari, M., Noy-Meir, I., and Goodall, D.W. (eds.), *Hot Deserts and Arid Shrublands, A*. Ecosystems of the World, vol. 12A. Elsevier, Amsterdam, pp. 1–22.

Evenari, M. (1985b) Adaptations of plants and animals to the desert environment. In: Evenari, M., Noy-Meir, I., and Goodall, D.W. (eds.), *Hot Deserts and Arid Shrublands, A*. Ecosystems of the World, vol. 12A. Elsevier, Amsterdam, pp. 79–92.

Evenari, M., Shanan, L., and Tadmor, A. (1971) *The Negev-The Challenge of a Desert*. Harvard University Press, Cambridge, MA.

Evenari, M., Yaalon, D.H., and Gutterman, Y. (1974) Notes on soils with vesicular structure in deserts. *Zeitschrift für Geomorphologie* **18**(2), 162–72.

Everard, M. (1996) The importance of periodic droughts for maintaining diversity in the freshwater environment. *Freshwater Forum* **7**, 33–50.

Everitt, B.L. (1979) Fluvial adjustments to the spread of tamarisk in the Colorado Plateau: Discussion. *Geological Society of America Bulletin* **90**, 1183.

Ewing, S.A., Sutter, B., Owen, J. et al. (2006) A threshold in soil formation at Earth's arid-hyperarid transition. *Geochimica et Cosmochimica Acta* **70**, 5293–322.

Ezcurra, E. and Rodrigues, V. (1986) Rainfall patterns in the Gran Desierto, Sonora, Mexico. *Journal of Arid Environments* **10**, 13–28.

Ezcurra, E., Felger, R.S., Russell, A.D., and Equihua, M. (1988) Freshwater Islands in a Desert Sand Sea: the hydrology, flora, and phytogeography of the Gran Desierto Oases of northwestern Mexico. *Desert Plants* **9**, 36–63.

Fall, P.L., Lines, L., and Falconer, S.E. (1998) Seeds of civilization: Bronze Age rural economy and ecology in the Southern Levant. *Annals of the Association of American Geographers* **88**, 107–25.

Fenneman, N.M. (1931) *Physiography of the Western United States*. McGraw-Hill, New York.

Fensham, R.J. and Fairfax, R.J. (2003) Spring wetlands of the Great Artesian Basin, Queensland, Australia. *Wetland Ecology and Management* **11**, 343–62.

Ferri, F., Smith, P.H., Lemmon, M., and Nilton, O.R. (2003) Dust devils as observed by Mars Pathfinder. *Journal of Geophysical Research* **108**, no. E12, 5133, doi:10.1029/2000JE001421.

Fierer, N., Craine, J.M., McLauchlan, K., and Schimel, J.P. (2005) Litter quality and the temperature sensitivity of decomposition. *Ecology* **86**, 320–6.

Fiol, L.A., Fornós, J.J., Gelabert, B., and Guijarro, J.A. (2005) Dust rains in Mallorca (Western Mediterranean): their occurrence and role in some recent geological processes. *Catena* **63**, 64–84.

Fisher, S.G. and Minckley, W.L. (1978) Chemical characteristics of a desert stream in flash flood. *Journal of Arid Environments* **1**, 25–33.

Fitzpatrick, E.A. (1979) Radiation. In: Goodall, D.W., Perry, R.A., and Howes, K.M.W. (eds.), *Arid-Land Ecosystems: Structure, Functioning and Management*. Cambridge University Press, Cambridge, pp. 347–71.

Foley, J.A., Coe, M.T., Scheffer, M., and Wang, G. (2003) Regime shifts in the Sahara and Sahel: interactions between ecological and climatic systems in Northern Africa. *Ecosystems* **6**, 524–32.

Foley, J.C. (1956) 500 mb contour patterns associated with the occurrence of widespread rains in Australia. *Australian Meteorological Magazine* **13**, 1–17.

Fontugne, M., Usselmann, P., Lavallée, D., and Julien, M. (1999) El Niño variability in the coastal desert of southern Peru during the mid-Holocene. *Quaternary Research* **52**, 171–9.

Fornari, M., Risacher, F., and Féraud, G. (2001) Dating of paleolakes in the central Altiplano of Bolivia. *Palaeogeography, Palaeoclimatology, Palaeoecology* **172**, 269–82.

Frank, A. and Kocurek, G. (1996) Toward a model for airflow on the lee side of aeolian dunes. *Sedimentology* **43**, 451–8.

Fredrickson, E., Havstad, K.M., Estell, R., and Hyder, P. (1998) Perspectives on desertification: south-western United States. *Journal of Arid Environments* **39**, 191–207.

Froend, R.H., Heddle, E.M., Bell, D.T., and McComb, A.J. (1987) Effects of salinity and waterlogging on the vegetation of Lake Toolibin, Western Australia. *Australian Journal of Ecology* **12**, 281–98.

Fryberger, S.G. (1979) Dune forms and wind regime. In: McKee, E.D. (ed.), *A Study of Global Sand Seas*. US Geological Survey Professional Paper 1052. US Government Printing Office, Washington DC, pp. 137–69.

Fryberger, S.G. and Goudie, A.S. (1981) Arid geomorphology. *Progress in Physical Geography* **5**, 420–8.

Fullen, M.A. and Mitchell, D.J. (1991) Taming the Shamo 'Dragon'. *Geographical Magazine* **63**, 26–9.

Gabriel, A. (1938) The southern Lut and Iranian Baluchistan. *Geographical Journal* **92**, 193–210.

Galloway, R.W. (1970) The full-glacial climate in the southwestern United States. *Annals of the Association of American Geographers* **60**, 245–56.

Gardner, G.J., Mortlock, A.J., Price, D.M., Readhead, M.L., and Wasson, R.J. (1987) Thermoluminescence and radiocarbon dating of Australian desert dunes. *Australian Journal of Earth Sciences* **34**, 343–57.

Garratt, J.R. (1992) Extreme maximum land surface temperatures. *Journal of Applied Meteorology* **31**, 1096–1105.

Garreaud, R.D. and Aceituno, P. (2001) Interannual rainfall variability over the South American Altiplano. *Journal of Climate* **14**, 2779–89.

Gawthorpe, R.L. and Leeder, M.R. (2000) Tectono-sedimentary evolution of active extensional basins. *Basin Research* **12**, 195–218.

Gay, S.P. (2005) Blowing sand and surface winds in the Pisco to Chala area, Southern Peru. *Journal of Arid Environments* **61**, 101–17.

Gee, G.W., Wierenga, P.J., Andraski, B.J., Young, M.H., Fayer, M.J., and Rockhold, M.L. (1994) Variations in water balance and recharge potential at three western desert sites. *Soil Science Society of America Journal* **58**, 63–72.

Geist, H. (2005) *The Causes and Progression of Desertification.* Ashgate Publishing, Aldershot.

Gellis, A., Hereford, R., Schumm, S.A., and Hayes, B.R. (1991) Channel evolution and hydrologic variations in the Colorado River basin: factors influencing sediment and salt loads. *Journal of Hydrology* **124**, 317–44.

Gellis, A.C., Pavich, M.J., Bierman, P.R., Clapp, E.M., Ellevein, A., and Aby, S. (2004) Modern sediment yield compared to geologic rates of sediment production in a semi-arid basin, New Mexico: assessing the human impact. *Earth Surface Processes and Landforms* **29**, 1359–72.

Genxu, W. and Guodong, C. (1999) Water resource development and its influence on the environment in arid areas of China – the case of the Hei River basin. *Journal of Arid Environments* **43**, 121–31.

Ghassemi, F., Jakeman, A.J., and Nix, H.A. (1995) *Salinisation of Land and Water Resources: Human Causes, Extent, Management and Case Studies.* CAB International, Oxon.

Gheith, H. and Sultan, M. (2002) Construction of a hydrologic model for estimating Wadi runoff and groundwater recharge in the Eastern Desert, Egypt. *Journal of Hydrology* **263**, 36–55.

Gibling, M.R., Nanson, G.C., and Maroulis, J.C. (1998) Anastomosing river sedimentation in the Channel Country of central Australia. *Sedimentology* **45**, 595–619.

Gilbert, G.K. (1875) *Report on the Geology of Portions of Nevada, Utah, California, and Arizona.* Geographical and Geological Surveys West of the 100th Meridian, vol. III. US Government Printing Office, Washington DC.

Gilbert, G.K. (1877) *Report on the Geology of the Henry Mountains: United States Geographical and Geological Survey of the Rocky Mountain Region.* US Government Printing Office, Washington DC.

Gilbert, G.K. (1890) *Lake Bonneville.* US Geological Survey Monograph 1. US Geological Survey, Washington DC.

Gilbert, G.K. (1909) The convexity of hilltops. *Journal of Geology* **17**, 344–51.

Gile, L.H., Peterson, F.F., and Grossman, R.B. (1965) The K horizon – a master soil horizon of carbonate accumulation. *Soil Science* **99**, 74–82.

Gill, A.E. and Rasmusson, E.M. (1983) The 1982–83 climate anomaly in the equatorial Pacific. *Nature* **306**, 229–34.

Gill, T.E. (1996) Eolian sediments generated by anthropogenic disturbance of playas: human impacts on the geomorphic system and geomorphic impacts on the human system. *Geomorphology* **17**, 207–28.

Gillette, D.A. (1999) A qualitative geophysical explanation for "hot spot" dust emission source regions. *Contributions to Atmospheric Physics* **72**, 67–77.

Gillette, D.A. and Hanson, K.J. (1989) Spatial and temporal variability of dust production caused by wind erosion in the United States. *Journal of Geophysical Research* **94**(D2), 2197–206.

Gillette, D.A. and Dobrowolski, J.P. (1993) Soil crust formation by dust deposition at Shaartuz, Tadzhik, S.S.R. *Atmospheric Environment* **27A**, 2519–25.

Gillette, D.A., Adams, J., Endo, A., Smith, D., and Kihl, R. (1980) Threshold velocities for input of soil particles into the air by desert soils, *Journal of Geophysical Research* **85**, 5621–30.

Gillette, D.A., Adams, J., Muhs, D., Smith, D., and Kihl, R. (1982) Threshold friction velocities and rupture moduli for crusted desert soils for the input of soil particles into the air. *Journal of Geophysical Research* **87**, 9003–15.

Glenn, E., Tanner, R., Mendez, S. et al. (1998) Growth rates, salt tolerance and water use characteristics of native and invasive riparian plants from the delta of the Colorado River, Mexico. *Journal of Arid Environments* **40**, 281–94.

Glennie, K.W. (1970) *Desert Sedimentary Environments.* Elsevier, Amsterdam.

Glennie, K.W. (1987) Desert sedimentary environments, past and present – a summary. *Sedimentary Geology* **50**, 135–66.

Glennie, K.W. (1998) The desert of southeast Arabia: a product of Quaternary climatic change. In: Alsharhan, A.S., Glennie, K.W., Whittle, G.L., and Kendall, C.G.St. C. (eds.), *Quaternary Deserts and Climate Change.* Balkema, Rotterdam, pp. 279–91.

Gomes, L., Bergametti, G., Coudé-Gaussen, G., and Rognon, P. (1990) Submicron desert dusts: a sandblasting process. *Journal of Geophysical Research* **95**, 13927–35.

Gonzalez, M.A. (2001) Recent formation of arroyos in the Little Missouri badlands of southwestern North Dakota. *Geomorphology* **38**, 63–84.

Goossens, D. (1995) Field experiments of aeolian dust accumulation on rock fragment substrata. *Sedimentology* **42**, 391–402.

Goudie, A.A. (2007) Mega-yardangs: a global analysis. *Geography Compass* **1**, 65–81.

Goudie, A.S. (1974) Further experimental investigation of rock weathering by salt crystallisation and other mechanical processes. *Zeitschrift für Geomorphologie* suppl. 21(**11**), 1–13.

Goudie, A.S. (1977) Sodium sulphate weathering and the disintegration of Mehenjo-Daro, Pakistan. *Earth Surface Processes* **2**, 75–86.

Goudie, A.S. (1983a) Calcrete. In: Goudie, A.S. and Pye, K. (eds.), *Chemical Sediments and Geomorphology: Precipitates and Residua in the Near Surface Environment.* Academic Press, New York, pp. 93–131.

Goudie, A.S. (1983b) Dust storms in space and time. *Progress in Physical Geography* **7**, 502–30.

Goudie, A.S. (1989) Wind erosion in deserts. *Proceedings of the Geologists' Association* **100**, 89–92.

Goudie, A.S. (1991) The climatic sensitivity of desert margins. *Geography* **76**, 73–6.

Goudie, A.S. (1996) Climate: past and present. In: Adams, W.M., Goudie, A.S., and Orme, A.R. (eds.), *The Physical Geography of Africa.* Oxford University Press, Oxford, pp. 34–59.

Goudie, A.S. (1997) Weathering processes. In: Thomas, D.S.G. (ed.), *Arid Zone Geomorphology: Process, Form and Change in Drylands,* 2nd edn. John Wiley & Sons, Chichester, pp. 25–39.

Goudie, A.S. (1999) Wind erosional landforms; yardangs and pans. In: Goudie, A.S., Livingstone, I., and Stokes, S. (eds.), *Aeolian Environments, Sediments and Landforms.* John Wiley & Sons, Chichester, pp. 167–80.

Goudie, A.S. and Sperling, C.H.B. (1977) Long distance transport of foraminiferal tests by the wind in the Thar Desert, north-west India. *Journal of Sedimentary Petrology* **47**, 630–3.

Goudie, A.S. and Day, M.J. (1980) Disintegration of fan sediments in Death Valley, California, by salt weathering. *Physical Geography* **1**, 126–37.

Goudie, A.S. and Middleton, N.J. (1992) The changing frequency of dust storms through time. *Climatic Change* **18**, 197–225.

Goudie, A.S. and Wells, G.L. (1995) The nature, distribution and formation of pans in arid zones. *Earth-Science Reviews* **38**, 1–69.

Goudie, A.S. and Viles, H. (1997) *Salt Weathering Hazards.* John Wiley & Sons, Chichester.

Goudie, A.S. and Middleton, N.J. (2001) Saharan dust storms: nature and consequences. *Earth-Science Reviews* **56**, 179–204.

Goudie, A.S., Warren, A., Jones, D.K.C., and Cooke, R.U. (1987) The character and possible origins of the aeolian sediments of the Wahiba Sand Sea, Oman. *Geographical Journal* **153**, 231–56.

Goudie, A.S., Viles, H.A., and Parker, A.G. (1997) Monitoring of rapid salt weathering in the central Namib desert using limestone blocks. *Journal of Arid Environments* **37**, 581–98.

Goutorbe, J.-P., Lebel, T., Tinga, A. et al. (1994) HAPEX-Sahel: a large-scale study of land-atmosphere interactions in the semi-arid tropics. *Annales Geophysicae* **12**, 53–64.

Govers, G. and Poesen, J. (1998) Field experiments on the transport of rock fragments by animal trampling on scree slopes. *Geomorphology* **23**, 193–203.

Graetz, R.D. (1991) Desertification: a tale of two feedbacks. In: Mooney, H.A., Medina, E., Schindler, D.W., Schulze E.-D., and Walker, B.W. (eds.), *Ecosystem Experiments: Scope Forty-Five.* John Wiley & Sons, Hoboken, pp. 59–87.

Graf, W.L. (1978) Fluvial adjustments to the spread of tamarisk in the Colorado Plateau region. *Geological Society of America Bulletin* **89**, 1491–1501.

Graf, W.L. (1979) Fluvial adjustments to the spread of tamarisk in the Colorado Plateau region: Reply. *Geological Society of America Bulletin* **90**, 1183–4.

Graf, W.L. (1983a) Downstream changes in stream power in the Henry Mountains, Utah. *Annals of the Association of American Geographers* **73**, 373–87.

Graf, W.L. (1983b) The arroyo problem – palaeohydrology and palaeohydraulics in the short term. In: Gregory, K.J. (ed.), *Background to Palaeohydrology.* John Wiley & Sons, Chichester, pp. 279–302.

Graf, W.L. (1988) *Fluvial Processes in Dryland Rivers.* Springer-Verlag, New York.

Graf, W.L. (1999) Dam nation: a geographic census of American dams and their large-scale hydrologic impacts. *Water Resources Research* **35**, 1305–12.

Greeley, R. (1994) Geology of terrestrial planets with dynamic atmospheres. *Earth, Moon, and Planets* **67**, 13–29.

Greeley, R. and Iversen, J. (1985) *Wind as a Geological Process on Earth, Mars, Venus and Titan.* Cambridge University Press, Oxford.

Greeley, R., Bridges, N., Kuzmin, R.O., and Laity, J.E. (2002) Terrestrial analogs to wind-related features at the Viking and Pathfinder landing sites on Mars. *Journal of Geophysical Research* **107**, 5-1–5-21.

Greeley, R., Balme, M.R., Iversen, J.D., Metzger, S., Mickelson, R. Phoreman, J., and White, B. (2003) Martian dust devils: Laboratory simulations of particle threshold. *Journal of Geophysical Research* **108**, 1–11, No. E5, 5041, doi:10.1029/2002JE001987.

Gregory, H.E. (1917) *Geology of the Navajo Country.* US Geological Survey Professional Paper 93. US Government Printing Office, Washington DC.

Griffin, D.W., Garrison, V.H., Herman, J.R., and Shinn, E.A. (2001) African desert dust in the Caribbean atmosphere: microbiology and public health. *Aerobiologia* **17**, 203–13.

Griffiths, J.F. (1972) Climates of Africa. Landsberg, H.E. (ed), *World Survey of Climatology,* vol. 10. Elsevier, New York.

Griggs, D.T. (1936) The factor of fatigue in rock exfoliation. *Journal of Geology* **74**, 733–96.

Grolier, M.J., McCauley, J.F., Breed, C.S., and Embabi, N.S. (1980) Yardangs of the Western Desert. *Geographical Journal* **146**, 86–7.

Grosjean, M., Núñez, L., Cartajena, I., and Messerli, B. (1997) Mid-Holocene climate and culture change in the Atacama Desert, Northern Chile. *Quaternary Research* **48**, 239–46.

Grove, A.T. (1977) The geography of semi-arid lands. *Philosophical Transactions of the Royal Society of London, Series B* **278**, 457–75.

Gulick, V.C. (2001) Origin of the valley networks on Mars: a hydrological perspective. *Geomorphology* **37**, 241–68.

Guo, Z., Ruddiman, W., Hao, Q. et al. (2002) Onset of Asian desertification by 22 myr ago inferred from loess deposits in China. *Nature* **416**, 1159–63.

Gupta, R.K. (1979) Integration. In: Goodall, D.W. and Perry, R.A. (eds.), *Arid-Lands Ecosystems: Their Structure, Functioning and Management*, vol. 1. Cambridge University Press, Cambridge, pp. 661–75.

Gupta, R.K. (1986) The Thar Desert. In: Evenari, M., Noy-Meir, I., and Goodall, D.W. (eds.), *Hot Deserts and Arid Shrublands, B*. Elsevier, Amsterdam, pp. 55–99.

Gutiérrez, J.R., Arancio, G., and Jacsik, F.M. (2000) Variation in vegetation and seed bank in a Chilean semi-arid community affected by ENSO 1997. *Journal of Vegetation Science* **11**, 641–8.

Gutiérrez, M., Benito, G., and Rodríguez, J. (1988) Piping in badland areas of the Middle Ebro Basin, Spain. In: Harvey, A.M. and Sala, M. (eds.), *Geomorphic Processes in Environments with Strong Seasonal Contrasts, vol. II. Geomorphic Systems*. Catena Supplement 13. Catena-Verlag, Cremlingen-Destedt, pp. 49–60.

Gutiérrez-Elorza, M., Desir, G., and Gutiérrez-Santolalla, F. (2002) Yardangs in the semiarid central sector of the Ebro Depression (NE Spain). *Geomorphology* **44**, 155–70.

Habermehl, M.A. (1982) *Springs in the Great Artesian Basin, Australia – Their Origin and Nature*. Bureau of Mineral Resources, Geology and Geophysics Report 235. Bureau of Mineral Resources, Canberra.

Hack, J.T. (1941) Dunes of the Western Navajo country. *Geographical Review* **31**, 240–63.

Hadley, R.F. (1961) *Influence of Riparian Vegetation on Channel Shape, Northeastern Arizona*. US Geological Survey Professional Paper 424-C. US Government Printing Office, Washington DC, pp. 30–1.

Hadley, R.F. (1967) Pediments and pediment-forming processes. *Journal of Geological Education* **15**, 83–9.

Haff, P.K. and Werner, B.T. (1996) Dynamical processes on desert pavements and the healing of surficial disturbances. *Quaternary Research* **45**, 38–46.

Hagedorn, H. (1971) Untersuchungen über Relieftypen arider Räume an Beispielen aus dem Tibesti-Gebirge und seiner Umgebung. *Zeitschrift für Geomorphologie Supplementband* **11**, 1–271.

Halimov, M. and Fezer, F. (1989) Eight yardang types in central Asia. *Zeitschrift für Geomorphologie NF* **33**, 205–17.

Hall, F.F. (1981) Visibility reductions from soil dust in the western U.S. *Atmospheric Environment* **15**, 1929–33.

Hallmark, C.T. and Allen, B.L. (1975) The distribution of creosotebush in West Texas and eastern New Mexico as affected by selected soil properties. *Soil Science Society of America Proceedings* **39**, 120–4.

Halsey, D.P., Mitchell, D.J., and Dews, S.J. (1998) Influence of climatically induced cycles in physical weathering. *Quarterly Journal of Engineering Geology and Hydrology* **31**, 359–67.

Hara, Y., Uno, I., and Wang, Z. (2006) Long-term variation of Asian dust and related climate factors. *Atmospheric Environment* **40**, 6730–40.

Harbeck, G.E. (1958) *Water-loss Investigations: Lake Mead Studies*. US Geological Survey Professional Paper 298. US Government Printing Office, Washington DC.

Harden, J.W. (1990) Soil development on stable landforms and implications for landscape studies. *Geomorphology* **3**, 391–8.

Hare, F.K. (1980) Long-term annual surface heat and water balances over Canada and the United States south of 60°N: reconciliation of precipitation, runoff and temperature fields. *Atmospheres and Oceans* **18**, 127–53.

Harrison, J.B.J. and Yair, A. (1998) Late Pleistocene aeolian and fluvial interactions in the development of the Nizzana dune field, Negev Desert, Israel. *Sedimentology* **45**, 507–18.

Hartley, A. (2003) Andean uplift and climate change. *Journal of the Geological Society* **160**, 7–10.

Hartley, A.J., Chong, G., Houston, J., and Mather, A.E. (2005) 150 million years of climatic stability: evidence from the Atacama Desert, northern Chile. *Journal of the Geological Society* **162**, 421–4.

Harvey, A. (1982) The role of piping in the development of badlands and gully systems in south-east Spain. In: Bryan, R. and Yair, A. (eds.), *Badland Geomorphology and Piping*. Geo Books, Norwich, pp. 317–25.

Harvey, A.M. (1988) Controls of alluvial fan development: the alluvial fans of the Sierra de Carrascoy, Murcia, Spain. In: Harvey, A.M. and Sala, M. (eds.), *Geomorphic Processes in Environments with Strong Seasonal Contrasts, vol. II. Geomorphic Systems*. Catena Supplement 13. Catena-Verlag, Cremlingen-Destedt, pp. 123–37.

Harvey, A.M. (1997) The role of alluvial fans in arid zone fluvial systems. In: Thomas, D.S.G. (ed.), *Arid Zone Geomorphology: Process, Form and Change in Drylands*, 2nd edn. John Wiley & Sons, Chichester, pp. 231–59.

Hastings, J.R. and Turner, R.M. (1965) *The Changing Mile: an Ecological Study of Vegetation Change With Time in the Lower Mile of an Arid and Semiarid Region*. University of Arizona Press, Tucson, AZ.

Haynes, Jr, C.V. (1980) Geologic evidence of pluvial climates in the El Nabta area of the Western Desert, Egypt. In: Wendorf, F. and Schild, R. (eds.), *Prehistory of the Eastern Sahara*. Academic Press, New York, pp. 353–71.

Haynes, Jr, C.V. (1982) The Darb El-Arba'in desert: a product of Quaternary climatic change. In: El-Baz, F. and Maxwell, T.A. (eds.), *Desert Landforms of Southwest Egypt: a Basis for Comparison with Mars*. Contractor report 3611. National Aeronautics and Space Administration, Washington DC, pp. 91–117.

Haynes, Jr, C.V. (1985) Quaternary Studies, Western Desert, Egypt and Sudan, 1979–1983 field studies. *National Geographic Society Research Reports* 19, 269–341.

Haynes, Jr, C.V. (2001) Geochronology and climate change of the Pleistocene-Holocene transition in the Darb el Arba'in Desert, eastern Sahara. *Geoarchaeology: An International Journal* 16, 119–41.

Hazan, N., Stein, M., Agnon, A. et al. (2005) The late Quaternary limnological history of Lake Kinneret (Sea of Galilee), Israel. *Quaternary Research* 63, 60–77.

Hedin, S. (1903) *Central Asia and Tibet*. Greenwood Press, New York.

Helgren, D.M. and Prospero, J.M. (1987) Wind velocities associated with dust deflation events in the Western Sahara. *Journal of Applied Meteorology* 26, 1147–51.

Hendrickson, D.A. and Minckley, W.L. (1984) Cienegas-Vanishing climax communities of the American Southwest. *Desert Plants* 6, 131–75.

Hennessy, J.T., Gibbens, R.P., Tromble, J.M., and Cardenas, M. (1985) Mesquite (*Prosopis glandulosa* Torr.) dunes and interdunes in southern New Mexico: a study of soil properties and soil water relations. *Journal of Arid Environments* 9, 27–38.

Hereford, R. (2002) Valley-fill alluviation during the Little Ice Age (ca. A.D. 1400–1880), Paria River basin and southern Colorado Plateau, United States. *Geological Society of America Bulletin* 114, 1550–63.

Hereford, R. and Huntoon, P.W. (1990) Rock movement and mass wastage in the Grand Canyon. In: Beuss, S.S. and Morales, M. (eds.), *Grand Canyon Geology*. Oxford University Press, New York, pp. 433–59.

Herrmann, S.M. and Hutchinson, C.F. (2005) The changing contexts of the desertification debate. *Journal of Arid Environments* 63, 538–55.

Hesp, P.A. (1981) Formation of shadow dunes. *Journal of Sedimentary Research* 51, 101–12.

Hesp, P.A., Hyde, R., Hesp, V.J., and Zhengyu, Q. (1989) Longitudinal dunes can move sideways. *Earth Surface Processes and Landforms* 14, 447–51.

Hesse, P. and McTainsh, G.H. (2003) Australian dust deposits: modern processes and the Quaternary record. *Quaternary Science Reviews* 22, 2007–35.

Hesse, P.P. and Simpson, R.L. (2006) Variable vegetation cover and episodic sand movement on longitudinal desert sand dunes. *Geomorphology* 81, 276–91.

Hidore, J.J. and Oliver, J.E. (1993) *Climatology: an Atmospheric Science*. Macmillan, New York.

Higgins, C.G. (1990) Gully development. In: Higgins, C.G. and Coates, D.R. (eds.), *Groundwater Geomorphology: the Role of Subsurface Water in Earth-Surface Processes and Landforms*. Geological Society of America Special Paper 252, 139–55.

Higgins, R.W., Chen, Y., and Douglas, A.V (1999) Interannual variability of the North American warm season precipitation regime. *Journal of Climate* 12, 653–80.

Hilton-Taylor, C. (2000) *2000 IUCN Red List of Threatened Species*. IUCN, Gland.

Hobbs, W.H. (1917) The erosional and degradational processes of deserts, with especial reference to the origin of desert depressions. *Annals of the Association of American Geographers* 7, 25–60.

Hoke, G.D., Isacks, B.L., Jordan, T.E., and Yu, J.S. (2004) Groundwater-sapping origin for the giant quebradas of northern Chile. *Geology* 32, 605–8.

Holden, J., Burt, T.P., and Vilas, M. (2002) Application of ground-penetrating radar to the identification of subsurface piping in blanket peat. *Earth Surface Processes and Landforms* 27, 235–49.

Hollands, C.B., Nanson, G.C., Jones, B.G., Bristow, C.S., Price, D.M., and Pietsch, T.J. (2006) Aeolian-fluvial interaction: evidence for late Quaternary channel change and wind-rift linear dune formation in the northwestern Simpson Desert, Australia. *Quaternary Science Reviews* 25, 142–62.

Holliday, V.T. (1997) Origin and evolution of lunettes on the High Plains of Texas and New Mexico. *Quaternary Research* 47, 54–69.

Holmgren, M., Scheffer, M., Ezcurra, E., Gutiérrez, J.R., and Mohren, G.M.J. (2001) El Niño effects on the dynamics of terrestrial ecosystems. *Trends in Ecology and Evolution* 16, 89–94.

Honda, M. and Shimizu, H. (1998) Geochemical, mineralogical and sedimentological studies on the Taklimakan Desert sands. *Sedimentology* 45, 1125–43.

Hooke, R.L. (1967) Processes on arid region alluvial fans. *Journal of Geology* 75, 438–60.

Hooke, R.L. (1972) Geomorphic evidence for late-Wisconsin and Holocene tectonic deformation, Death Valley, California. *Geological Society of America Bulletin* 85, 2073–98.

Hooke, R.L. (1999) Lake Manly (?) shorelines in the eastern Mojave Desert. *Quaternary Research* 52, 528–36.

Hooke, R.L. (2002) Is there any evidence of Mega-Lake Manly in the eastern Mojave Desert during Oxygen Isotope Stage 5e/6?: Reply. *Quaternary Research* 57, 177–9.

Houston, J. (2006a) Evaporation in the Atacama Desert: An empirical study of spatio-temporal variations and their causes. *Journal of Hydrology* 330, 402–12.

Houston, J. (2006b) Variability of precipitation in the Atacama Desert: Its causes and hydrologic impact. *International Journal of Climatology* 26, 2181–98.

Howard, A.D. (1942) Pediment passes and the pediment problem. *Journal of Geomorphology* 5, 3–31, 95–136.

Howard, A.D. (1994) Badlands. In: Abrahams, A.D. and Parsons, A.J. (eds.), *Geomorphology of Desert Environments*. Chapman & Hall, London, pp. 211–42.

Howard, A.D. (1995) Simulation modeling and statistical classification of escarpment planforms. *Geomorphology* **12**, 187–214.

Howard, A.D. and Kochel, R.C. (1988) Introduction to cuesta landforms and sapping processes on the Colorado Plateau. In: Howard, A.D., Kochel, R.C., and Holt, H.E. (eds.), *Sapping Features on the Colorado Plateau*. Proceedings and Field Guide for the NASA Groundwater Sapping Conference. National Aeronautics and Space Administration Special Publication SP-491. National Aeronautics and Space Administration, Washington DC, pp. 6–56.

Howard, A.D. and Selby, M.J. (1994) Rock slopes. In: Abrahams, A.D. and Parsons, A.J. (eds.), *Geomorphology of Desert Environments*. Chapman & Hall, London, pp. 123–72.

Hubert, J.F. and Filipov, A.J. (1989) Debris–flow deposits in alluvial fans on the west flank of the White Mountains, Owens Valley, California, USA. *Sedimentary Geology* **61**, 177–205.

Huinink, H.P., Pel, L., and Kopinga, K. (2004) Simulating the growth of tafoni. *Earth Surface Processes and Landforms* **29**, 1225–33.

Hull, Jr, A.C. (1972) Rainfall and snowfall interception of big sagebrush. *Utah Academy of Science, Arts and Letters* **49**, 64.

Hulme, M. (1996) Recent climatic change in the World's drylands. *Geophysical Research Letters* **23**, 61–4.

Hulme, M., Doherty, R., Ngara, T., New, M., and Lister, D. (2001) African climate change: 1900–2100. *Climate Research* **17**, 145–68.

Hume, W.F. (1925) *Geology of Egypt, vol. I., The Surface Features of Egypt, their Determining Causes and Relation to Geological Structure*. Government Press, Cairo.

Hunt, C.B. (1974) *Natural Regions of the United States and Canada*. W.H. Freeman, San Francisco.

Hunt, C.B., Robinson, T.W., Bowles, W.A., and Washburn, A.L. (1966) *Hydrologic Basin, Death Valley, California*. US Geological Survey Professional Paper 494-B. US Government Printing Office, Washington DC.

Hunt, H.A. (1929) *Results of Rainfall Observations in Western Australia*. Commonwealth Bureau of Meteorology, Melbourne.

Huntington, E. (1914) *The Climatic Factor as Illustrated in Arid America*. Publication 192. Carnegie Institute of Washington, Washington DC.

Husar, R.B., Prospereo, J.M., and Stowe, L.L. (1997) Characterization of tropospheric aerosols over the oceans with the NOAA advanced very high resolution radiometer optical thickness operational product. *Journal of Geophysical Research* **102**, 16899–909.

Husar, R.B., Tratt, D.M., Schichtel, B.A. et al. (2001) Asian dust of April 1998. *Journal of Geophysical Research* **106**, 18317–30.

Idso, S.B. (1967) Dust storms. *Scientific American* **235**, 108–14.

Ikehara, M.E. and Phillips, S.P. (1994) *Determination of Land Subsidence Related to Groundwater-level Declines using Global Positioning System and Leveling Surveys in Antelope Valley, Los Angeles and Kern Counties, California*. US Geological Survey Water Resources Investigations Report 94-4184. US Geological Survey, Denver, CO.

Imeson, A.C. and Verstraten, J.M. (1981) Suspended solids concentrations and river water chemistry. *Earth Surface Processes and Landforms* **6**, 251–63.

Imeson, A.C. and Verstraten, J.M. (1988) Rills on badland slopes: a physico-chemically controlled phenomenon. In: Imeson, A.C. and Sala, M. (eds.), *Geomorphic Processes in Environments with Strong Seasonal Contrasts: vol. I. Hillslope Processes*. Catena Supplement 12. Catena-Verlag, Cremlingen-Destedt, pp. 139–50.

Inbar, M. and Risso, C. (2001) Holocene yardangs in volcanic terrains in the southern Andes, Argentina. *Earth Surface Processes and Landforms* **26**, 657–66.

Inman, D.L., Ewing, G.C., and Corliss, J.B. (1966) Coastal sand dunes of Guerrero Negro, Baja California, Mexico. *Geological Society of America Bulletin* **77**, 787–802.

Inouye, R.S. (1991) Population biology of desert annual plants. In: Polis, G.A. (ed.), *The Ecology of Desert Communities*. University of Arizona Press, Tucson, AZ, pp. 27–54.

Iverson, R.M. (1997) The physics of debris flows. *Reviews of Geophysics* **35**, 245–96.

Izbicki, J.A., Martin, P., and Michel, R.L. (1995) Source, movement and age of groundwater in the upper part of the Mojave River basin, California, USA. In: Adar, E.M. and Leibundgut, C. (eds.), *Application of Tracers in Arid Zone Hydrology (Proceedings of the Vienna Symposium)*. IAHS Publication 232. IAN Press, CEH, Wallingford, pp. 43–56.

Izbicki, J.A., Radyk, J., and Michel, R.L. (2000) Water movement through a thick unsaturated zone underlying an intermittent stream in the western Mojave Desert, southern California, USA. *Journal of Hydrology* **238**, 194–217.

Jackson, E.C., Krogh, S.N., and Whitford, W.G. (2003) Desertification and biopedturbation in the northern Chihuahuan Desert. *Journal of Arid Environments* **53**, 1–14.

Jackson, P.S. and Hunt, J.C.R. (1975) Turbulent wind flow over a low hill. *Quarterly Journal of the Royal Meteological Society* **101**, 929–55.

Jacsik, F.M. (2001) Ecological effects of El Niño in terrestrial ecosystems of western South America. *Ecography* **24**, 241–50.

Jain, M. and Tandon, S.K. (2003) Fluvial response to Late Quaternary climate changes, western India. *Quaternary Science Reviews* **22**, 2223–35.

Jankowski, J. and Jacobson, G. (1989) Hydrochemical evolution of regional groundwaters to playa brines in central Australia. *Journal of Hydrology* **108**, 123–73.

Jayko, A.S. (2005) Late Quaternary denudation, Death and Panamint Valleys, eastern California. *Earth-Science Reviews* **73**, 271–89.

Jefferson, G.T. (1991) The Camp Cady local fauna: stratigraphy and paleontology of the Lake Manix basin. *San Bernardino County Museum Association Quarterly* **38**, 93–9.

Jennings, J.N. (1968) A revised map of the desert dunes of Australia. *Australian Geographer* **10**, 408–9.

Jessup, R.W. (1960) The stony tableland soils of the southeastern portion of the Australian arid zone and their evolutionary history. *Journal of Soil Science* **11**, 188–97.

Jimenez, J.A., Maia, L.P., Serra, J., and Morais, J. (1999) Aeolian dune migration along the Ceará coast, northeastern Brazil. *Sedimentology* **46**, 689–701.

Jones, C.J., Lawton, J.H., and Shachak, M. (1994) Organisms as ecosystem engineers. *Oikos* **69**, 373–86.

Jones, J.A. (1971) Soil piping and channel initiation. *Water Resources Research* **7**, 602–10.

Jones, J.A.A. (1990) Piping effects in humid lands. In: Higgins, C.G. and Coates, D.R. (eds.), *Groundwater Geomorphology: the Role of Subsurface Water in Earth-Surface Processes and Landforms*. Geological Society of America Special Paper **252**, 111–38.

Jordan, P.W. and Nobel, P.S. (1979) Infrequent establishment of seedlings of Agave deserti (Agavaceae) in the northwestern Sonoran Desert. *American Journal of Botany* **66**, 1079–84.

Jürgens, N. (1997) Floristic biodiversity and history of African arid regions. *Biodiversity and Conservation* **6**, 795–814.

Juyal, N., Singhvi, A.K., and Glennie, K.W. (1997) Chronology and paleoenvironmental significance of Quaternary desert sediment in southeastern Arabia. In: Alsharan, A.S., Glennie, K.W., Whittle, G.L., and Kendall, C. (eds.), *Quaternary Deserts and Climate Change*. Balkema, Rotterdam, pp. 315–25.

Juyal, N., Chamyal, L.S., Bhandari, S., Bhushan, R., and Singhvi, A.K. (2006) Continental record of the southwest monsoon during the last 130 ka: evidence from the southern margin of the Thar Desert, India. *Quaternary Science Reviews* **25**, 2632–50.

Kajale, M.D. and Deotare, B.C. (1997) Late Quaternary environmental studies on salt lakes in western Rajasthan, India: a summarised view. *Journal of Quaternary Science* **12**, 405–12.

Kale, V.S., Singhvi, A.K., Mishra, P.K., and Banerjee, D. (2000) Sedimentary records and luminescence chronology of Late Holocene palaeofloods in the Luni River, Thar Desert, northwest India. *Catena* **40**, 337–58.

Kampf, S.K., Tyler, S.W., Ortiz, C.A., Muñoz, J.F., and Adkins, P.L. (2005) Evaporation and land surface energy budget at the Salar de Atacama, Northern Chile. *Journal of Hydrology* **310**, 236–52.

Kar, A. (1993) Aeolian processes and bedforms in the Thar Desert. *Journal of Arid Environments* **25**, 83–96.

Kar, A. and Takeuchi, K. (2004) Yellow dust: an overview of research and felt needs. *Journal of Arid Environments* **59**, 167–87.

Kar, A., Felix, C., Rajaguru, S.N., and Singhvi, A.K. (1998) Late Holocene growth and mobility of a transverse dune in the Thar Desert. *Journal of Arid Environments* **38**, 175–85.

Kar, A., Singhvi, A.K., Rajaguru, S.N. et al. (2001) Reconstruction of the late Quaternary environment of the lower Luni Plains, Thar Desert, India. *Journal of Quaternary Science* **16**, 61–8.

Kariuki, P.C., Woldai, T., and van der Meer, F. (2004) The role of remote sensing in mapping swelling soils. *Asian Journal of Geoinformatics* **5**, 43–54.

Kawamura, R. (1951) *Study of Sand Movement by Wind*. Report 5. Institute of Science and Technology, Tokyo, pp. 95–112.

Kellogg, C.A. and Griffin, D.W. (2006) Aerobiology and the global transport of desert dust. *Trends in Ecology and Evolution* **21**, 638–44.

Kershaw, P., Moss, P., and Van Der Kaars, S. (2003) Causes and consequences of long-term climatic variability on the Australian continent. *Freshwater Biology* **48**, 1274–83.

Key, L.J., Delph, L.F., Thompson, D.B., and Van Hoogenstyn, E.P. (1984) Edaphic factors and the perennial plant community of a Sonoran Desert bajada. *Southwestern Naturalist* **29**, 211–22.

Kidron, G.J. (2005) Angle and aspect dependent dew and fog precipitation in the Negev desert. *Journal of Hydrology* **301**, 66–74.

Kidron, G.J. and Pick, K. (2000) The limited role of localized convective storms in runoff production in the western Negev Desert. *Journal of Hydrology* **229**, 281–89.

Kim, J. and Sultan, M. (2002) Assessment of the long-term hydrologic impacts of Lake Nasser and related irrigation projects in Southwestern Egypt. *Journal of Hydrology* **262**, 68–83.

Kim, Y.J., Kim, K.W., and Oh, S.J. (2001) Seasonal characteristics of haze observed by continuous visibility monitoring in the urban atmosphere of Kwangju, Korea. *Environment Monitoring and Assessment* **70**, 35–46.

King, G.C.P. and Vita-Finzi, C. (1981) Active folding in the Algerian earthquake of 10 October 1980. *Nature* **292**, 22–6.

Kinlaw, A. (1999) A review of burrowing by semi-fossorial vertebrates in arid environments. *Journal of Arid Environments* **41**, 127–45.

Klausmeier, C.A. (1999) Regular and irregular patterns in semiarid vegetation. *Science* **284**, 1826–8.

Klemmedson, J.O. and Smith, J.G. (1964) Cheatgrass (Bromus tectorum L.) *The Botanical Review* **30**, 226–62.

Knapp, P.A. (1992) Secondary plant succession and vegetation recovery in two western Great Basin Desert ghost towns. *Biological Conservation* **60**, 81–9.

Knetsch, G. and Yallouze, M. (1955) Remarks on the origin of the Egyptian oasis depressions. *Bulletin Societe Géographique Egypte* **28**, 21–33.

Knight, A.W., McTainsh, G.H., and Simpson, R.W. (1995) Sediment loads in an Australian dust storm: implications for present and past dust processes. *Catena* **25**, 195–213.

Knighton, A.D. and Nanson, G.C. (1994) Flow transmission along an arid zone anastomosing river, Cooper Creek, Australia. *Hydrological Processes* **8**, 137–54.

Knighton, D. and Nanson, G. (1997) Distinctiveness, diversity and uniqueness in arid zone river systems. In: Thomas, D.S.G. (ed.), *Arid Zone Geomorphology: Process, Form and Change in Drylands*, 2nd edn, John Wiley & Sons, Chichester, pp. 185–203.

Kocurek, G. (1998) Aeolian system response to external forcing factors – a sequence stratigraphic view of the Sahara Region. In: Alsharhan, A.S., Glennie, K.W., Whittle, G.L., and Kendall, C.G. St. C. (eds.), *Quaternary Deserts and Climate Change*. Balkema, Rotterdam, pp. 327–37.

Kocurek, G. and Nielson, J. (1986) Conditions favorable for the formation of warm-climate aeolian sand sheets. *Sedimentology* **33**, 795–816.

Kocurek, G. and Lancaster, N. (1999) Aeolian system sediment state: theory and Mojave Desert Kelso dune field example. *Sedimentology* **46**, 505–15.

Kocurek, G. and Ewing, R.C. (2005) Aeolian dune field self-organization – implications for the formation of simple versus complex dune-field patterns. *Geomorphology* **72**, 94–105.

Kohfeld, K.E. and Harrison, S.P. (2001) DIRTMAP: the geological record of dust. *Earth-Science Reviews* **54**, 81–114.

Kolla, V. and Biscaye, P.C. (1977) Distribution and origin of quartz in the sediments of the Indian Ocean. *Journal of Sedimentary Petrology* **47**, 642–49.

Koons, D. (1955) Cliff retreat in southwestern United States. *American Journal of Science* **253**, 44–52.

Koren, I., Kaufman, Y.J., Washington, R. et al. (2006) The Bodélé depression: a single spot in the Sahara that provides most of the mineral dust to the Amazon forest. *Environmental Research Letters* **1**, 014005, doi:10.1088/1748-9326/1/1/014005.

Kotlyakov, V.M. (1991) The Aral Sea Basin, a critical environmental zone. *Moscow Environment* **33**, 4–9, 36–8.

Kotwicki, V. and Allan, R. (1998) La Niña de Australia – contemporary and palaeo-hydrology of Lake Eyre. *Palaeogeography, Palaeoclimatology, Palaeoecology* **144**, 265–80.

Krapf, C.B.E., Stollhofen, H., and Stanistreet, I.G. (2003) Contrasting styles of ephemeral river systems and their interaction with dunes of the Skeleton Coast erg (Namibia). *Quaternary International* **104**, 41–52.

Kresan, P.L. (1988) Tucson, Arizona, Flood of October 1983. In: Baker, V.R., Kochel, R.C., and Patton, P.C. (eds.), *Flood Geomorphology*. John Wiley & Sons, New York, pp. 465–89.

Krinsley, D.B. (1970) A geomorphological and palaeoclimatological study of the playas of Iran. *US Geological Survey Final Report*, contract no. PRO CP 70–800. US Government Printing Office, Washington DC.

Krumbein, W.E. and Jens, K. (1981) Biogenic rock varnishes of the Negev Desert (Israel): an ecological study of iron and manganese transformation by cyanobacteria and fungi. *Oecologia* **50**, 25–38.

Kuenen, P.H. (1960) Sand. *Scientific American* **202**, 94–110.

Kuhn, N.J., Yair, A., and Grubin, M.K. (2004) Spatial distribution of surface properties, runoff generation and landscape development in the Zin Valley Badlands, Northern Negev, Israel. *Earth Surface Processes and Landforms* **29**, 1417–30.

Kumar, M. and Bhandari, M.M. (1992) Impact of protection and free grazing on sand dune vegetation in the Rajasthan Desert, India. *Land Degradation and Rehabilitation* **3**, 215–27.

Kutzbach, J.E. and Wright, Jr, H.E. (1985) Simulation of the climate of 18,000 B.P.: results for the North American/North Atlantic/European Sector and comparison with the geologic record of North America. *Quaternary Science Reviews* **4**, 147–87.

Kutzbach, J.E. and Guetter, P.J. (1986) The influence of changing orbital parameters and surface boundary conditions on climate simulations for the past 18,000 years. *Journal of Atmospheric Science* **43**, 1726–59.

LADWP 2005. *Mitigation measures.* http://wsoweb.ladwp.com/Aqueduct/EnvironmentalProjects/owenslakedustmitigation/mitigation.htm (accessed 12 September 2005).

Laity, J.E. (1983) Diagenetic controls on groundwater sapping and valley formation, Colorado Plateau, revealed by optical and electron microscopy. *Physical Geography* **4**, 103–25.

Laity, J.E. (1987) Topographic effects on ventifact formation, Mojave Desert, California. *Physical Geography* **2**, 113–32.

Laity, J.E. (1988) The role of groundwater sapping in valley evolution on the Colorado Plateau. In: Howard, A.D., Kochel, R.C., and Holt, H.E. (eds.), *Sapping features of the Colorado Plateau, a Comparative Planetary Geology Field Guide*. National Aeronautics and Space Administration Special Publication 491. National Aeronautics and Space Administration, Washington DC, pp. 63–70.

Laity, J.E. (1992) Ventifact evidence for Holocene wind patterns in the east-central Mojave Desert. *Zeitschrift für Geomorphologie NF Supplementband* **84**, 1–16.

Laity, J.E. (1994) Landforms of aeolian erosion. In: Abrahams, A.D. and Parsons, A.J. (eds.), *Geomorphology of Desert Environments*. Chapman & Hall, London, pp. 506–35.

Laity, J.E. (1995) Wind abrasion and ventifact formation in California. In: Tchakerian, V.P. (ed.), *Desert Aeolian Processes*. Chapman & Hall, London, pp. 296–321.

Laity, J.E. (2002) Desert Environments. In: Orme, A.R. (ed.), *The Physical Geography of North America*. Oxford University Press, New York, pp. 380–401.

Laity, J.E. (2003) Aeolian destabilization along the Mojave River, Mojave Desert, California: linkages among fluvial, groundwater, and aeolian systems. *Physical Geography* **24**, 196–221.

Laity, J.E. and Malin, M.C. (1985) Sapping processes and the development of theater-headed valley networks on the Colorado Plateau. *Geological Society of America Bulletin* **96**, 203–17.

Lake, P.S. (2003) Ecological effects of perturbation by drought in flowing waters. *Freshwater Biology* **48**, 1161–72.

Lakin, H.W., Hunt, C.B., Davidson, D.F., and Oda, U. (1963) *Variations in Minor-Element Content of Desert Varnish*. US Geological Survey Professional Paper 475B. US Government Printing Office, Washington DC, pp. 28–31.

Lamb, H.F. (2001) Multi-proxy records of Holocene climate and vegetation change from Ethiopian Crater Lakes. *Biology and Environment: Proceedings of the Royal Irish Academy*, **101B**, 35–46.

Lamb, P.J. and Peppler, R.A. (1987) North Atlantic Oscillation: concept and an application. *Bulletin of the American Meteorological Society* **68**, 1218–25.

Lamplugh, G.H. (1902) Calcrete. *Geological Magazine* **9**, 75.

Lamplugh, G.H. (1907) The geology of the Zambezi Basin around the Batoka Gorge (Rhodesia). *Quarterly Journal of the Geological Society of London* **63**, 162–216.

Lancaster, J., Lancaster N., and Seely, M.K. (1984) Climate of the central Namib Desert. *Madoqua* **14**, 5–61.

Lancaster, N. (1982) Linear dunes. *Progress in Physical Geography* **6**, 476–504.

Lancaster, N. (1983) Linear dunes of the Namib sand sea. *Zeitschrift für Geomorphologie NF Supplementband* **45**, 27–49.

Lancaster, N. (1985) Variations in wind velocity and sand transport on the windward flanks of desert sand dunes. *Sedimentology* **32**, 581–93.

Lancaster, N. (1988) Development of linear dunes in the southwestern Kalahari, southern Africa. *Journal of Arid Environments* **14**, 233–44.

Lancaster, N. (1989) Star dunes. *Progress in Physical Geography* **13**, 67–92.

Lancaster, N. (1990) Palaeoclimatic evidence from sand seas. *Palaeogeography, Palaeoclimatology, Palaeoecology* **76**, 279–90.

Lancaster, N. (1995) *Geomorphology of Desert Dunes*. Routledge, London.

Lancaster, N. (2002) How dry was dry? Late Pleistocene palaeoclimates in the Namib Desert. *Quaternary Science Reviews* **21**, 769–82.

Lancaster, N. and Teller, J.T. (1988) Interdune deposits of the Namib Sand Sea. *Sedimentary Geology* **55**, 91–108.

Lancaster, N. and Nickling, W.G. (1994) Aeolian sediment transport. In: Abrahams, A.D. and Parsons, A.J. (eds.), *Geomorphology of Desert Environments*. Chapman & Hall, London, pp. 447–73.

Lancaster, N. and Tchakerian, V.P. (1996) Geomorphology and sediments of sand ramps in the Mojave Desert. *Geomorphology* **17**, 151–65.

Lancaster, N. and Baas, A. (1998) Influence of vegetation cover on sand transport by wind: field studies at Owens Lake, California. *Earth Surface Processes and Landforms* **23**, 69–82.

Lancaster, N., Nickling, W.G., McKenna Neumann, C.K., and Wyatt, V.E. (1996) Sediment flux and airflow on the stoss slope of a barchan dune. *Geomorphology* **17**, 55–62.

Lancaster, N., Kocurek, G., Singhvi, A., Pandey, V., Deynoux, M., Ghienne, J.-F., and Khalidou, L. (2002) Late Pleistocene and Holocene dune activity and wind regimes in the western Sahara Desert of Mauritania. *Geology* **30**, 991–4.

Landsberg, H.E. (1965) Global distribution of solar and sky radiation. In: Rodenwaldt, E. and Jusatz, H.J. (eds.), *World Maps of Climatology*, 2nd edn. Springer-Verlag, Berlin, pp. 1–6.

Langbein, W.B. and Schumm, S.A. (1958) Yield of sediment in relation to mean annual precipitation. *Transactions of the American Geophysical Union* **39**, 1076–84.

Lange, J. (2005) Dynamics of transmission losses in a large arid stream channel. *Journal of Hydrology* **306**, 112–26.

Lange, O.L. (1990) Twenty-three years of growth measurements on the crustose lichen *Caloplaca aurantia* in the central Negev Desert. *Israel Journal of Botany* **39**, 383–94.

Langford, R.P. (2003) The Holocene history of the White Sands dune field and influences on eolian deflation and playa lakes. *Quaternary International* **104**, 31–9.

Langford-Smith, T. (1978) Review of silcrete research in Australia. In: Langford-Smith, T. (ed.), *Silcrete in Australia*. Department of Geography, University of New England, Armidale, NSW, pp. 1–11.

Larmuth, J. (1984) Microclimates. In: Cloudsley-Thompson, J.L. (ed.), *Key Environments: Sahara Desert*. Pergamon Press, Oxford, pp. 57–66.

Laronne, J.B. and Reid, I. (1993) Very high rates of bedload sediment transport by ephemeral desert rivers. *Nature* **366**, 148–50.

Laronne, J.B., Reid, I., Yitshak, Y., and Frostick, E. (1994) The nonlayering of gravel streambeds under ephemeral flood regimes. *Journal of Hydrology* **159**, 353–63.

Larrain, H., Velásquez, F., Cereceda, P. et al. (2002) Fog measurements at the site "Falda Verde" north of Chañaral compared with other fog stations of Chile. *Atmospheric Research* **64**, 273–84.

Latorre, C., Betancourt, J.L., Rylander, K.A., and Quade, J. (2002) Vegetation invasions into absolute desert: a 45000 yr rodent midden record from the Calama-Salar de Atacama basins, northern Chile (lat 22°–24°S). *Geological Society of America Bulletin* **114**, 349–66.

Latorre, C., Betancourt, J.L., Rech, J.A. et al. (2005) Late Quaternary history of the Atacama Desert. In: Smith, M. and Hesse, P. (eds.), *23°S: Archaeology and Environmental History of the Southern Deserts*. National Museum of Australia Press, Canberra.

Lattman, L.H. (1973) Calcium carbonate cementation of alluvial fans in southern Nevada. *Bulletin of the Geological Society of America* **84**, 3013–28.

Laudermilk, J.D. (1931) On the origin of desert varnish. *American Journal of Science, Series 5* **21**, 51–66.

Lavauden, L. (1927) Les forêts du Sahara. *Revue des Eaux et Forêts* **65**(6), 265–77; **65**(7), 329–41.

Lawson, A.C. (1915) The epigene profiles of the desert. *University of California Department of Geology Bulletin* **9**, 23–48.

Lawson, M.P. and Thomas, D.S.G. (2002) Late Quaternary lunette dune sedimentation in the southwestern Kalahari desert, South Africa: luminescence based chronologies of aeolian activity. *Quaternary Science Reviews* **21**, 825–36.

Lee, J.A., Wigner, K.A., and Gregory, J.M. (1993) Drought, wind, and blowing dust on the Southern High Plains of the United States. *Physical Geography* **14**, 56–67.

Leeder, M.R. (1999) *Sedimentology and Sedimentary Basins*. Blackwell Science, Oxford.

Leeder, M.R. and Jackson, J.A. (1993) The interaction between normal faulting and drainage in active extensional basins, with examples from the western United States and central Greece. *Basin Research* **5**, 79–102.

Le Floc'h, E., Le Houérou, H.N., and Mathez, J. (1990) History and patterns of plant invasion in Northern Africa. In: di Castri, F., Hansen, A.J., and Debussche, M. (eds.), *Biological Invasions in Europe and the Mediterranean Basin*. Kluwer Academic Publishers, Dordrecht, pp. 105–33.

Le Houérou, H.N. (1980) The Rangelands of the Sahel. *Journal of Range Management* **33**, 41–6.

Le Houérou, H.N. (1986) The desert and arid zones of Northern Africa. In: Evenari, M., Noy-Meir, I., and Goodall, D.W. (eds.), *Hot Deserts and Arid Shrublands, B.* Elsevier, Amsterdam, pp. 101–47.

Le Houérou, H.N. (2002) Man-made deserts: desertization processes and threats. *Arid Land Research and Management* **16**, 1–36.

Leopold, L.B. (1951a) Pleistocene climate in New Mexico. *American Journal of Science* **249**, 152–68.

Leopold, L.B. (1951b) Rainfall frequency: an aspect of climatic variation. *American Geophysical Union Transactions* **32**, 347–57.

Leopold, L.B. and Maddock, T. (1953) *Hydraulic Geometry of Stream Channels and Some Physiographic Implications*. US Geological Survey Professional Paper 252. US Government Printing Office, Washington DC.

Leopold, L.B. and Miller, J.P. (1956) *Ephemeral Streams – Hydraulic Factors and their Relation to the Drainage Net*. US Geological Survey Professional Paper 282-A. US Government Printing Office, Washington DC.

Leopold, L.B., Wolman, M.G., and Miller, J.P. (1964) *Fluvial Processes in Geomorphology*. W.H. Freeman, San Francisco, CA.

Leroy, S.A.G., Marret, F., Gibert, E., Chalié, F., Reyss, J.-L., and Arpe, K. (2007) River inflow and salinity changes in the Caspian Sea during the last 5500 years. *Quaternary Science Reviews* **26**, 3359–83.

Lézine, A.M. (1989) Late Quaternary vegetation and climate of the Sahel. *Quaternary Research* **32**, 317–34.

Li, F.-R., Kang, L.-F., Zhang, H., Zhao, L.-Y., Shirato, Y., and Taniyama, I. (2005) Change in intensity of wind erosion at different stages of degradation development in grasslands of Inner Mongolia, China. *Journal of Arid Environments* **62**, 567–85.

Li, X.Y., Liu, L.Y., and Wang, J.H. (2004) Wind tunnel simulation of aeolian sandy soil erodibility under human disturbance. *Geomorphology* **59**, 3–11.

Lines, G.C. (1999) *Ground-water and Surface-water Relations along the Mojave River, Southern California*. US Geological Survey Water Resources Investigations Report 99-4112. US Geological Survey, Sacramento, CA, pp. 1–43.

Liu, M., Westphal, D., Wang, S. et al. (2003) A high-resolution numerical study of the Asian dust storms of April 2001. *Journal of Geophysical Research* **108**, no. D23, 8653, doi:10.1029/2002JD003178.

Liu, T. (2003) Blind testing of rock varnish microstratigraphy as a chronometric indicator: results on late Quaternary lava flows in the Mojave Desert, California. *Geomorphology* **53**, 209–34.

Liu, T. and Broecker, W.S. (2000) How fast does rock varnish grow? *Geology* **28**, 183–6.

Liu, T. and Broecker, W.S. (2008) Rock varnish microlamination dating of late Quaternary geomorphic features in the drylands of western USA. *Geomorphology* **93**, 501–23.

Livingstone, I. (1988) New models for the formation of linear sand dunes. *Geography* **73**, 105–15.

Livingstone, I. and Warren, A. (1996) *Aeolian Geomorphology: an Introduction*. Longman, Harlow.

Lomax, J., Hilgers, A., Wopfner, H., Grün, R., Twidale, C.R., and Radtke, U. (2003) The onset of dune formation in the Strzelecki Desert, South Australia. *Quaternary Science Reviews* **22**, 1067–76.

Losno, R., Bergametti, G., Carlier, P., and Mouvier, G. (1991) Major ions in marine rainwater with attention to sources of alkaline and acidic species. *Atmospheric Environment, Part A* **25A**, 763–70.

Louw, G.N. and Seely, M.K. (1982) *Ecology of Desert Organisms*. Longman, New York.

Lowe, C.H. (1955) The eastern limit of the Sonoran Desert in the United States with additions to the known herpetofauna of New Mexico. *Ecology* **36**, 343–5.

Lu, H. and Sun, D. (2000) Pathways of dust input to the Chinese Loess Plateau during the last glacial and interglacial periods. *Catena* **40**, 251–61.

Lu, H., van Huissteden, K., Zhou, J., Vandenberghe, J., Liu, X., and An, Z. (2000) Variability of East Asian winter monsoon in Quaternary climatic extremes in North China. *Quaternary Research* **54**, 321–7.

Ludwig, J.A. (1987) Primary productivity in arid lands: myths and realities. *Journal of Arid Environments* **13**, 1–7.

Ludwig, J.A. and Tongway, D.J. (1997) A landscape approach to rangeland ecology. In: Belnap, J. and Lange, O.L. (eds.), *Biological Soil Crusts: Structure, Function, and Management*. Springer-Verlag, Berlin, pp. 217–40.

Ludwig, J.A., Wilcox, B.P., Breshears, D.D., Tongway, D.J., and Imeson, A.C. (2005) Vegetation patches and runoff-erosion as interacting ecohydrological processes in semiarid landscapes. *Ecology* **86**, 288–97.

Luo, W., Arvidson, R.E., Sultan, M. et al. (1997) Groundwater sapping processes, Western Desert, Egypt. *Geological Society of America Bulletin* **109**, 43–62.

Lustig, L.K. (1969) *Trend Surface Analysis of the Basin and Range Province, and Some Geomorphic Implications.* US Geological Survey Professional Paper 500-D. US Government Printing Office, Washington DC.

Mabbutt, J.A. (1965) Stone distribution in a stony tableland soil. *Australian Journal of Soil Research* **3**, 131–42.

Mabbutt, J.A. (1977a) Desert lands. In: Jeans, D.N. (ed.), *Australia, a Geography.* Sydney University Press, Sydney, pp. 113–33.

Mabbutt, J.A. (1977b) *Desert landforms, vol. 2. An Introduction to Systematic Geomorphology.* Australian National University Press, Canberra.

Mabbutt, J.A. (1984) Landforms of the Australian deserts. In: El Baz, F. (ed.), *Deserts and Arid Lands.* Martinus Nijhoff Publishers, The Hague, pp. 79–94.

Machette, M.N. (1985) Calcic soils of the southwestern United States. In: Weide, D. (ed.), *Soils and Quaternary Geology of the Southwestern United States. Geological Society of America Special Paper* **203**, 1–21.

MacKinnon, D.J., Elder, D.F., Helm, P.J., Tuesink, M.F., and Nist, C.A. (1990) A method of evaluating effects of antecedent precipitation on duststorms and its application to Yuma, Arizona, 1981–1988. *Climatic Change* **17**, 331–60.

MacMahon, J.A. and Wagner, F.H. (1985) The Mojave, Sonoran and Chihuahuan deserts of North America. In: Evenari, M., Noy-Meir, I., and Goodall, D.W. (eds.), *Hot Deserts and Arid Shrublands, A.* Ecosystems of the World, vol. 12A, Elsevier, Amsterdam, pp. 105–202.

Magee, J.W. and Miller, G.H. (1998) Lake Eyre palaeohydrology from 60 ka to the present: beach ridges and glacial maximum aridity. *Palaeogeography, Palaeoclimatology, Palaeoecology* **144**, 307–29.

Mahowald, N.M., Bryant, R.G., del Corral, J., and Steinberger, L. (2003) Ephemeral lakes and desert dust sources. *Geophysical Research Letters* **30**, 46-1–4.

Mainguet, M. (1968) Le Bourkou – aspects d'un modelé éolien. *Annales de Géographie* **77**, 296–322.

Mainguet, M. (1970) Un étonnant paysage: les cannelures gréseuses du Bembéché (N. du Tchad). Essai d'explication géomorphologique. *Annales de Géographie* **79**, 58–66.

Mainguet, M. (1972) *Le Modelé des Grès.* Institut Geographique National, Paris.

Mainguet, M. (1984) A classification of dunes based on aeolian dynamics and the sand budget. In: El-Baz, F. (ed.), *Deserts and Arid Lands.* Martinus Nijhoff Publishers, The Hague, pp. 31–58.

Mainguet, M. (1991) *Desertification, Natural Background and Human Mismanagement.* Springer-Verlag, New York.

Mainguet, M., Canon, L., and Chemin, M.C. (1980) Le Sahara: géomophologie et paléogéomorphologie éoliennes. In: Williams, M.A.J. and Faure, H. (eds.), *The Sahara and the Nile; Quaternary Environments and Prehistoric Occupation in Northern Africa.* Balkema, Rotterdam, pp. 17–35.

Malek, E., Bingham, G.E., Or, D., and McCurdy, G. (1997) Annual mesoscale study of water balance in a Great Basin heterogeneous desert valley. *Journal of Hydrology* **191**, 223–44.

Maley, J. (1982) Dust, clouds, rain types, and climatic variations in tropical north Africa. *Quaternary Research* **18**, 1–16.

Manabe, S. and Broccoli, A.J. (1985) The influence of continental ice sheets on the climate of an ice age. *Journal of Geophysical Research* **90**, 2167–90.

Mann, H.S., Shankaranarayan, K.A., and Dhir, R.P. (1984) Natural resource survey and environmental monitoring in arid-Rajasthan using remote sensing. In: El-Baz, F. (ed.), *Deserts and Arid Lands.* Martinus Nijhoff Publishers, The Hague, pp. 157–70.

Mares, M.A. (1999) *The Encyclopedia of Deserts.* University of Oklahoma Press, Norman, OK.

Mares, M.A., Morello, J., and Goldstein, G. (1985) The Monte desert and other subtropical semi-arid biomes of Argentina, with comments on their relation to North American arid areas. In: Evenari, M., Noy-Meir, I., and Goodall, D.W. (eds.), *Hot Deserts and Arid Shrublands, A.* Elsevier, Amsterdam, pp. 203–37.

Marín, L., Forman, S.L., Valdez, A., and Bunch, F. (2005) Twentieth century dune migration at the Great Sand Dunes National Park and Preserve, Colorado, relation to drought variability. *Geomorphology* **70**, 163–83.

Marker, M.E. and Holmes, P.J. (1995) Lunette dunes in the northeast Cape, South Africa, as geomorphic indicators of palaeoenvironmental change. *Catena* **24**, 259–73.

Marston, R.A. (1986) Maneuver-caused wind erosion impacts, South Central New Mexico. In: Nickling, W.G. (ed.), *Aeolian Geomorphology.* Allen & Unwin, Boston, MA, pp. 273–90.

Martin, P. (1994) Southern California basins regional aquifer. In: McGill, S.F. and Ross, T.M. (eds.), *Geological Investigations of an Active Margin.* Cordilleran Section Guidebook, 27th Annual Meeting. Geological Society of America, Boulder, CO, pp. 166–9.

Martinsen, G.D., Hall Cushman, J., and Whitham, T.G. (1990) Impact of pocket gopher disturbance on plant species diversity in a shortgrass prairie community. *Oecologia* **83**, 132–8.

Mason, S.J. (2001) El Niño, climate change, and Southern African climate. *Environmetrics* **12**, 327–45.

Matthews, W.J. (1998) *Patterns in Freshwater Fish Ecology.* Chapman & Hall, New York.

Maxey, G.B. and Jameson, C.H. (1948) Geology and water resources of the Las Vegas, Pahrump, and Indian Springs Valley, Clark and Nye Counties, Nevada. *Nevada Department of Conservation and Natural Resources, Water Resources Bulletin* **5**.

McCarthy, T.S. (1993) The great inland deltas of Africa. *Journal of African Earth Sciences* **17**, 275–91.

McCauley, J.F., Breed, C.S., and Grolier, M.J. (1977) Yardangs. In: Doehring, D.O. (ed.), *Geomorphology in Arid Regions*. Allen & Unwin, Boston, MA, pp. 233–69.

McCauley, J.F., Schaber, G.G., Breed, C.S. et al. (1982) Subsurface valleys and geoarchaeology of the eastern Sahara revealed by Shuttle radar. *Science* **218**, 1004–20.

McDonald, G.M. (2003) *Biogeography: Space, Time and Life*. John Wiley & Sons, New York.

McFadden, L.D., Wells, S.G., and Jercinovich, M.J. (1987) Influences of eolian and pedogenic processes on the origin and evolution of desert pavements. *Geology* **15**, 504–8.

McFadden, L.D., McDonald, E.V., Wells, S.G., Anderson, K., Quade, J., and Forman, S. (1998) The vesicular layer and carbonate collars of desert soils and pavements: formation, age and relation to climate change. *Geomorphology* **24**, 101–45.

McFadden, L.D., Eppes, M.C., Gillespie, A.R., and Hallet, B. (2005) Physical weathering in arid landscapes due to diurnal variation in the direction of solar heating. *Geological Society of America Bulletin* **117**, 161–73.

McGee, W.J. (1897) Sheetflood erosion. *Geological Society of America Bulletin* **8**, 87–112.

McGill, G.E. and Stromquist, A.W. (1974) The grabens of Canyonlands National Park, Utah; geometry, mechanics, and kinematics. *Journal of Geophysical Research* **84**, 4547–63.

McIlveen, R. (1992) *Fundamentals of Weather and Climate*. Chapman & Hall, London.

McIntyre, D.S. (1958a) Permeability measurements of soil crusts formed by raindrop impact. *Soil Science* **85**, 261–6.

McIntyre, D.S. (1958b) Soil splash and the formation of surface crusts by raindrop impact. *Soil Science* **85**, 185–9.

McKay, C.P., Friedmann, E., Gómez-Silva, B., Cáceres-Villanueva, L., Andersen, D.T., and Landheim, R. (2003) Temperature and moisture conditions for life in the extreme arid region of the Atacama Desert: four years of observations including the El Niño of 1997–1998. *Astrobiology* **3**, 393–406.

McKee, E.D. (1966) Structures of dunes at White Sands National Monument, New Mexico (and a comparison with structures of dunes from other selected areas). *Sedimentology* **7**, 1–69.

McKee, E.D. (1979) Introduction to a study of global sand seas. In: McKee, E.D. (ed.), *A Study of Global Sand Seas*. US Geological Survey Professional Paper 1052. US Government Printing Office, Washington DC.

McKee, E.D. (1983) Eolian sand bodies of the world. In: Brookfield, M.E. and Ahlbrandt, T.S. (eds.), *Eolian Sediments and Processes*. Developments in Sedimentology vol. 38. Elsevier, Amsterdam, pp. 1–25.

McKenna Neuman, C. (2004) Effects of temperature and humidity upon the transport of sedimentary particles by wind. *Sedimentology* **51**, 1–17.

McKenna Neuman, C., Lancaster, N., and Nickling, W.G. (1997) Relations between dune morphology, air flow, and sediment flux on reversing dunes, Silver Peak, Nevada. *Sedimentology* **44**, 1103–13.

McMahon, J.A. and Wagner, F.H. (1985) The Mojave, Sonoran, and Chihuahuan Deserts of North America. In: Evans, D.D. and Thames, J.L. (eds.), *Hot Deserts and Arid Shrublands*. Elsevier, Amsterdam, pp. 105–202.

McMahon, T.A. and Finlayson, B.L. (2003) Droughts and anti-droughts: the low flow hydrology of Australian rivers. *Freshwater Biology* **48**, 1147–60.

McTainsh, G.H. (1989) Quaternary aeolian dust processes and sediments in the Australian region. *Quaternary Science Reviews* **8**, 235–53.

McTainsh, G.H. and Leys, J.F. (1993) Soil erosion by wind. In: McTainsh, G.H. and Boughton, W.C. (eds.), *Land Degradation Processes in Australia*. Longman Cheshire, Melbourne, pp. 188–230.

Meadows, D.G., Young, M.H., and McDonald, E.V. (2007) Influence of relative surface age on hydraulic properties and infiltration on soils associated with desert pavements. *Catena* **72**, 169–78.

Meadows, M.E. (1996) Biogeography. In: Adams, W.M., Goudie, A.S., and Orme, A.R. (eds.), *The Physical Geography of Africa*. Oxford University Press, Oxford, pp. 161–72.

Meigs, P. (1953) World distribution of arid and semi-arid homoclimates. *Review of Research on Arid Zone Hydrology*. Arid Zone Programme 1. UNESCO, Paris, pp. 203–10.

Meigs, P. (1957) Arid and semiarid climate types of the world. In: *Proceedings, International Geographical Union, 17th Congress, 8th General Assembly*. International Geographical Union, Washington DC, pp. 135–8.

Meinzer, O.E. (1927) *Plants as Indicators of Ground Water*. US Geological Survey Water-Supply Paper 577. US Government Printing Office, Washington DC, pp. 1–95.

Melegy, A.A. (2005) Relationship of environmental geochemistry to soil degradation in Helwan catchment, Egypt. *Environmental Geology* **48**, 524–30.

Meng, Z. and Lu, B. (2007) Dust events as a risk factor for daily hospitalization for respiratory and cardiovascular diseases in Minqin, China. *Atmospheric Environment* **41**, 7048–58.

Menking, K.M., Anderson, R.Y., Shafike, N.G., Syed, K.H., and Allen, B.D. (2004) Wetter or colder during the Last Glacial Maximum? Revisiting the pluvial lake question in southwestern North America. *Quaternary Research* **62**, 280–8.

Merrill, G.P. (1898) Desert varnish. *Bulletin of the United States Geological Survey* **150**, 389–91.

Merritt, D.M. and Cooper, D.J. (2000) Riparian vegetation and channel change in response to river regulation: a comparative study of regulated and unregulated streams in the Green River basin, USA. *Regulated Rivers: Research & Management* **16**, 543–64.

Mésochina, P., Bedin, E., and Ostrowski, S. (2003) Re-introducing antelopes into arid areas: lessons learnt from the oryx in Saudi Arabia. *Comptes Rendus Biologies* **326**, 158–65.

Messina, P., Stoffer, P., and Smith, W.C. (2005) Macropolygon morphology, development, and classification on North Panamint and Eureka playa, Death Valley National Park CA. *Earth-Science Reviews* **73**, 309–22.

Metternicht, G.I. and Zinck, J.A. (2003) Remote sensing of soil salinity: potentials and constraints. *Remote Sensing of Environment* **85**, 1–20.

Michalek, J.L., Colwell, J.E., Roller, N.E.G., Miller, N.A., Kasischke, E.S., and Schlesinger, W.H. (2001) Satellite measurements of albedo and radiant temperature from semi-desert grasslands along the Arizona/Sonora border. *Climatic Change* **48**, 417–25.

Michel, R.L. (1976) Tritium inventories in the world's oceans and their implications. *Nature* **263**, 103–6.

Micklin, P.P. (1988) Desiccation of the Aral Sea: a water management disaster in the Soviet Union. *Science* **241**, 1170–6.

Middleton, N.J. (1986) Dust storms in the Middle East. *Journal of Arid Environments* **10**, 83–96.

Middleton, N.J. (1997) Desert dust. In: Thomas, D.S.G. (ed.), *Arid Zone Geomorphology: Process, Form and Change in Drylands*, 2nd edn. John Wiley & Sons, Chichester, pp. 413–36.

Middleton, N.J. and Goudie, A.S. (2001) Saharan dust: sources and trajectories. *Transactions of the Institute of British Geographers* **26**, 165–81.

Middleton, N.J., Goudie, A.S., and Wells, G.L. (1986) The frequency and source areas of dust storms. In: Nickling, W.G. (ed.), *Aeolian Geomorphology*. Allen & Unwin, Boston, MA, pp. 237–59.

Mignon, P., Goudie, A., Allison, R., and Rosser, N. (2005) The origin and evolution of footslope ramps in the sandstone desert environment of south-west Jordan. *Journal of Arid Environments* **60**, 303–20.

Milana, J.P. and Ruzycki, L. (1999) Alluvial-fan slope as a function of sediment transport efficiency. *Journal of Sedimentary Research* **69**, 553–62.

Miller, R.L. and Tegen, I. (1998) Climate response to soil dust aerosols. *Journal of Climate* **11**, 3247–67.

Mitchell, D.J. and Fullen, M.A. (1994) Soil-forming processes on reclaimed desertified land in north-central China. In: Millington, A.C. and Pye, K. (eds.), *Environmental Change in Drylands*. John Wiley & Sons, Chichester, pp. 393–412.

Moffett, M.W. (1985) An Indian ant's novel method for obtaining water. *National Geographic Research* **1**, 146–9.

Monod, T. (1986) The Sahel zone north of the Equator. In: Evenari, M., Noy-Meir, I. and Goodall, D.W. (eds.), *Hot Deserts and Arid Shrublands, B*. Elsevier, Amsterdam, pp. 203–43.

Montaña, C., Ezcurra, E., Carrillo, A., and Delhoume, J.P. (1988) The decomposition of litter in grasslands of northern Mexico: a comparison between arid and non-arid environments. *Journal of Arid Environments* **14**, 55–60.

Morafka, D.J. (1977) *A Biogeographical Analysis of the Chihuahuan Desert Through its Herpetofauna*. Biogeographica **9**.

Morris, R.L., Devitt, D.A., Crites, A.M., Borden, G., and Allen, L.N. (1997) Urbanization and water conservation in Las Vegas Valley, Nevada. *Journal of Water Resources Planning and Management* **123**, 189–95.

Morrison, R.B. (1991) Quaternary stratigraphic, hydrologic, and climatic history of the Great Basin, with emphasis on Lakes Lahontan, Bonneville, and Tecopa. In: Morrison, R.B. (ed.), *Quaternary Nonglacial Geology: Conterminous U.S.* The Geology of North America K-2. Geological Society of America, Boulder, CO, pp. 283–320.

Morton, S.R. and MacMillen, R.E. (1982) Seeds as sources of preformed water for desert-dwelling granivores. *Journal of Arid Environments* **5**, 61–7.

Moseley, M.P. (1973) Rainsplash and the convexity of badland divides. *Zeitschrift für Geomorphologie Supplementband* **18**, 10–25.

Moss, J.H. (1977) The formation of pediments: scarp backwearing or surface downwasting? In: Doehring, D.O. (ed.), *Geomorphology in Arid Regions*. The Binghamton Symposia in Geomorphology, International Series 8. SUNY Binghamton, New York, pp. 51–78.

Mourelle, C. and Ezcurra, E. (1996) Species richness of Argentine cacti: a test of biogeographic hypotheses. *Journal of Vegetation Science* **7**, 667–80.

Muhs, D.R. and Maat, P.B. (1993) The potential response of aeolian sands to greenhouse warming and precipitation reduction on the Great Plains of the USA. *Journal of Arid Environments* **25**, 351–61.

Muhs, D.R. and Holliday, V.T. (2001) Origin of late Quaternary dune fields on the Southern High Plains of Texas and New Mexico. *Geological Society of America Bulletin* **113**, 75–87.

Muhs, D.R., Bush, C.A., Stewart, K.C., Rowland, T.R., and Crittenden, R.C. (1990) Geochemical evidence of Saharan dust parent material for soils developed on Quaternary limestones of Caribbean and western Atlantic islands. *Quaternary Research* **33**, 157–77.

Muhs, D.R., Reynolds, R.L., Been, J., and Skipp, G. (2003) Eolian sand transport pathways in the southwestern United States: importance of the Colorado River and local sources. *Quaternary International* **104**, 3–18.

Mulligan, K.R. (1988) Velocity profiles measured on the windward slope of a transverse dune. *Earth Surface Processes and Landforms* **13**, 573–82.

Muñoz-Schick, M., Pinto, R., Mesa, A., and Moreira-Muñoz, A. (2001) Fog oases during the El Niño Southern Oscillation 1997–1998, in the coastal hills south of Iquique, Tarapacá region, Chile. *Revista Chilena de Historia Natural* **74**, 389–405.

Munyikwa, K. (2005) Synchrony of Southern Hemisphere Late Pleistocene arid episodes: a review of luminescence chronologies from arid aeolian landscapes south of the Equator. *Quaternary Science Reviews* **24**, 2555–83.

Mustoe, G.E. (1983) Cavernous weathering in the Capitol Reef desert, Utah. *Earth Surface Processes and Landforms* **8**, 517–26.

Nagel, J.F. (1962) Fog precipitation measurements of Africa's southwest coast. *South Africa Weather Bureau, Notos* **11**, 51–60.

Nagy, K.A. (2004) Water economy of free-living desert animals. *International Congress Series* **1275**, 291–7.

Nanson, G.C. and Price, D.M. (1998) Quaternary change in the Lake Eyre basin of Australia: an introduction. *Palaeogeography, Palaeoclimatology, Palaeoecology* **144**, 235–7.

Nanson, G.C., Young, R.W., Price, D.M., and Rust, B.R. (1988) Stratigraphy, sedimentology and late-Quaternary chronology of the Channel Country of western Queensland. In: Warner, R.F. (ed.), *Fluvial Geomorphology of Australia*. Academic Press, Sydney, pp. 151–75.

Nanson, G.C., Chen, X.Y., and Price, D.M. (1995) Aeolian and fluvial evidence of changing climate and wind patterns during the past 100 ka in the western Simpson Desert, Australia. *Palaeogeography, Palaeoecology, Palaeoclimatoloy* **113**, 87–102.

Nanson, G.C., Callen, R.A., and Price, D.M. (1998) Hydroclimatic interpretation of Quaternary shorelines on South Australian playas. *Palaeogeography, Palaeoclimatology, Palaeoecology* **144**, 181–305.

Nash, D.J. and Shaw, P.A. (1998) Silica and carbonate relationships in silcrete-calcrete intergrade duricrusts from the Kalahari of Botswana and Namibia. *Journal of African Earth Sciences* **27**, 11–25.

Nash, III, T.H., Nebeker, G.T., Moser, T.J., and Reeves, T. (1979) Lichen vegetational gradients in relation to the Pacific Coast of Baja California: the maritime influence. *Madroño* **26**, 149–63.

Nasrallah, H.A., Brazel, J., and Balling, Jr, R.C. (1990) Analysis of the Kuwait City urban heat island. *International Journal of Climatology* **10**, 401–5.

Nasrallah, H.A., Nieplova, E., and Ramadan, E. (2004) Warm season extreme temperature events in Kuwait. *Journal of Arid Environments* **56**, 357–71.

Nativ, R., Adar, E., Dahan, O., and Nissim, I. (1997) Water salinization in arid regions – observations from the Negev desert, Israel. *Journal of Hydrology* **196**, 271–96.

Naylor, L.A., Viles, H.A., and Carter, N.E.A. (2002) Biogeomorphology revisited: looking towards the future. *Geomorphology* **47**, 3–14.

Neal, J.T. and Motts, W.S. (1967) Recent geomorphic changes in playas of western United States. *Journal of Geology* **75**, 511–25.

Neave, M. and Abrahams, A.D. (2001) Impact of small mammal disturbances on sediment yield from grassland and shrubland ecosystems in the Chihuahuan Desert. *Catena* **44**, 285–303.

Nicholls, N. (1992) Historical El Niño/Southern Oscillation variability in the Australasian region. In: Diaz, H.F. and Markgraf, V. (eds.), *El Niño: Historical and Paleoclimatic Aspects of the Southern Oscillation*, Cambridge University Press, Cambridge, pp. 151–74.

Nicholson, S.E. (1978) Climatic variations in the Sahel and other African regions during the past five centuries. *Journal of Arid Environments* **1**, 3–24.

Nicholson, S.E. (1985) Sub-Saharan rainfall 1981–84. *Journal of Climate and Applied Meteorology* **24**, 1388–91.

Nicholson, S.E. (2001) Climatic and environmental change in Africa during the last two centuries. *Climate Research* **17**, 123–44.

Nicholson, S. (2005) On the question of the "recovery" of the rains in the West African Sahel. *Journal of Arid Environments* **63**, 615–41.

Nicholson, S.E. and Selato, J.C. (2000) The influence of La Nina on African rainfall. *International Journal of Climatology* **20**, 1761–76.

Nicholson, S.E., Tucker, C.J., and Ba, M.B. (1998) Desertification, drought and surface vegetation: an example from the West African Sahel. *Bulletin of the American Meteorological Society* **79**, 815–29.

Nickling, W.G. and Wolfe, S.A. (1994) The morphology and origin of Nabkhas, Region of Mopti, Mali, West Africa. *Journal of Arid Environments* **28**, 13–30.

Nickling, W.G., McKenna Neuman, C., and Lancaster, N. (2002) Grainfall processes in the lee of transverse dunes, Silver Peak, Nevada. *Sedimentology* **49**, 191–209.

Nicoll, K. (2004) Recent environmental change and prehistoric human activity in Egypt and Northern Sudan. *Quaternary Science Reviews* **23**, 561–81.

Nielson, J. and Kocurek, G. (1987) Surface processes, deposits, and development of star dunes: Dumont dune field, California. *Geological Society of America Bulletin* **99**, 177–86.

Nobel, P.S. (1980) Morphology, surface temperatures, and northern limits of columnar cacti in the Sonoran Desert. *Ecology* **61**, 1–7.

Norwick, S.A. and Dexter, L.R. (2002) Rates of development of tafoni in the Moenkopi and Kaibab Formations in Meteor Crater and on the Colorado Plateau, northeastern Arizona. *Earth Surface Processes and Landforms* **27**, 11–26.

Noy-Meir, I. (1973) Desert ecosystems: environment and producers. *Annual Review of Ecology and Systematics* **4**, 25–51.

Noy-Meir, I. (1974) Desert ecosystems: higher trophic levels. *Annual Review of Ecology and Systematics* **5**, 195–214.

Noy-Meir, I. (1985) Desert ecosystem structure and function. In: Evenari, M., Noy-Meir, I., and Goodall, D.W. (eds.), *Hot Deserts and Arid Shrublands*. Ecosystems of the World, vol. 12A. Elsevier, Amsterdam, pp. 93–103.

N'Tchayi, G.M., Bertrand, J., Legrand, M., and Baudet, J. (1994) Temporal and spatial variations of the atmospheric dust loading throughout West Africa over the last thirty years. *Annales Geophysicae* **12**, 265–73.

Oba, G., Post, E., and Stenseth, N.C. (2001) Sub-saharan desertification and productivity are linked to hemispheric climate variability. *Global Change Biology* **7**, 241–6.

Oberlander, T.M. (1972) Morphogenesis of granitic boulder slopes in the Mojave Desert, California. *Journal of Geology* **80**, 1–20.

Oberlander, T.M. (1974) Landscape inheritance and the pediment problem in the Mojave Desert of southern California. *American Journal of Science* **274**, 849–75.

Oberlander, T.M. (1994) Global deserts: a geomorphic comparison. In: Abrahams, A.D. and Parsons, A.J. (eds.), *Geomorphology of Desert Environments*. Chapman & Hall, London, pp. 13–35.

Oberlander, T.M. (1997) Slope and pediment systems. In: Thomas, D.S.G. (ed.), *Arid Zone Geomorphology: Process, Form and Change in Drylands*, 2nd edn. John Wiley & Sons, Chichester, pp. 135–63.

O'Brien, M.P. and Rindlaub, B.D. (1936) The transportation of sand by wind. *Civil Engineering* **6**, 325–7.

Offer, Z.Y. and Goossens, D. (2001) Ten years of aeolian dust dynamics in a desert region (Negev desert, Israel): analysis of airborne dust concentration, dust accumulation and the high-magnitude dust events. *Journal of Arid Environments* **47**, 211–49.

O'Hara, S.L., Wiggs, G.F.S., Mamedov, B., Davidson, G., and Hubbard, R.B. (2000) Exposure to airborne dust contaminated with pesticide in the Aral Sea region. *The Lancet* **355**, 627–8.

Oke, T.R. (1978) *Boundary Layer Climates*. Methuen, London.

Okin, G.S. and Gillette, D.A. (2004) Modelling wind erosion and dust emission on vegetated surfaces. In: Kelly, R., Drake, N., and Barr, S. (2004) *Spatial Modelling of the Terrestrial Environment*. John Wiley & Sons, Chichester, pp. 137–56.

Olivier, J. (2004) Fog harvesting: an alternative source of water supply on the West Coast of South Africa. *GeoJournal* **61**, 203–14.

Olivier, J. and de Rautenbach, C.J. (2002) The implementation of fog water collection systems in South Africa. *Atmospheric Research* **64**, 227–38.

Oostwoud Wijdenes, D.J., Poesen, J., Vandekerckhove, L., Nachtergaele, J., and De Baerdemaeker, J. (1999) Gully-head morphology and implications for gully development on abandoned fields in a semi-arid environment, Sierra de Gata, southeast Spain. *Earth Surface Processes and Landforms* **24**, 585–603.

Orlovsky, L., Orlovsky, N., and Durdyev, A. (2005) Dust storms in Turkmenistan. *Journal of Arid Environments* **60**, 83–97.

Orme, A.J. and Orme, A.R. (1991) Relict barrier beaches as paleoenvironmental indicators in the California desert. *Physical Geography* **12**, 334–46.

Orme, A.R. (2002) The Pleistocene legacy: beyond the ice front. In: Orme, A.R. (ed.), *The Physical Geography of North America*. Oxford University Press, New York, pp. 380–401.

Orme, A.R. (2004) *Lake Thompson, Mojave Desert, California: A Desiccating Late Quaternary Lake System*. Engineer Research and Development Center, Corps of Engineers, Cold Regions Research and Engineering Laboratory, ERDC/CCRREL TR-04–1. Cold Regions Research and Engineering Laboratory, Hanover, NH.

Orshan, G. (1986) The deserts of the Middle East. In: Evenari, M., Noy-Meir, I., and Goodall, D.W. (eds.), *Hot Deserts and Arid Shrublands*. Ecosystems of the World, vol. 12B. Elsevier, Amsterdam, pp. 1–28.

Ortega-Ramírez, J., Maillol, J.M., Bandy, W. et al. (2004) Late Quaternary evolution of alluvial fans in the Playa, El Fresnal region, northern Chihuahua desert, Mexico: palaeoclimatic implications. *Geofísica Internacional* **43**, 445–66.

Osterkamp, W.R. and Friedman, J.M. (2000) The disparity between extreme rainfall events and rare floods – with emphasis on the semi-arid American West. *Hydrological Processes* **14**, 2817–29.

Osterkamp, W.R. and Wood, W.W. (1987) Playa-lake basins on the southern High Plains of Texas and New Mexico: Part 1 – Hydrologic, geomorphic, and geologic evidence for their development. *Bulletin of the Geological Society of America* **99**, 215–23.

Otterman, J. (1974) Baring high-albedo soils by overgrazing: a hypothesized desertification mechanism. *Science* **186**, 531–3.

Otterman, J. and Tucker, C.J. (1985) Satellite measurements of surface albedo and temperatures in a semi-desert. *Journal of Climate and Applied Meteorology* **24**, 228–35.

Owen, L.A., Windley, B.F., Cunningham, W.D., Badamgarav, J., and Dorjnamjaa, D. (1997) Quaternary alluvial fans in the Gobi of southern Mongolia: evidence for neotectonics and climate change. *Journal of Quaternary Science* **12**, 239–52.

Paine, J.G. (1994) Subsidence beneath a playa basin on the Southern High Plains, USA: evidence from shallow seismic data. *Bulletin of the Geological Society of America* **106**, 233–42.

Palmer, W.C. (1965) *Meteorological Drought*. Research Paper 45. US Department of Commerce Weather Bureau, San Diego, CA.

Parker, G.G. (1963) Piping, a geomorphic agent in landform development of the drylands. *International Association of Scientific Hydrology Publication* **65**, 103–13.

Parker, Sr, G.G., Higgins, C.G., and Wood, W.W. (1990) Piping and pseudokarst in drylands. In: Higgins, C.G. and Coates, D.R. (eds.), *Groundwater Geomorphology: the Role of Subsurface Water in Earth-Surface Processes and Landforms*. Geological Society of America Special Paper **252**, 77–110.

Parker, K.C. (1991) Topography, substrate, and vegetation patterns in the northern Sonoran Desert. *Journal of Biogeography* **18**, 151–63.

Patton, P.C. and Schumm, S.A. (1975) Gully erosion, northwestern Colorado: a threshold phenomenon. *Geology* **3**, 88–9.

Patton, P.C. and Boison, P.J. (1986) Processes and rates of formation of Holocene alluvial terraces in Harris Wash, Escalante River basin, south-central Utah. *Geological Society of America Bulletin* **97**, 369–78.

Patton, P.C., Biggar, N., Condit, C.D. et al. (1991) Quaternary geology of the Colorado Plateau. In: Morrison, R.B. (ed.), *Quaternary Nonglacial Geology; Conterminous U.S.* The Geology of North America, K-2. Geological Society of America, Boulder, CO, pp. 373–406.

Pavlik, B.M. (1985) Sand dune flora of the Great Basin and Mojave deserts, California, Nevada and Oregon. *Madrono* **32**, 197–213.

Pease, P.P. and Tchakerian, V.P. (2003) Geochemistry of sediments from Quaternary sand ramps in the southeastern Mojave Desert, California. *Quaternary International* **104**, 19–29.

Peel, R.G. (1941) Denudational landforms of the central Libyan Desert. *Journal of Geomorphology* **4**, 3–23.

Peel, R.R. (1974) Insolation weathering: some measurements of diurnal temperature change in exposed rocks in the Tibesti region, central Sahara. *Zeitschrift für Geomorphologie NF Supplementband* **21**, 19–28.

Pell, S.D., Chivas, A.R., and Williams, I.S. (2000) The Simpson, Strzelecki and Tirari Deserts: development and sand provenance. *Sedimentary Geology* **130**, 107–30.

Pelletier, J.D., Cline, M., and Long, S.B. (2007) Desert pavement dynamics: numerical modeling and field-based calibration. *Earth Surface Processes and Landforms* **32**(13), 1913–27.

Peneva, E.L., Stanev, E.V., Stanychni, S.V., Salokhiddinov, A., and Stulina, G. (2004) The recent evolution of the Aral Sea level and water properties: analysis of satellite, gauge and hydrometeorological data. *Journal of Marine Systems* **47**, 11–24.

Pérez, F.L. (1987) Soil moisture and the upper altitudinal limit of giant paramo rosettes. *Journal of Biogeography* **14**, 173–86.

Perry, R.S. and Adams, J. (1978) Desert varnish: evidence of cyclic deposition of manganese. *Nature* **276**, 489–91.

Perry, R.S., Engel, M.H., Botta, O., and Staley, J.T. (2003) Amino acid analyses of desert varnish from the Sonoran and Mojave Deserts. *Geomicrobiology Journal* **20**, 427–38.

Peterson, T.C. (2003) Assessment of urban versus rural in situ surface temperatures in the contiguous United States: no difference found. *Journal of Climate* **16**, 2941–59.

Philander, S.G.H. (1983) El Niño Southern Oscillation phenomena. *Nature* **302**, 295–301.

Philander, S.G.H. (1985) El Niño and La Niña. *Journal of the Atmospheric Sciences* **42**, 2652–62.

Phillips, D.L. and McMahon, J.A. (1978) Gradient analysis of a Sonoran Desert bajada. *Southwestern Naturalist* **23**, 669–80.

Pickard, J. (1994) Post-European changes in creeks of semi-arid rangelands, "Polpah Station", New South Wales. In: Millington, A.C. and Pye, K. (eds.), *Environmental Change in Drylands: Biogeographical and Geomorphological Perspectives*. John Wiley & Sons, Hoboken, pp. 271–83.

Pima Association of Governments (2005) *Population Data and Information*. www.pagnet.org/Population/.

Polis, G.A. (1991) *The Ecology of Desert Communities*. University of Arizona Press, Tucson, AZ.

Polis, G.A., Hurd, S.D., Jackson, C.T., and Pinero, F.S. (1997) El Niño effects on the dynamics and control of an island ecosystem in the Gulf of California. *Ecology* **78**, 1884–97.

Ponder, W.F. (2002) Desert springs of the Australian Great Artesian Basin. In: Sada, D.W. and Sharpe, S.E. (eds.), Conference Proceedings, *Spring-fed Wetlands: Important Scientific and Cultural Resources of the Intermountain Region*, 7–9 May 2002, Las Vegas, NV. DHS Publication no. 41210. www.wetlands.dri.edu (accessed 21 June 2005).

Porter, M.L. (1986) Sedimentary record of erg migration. *Geology* **14**, 497–500.

Portnov, B.A. and Safriel, U.N. (2004) Combating desertification in the Negev: dryland agriculture vs. dryland urbanization. *Journal of Arid Environments* **56**, 659–80.

Prospero, J.M. (1981) Arid regions as sources of mineral aerosols in the marine atmosphere. In: Péwé, T.L. (ed.), *Desert Dust: Origin, Characteristics and Effect on Man*. Geological Society of America Special Paper **186**, 71–86.

Prospero, J.M. (1996) Saharan dust transport over the North Atlantic Ocean and Mediterranean; an overview. In: Guerzoni, S. and Chester, R. (eds.), *The Impact of Dust Across the Mediterranean*. Kluwer Academic Publishers, Dordrecht, pp. 133–52.

Prospero, J.M. and Lamb, P.J. (2003) African droughts and dust transport to the Caribbean; climate change implications. *Science* **302**, 1024–7.

Prospero, J.M., Ginoux, P., Torres, O., Nicholson, S.E., and Gill, T.E. (2002) Environmental characterization of global sources of atmospheric soil dust identified with the Nimbus 7 Total Ozone Mapping Spectrometer (TOMS) absorbing aerosol product. *Reviews of Geophysics* **40**, 2-1–31.

Prudic, D.E. (1994) *Estimates of Percolation Rates and Ages of Water in Unsaturated Sediments at two Mojave Desert Sites, California-Nevada*. US Geological Survey Water Resources Investigations Report 94-4160. US Geological Survey, Denver, CO.

Pye, K. (1987) *Aeolian Dust and Dust Deposits*. Academic Press, London.

Pye, K. (1993) Late Quaternary development of coastal parabolic megadune complexes in northeastern Australia. *Special Publication of the International Association of Sedimentologists* **16**, 23–44.

Pye, K. and Tsoar, H. (1987) The mechanics and geological implications of dust transport and deposition in deserts, with particular reference to loess formation and dune sand diagenesis in the northern Negev, Israel. In: Frostick, L. and Reid, I. (eds.), *Desert Sediments Ancient and*

Modern. Geological Society of London, Special Publication 35. Blackwell, Oxford, pp. 139–56.

Pye, K. and Tsoar, H. (1990) *Aeolian Sand and Sand Dunes*. Unwin Hyman, London.

Qin, B. and Yu, G. (1998) Implications of lake level variations at 6 ka and 18 ka in mainland Asia. *Global and Planetary Change* **18**, 59–72.

Quade, J. (2001) Desert pavements and associated rock varnish in the Mojave Desert: how old can they be? *Geology* **29**, 855–8.

Quade, J., Rech, J.A., Latorre, C., Betancourt, J.L., Gleeson, E., and Kalin, M.T.K. (2007) Soils at the hyperarid margin: the isotopic composition of soil carbonate from the Atacama Desert, Northern Chile. *Geochimica et Cosmochimica Acta* **71**, 3772–95.

Quirk, J.P. and Blackmore, A.V. (1955) *Effect of Electrolyte Concentration of the Irrigation Water on the Infiltration of Several Riverina Soils*. Divisional Report, no. 12/55. CSIRO Australia, Division of Soils, Melbourne.

Radies, D., Preusser, F., Matter, A., and Mange, M. (2004) Eustatic and climatic controls on the development of the Wahiba Sand Sea, Sultanate of Oman. *Sedimentology* **51**, 1359–85.

Rajot, J.L., Alfaro, S.C., Gomes, L., and Gaudichet, A. (2003) Soil crusting on sandy soils and its influence on wind erosion. *Catena* **53**, 1–16.

Ramesh, R., Jani, R.A., and Bhushan, R. (1993) Stable isotopic evidence for the origin of salt lakes in the Thar Desert. *Journal of Arid Environments* **25**, 117–23.

Rapport, D.J. and Whitford, W.G. (1999) How ecosystems respond to stress: common properties of arid and aquatic systems. *Bioscience* **49**, 193–203.

Rasmussen, K.R. and Mikkelsen, H.E. (1998) On the efficiency of vertical array aeolian field traps. *Sedimentology* **45**, 789–800.

Rauh, W. (1985) The Peruvian-Chilean Deserts. In: Evenari, M., Noy-Meir, I., and Goodall, D.W. (eds.), *Hot Deserts and Arid Shrublands, A*. Elsevier, Amsterdam, pp. 239–67.

Rauh, W. (1986) The arid region of Madagascar. In: Evenari, M., Noy-Meir, I., and Goodall, D.W. (eds.), *Hot Deserts and Arid Shrublands, B*. Elsevier, Amsterdam, pp. 361–77.

Raupach, M., McTainsh, G., and Leys, J. (1994) Estimates of dust mass in recent major Australian dust storms. *Australian Journal of Soil and Water Conservation* **7**, 20–4.

Reheis, M.C., Goodmacher, J.C., Harden, J.W. et al. (1995) Quaternary soils and dust deposition in southern Nevada and California. *Geological Society of America Bulletin* **107**, 1003–22.

Reheis, M.C., Slate, J.L., Throckmorton, C.K., McGeehin, J.P., Sarna-Wojcicki, A.M., and Dengler, L. (1996) Late Quaternary sedimentation on the Leidy Creek fan, Nevada-California: geomorphic responses to climate change. *Basin Research* **8**, 279–99.

Reiche, P. (1937) The Toreva block, a distinctive landform type. *Journal of Geology* **45**, 538–48.

Reid, I. and Laronne, J.B. (1995) Bed load sediment transport in an ephemeral stream and a comparison with

seasonal and perennial counterparts. *Water Resources Research* **31**, 773–82.

Reid, I. and Frostick, L.E. (1997) Channel form, flows, and sediments in deserts. In: Thomas, D.S.G. (ed.), *Arid Zone Geomorphology: Process, Form and Change in Drylands*, 2nd edn. John Wiley & Sons, Chichester, pp. 205–29.

Reid, J.S., Flocchini, R.G., Cahill, T.A., Ruth, R.S., and Salgado, D.P. (1994) Local meteorological, transport, and source aerosol characteristics of late autumn Owens Lake (dry) dust storms: *Atmospheric Environment* **28**, 1699–1706.

Reineck, H.-E. and Singh, I.B. (1975) *Depositional Sedimentary Environments*. Springer-Verlag, Heidelberg.

Rendell, H. (1997) Tectonic frameworks. In: Thomas, D.S.G. (ed.), *Arid Zone Geomorphology: Process, Form and Change in Drylands*, 2nd edn. John Wiley & Sons, Chichester, pp. 13–22.

Rendell, H.M. and Sheffer, N.L. (1996) Luminescence dating of sand ramps in the Eastern Mojave Desert. *Geomorphology* **17**, 187–97.

Reneau, S.L. and Raymond, Jr, R. (1991) Cation-ratio dating of rock varnish: why does it work? *Geology* **19**, 937–40.

Repka, J.L., Anderson, R.S., and Frinkel, R.C. (1997) Cosmogenic dating of fluvial terraces, Fremont River, Utah. *Earth and Planetary Science Letters* **152**, 59–73.

Reynolds, R., Belnap, J., Reheis, M., Lamothe, P., and Luiszer, F. (2001) Aeolian dust in Colorado Plateau soils: nutrient inputs and recent change in source. *Proceedings of the National Academy of Sciences USA* **98**, 7123–7.

Reynolds, R., Reheis, M.C., Neff, J.C., Goldstein, H., and Yount, J. (2006) Late Quaternary eolian dust in surficial deposits of a Colorado Plateau grassland: controls on distribution and ecologic effects. *Catena* **66**, 251–66.

Reynolds, R.L., Yount, J.C., Reheis, M. et al. (2007) Dust emission from wet and dry playas in the Mojave Desert, USA. *Earth Surface Processes and Landforms* **32**, 1811–27.

Rice, M.A., Mullins, C.E., and McEwan, I.K. (1997) An analysis of soil crust strength in relation to potential abrasion by saltating particles. *Earth Surface Processes and Landforms* **22**, 869–83.

Richardson, C.J., Reiss, P., Hussain, N.A., Alwash, A.J., and Pool, D.J. (2005) The restoration potential of the Mesopotamian marshes of Iraq. *Science* **307**, 1307–11.

Richter, B.D. and Richter, H.E. (2000) Prescribing flood regimes to sustain riparian ecosystems along meandering rivers. *Conservation Biology* **14**, 1467–78.

Risacher, F., Alonso, H., and Salazar, C. (2003) The origin of brines and salts in Chilean salars: a hydrochemical review. *Earth-Science Reviews* **63**, 249–93.

Ritchie, J.C. and Haynes, Jr, C.V. (1987) Holocene vegetation zonation in the eastern Sahara. *Nature* **330**, 645–7.

Roberts, D.L. (2003) *Age, Genesis and Significance of South African Coastal Belt Silcretes*. Memoir 95. Council for Geoscience, Pretoria.

Roberts, S.M., Spencer, R.J., Yang, W., and Krouse, H.R. (1997) Deciphering some unique paleotemperature indicators in halite-bearing saline lake deposits from Death Valley, California, USA. *Journal of Paleolimnology* **17**, 101–30.

Robinson, T.W. (1958) *Phreatophytes*. US Geological Survey Water-Supply Paper 1423. US Government Printing Office, Washington DC, pp. 1–84.

Robinson, T.W. (1965) *Introduction, Spread and Aerial Extent of Saltcedar (Tamarix) in the Western States*. US Geological Survey Professional Paper 491-A. US Government Printing Office, Washington DC.

Rodier, J.A. (1985) Aspects of arid zone hydrology. In: Rodda, J.C. (ed.), *Facets of Hydrology*, vol. II. John Wiley & Sons, Chichester, pp. 205–47.

Rodrigues, D., Abell, P.I., and Kröpelin, S. (2000) Seasonality in the early Holocene climate of Northwest Sudan: interpretation of *Etheria elliptica* shell isotopic data. *Global and Planetary Change* **26**, 181–7.

Rodriguez-Iturbe, I. (2000) Ecohydrology: a hydrologic perspective of climate-soil-vegetation dynamics. *Water Resources Research* **36**, 3–9.

Rodriguez-Navarro, C. and Doehne, E. (1999) Salt weathering: influence of evaporation rate, supersaturation and crystallization pattern. *Earth Surface Processes and Landforms* **24**, 191–209.

Rosen, M.R. (1994) The importance of groundwater in playas: a review of playa classifications and the sedimentology and hydrology of playas. In: Rosen, M.R. (ed.), *Paleoclimate and Basin Evolution of Playa Systems*. Geological Society of America Special Paper **289**, 1–18.

Rosenfeld, D. (1997) Comments on "A New Look at the Israeli Cloud Seeding Experiments." *Journal of Applied Meteorology* **36**, 260–71.

Rosenfeld, D., Rudich, Y. and Lahav, R. (2001) Desert dust suppressing precipitation: a possible desertification feedback loop. *Proceedings of the National Academy of Sciences USA* **98**, 5975–80.

Roshier, D.A., Robertson, A.I., and Kingsford, R.T. (2002) Responses of waterbirds to flooding in an arid region of Australia and implications for conservation. *Biological Conservation* **106**, 399–411.

Ross, G.M. (1983) Bigbear erg: a Proterozoic intermontane eolian sand sea in the Hornby Bay Group, Northwest Territories, Canada. In: Brookfield, M.E. and Ahlbrandt, T.A. (eds.), *Eolian Sediments and Processes*. Elsevier, Amsterdam, pp. 483–519.

Roth, E.S. (1965) Temperature and water content as factors in desert weathering. *Journal of Geology* **73**, 454–68.

Rowe, H.D., Dunbar, R.B., Mucciarone, D.A., Seltzer, G.O., Baker, P.A., and Fritz, S. (2002) Insolation, moisture balance and climate change on the South American Altiplano since the last glacial maximum. *Climatic Change* **52**, 175–99.

Rowell, D.P. and Milford, J.R. (1993) On the generation of African squall lines. *Journal of Climate* **6**, 1181–93.

Rowlands, P.H., Johnson, H., Ritter, E., and Endo, A. (1982) The Mojave Desert. In: Bender, G.L. (ed.), *Reference Handbook on the Deserts of North America*. Greenwood Press, Westport, CT, pp. 103–45.

Rubin, D.M. (1990) Lateral migration of linear dunes in the Strzlecki Desert, Australia. *Earth Surface Processes and Landforms* **15**, 1–14.

Rundel, P.W. (1978) Ecological relationships of desert fog zone lichens. *Bryologist* **81**, 277–93.

Russell, I.C. (1885) Geological history of Lake Lahontan, a Quaternary lake of northwestern Nevada. US Geological Survey Monograph 11. US Geological Survey, Washington DC.

Said, R. (1960) New light on the origin of the Qattara depression. *Bulletin Société Géographique Egypte* **33**, 37–44.

Said, R. (1981) *The Geological Evolution of the River Nile*. Springer, Berlin.

Said, R. (1983) Remarks on the origin of the landscape of the eastern Sahara. *Journal of African Earth Sciences* **1**, 153–8.

Sala, A., Smith, S.D., and Devitt, D.A. (1996) Water use by *Tamarix ramosissima* and associated phreatophytes in a Mojave Desert floodplain. *Ecological Applications* **6**, 888–98.

Sarnthein, M. (1978) Sand deserts during glacial maximum and climatic optimum. *Nature* **272**, 43–6.

Savard, C.S. (1998) *Estimated Ground-water Recharge from Streamflow in Fortymile Wash near Yucca Mountain, Nevada*. US Geological Survey Water Resources Investigations Report 97-4273. US Geological Survey, Denver, CO.

Savoie, D.L. and Prospero, J.M. (1977) Aerosol concentration statistics for the northern tropical Atlantic. *Journal of Geophysical Research* **82**, 5954–64.

Schemenauer, R.S., Fuenzalida, H., and Cereceda, P. (1988) A neglected water resource: the Camanchaca of South America. *Bulletin of the American Meteorological Society* **69**, 138–47.

Schick, A. (1988) Hydrologic aspects of floods in extreme arid environments. In: Baker, V.R., Kockel, R.C., and Patton, P.C. (eds.), *Flood Geomorphology*. John Wiley & Sons, New York, pp. 189–203.

Schlesinger, W.H. and Pilmanis, A.M. (1998) Plant-soil interactions in deserts. *Biogeochemistry* **42**, 169–87.

Schlesinger, W.H., Reynolds, J.F., Cunningham, G.L. et al. (1990) Biological feedbacks in global desertification. *Science* **247**, 1043–7.

Schlesinger, W.H., Raikes, J.A., Hartley, A.E., and Cross, A.F. (1996) On the spatial pattern of soil nutrients in desert ecosystems. *Ecology* **77**, 364–74.

Schmidt, K.-H. (1989a) Talus and pediment flatirons-erosional and depositional features of dryland cuesta scarps. In: Yair, A. and Berkowicz, S. (eds.), *Arid and Semi-arid Environments: Geomorphological and Pedological Aspects*. Catena Supplement 14. Catena-Verlag, Cremlingen, pp. 107–18.

Schmidt, K.-H. (1989b) The significance of scarp retreat for Cenozoic landform evolution on the Colorado Plateau, USA. *Earth Surface Processes and Landforms* **14**, 93–105.

Schmidt, Jr, R.H. (1979) A climatic delineation of the "real" Chihuahuan Desert. *Journal of Arid Environments* **2**, 243–50.

Schmidt, Jr, R.H. (1989) The arid zones of Mexico: climatic extremes and conceptualization of the Sonoran Desert. *Journal of Arid Environments* **16**, 241–56.

Schulten, J.A. (1985) Soil aggregation by cryptogams of a sand prairie. *American Journal of Botany* **72**, 1657–61.

Schulz, E. and Whitney, J.W. (1986) Upper Pleistocene and Holocene lakes in the An Nafud, Saudi Arabia. *Hydrobiologia* **143**, 175–90.

Schumm, S.A. (1956) Evolution of drainage systems and slopes in badlands at Perth Amboy, New Jersey. *Geological Society of America Bulletin* **67**, 597–646.

Schumm, S.A. (1961) *The Effect of Sediment Characteristics on Erosion and Deposition in Ephemeral Stream Channels*. US Geological Survey Professional Paper 352-C. US Government Printing Office, Washington DC, pp. 31–70.

Schumm, S.A. (1973) Geomorphic thresholds and complex response of drainage systems. In: Morisawa, M. (ed.), *Fluvial Geomorphology*. Publications in Geomorphology. State University of New York, Binghamton, NY, pp. 299–310.

Schumm, S.A. and Hadley, R.F. (1957) Arroyos and the semi-arid cycle of erosion. *American Journal of Science* **255**, 161–74.

Schumm, S.A. and Lusby, G.C. (1963) Seasonal variation of infiltration capacity and runoff on hillslopes in western Colorado. *Journal of Geophysical Research* **68**, 3655–66.

Schumm, S.A. and Chorley, R.J. (1964) The fall of Threatening Rock. *American Journal of Science* **262**, 1041–54.

Schumm, S.A. and Chorley, R.J. (1966) Talus weathering and scarp recession in the Colorado Plateau. *Zeitschrift für Geomorphologie* **10**, 11–36.

Schumm, S.A., Mosley, M.P., and Weaver. W.E. (1987) *Experimental Fluvial Geomorphology*. John Wiley & Sons, New York.

Schumm, S.A., Boyd, K.F., Wolff, C.G., and Spitz, W.J. (1995) A ground-water sapping landscape in the Florida Panhandle. *Geomorphology* **12**, 281–97.

Schuster, M., Duringer, P., Ghienne, J.-F. et al. (2003) Coastal conglomerates around the Hadjer el Khamis inselbergs (western Chad, central Africa): new evidence for Lake Mega-Chad episodes. *Earth Surface Processes and Landforms* **28**, 1059–69.

Schwarz, H.E., Emel, J., Dickens, W.J., Rogers, P., and Thompson, J. (1991) Water quality and flows. In: Turner, B.L. (ed.), *The Earth as Transformed by Human Action*. Cambridge University Press, Cambridge, pp. 253–70.

Seely, M.K. and Louw, G.N. (1980) First approximation of the effects of rainfall on the ecology and energetics of a Namib dune ecosystem. *Journal of Arid Environments* **3**, 25–54.

Sefe, F., Ringrose, S., and Matheson, W. (1966) Desertification in north-central Botswana: causes, processes, and impacts. *Journal of Soil and Water Conservation* **51**, 241–8.

Seghieri, J. and Galle, S. (1999) Run-on contribution to a Sahelian two-phase mosaic system: soil water regime and vegetation life cycles. *Acta Oecologica* **20**, 209–17.

Sena, G., Connell, S., Wells, S., and Anderson, K. (1994) Investigation of surficial processes active on fan pavement surfaces using tilted carbonate collars, Providence Mountains, California. In: McGill, S.F. and Ross, T.M. (eds.), *Geological Investigations of an Active Margin*. Cordilleran Section Guidebook, 27th Annual Meeting. Geological Society of America, Boulder, CO, pp. 210–13.

Seymour, R.S. and Seely, M.K. (1996) The respiratory environment of the Namib Desert Golden Mole. *Journal of Arid Environments* **32**, 453–61.

Shafer, D.S., Young, M.H., Zitzer, S.F., Caldwell, T.G., and McDonald, E.V. (2007) Impacts of interrelated biotic and abiotic processes during the past 125,000 years of landscape evolution in the Northern Mojave Desert, Nevada, USA. *Journal of Arid Environments* **69**, 633–57.

Shaltout, M.A. and El Housry, T. (1997) Estimating the evaporation over Nasser Lake in the Upper Egypt from Meteosat Observations. *Advances in Space Research* **19**, 515–18.

Shantz, H.L. (1927) Drought resistance and soil moisture. *Ecology* **8**, 145–57.

Shanyengana, E.S., Henschel, J.R., Seely, M.K., and Sanderson, R.D. (2002) Exploring fog as a supplementary water source in Namibia. *Atmospheric Research* **64**, 251–9.

Shao, Y.P. and Zhao, S. (2001) Wind erosion and wind erosion research in China: a review. *Annals of Arid Zone* **40**, 317–36.

Shao, Y., Raupach, M.R., and Findlater, P.A. (1993) The effect of saltation bombardment on the entrainment of dust by wind. *Journal of Geophysical Research* **98D**, 12719–26.

Sharma, K.D. and Murthy, J.S.R. (1996) Ephemeral flow modeling in arid regions. *Journal of Arid Environments* **33**, 161–78.

Sharon, D. (1962) On the nature of hamadas in Israel. *Zeitschrift für Geomorphologie* **6**, 129–47.

Sharon, D. (1972) The spottiness of rainfall in a desert area. *Journal of Hydrology* **17**, 161–75.

Sharp, R.P. (1949) Pleistocene ventifacts east of the Big Horn Mountains, Wyoming. *Journal of Geology* **57**, 173–95.

Sharp, R.P. (1963) Wind ripples. *Journal of Geology* **71**, 617–36.

Sharp, R.P. (1966) Kelso Dunes, Mojave Desert, California. *Geological Society of America Bulletin* **77**, 1045–74.

Sharp, R.P. (1980) Wind-driven sand in Coachella Valley, California: further data. *Geological Society of America Bulletin* **91**, 724–30.

Shata, A.A. (1992) Water resources and the future of arid lands. *Water Resources Development* **8**, 87–97.

Shaw, P.A. (1997) Africa and Europe. In: Thomas, D.S.G. (ed.), *Arid Zone Geomorphology: Process, Form and Change in Drylands*, 2nd edn. John Wiley & Sons, Chichester, pp. 467–85.

Shaw, P.A. and Thomas, D.S.G. (1997) Pans, playas and salt lakes. In: Thomas, D.S.G. (ed.), *Arid Zone Geomorphology: Process, Form and Change in Drylands*, 2nd edn. John Wiley & Sons, Chichester, pp. 293–317.

Sheehan, P.M. (2001) The Late Ordovician Mass Extinction. *Annual Review of Earth and Planetary Sciences* **29**, 331–64.

Shi, P., Yan, P., Yuan, Y., and Nearing, M.A. (2004) Wind erosion research in China: past, present and future. *Progress in Physical Geography* **28**, 366–86.

Shinn, E.A., Smith, G.W., Prospero, J.M. et al. (2000) African dust and the demise of Caribbean coral reefs. *Geophysical Research Letters* **27**, 3029.

Shlemon, R.J. (1978) Quaternary soil-geomorphic relationships, southeastern Mojave Desert, California and Arizona. In: Mahaney, W.C. (ed.), *Quaternary Soils*. Geo Books, Norwich, pp. 187–207.

Shmida, A. and Ellner, S. (1984) Coexistence of plant species with similar niches. *Vegetatio* **58**, 29–55.

Shreve, F. (1911) The influence of low temperature on the distribution of the giant cactus. *Plant World* **14**, 136–46.

Shreve, F. (1942) The desert vegetation of North America. *Botanical Review* **8**, 195–246.

Shreve, F. (1951) *Vegetation of the Sonoran Desert*. Carnegie Institute Washington Publication, Washington DC.

Shreve, F. (1964) Vegetation of the Sonoran Desert. In: Shreve, F. and Wiggins, I.L. (eds.), *Vegetation and Flora of the Sonoran Desert*, vol. 1, Part 1. Stanford University Press, Stanford, CA, pp. 9–186.

Shulmeister, J., Goodwin, I., Renwick, J. et al. (2004) The Southern Hemisphere westerlies in the Australasian sector over the last glacial cycle: a synthesis. *Quaternary International* **118–19**, 25–53.

Singh, V.P. (1992) *Elementary Hydrology*. Prentice Hall, Upper Saddle River, NJ.

Slattery, M.C. and Bryan, R.B. (1992) Laboratory experiments on surface seal development and its effect on interrill erosion processes. *European Journal of Soil Science* **43**, 517–29.

Slatyer, R.O. (1965) Measurements of precipitation interception by an arid zone plant community (*Acacia aneura* F. Muell). In: Eckardt, F.E. (ed.), *Proceedings of the Montpellier Symposium on Methodology of Plant Eco-physiology*. UNESCO Arid Zone Research 25. UNESCO, Paris, pp. 181–92.

Slatyer, R.O. and Mabbutt, J.A. (1964) Hydrology of arid and semi-arid regions. In: Chow, V.T. (ed.), *Handbook of Applied Hydrology*. McGraw-Hill, New York, pp. 241–46.

Small, E.E., Sloan, L.C., and Nychka, D. (2001) Changes in surface air temperature caused by desiccation of the Aral Sea. *Journal of Climate* **14**, 284–99.

Small, I., Falzon, D., van der Meer, J., Ford, N., and Upshur, R. (2001a) Not a drop to drink in the Aral Sea. *Lancet* **358**, 1649.

Small, I., van der Meer, J., and Upshur, R.E.G. (2001b) Acting on an environmental health disaster: the case of the Aral Sea. *Environmental Health Perspectives* **109**, 547–9.

Smith, B.J. (1994) Weathering processes and forms. In: Abrahams, A.D. and Parsons, A.J. (eds.), *Geomorphology of Desert Environments*. Chapman & Hall, London, pp. 39–63.

Smith, B.J. and Whalley, W.B. (1988) A note on the characteristics and possible origins of desert varnish from southeast Morocco. *Earth Surface Processes and Landforms* **13**, 251–8.

Smith, B.J., Warke, P.A., McGreevy, J.P., and Kane, H.L. (2004). Salt-weathering simulations under hot desert conditions: agents of enlightenment or perpetuators of preconceptions? *Geomorphology* **67**, 211–27.

Smith, E.A. (1986) The structure of the Arabian heat low. Part 1: Surface energy budget. *Monthly Weather Review* **114**, 1067–83.

Smith, G. (1984) Climate. In: Cloudsley-Thompson, J.L. (ed.), *Key Environments: Sahara Desert*. Pergamon Press, Oxford, pp. 17–30.

Smith, G.I. (1984) Paleohydrologic regimes in the southeast Great Basin, 0.32 my ago, compared with other long records of "global" climate. *Quaternary Research* **22**, 1–17.

Smith, G.I. and Bischoff, J.L. (eds.) (1997) *An 800,000-year paleoclimatic record from core OL-92, Owens Lake, southeast California*. Geological Society of America Special Paper **317**.

Smith, H.T.U. (1967) Past versus present wind action in the Mojave Desert region, California. *Air Force Cambridge Research Laboratories* 67-0683, pp. 1–26.

Smith, R.S.U. (1982) Sand dunes in North American deserts. In: Bender, G.L. (ed.), *Reference Handbook on the Deserts of North America*. Greenwood Press, Westport, CT, pp. 481–526.

Smith, R.S.U. (1984) Eolian geomorphology of the Devils Playground, Kelso Dunes and Silurian Valley, California. In: Lintz, J. (ed.), *Western Geological Excursions* 1. 97th Annual Meeting Field Trip Guidebook. Geological Society of America, Reno, NV, pp. 239–51.

Smith, S.I., Monson, R.K., and Anderson, J.E. (1997) *Physiological Ecology of North American Desert Plants*. Springer-Verlag, Berlin.

Soliman, K.H. (1953) Rainfall over Egypt. *Quarterly Journal Royal Meteorological Society* **79**, 389.

Songqiao, Z. and Xuncheng, X. (1984) Evolution of the Lop Desert and the Lop Nor. *Geographical Journal* **150**, 311–21.

Springer, M.E. (1958) Desert pavement and vesicular layer of some soils of the desert of the Lahontan Basin, Nevada. *Proceedings of the Soil Science Society of America* **22**, 63–6.

St. Amand, P., Mathews, l., Gaines, C., and Reinking, R. (1986) *Dust Storms from Owens and Mono Lakes*. Technical Publication 6731. Naval Weapons Center, China Lake, CA.

Stanistreet, I.G. and Stollhofen, H. (2002) Hoanib River flood deposits of Namib Desert interdunes as analogues for thin permeability barrier mudstone layers in aeolianite reservoirs. *Sedimentology* **49**, 719–36.

Stanley, E.H., Buschman, D.L., Boulton, A.J., Grimm, N.B., and Fisher, S.G. (1994) Invertebrate resistance and resilience to intermittency in a desert stream. *American Midland Naturalist* **131**, 288–300.

Stapff, F.M. (1887) Karte des unteren Khuisebtals. *Petermanns Geographische Mitteilungen* **33**, 202–14.

St. Clair, L.L., Johansen, J.R., and Rushforth, S.R. (1993) Lichens of soil crust communities in the Intermountain Area of the western United States. *Great Basin Naturalist* **53**, 5–12.

Stebbing, E.P. (1935) The encroaching Sahara. *Geographical Journal* **86**, 509–10.

Steinberger, Y., Loboda, I., and Gamer, W. (1989) The influence of autumn dewfall on spatial and temporal distribution of nematodes in the desert ecosystem. *Journal of Arid Environments* **16**, 177–83.

Stephens, T. (1997) A river runs through desert agriculture. *California Agriculture* **51**, 6–10.

Stokes, S., Thomas, D.S.G., and Washington, R. (1997) Multiple episodes of aridity in southern Africa since the last interglacial period. *Nature* **388**, 154–8.

Stone, T. (2006) Late glacial cycle hydrological change at Lake Tyrrell, southeast Australia. *Quaternary Research* **66**, 176–81.

Storie, R.E. and Trussel, D.F. (1933) *Soil survey of the Barstow Area, California*. US Department of Agriculture, Bureau of Chemistry and Soils, Washington DC.

Stulina, G. and Sektimenko, V. (2004) The change in soil cover on the exposed bed of the Aral Sea. *Journal of Marine Systems* **47**, 121–5.

Summerfield, M.A. (1983) Silcrete. In: Goudie, A.S. and Pye, K. (eds.), *Chemical Sediments and Geomorphology: Precipitates and Residua in the Near Surface Environment*. Academic Press, New York, pp. 59–91.

Suppiah, R. and Hennessy, K.J. (1998) Trends in total rainfall, heavy rain events and number of dry days in Australia, 1910–1990. *International Journal of Climatology* **18**, 1141–64.

Svendsen, J.B. (2003) Parabolic halite dunes on the Salar de Uyuni, Bolivia. *Sedimentary Geology* **155**, 147–56.

Swap, R., Garstang, M., Greco, S., Talbot, R., and Kaallberg, P. (1992) Saharan dust in the Amazon basin. *Tellus, Series B* **44**, 133–49.

Ta, W., Mao, H., and Dong, Z. (2008) Long-term morphodynamic changes of a desert reach of the Yellow River following upstream large reservoirs' operation. *Geomorphology* **97**, 249–59.

Tator, B.A. (1952) Pediment characteristics and terminology. *Annals of the Association of American Geographers* **42**, 295–317.

Taylor, C.M. and Lebel, T. (1998) Observational evidence of persistent convective-scale rainfall patterns. *Monthly Weather Review* **126**, 1597–607.

Taylor, P.A. (1977) Numerical studies of neutrally stratified planetary boundary-layer flow above gentle topography. *Boundary-Layer Meteorology* **12**, 37–60.

Tchakerian, V.P. (1991) Lake Quaternary aeolian geomorphology of the Dale Lake sand sheet, southern Mojave Desert, California. *Physical Geography* **12**, 347–69.

Tegen, I., Lacis, A.A., and Fung, I. (1996) The influence on climate forcing of mineral aerosols from disturbed soils. *Nature* **380**, 419–22.

Thiagarajan, N. and Lee, C.-T.A. (2004) Trace-element evidence for the origin of desert varnish by direct aqueous atmospheric deposition. *Earth and Planetary Science Letters* **224**, 131–41.

Thiéry, J.M., d'Herbès, J.-M., and Valentin, C. (1995) A model simulating the genesis of banded vegetation patterns in Niger. *Journal of Ecology* **83**, 497–507.

Thomas, D.S.G. (1988) The biogeomorphology of arid and semi-arid environments. In: Viles, H.A. (ed.), *Biogeomorphology*. Blackwell, Oxford, pp. 193–221.

Thomas, D.S.G. (1997) Arid environments: their nature and extent. In: Thomas, D.S.G. (ed.), *Arid Zone Geomorphology: Process, Form and Change in Drylands*, 2nd edn. John Wiley & Sons, Chichester, pp. 3–12.

Thomas, D.S.G. and Shaw, P.A. (1991) *The Kalahari Environment*. Cambridge University Press, Cambridge.

Thomas, D.S.G. and Middleton, N.J. (1994) *Desertification: Exploding the Myth*. John Wiley & Sons, Chichester.

Thomas, D.S.G. and Leason, H.C. (2005) Dunefield activity response to climate variability in the southwest Kalahari. *Geomorphology* **64**, 117–32.

Thompson, D.G. (1929) *The Mojave Desert Region, California. A Geographic, Geologic, and Hydrologic Reconnaissance*. US Geological Survey Water Supply Paper 578. US Government Printing Office, Washington, DC, pp. 1–759.

Thompson, L., Mosley-Thompson, E., and Thompson, P. (1992) Reconstructing interannual climate variability from tropical and subtropical ice cores. In: Diaz, H. and Markgraf, V. (eds.), *El Niño, Historical and Paleoclimatic Aspects of the Southern Oscillation*. Cambridge University Press, New York, pp. 295–322.

Thornes, J.B. (1994) Catchment and channel hydrology. In: Abrahams, A.D. and Parsons, A.J. *Geomorphology of Desert Environments*. Chapman & Hall, London, pp. 257–87.

Thornthwaite, C.W. (1948) An approach toward a rational classification of climate. *Geographical Review* **38**, 55–94.

Thorweihe, U. (1990) Nubian aquifer system. In: Said, R. (ed.), *Geology of Egypt*. Balkema, Rotterdam, pp. 601–11.

Tivy, J. (1993) *Biogeography: a Study of Plants in the Ecosphere*. John Wiley & Sons, New York.

Todorov, A.V. (1985) Sahel: the changing rainfall regime and the "normals" used for its assessment. *Journal of Climate and Applied Meteorology* **24**, 97–107.

Tongway, D.J. and Ludwig, J.A. (1990) Vegetation and soil patterning in semi-arid mulga lands of Eastern Australia. *Austral Ecology* **15**, 23–34.

Tooth, S. (2000) Process, form and change in dryland rivers: a review of recent research. *Earth-Science Reviews* **51**, 67–107.

Tooth, S. (2007) Arid geomorphology: investigating past, present and future changes. *Progress in Physical Geography* **31**, 319–35.

Tooth, S. and Nanson, G.C. (2000) The role of vegetation in the formation of anabranching channels in an ephemeral river, Northern plains, arid central Australia. *Hydrological Processes* **14**, 3099–117.

Tooth, S. and Nanson, G.C. (2004) Forms and processes of two highly contrasting rivers in arid central Australia, and the implications for channel-pattern discrimination and prediction. *Geological Society of America Bulletin* **116**, 802–16.

Tsoar, H. (1983) Wind tunnel modeling of echo and climbing dunes. In: Brookfield, M.E. and Ahlbrandt, T.S. (eds.), *Eolian Sediments and Processes*. Elsevier, Amsterdam, pp. 247–59.

Tsoar, H. (1985) Profiles analysis of sand dunes and their steady state signification. *Geografiska Annaler* **67A**, 47–61.

Tsoar, H. (1989) Linear dune-forms and formation. *Progress in Physical Geography* **13**, 507–28.

Tsoar, H. (2001) Types of aeolian sand dunes and their formation. In: Balmforth, N.J. and Provenzale, A. (eds.), *Geomorphological Fluid Mechanics*. Lecture Notes in Physics Series, vol. 582. Springer-Verlag, Berlin, pp. 403–29.

Tsoar, H. and Møller, J.T. (1986) The role of vegetation in the formation of linear sand dunes. In: Nickling, W.G. (ed.), *Aeolian Geomorphology*. Allen & Unwin, Boston, MA, pp. 75–95.

Tsoar, H. and Blumberg, D.G. (2002) Formation of parabolic dunes from barchan and transverse dunes along Israel's Mediterranean coast. *Earth Surface Processes and Landforms* **27**, 1147–61.

Tsoar, H., Goldsmith, V., Schoenhaus, S., Clarke, K., and Karnieli, A. (1995) Reversed desertification on sand dunes along the Sinai/Negev border. In: Tchakerian, V.P. (ed.), *Desert Aeolian Processes*. Chapman & Hall, London, pp. 251–67.

Tucker, C.J. and Nicholson, S.E. (1999) Variations in the size of the Sahara Desert from 1980 to 1997. *Ambio* **28**, 587–91.

Tucker, C.J., Dregne, H.E., and Newcomb, W.W. (1991) Expansion and contraction of the Sahara Desert from 1980 to 1990. *Science* **253**, 299–301.

Tucker, M.E. (1978) Gypsum crusts (gypcrete) and patterned ground from northern Iraq. *Zeitschrift für Geomorphologie NF* **22**, pp. 89–100.

Turner, B.R. and Makhlouf, I. (2002) Recent colluvial sedimentation in Jordan: fans evolving into sand ramps. *Sedimentology* **49**, 1283–98.

Turner, R.M. (1990) Long-term vegetation change at a fully protected Sonoran Desert site. *Ecology* **71**, 464–77.

Turner, R.M. and Brown, D.E. (1982) Sonoroan desertscrub. Biotic Communities of the American Southwest-United States and Mexico. *Desert Plants* **4**, 181–221.

Twidale, C.R. (1967) Hillslopes and pediments in the Flinders ranges, South Australia. In: Jennings, J.N. and Mabbutt, J.A. (eds.), *Landform Studies from Australia and New Guinea*. Cambridge University Press, Cambridge, pp. 95–117.

Twidale, C.R. (1972) Landform development in the Lake Eyre region, Australia. *Geographical Review* **62**, 40–70.

Twidale, C.R. (1978a) On the origin of Ayres Rock, Central Australia, *Zeitschrift für Geomorphologie NF Supplementband* **31**, 177–206.

Twidale, C.R. (1978b) On the origin of pediments in different structural settings. *American Journal of Science* **278**, 1138–76.

Twidale, C.R. and Corbin, E.M. (1963) Gnammas. *Revue de Géomorphologie Dynamique* **14**, 1–20.

Twidale, C.R., Prescott, J.R., Bourne, J.A., and Williams, F.M. (2001) Age of desert dunes near Birdsville, southwest Queensland. *Quaternary Science Reviews* **20**, 1355–64.

Tyler, S.W., Kranz, S., Parlange, M.B. et al. (1997) Estimation of groundwater evaporation and salt flux from Owens Lake, California, USA. *Journal of Hydrology* **200**, 110–35.

Ullman, W.J. (1985) Evaporation rate from a salt pan: estimates from chemical profiles in near-surface groundwaters. *Journal of Hydrology* **79**, 365–73.

United Nations Environment Program (1992) *World Atlas of Desertification*. Edward Arnold, London.

Utah Division of Water Resources (2005) *Utah Cloud Seeding Activities Water Year 2005*. Utah Division of Water Resources. http://water.utah.gov/cloudseeding/currentprojects/ (accessed 23 June 2005).

Vale, T.R. (1988) Clearcut logging, vegetation dynamics, and human wisdom. *Geographical Review* **78**, 375–86.

Valentine, G.A. and Harrington, C.D. (2006) Clast size controls and longevity of Pleistocene desert pavements at Lathrop Wells and Red Cone volcanoes, southern Nevada. *Geology* **34**, 533–6.

Valero-Garcés, B.L., Delgado-Huertas, A., Ratto, N., Navas, A., and Edwards, L. (2000) Paleohydrology of Andean saline lakes from sedimentological and isotopic records, Northwestern Argentina. *Journal of Paleolimnology* **24**, 343–59.

Van den Ancker, J.A.M., Jungerius, P.D., and Mur, L.R. (1985) The role of algae in the stabilization of coastal dune blowouts. *Earth Surface Processes and Landforms* **10**, 189–92.

Van Devender, T.R. (1990) Late Quaternary vegetation and climate of the Sonoran Desert, United States and Mexico. In: Betancourt, J.L., Van Devender, T.R., and Martin, P.S. (eds.), *Packrat Middens: The Last 40,000 Years of Biotic Change*. University of Arizona Press, Tucson, AZ, pp. 134–66.

Viles, H.A. and Goudie, A.S. (2007) Rapid salt weathering in the coastal Namib desert: implications for landscape development. *Geomorphology* **85**, 49–62.

Villa, N., Dorn, R.I., and Clark, J. (1995) Fine material in rock fractures: aeolian dust or weathering? In: Tchakerian, V.P. (ed.), *Desert Aeolian Processes*. Chapman & Hall, London, pp. 219–31.

Wainwright, J., Parsons, A.J., and Abrahams, A.D. (1999) Field and computer simulation experiments on the formation of desert pavement. *Earth Surface Processes and Landforms* **24**, 1025–37.

Walker, A.S. (1986) Eolian landforms. In: Short, N.M. and Blair, Jr, R.W. (eds.), *Geomorphology from Space: a Global Overview of Regional Landforms*. Special Publication 486. NASA Scientific and Technical Branch, Washington DC, pp. 447–520.

Walker, I.J. and Nickling. W.G. (2002) Dynamics of secondary airflow and sediment transport over and in the lee of transverse dunes. *Progress in Physical Geography* **26**, 47–75.

Walling, D.E. (1996) Hydrology and rivers. In: Adams, W. M., Goudie, A.S., and Orme, A.R. (eds.), *The Physical Geography of Africa*. Oxford University Press, Oxford, pp. 103–21.

Walling, D.E. and Webb, B.W. (1983) Patterns of sediment yield. In: Gregory, K.G. (ed.), *Background to Palaeohydrology: a Perspective*. John Wiley & Sons, Chichester, pp. 69–100.

Walsh, P.A. and Hoffer, T.E. (1991) The changing environment of a desert boomtown. *The Science of the Total Environment* **105**, 233–58.

Walter, H. (1986) The Namib Desert. In: Evenari, M., Noy-Meir, I., and Goodall, D.W. (eds.), *Hot Deserts and Arid Shrublands, B*. Elsevier, Amsterdam, pp. 245–82.

Walter, H. and Box, E.O. (1983a) Caspian lowland biome. In: West, N.E. (ed.), *Temperate Deserts and Semi-Deserts*. Ecosystems of the World, vol. 5. Elsevier, Amsterdam, pp. 9–41.

Walter, H. and Box, E.O. (1983b) Middle Asian Deserts. In: West, N.E. (ed.), *Temperate Deserts and Semi-Deserts*. Ecosystems of the World, vol. 5. Elsevier, Amsterdam, pp. 79–104.

Walter, H. and Box, E.O. (1983c) Semi-deserts and deserts of central Kazakhstan. In: West, N.E. (ed.), *Temperate Deserts and Semi-Deserts*. Ecosystems of the World, vol. 5. Elsevier, Amsterdam, pp. 43–78.

Walter, H. and Box, E.O. (1983d) Overview of Eurasian continental deserts and semi-deserts. In: West, N.E. (ed.), *Temperate Deserts and Semi-Deserts*. Ecosystems of the World, vol. 5. Elsevier, Amsterdam, pp. 3–7.

Walter, H. and Box, E.O. (1983e) The Karakum Desert, an example of a well-studied eu-biome. In: West, N.E. (ed.), *Temperate Deserts and Semi-Deserts*. Ecosystems of the World, vol. 5. Elsevier, Amsterdam, pp. 105–59.

Waltham, T. and Sholji, I. (2001) The demise of the Aral Sea – an environmental disaster. *Geology Today* **17**, 206.

Walther, J. (1891) Die Denudationin der Wüste und ihre geologische Bedeutung. *Abhandlungen Sächsische Gesellschaft Wisssenshaft* **16**, 345–570.

Walther, J. (1893) The North American deserts. *National Geographic Magazine* **4**, 163–208.

Walvoord, M.A. and Phillips, F.M. (2004) Identifying areas of basin-floor recharge in the Trans-Pecos region and the link to vegetation. *Journal of Hydrology* **292**, 59–74.

Wang, J. and Mitsuta, Y. (1992) Evaporation from the desert: some preliminary results of HEIFE. *Boundary-Layer Meteorology* **59**, 413–18.

Wang, T., Zhang, W., Dong, Z. et al. (2005) The dynamic characteristics and migration of a pyramid dune. *Sedimentology* **52**, 429–40.

Wang, X., Dong, Z., Zhang, J., and Chen, G. (2002) Geomorphology of sand dunes in the Northeast Taklimakan Desert. *Geomorphology* **42**, 183–95.

Wang, X., Dong, Z., Liu, L., and Qu, J. (2004a) Sand sea activity and interactions with climatic parameters in the Taklimakan Sand Sea, China. *Journal of Arid Environments* **57**, 85–98.

Wang, X., Dong, Z., Zhang, J., and Liu, L. (2004b) Modern dust storms in China: an overview. *Journal of Arid Environments* **58**, 559–74.

Wang, X., Dong, Z., Yan, P., Zhang, J., and Qian, G. (2005) Wind energy environments and dunefield activity in the Chinese deserts. *Geomorphology* **65**, 33–48.

Wang, Y., McDonald, E., Amundson, R., McFadden, L., and Chadwick, O. (1996) An isotopic study of soils in chronological sequences of alluvial deposits, Providence Mountains, California. *Geological Society of America Bulletin* **108**, 379–91.

Wang, Y., Xiao, D., and Li, Y. (2007) Temporal-spatial change in soil degradation and its relationship with landscape types in a desert-oasis ecotone: a case study in the Fubei region of Xinjiang Province, China. *Environmental Geology* **51**, 1019–28.

Ward, A.W. (1979) Yardangs on Mars; evidence of recent wind erosion. *Journal of Geophysical Research* **84**, 8147–66.

Ward, A.W. and Greeley, R. (1984) Evolution of the yardangs at Rogers Lake, California. *Geological Society of America Bulletin* **95**, 829–37.

Warner, T.T. (2004) *Desert Meteorology*. Cambridge University Press, Cambridge.

Warren, A. (1996) Desertification. In: Adams, W.M., Goudie, A.S., and Orme, A.R. (eds.), *The Physical Geography of Africa*. Oxford University Press, Oxford, pp. 342–55.

Warren-Rhodes, K.A., Rhodes, K.L., Pointing, S.B. et al. (2006) Hypolithic cyanobacteria, dry limit of photosynthesis, and microbial ecology in the hyperarid Atacama Desert. *Microbial Ecology* **52**, 389–98.

Washington, R. and Todd, M.C. (2005) Atmospheric controls on mineral dust emission from the Bodélé Depression, Chad: the role of the low level jet. *Geophysical Research Letters* **32**, L17701.

Washington, R., Todd, M.C., Lizcano, G. et al. (2006) Links between topography, wind, deflation, lakes and dust: the cast of the Bodélé Depression, Chad. *Geophysical Research Letters* **33**, L09401.

Wasson, R.J. (1983a) Dune sediment types, sand colour, sediment provenance and hydrology in the Simpson-Strzelecki dunefield, Australia. In: Brookfield, M.E. and Ahlbrandt, T.S. (eds.), *Eolian Sediments and Processes*. Elsevier, Amsterdam, pp. 165–95.

Wasson, R.J. (1983b) The Cainozoic history of the Strzelecki and Simpson dunefields (Australia), and the origin of the desert dunes. *Zeitschrift für Geomorphologie NF Supplementband* **45**, 85–115.

Wasson, R.J. and Hyde, R. (1983) Factors determining desert dune type. *Nature* **304**, 337–9.

Wasson, R.J. and Nanninga, P.M. (1986) Estimating transport of sand on vegetated surfaces. *Earth Surface Processes and Landforms* **11**, 505–14.

Wasson, R.J., Rajaguru, S.N., Misra, V.N. et al. (1983) Geomorphology, late Quaternary stratigraphy and palaeoclimatology of the Thar dunefield. *Zeitschrift für Geomorphologie NF Supplementband* **45**, 117–51.

Wasson, R.J., Fitchett, K., Mackey, B., and Hyde, R. (1988) Large-scale patterns of dune type, spacing and orientation in the Australian Continental Dunefield. *Australian Geographer* **19**, 89–104.

Waters, M.R. (1988) Holocene alluvial geology and geoarchaeology of the San Xavier reach of the Santa Cruz River, Arizona. *Geological Society of America Bulletin* **100**, 479–91.

Waters, M.R. (1991) The geoarchaeology of gullies and arroyos in southern Arizona. *Journal of Field Archaeology* **18**, 141–59.

Waters, M.R. and Haynes, C.V. (2001) Late Quaternary arroyo formation and climate change in the American Southwest. *Geology* **29**, 399–402.

Watson, A. (1983) Gypsum crusts. In: Goudie, A.S. and Pye, K. (eds.), *Chemical Sediments and Geomorphology: Precipitates and Residua in the Near Surface Environment*. Academic Press, New York, pp. 133–6.

Watson, A. (1985) Structure, chemistry and origins of gypsum crusts in southern Tunisia and the central Namib Desert. *Sedimentology* **32**, 855–75.

Watson, A. (1988) Desert gypsum crusts as palaeoenvironmental indicators: a micropetrographic study of crusts from southern Tunisia and the central Namib Desert. *Journal of Arid Environments* **15**, 19–42.

Watson, A. and Nash, D.J. (1997) Desert crusts and varnishes. In: Thomas, D.S.G. (ed.), *Arid Zone Geomorphology: Process, Form and Change in Drylands*, 2nd edn. Wiley & Sons, Chichester, pp. 69–107.

Waylen, P. and Poveda, G. (2002) El Niño-Southern Oscillation and aspects of western South American hydroclimatology. *Hydrological Processes* **16**, 1247–60.

Wells, S.G., Dohrenwend, J.C., McFadden, L.D., Turrin, B.D., and Mahrer, K.D. (1985) Late Cenozoic landscape evolution on lava flow surfaces of the Cima volcanic field, Mojave Desert, California. *Geological Society of America Bulletin* **96**, 1518–29.

Wells, S.G., McFadden, L.D., and Dohrenwend, J.C. (1987) Influence of Late Quaternary climatic change on geomorphic and pedogenic processes on a desert piedmont, eastern Mojave Desert, California. *Quaternary Research* **27**, 130–46.

Wells, S.G., McFadden, L.D., and Schultz, J.D. (1990) Eolian landscape evolution and soil formation in the Chaco dune field, southern Colorado Plateau, New Mexico. *Geomorphology* **3**, 517–46.

Wells, S.G., McFadden, L.D., Olinger, C.T., and Poths, J. (1994) Use of cosmogenic ^{3}HE to understand desert pavement formation. In: McGill, S.F. and Ross, T.M. (eds.), *Geological Investigations of an Active Margin*. Cordilleran Section Guidebook, 27th Annual Meeting. Geological Society of America, Boulder, CO, pp. 201–5.

Wendler, G. and Eaton, F. (1983) On the desertification of the Sahel Zone. Part 1: Ground observations. *Climatic Change* **5**, 365–80.

Went, F.W. (1979) Germination and seedling behavior of desert plants. In: Goodall, D.W., Perry, R.A., and Howes, K.M.W. (eds.), *Arid-Land Ecosystems: Structure, Functioning and Management*. Cambridge University Press, Cambridge, pp. 477–89.

Werner, B.T. (1995) Eolian dunes: computer simulations and attractor interpretation. *Geology* **23**, 1107–10.

West, N.E. and Gifford, G.F. (1976) Rainfall interception by cool-desert shrubs. *Journal of Range Management* **29**, 171–2.

White, F. (1983) *The Vegetation of Africa*. UNESCO, Paris.

White, K., McLaren, S., Black, S., and Parker, A. (2000) Evaporite minerals and organic horizons in sedimentary sequences in the Libyan Fezzan: implications for palaeoenvironmental reconstruction. In: McLaren, S.J. and Kniveton, D.R. (eds.), *Linking Climate Change to Land Surface Change*. Kluwer Academic, Dordrecht, pp. 193–208.

Whitford, W.G. and Kay, F.R. (1999) Biopedturbation by mammals in deserts: a review. *Journal of Arid Environments* **41**, 203–30.

Whitney, M.I. (1983) Eolian features shaped by aerodynamic and vorticity processes. In: Brookfield, M.E. and Ahlbrandt, T.S. (eds.), *Eolian Sediments and Processes*. Elsevier, Amsterdam, pp. 223–45.

Whitney, M.I. (1985) Yardangs. *Journal of Geological Education* **33**, 93–6.

Wiegand, J.P. (1977) *Dune Morphology and Sedimentology at Great Sand Dunes National Monument*. MSc Thesis, Colorado State University, Fort Collins, CO.

Wiggs, G.F.S. (1997) Sediment mobilisation by the wind. In: Thomas, D.S.G. (ed.), *Arid Zone Geomorphology: Process, Form and Change in Drylands*, 2nd edn. John Wiley & Sons, Chichester, pp. 352–72.

Wiggs, G.F.S. (2001) Desert dune processes and dynamics. *Progress in Physical Geography* **25**, 53–79.

Wiggs, G.F.S., Thomas, D.S.G., Bullard, J.E., and Livingstone, I. (1995) Dune mobility and vegetation cover in the southwest Kalahari Desert. *Earth Surface Processes and Landforms* **20**, 515–30.

Wiggs, G.F.S., O'Hara, S., Wegerdt, J., Van Der Meers, J., Small, I., and Hubbard, R. (2003) The dynamics and characteristics of aeolian dust in dryland Central Asia: possible impacts on human exposure and respiratory health in the Aral Sea basin. *Geographical Journal* **169**, 142–57.

Wilby, A., Shachak, M., and Boeken, B. (2001) Integration of ecosystem engineering and tropic effects of herbivores. *Oikos* **92**, 436–44.

Williams, G. (1964) Some aspects of the eolian saltation load. *Sedimentology* **3**, 257–87.

Williams, J.D., Dobrowolski, J.P., and West, N.E. (1995a) Microphytic crust influence on interrill erosion and infiltration capacity. *Transactions of the ASAE* **38**, 139–46.

Williams, J.D., Dobrowolski, J.P., West, N.E., and Gillette, D.A. (1995b) Microphytic crust influence on wind erosion. *Transactions of the ASAE* **38**, 131–7.

Williams, M.A.J. (1982) Quaternary environments in North Africa. In: Williams, M.A.J. and Adamson, D.A. (eds.), *A Land Between Two Niles. Quaternary Geology and Biology of the Central Sudan*. Balkema, Rotterdam, pp. 13–22.

Williams, M.A.J. (1994) Cenozoic climatic changes in deserts: a synthesis. In: Abrahams, A.D. and Parsons, A.J. (eds.), *Geomorphology of Desert Environments*. Chapman & Hall, London, pp. 644–70.

Williams, M.A.J., Dunkerley, D.L., De Deckker, P., Kershaw, A.P., and Stokes, T. (1993) *Quaternary Environments*. Edward Arnold, London.

Williams, O.B. and Calaby, J.H. (1985) The hot deserts of Australia. In: Evenari, M., Noy-Meir, I., and Goodall, D.W. (eds.), *Hot Deserts and Arid Shrublands, A*. Elsevier, Amsterdam, pp. 269–312.

Williams, S.H. and Zimbelman, J.R. (1994) Desert pavement evolution: an example of the role of sheetflood. *Journal of Geology* **102**, 243–8.

Williams, W.D. (2002) Environmental threats to salt lakes and the likely status of inland saline ecosystems in 2025. *Environmental Conservation* **29**, 154–67.

Wilson, I.G. (1973) Ergs. *Sedimentary Geology* **10**, 77–106.

Winkler, E.M. (1994) *Stone in Architecture*. Springer-Verlag, Berlin.

Winkler, E.M. and Wilhelm, E.J. (1970) Saltburst by hydration pressures in architectural stone in urban atmosphere. *Bulletin of the Geological Society of America* **81**, 567–72.

Wolf, B. (2000) Global warming and avian occupancy of hot deserts; a physiological and behavioral perspective. *Revista Chilena de Historia Natural* **73**, 395–400.

Wood, M.K., Jones, T.L., and Vera-Cruz, M.T. (1998) Rainfall interception by selected plants in the Chihuahuan Desert. *Journal of Range Management* **51**, 91–6.

Woodhouse, C.A. (1997) Winter climate and atmospheric circulation patterns in the Sonoran Desert region, USA. *International Journal of Climatology* **17**, 859–73.

Wopfner, H. (1978) Silcretes of northern south Australia and adjacent regions. In: Langford-Smith, T. (ed.), *Silcrete in Australia*. Department of Geography, University of New England, Armidale, NSW, pp. 93–141.

Wopfner, H. and Twidale, C.R. (1967) Geomorphological history of the Lake Eyre basin. In: Jennings, J.A. and Mabbutt, J.A. (eds.), *Landform Studies from Australia and New Guinea*. Australian University Press, Canberra, pp. 118–43.

Wopfner, H. and Twidale, C.R. (1988) Formation and age of desert dunes in the Lake Eyre depocentres in central Australia. *Geologische Rundschau* **77**, 815–34.

Wopfner, H. and Twidale, C.R. (1992) Response to R.A. Callen and G.C. Nanson: Reply. *Geologische Rundschau* **81**, 595–9.

Wright, H.A. and Klemmedson, J.O. (1965) Effect of fire on bunchgrasses of the sagebrush-grass region in southern Idaho. *Ecology* **46**, 680–8.

Xia, X.C. (1987) *A Scientific Expedition and Investigation to Lop Nor Area*. Scientific Press, Beijing.

Xu, J. (2006) Sand-dust storms in and around the Ordos Plateau of China as influenced by land use change and desertification. *Catena* **65**, 279–84.

Xuan, J. and Sokolik, I. (2002) Characterization of sources and emission rates of mineral dust in Northern China. *Atmospheric Environment* **36**, 4863–76.

Xuan, J., Liu, G.L., and Du, K. (2000) Dust emission inventory in northern China. *Atmospheric Environment* **34**, 4565–70.

Yaalon, D.H. (1997) Soils in the Mediterranean region: what makes them different? *Catena* **28**, 157–69.

Yair, A. (1973) Theoretical considerations on the evolution of convex hillslopes. *Zeitschrift für Geomorphologie Supplementband* **18**, 1–9.

Yair, A. and Rutin, J. (1981) Some aspects of the regional variation in the amount of available sediment produced by isopods and porcupines in northern Negev, Israel. *Earth Surface Processes and Landforms* **6**, 221–34.

Yair, A. and Lavee, H. (1985) Runoff generation in arid and semi-arid zones. In: Anderson, M.G. and Burt, T.P. (eds.), *Hydrological Forecasting*. John Wiley & Sons, Chichester, pp. 183–220.

Yair, A. and Kossovsky, A. (2002) Climate and surface properties: hydrological response of small arid and semi-arid watersheds. *Geomorphology* **42**, 43–57.

Yair, A., Lavee, H., Bryan, R.B., and Adar, E. (1980) Runoff and erosion processes and rates in the Zion Valley badlands, northern Negev, Israel. *Earth Surface Processes* **5**, 205–25.

Yang, W., Krouse, H.R., Spencer, R.J. et al. (1999) A 200,000-year record of change in oxygen isotope composition of sulfate in a saline sediment core, Death Valley, California. *Quaternary Research* **51**, 148–57.

Yang, X., Liu, T., and Xiao, H. (2003) Evolution of megadunes and lakes in the Badain Jaran Desert, Inner Mongolia, China during the last 31,000 years. *Quaternary International* **104**, 99–112.

Yang, X., Rost, K.T., Lehmkuhl, F., Zhenda, Z., and Dodson, J. (2004) The evolution of dry lands in northern China and in the Republic of Mongolia since the Last Glacial Maximum. *Quaternary International* **118–19**, 69–85.

Yeager, E.C. (1957) *The North American Deserts*. Stanford University Press, Stanford, CA.

Young, J.A. and Evans, R.A. (1978) Population dynamics after wildfires in sagebrush grasslands. *Journal of Range Management* **31**, 283–9.

Young, M.H., McDonald, E.V., Caldwell, T.G., Benner, S.G., and Meadows, D.G. (2004) Hydraulic properties of a desert soil chronosequence in the Mojave Desert, USA. *Vadose Zone Journal* **3**, 956–63.

Yu, B., Neil, D.T., and Hesse, P.P. (1992) Correlation between rainfall and dust occurrence at Mildura, Australia: the difference between local and source area rainfalls. *Earth Surfaces Processes and Landforms* **17**, 723–7.

Yu, G., Harrison, S.P., and Xue, B. (2001) *Lake Status Records from China: Data Base Documentation*. Technical Report no. 4. Max-Planck Institute, Jena.

Yu, Y.C., Wang, X.L., and Wintle, A.G. (2007) A new OSL chronology for dust accumulation in the last 130,000 yr for the Chinese Loess Plateau. *Quaternary Research* **67**, 152–60.

Zangvil, A. (1996) Six years of dew observations in the Negev Desert, Israel. *Journal of Arid Environments* **32**, 361–71.

Zárate, M.A. (2003) Loess of southern South America. *Quaternary Science Reviews* **22**, 1987–2006.

Zdanowicz, C., Hall, G., Vaive, J. et al. (2006) Asian dustfall in the St. Elias Mountains, Yukon, Canada. *Geochimica et Cosmochimica Acta* **70**, 3493–507.

Zehfuss, P.H., Bierman, P.R., Gillespie, A.R., Burke, R.M., and Caffee, M.W. (2001) Slip rates on the Fish Springs fault, Owens Valley, California, deduced from cosmogenic ^{10}Be and ^{26}Al and soil development on fan surfaces. *Geological Society of America Bulletin* **113**, 241–55.

Zehnder, K. and Arnold, A. (1989) Crystal growth in salt efflorescence. *Journal of Crystal Growth* **97**, 513–21.

Zeng, N., Neelin, J.D., Lau K.-M., and Tucker, C.J.A. (1999) Enhancement of interdecadal climate variability in the Sahel by vegetation interaction. *Science* **286**, 1537–40.

Zhang, H., Wünnemann, B., Ma, Y. et al. (2002) Lake level and climate changes between 42,000 and 18,000 ^{14}C yr

B.P. in the Tengger Desert, Northwestern China. *Quaternary Research* **58**, 62–72.

Zhang, X.Y., Arimoto, R., An, Z., Chen, T., Zhang, G., and Ray, B.J. (1994) Late Quaternary records of the atmospheric input of eolian dust to the center of the Chinese loess plateau. *Quaternary Research* **41**, 35–43.

Zhang, X.Y., Arimoto, R., and An, Z.S. (1997) Desert emission from Chinese desert sources linked to variations in atmospheric circulation. *Journal of Geophysical Research* **102**, 28041–7.

Zhou, B., Liu, T., and Zhang, Y. (2000) Microlaminations in rock varnish from northern Tian Shan and their paleoclimatic implications. *Chinese Science Newsletters* **45**, 372–5.

Zhou, M., Chen, Z., Huang, R. et al. (1994) Effects of two dust storms on solar radiation in the Beijing-Tianjin area. *Geophysical Research Letters* **21**, 2697–700.

Zhou, X., Fang, B., Wan, L., Cao, W., Wu, S., and Feng, W. (2006) Occurrence of soluble salts and moisture in the unsaturated zone and groundwater hydrochemistry along the middle and lower reaches of the Heihe river in northwest China. *Environmental Geology* **50**, 1085–93.

Zhu, C., Waddell, R.K., Star, I., and Ostrander, M. (1998) Responses of ground water in the Black Mesa Basin, northeastern Arizona, to paleoclimatic changes during the late Pleistocene and Holocene. *Geology* **26**, 127–30.

Zhu, Z. (1989) Advances in desertification research in China. *Journal of Desert Research* **9**, 1–13 [in Chinese, abstract in English].

Zhu, Z. and Liu, S. (1988) Desertification processes and their control in northern China. *Chinese Journal of Arid Land Research* **1**, 27–36.

Zimbelman, J.R., Williams, S.H., and Tchakerian, V.P. (1995) Sand transport paths in the Mojave Desert, southwestern United States. In: Tchakerian, V.P. (ed.), *Desert Aeolian Processes*. Chapman & Hall, London, pp. 101–30.

Zohary, M. (1973) *Geobotanical Foundation of the Middle East*. Gustav Fisher Verlag, Stuttgart.

Zou, X.K. and Zhai, P.M. (2004) Relationship between vegetation coverage and spring dust storms over northern China. *Journal of Geophysical Research* **109**, D03104, doi:10.1029/2003JD003913.

INDEX

Figures in *italic* refer to figures
Figures in **bold** refer to tables